MUNICIPAL SOLID WASTE TO ENERGY CONVERSION PROCESSES

MUNICIPAL SOLID WASTE TO ENERGY CONVERSION PROCESSES

ECONOMIC, TECHNICAL, AND RENEWABLE COMPARISONS

Gary C. Young, PhD., P.E.

A JOHN WILEY & SONS, INC., PUBLICATION

Published by John Wiley & Sons, Inc., Hoboken, New Jersey
Published simultaneously in Canada

For general information on our other products and services or for technical support, please contact our Customer Care Department within the United States at (800) 762-2974, outside the United States at (317) 572-3993 or fax (317) 572-4002.

Wiley also publishes its books in a variety of electronic formats. Some content that appears in print may not be available in electronic formats. For more information about Wiley products, visit our web site at www.wiley.com.

Library of Congress Catalogiitg-in-Publicatton Data:

Young, Gary C., 1943-
   Municipal solid waste to energy conversion processes : economic, technical,
and renewable comparisons / Gary C. Young.
       p. cm.
   Includes index.
   ISBN 978-0-470-53967-5 (cloth)
 1.   Refuse as fuel. 2.   Waste products as fuel. 3.   Refuse and refuse
disposal. 4.   Incineration. 5.   Resource recovery facilities. I.   Title.
   TD796.2.Y68 2010
   662'.87–dc22

                                                        2009043672

Printed in the United States of America

10 9 8 7 6 5 4 3 2 1

■■■■■ CONTENTS

◼◼◼◼◼ PREFACE

In the many years of practicing engineering as a profession from research and development to commercial plant operations, I observed over time that many good ideas never came to fruition. The good ideas were typically blocked by those who did not understand the technology behind the good idea and/or by those ideas supported by others with vested interests or agendas. Thus, the quotation presented here keeps the true professional aware of potential roadblocks in order to be successful.

"Show me an agenda and it likely will lead you astray.
Give me technology and it will lead you to prosperity."

In all my professional work experiences from research and development, process development and design, construction, start-up and operation of commercial plant operations, and waste management always played a role in the final outcome of a plant's commercial profitability. My exposure, interest, and professional knowledge in gasification of hydrocarbons began in the late 1960s while working in the energy industry. When my exposure to plasma gasification technology began in the early 2000s from listening to a seminar by Dr. Circeo, the practical process benefits of plasma gasification technology for the management of wastes became apparent and caught my professional interest. After several years of further scientific inquiry into plasma technology, it became obvious that a proven technology was available to manage a wide variety of solid wastes, such as municipal solid waste (MSW), to produce energy and valuable recyclable by-products with essentially no wastes. Further economic assessment of the plasma gasification process on MSW indicated a proven process of commercial viability.

Thus, this book was undertaken to inform the public, teachers, professors, public officials, city, state, federal governments, businessmen, and businesswomen on how the proven plasma gasification technology can be used to manage MSW and generate energy and revenues for local communities in an environmentally safe manner with essentially no wastes. Furthermore, currently generated MSW can be processed with this proven plasma gasification technology to eliminate air and water pollution from landfills. It is my professional hope that this proven and economic plasma gasification technology for the management of MSW will be understood and embraced by the reader so as to lead to prosperity.

PROFESSIONAL BIOGRAPHY

Dr. Gary C. Young, Ph.D., P.E., is a knowledgeable professional in research, development, economic assessment, and commercialization of industrial processes. Dr. Young has over 40+ years of industrial experience in processes involving energy, food, agricultural, chemical, and pharmaceutical businesses. He has done consulting in areas of research and development, troubleshooting plant operations and process bottlenecks, maintenance, engineering, and environmental challenges. Dr. Young has successfully commercialized new processes from the laboratory to process development, engineering, procurement, construction, training, and start-up with final management of the production operation. Furthermore, a new agricultural herbicide process was successfully completed from the laboratory to full-scale commercialization without the need for a pilot plant. Dr. Young is the inventor of many patents. He has work experiences with CONOCO, Stauffer Chemical Company, Beatrice Foods Company, Monsanto Company, and Carus Chemical Company.

Current "technical and commercial economic" interests are with the conversion of "waste solids to energy" such as electricity and fuels. In addition, Dr. Young recently developed a novel thermal process for converting carbon dioxide (CO_2) from a gaseous stream into energy and fuels.

Dr. Young holds B.S., M.S., and Ph.D. degrees in Chemical Engineering from the University of Nebraska. He is a licensed professional engineer in the states of California, Texas, Illinois, Iowa, and Wisconsin. Dr. Young is the founder and owner of Bio-Thermal-Energy, Inc. (B-T-E, Inc.).

CHAPTER 1

Introduction to Gasification/Pyrolysis and Combustion Technology(s)

HISTORICAL BACKGROUND AND PERSPECTIVE

Archeological studies demonstrate that Trash-Garbage-Waste was generated by Native Americans in Colorado about 6,500 BC in North America. Based upon archeological assessment of the waste site, the Native Americans in that ancient clan generated 5.3 pounds of waste per day as compared to 2.5 pounds per day for middle-class Americans today. The first municipal dump in the Western world is credited to the Athenians of Greece about 500 BC. In Jerusalem/Palestine, the New Testament of The Bible mentions Sheol was likely a dump outside the city of Jerusalem and became synonymous with "Hell." In 1388, the English parliament barred waste disposal in public waterways and ditches. Recycling was mentioned in 1690 when Rittenhouse Mill, Philadelphia, made paper from recycled fibers of waste paper and rags. In Nottingham, England about 1874, a new technology known as "the Destructor" was used to manage garbage; it involved systematic burning, i.e., incineration. The first garbage incinerator was built in the United States on Governor's Island, New York about 1885. It was reported in 1889 around Washington, D.C., that there was lack of places for refuse. Also, the first recycling/sorting of rubbish in the United States occurred in New York around 1898.[1]

Landfills became popular in the 1920s as a means of reclaiming swampland while disposing of trash. Then in 1965, the Federal government of the United States enacted the first Federal solid waste (SW) management laws. In 1976, the Resource Conservation and Recovery Act (RCRA) was created for stressing recycling and hazardous waste management, which likely was instigated by the discovery of Love Canal.[1]

This proves that since the creation of mankind, humans have generated waste. But waste disposal was not a problem when we had a nomadic existence; mankind simply moved away and left their waste behind. In addition, populations concentrating in urban areas necessitated better methods for management of waste. With the initiation of the industrial revolution, waste management became a critical issue. The population increase and migration of people to industrial towns and cities from rural areas resulted in a consequent increase in the domestic and industrial waste (IW), posing threat to human health and environmental issues of water quality, air pollution, and land toxicity issues. As the American population grew and people left the farms for life in the city, the amount of waste increased. But the method of getting rid of the waste needed to improve. We continue to dump it. Today, about 55% of our garbage is hauled off and buried in sanitary landfills.

Municipal solid waste (MSW) is garbage that comes from homes, businesses, and schools. Today, this garbage is disposed of in "municipal solid waste landfills" so the garbage does not harm the public health, or land, water, and air environment. MSW landfills are not dumps for the new landfills are required to have liners, leachate collection systems, gas collection equipment, groundwater monitoring, and environmental reporting requirements so as to protect the health and welfare of the community.

Our population is still growing and we are producing more garbage, even with the recycling efforts in full operation. We have come to the "place in time" where the momentum of TECHNOLOGY can help "protect human health and welfare," and thus the environment, by creating an infrastructure design, creation and building of sustainable MSW processes that can turn our WASTE PROBLEM into useful GREEN ENERGY for the betterment of ALL.

INTRODUCTION

The management/treatment of SWs by thermal pyrolysis/gasification technology is increasingly viewed as the best suitable and economically viable approach for the management of wastes such as: residential waste (RW), commercial waste (CW), IW, and MSW, which can be a mixture of these wastes. Various types of Thermal Processes using pyrolysis/gasification technology will be discussed and also why plasma arc gasification process was selected as most attractive for commercial viability.[2-4] The various types of thermal processes based upon pyrolysis/gasification technology are pyrolysis, pyrolysis/gasification, conventional gasification, and plasma arc gasification. One additional thermal process was also considered, which is based upon combustion technology and is known as mass burn (incineration). The key product from these thermal gasification technologies is the conversion of MSW into synthesis gas (syngas), which is predominantly carbon monoxide (CO) and hydrogen (H_2), which can be converted into energy (steam and/or electricity), other gases, fuels, and/or chemicals, and will be discussed in detail throughout this book.

One approach or option for the use of the key product from the conversion of MSW into syngas by a thermal process is for generation of steam and/or electricity in a powerhouse. This approach or "Power Option" will be discussed later in Chapters 2, 4, 5 and 7.

Another approach to the management of MSW is the "BioChemistry Option" (biochemical or biological technologies), which by necessity operates at conditions appropriate for living organisms/microbes. Consequently, the reaction rates are lower and these technologies require feedstock that is biodegradable. One, therefore, could conclude that these biochemical technologies have limitations for applicability for treating MSW compared to the thermal processes. Thermal processes are brute force chemical reaction approach to the management of MSW feedstock in comparison to the finesse of biochemical/biological reactions and consequent limitations of feedstock acceptance. However, the real niche for biochemical processes is to take the syngas (predominantly, CO and H_2), produced by a thermal process, and have the biochemical process (bacteria/microbes) convert the syngas into products such as fuels and chemicals, for example, ethanol, methanol, etc.[5,6] This approach or "BioChemistry Option" will be covered in Chapter 3 with a case study.

Another approach could be labeled the "Chemistry Option," which converts syngas into fuels and chemicals by catalytic chemistry. A catalyst that is used typically is called Fischer–Tropsch catalyst. Thus, a thermal process can be used to produce syngas from MSW and then convert the syngas into chemicals by Fischer–Tropsch chemistry. This "Chemistry Option" is also covered in Chapter 3 with a case study.

Lastly, one could consider landfill gas (LFG) as an approach, which involves the use of microorganisms to produce LFG *in situ* within the landfill. LFG is predominantly methane (CH_4) and carbon dioxide (CO_2) gas, i.e., approximately 50% CH_4 and CO_2. LFG is extracted from landfills with a system typically comprising gas collecting from wells at the landfill to a central point, a gas processing plant, and a gas delivery pipeline to customer(s). LFG could be used in a boiler, dryer, kiln, greenhouse, or other applications. A basic drawback of LFG facility is that the microorganisms producing the LFG leave behind *in situ* landfill leachate as a by-product of the microbiological process that can contaminate soil and groundwater. Even with the latest designs and use of liners in landfills, no LFG system is fail-safe. Another negative factor is that an LFG facility just depletes the energy value of the landfill wastes by using up the most easily biodegraded organics in the MSW. Thus, a lesser energy value of MSW remaining in the landfill after an LFG facility will make it more difficult economically to justify a future MSW management system to eliminate the landfill. In summary, an LFG process just skims off the energy leaving a degraded MSW mess behind to be dealt with later at a much greater cost to any future management system. Thus, this approach is not discussed further as a suitable approach both economically and environmentally.[7,8]

These basic approaches for the management of MSW are schematically shown in Fig. 1.1, whereby the options for the syngas are numerous.

Key Thermal Processes will be discussed next with emphasis upon the conversion of MSW to syngas and an assessment of each process with a thorough technical and economic analysis.

WHAT IS PYROLYSIS?

Pyrolysis can be defined as the thermal decomposition of carbon-based materials in an oxygen-deficient atmosphere using heat to produce syngas.[9] No air or oxygen is present and no direct burning takes place. The process is endothermic.

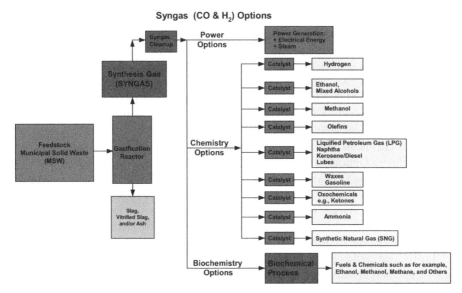

FIGURE 1.1   MSW to Energy, Gases, Fuels, and Chemicals.

Typically, most organic compounds are thermally unstable, and at high temperatures, the chemical bonds of organic molecules break, producing smaller molecules such as hydrocarbon gases and hydrogen gas. At high temperatures, the gaseous mixture produced comprises predominantly the thermodynamically stable small molecules of CO and H_2. This gaseous mixture of CO and H_2 is called "syngas." This latter stage of the thermal process is known as gasification.

A typical pyrolysis process is illustrated in Fig. 1.2.

As illustrated in Fig. 1.2, feedstock as MSW is preprocessed to remove profitable recyclables. Then the preprocessed material is fed into the pyrolysis reactor where an indirect source of heat elevates the contents to a temperature between 1,200 and 2,200°F to produce raw syngas overhead and a bottom ash, carbon char, and metals from the reactor. Some report the pyrolysis process to occur at a reactor temperature between 750 and 1,650°F.[9] The pyrolysis process occurs in an oxygen-deficient (starved) atmosphere.

The syngas cleanup step is designed to remove carry-over particulate matter from the reactor, sulfur, chlorides/acid gases (such as hydrochloric acid), and trace metals such as mercury.

Syngas is used in the power generation plant to produce energy, such as steam and electricity, for use in the process and export energy. The export energy is typically converted into electricity and supplied/sold to the grid.

The bottoms from the reactor are ash, carbon char, and metals. The carbon char and metals have use as recyclables in industry. However, the ash from the pyrolysis process is usually disposed of in a landfill, which is one of the major environmental shortcomings of the pyrolysis process when used for MSW management.

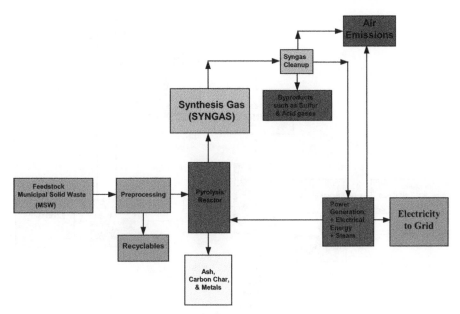

FIGURE 1.2   Process Schematic, MSW to Electricity via Pyrolysis.

WHAT IS PYROLYSIS/GASIFICATION?

Pyrolysis/gasification is a variation of the pyrolysis process. Another reactor is added whereby any carbon char or pyrolysis liquids produced from the initial pyrolysis step are further gasified in a close-coupled reactor, which may use air, oxygen, and/or steam for these gasification reactions. As shown in Fig. 1.3, a controlled amount of air/oxygen is fed into the pyrolysis/gasification reactor whereby some of the char and pyrolysis liquids react, i.e., there is combustion with oxygen. The combustion reactions (exothermic reactions) are controlled so as to supply sufficient heat for the pyrolysis reactions (endothermic reactions), yielding a temperature typically between 1,400 and 2,800°F. Sometimes the pyrolysis/gasifier conditions are stated as 750–1,650°F for the pyrolysis zone and 1,400–2,800°F for the gasification zone. In addition, steam is supplied to the reactor for the chemical reactions that yield CO and H_2.[9]

Pyrolysis/gasification reactor operates predominantly in an oxygen-starved environment, since the combustion reactions (exothermic reactions) quickly consume the oxygen producing heat sufficient for the pyrolysis reactions (endothermic reactions), resulting in a raw syngas exiting the reactor. The raw syngas is cleaned up of carry-over particulate matter from the reactor, sulfur, chlorides/acid gases (such as hydrochloric acid), and trace metals such as mercury. Syngas is used in the power generation plant to produce energy, such as steam and electricity, for use in the process and export energy. The export energy is typically converted into electricity and supplied/sold to the grid.

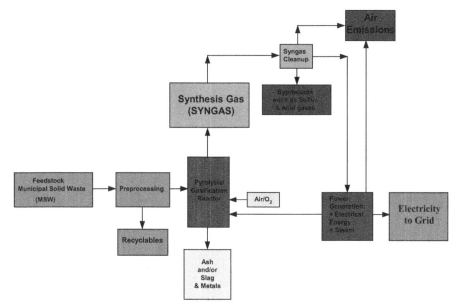

FIGURE 1.3 Process Schematic, MSW to Electricity via Pyrolysis/Gasification.

The bottoms from the reactor are typically ash, slag, and metals depending on the temperature of the pyrolysis/gasification reactor. The metals find use as recyclables in industry. However, the ash and/or slag is typically disposed of in a landfill, which is one of the major environmental shortcomings of the pyrolysis/gasification process when used for MSW management.

WHAT IS CONVENTIONAL GASIFICATION?

Conventional gasification is a thermal process, which converts carbonaceous materials, such as MSW, into syngas using a limited quantity of air or oxygen.

The conventional gasification conditions are sometimes between 1,450 and 3,000°F. Steam is injected into the conventional gasification reactor to promote CO and H_2 production.

For simplicity, some basic chemical reactions in the gasification process are:

$$C + O_2 \rightarrow CO_2 \tag{1.1}$$

$$C + H_2O \rightleftharpoons CO + H_2 \tag{1.2}$$

$$C + 2H_2 \rightleftharpoons CH_4 \tag{1.3}$$

$$C + CO_2 \rightleftharpoons 2CO \tag{1.4}$$

$$CO + H_2O \rightleftharpoons CO_2 + H_2 \tag{1.5}$$

$$C_nH_m + nH_2O \rightleftharpoons nCO + (n + \tfrac{1}{2}m)H_2 \tag{1.6}$$

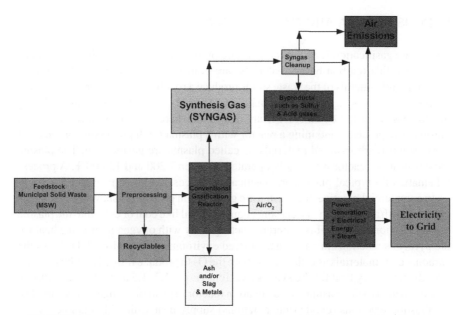

FIGURE 1.4   Process Schematic, MSW to Electricity via Conventional Gasification.

Thus, CO, H_2, and CH_4 are the basic components of the gasification process producing the gaseous mixture. Of these components, the gaseous mixture comprises predominantly of CO) and H_2. Equation (1.1) shows the carbonaceous components of the MSW as carbon (C) that reacts with oxygen (O_2) to produce limited combustion but with the necessary heat for the syngas reactions (Eqs. (1.2–1.5 and 1.6)).

Figure 1.4 illustrates a typical conventional gasification process. As shown, a controlled amount of air/oxygen is fed into the conventional gasification reactor whereby some feedstock material reacts, i.e., there is combustion with oxygen. The combustion reactions (exothermic reactions) are controlled so as to supply sufficient heat for the predominantly syngas reactions (endothermic reactions), yielding a temperature typically between 1,450 and 3,000°F. The raw syngas exits the reactor and is cleaned up of carry-over particulate matter from the reactor, sulfur, chlorides/acid gases (such as hydrochloric acid), and trace metals such as mercury. Syngas is sent to the power generation plant to produce energy, such as steam and electricity, for use in the process and export energy. The export energy is converted to electricity and supplied/sold to the grid.

The bottoms from the conventional gasification reactor are ash and/or slag and metals depending upon the temperature of the conventional gasification reactor. However, the ash and/or slag from the reactor bottoms is usually disposed off in a landfill which is one of the major environmental shortcomings when used for MSW management.

WHAT IS PLASMA ARC GASIFICATION?

Plasma arc gasification is a high-temperature pyrolysis process whereby the organics of waste solids (carbon-based materials) are converted into syngas and inorganic materials and minerals of the waste solids produce a rocklike glassy by-product called vitrified slag. The syngas is predominantly CO and H_2. The high temperature during the process is created by an electric arc in a torch whereby a gas is converted into a plasma. The process containing a reactor with a plasma torch processing organics of waste solids (carbon-based materials) is called plasma arc gasification. The plasma arc gasification reactor is typically operated between 7,200 and 12,600°F. A process schematic of a typical plasma arc gasification process is shown in Fig. 1.5.

In commercial practice, the plasma arc gasification process, as shown in Fig. 1.5, is operated with an injection of a carbonaceous material like coal or coke into the plasma arc gasification reactor. This material reacts quickly with oxygen to produce heat for the pyrolysis reactions in an oxygen-starved environment. Equation (1.1) shows the carbonaceous materials as C that reacts with the O_2 to produce limited combustion but with the necessary heat for the syngas reactions (Eqs. (1.2–1.5 and 1.6)). In addition, steam is added to the plasma arc gasification reactor to promote syngas reactions. The combustion reactions (exothermic reactions) supply heat with additional heat from the plasma arc torches for the pyrolysis reactions (endothermic reactions), yielding a temperature typically between 7,200 and 12,600°F.

The inorganic minerals of the waste solids (MSW) produce a rocklike by-product. Since operating conditions are very high (7,200–12,600°F), these minerals are

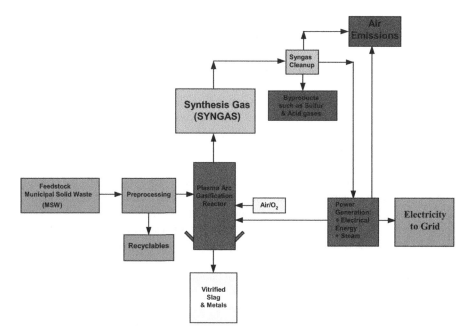

FIGURE 1.5 Process Schematic, MSW to Electricity via Plasma Arc Gasification.

converted into a vitrified slag typically comprising metals and silica glass. This vitrified slag is basically nonleaching and exceeds EPA standards. Metals can be recovered from the slag and the slag can be used to produce other by-products such as rock wool, floor tiles, roof tiles, insulation, and landscaping blocks, to mention a few.[5,10] The vitrified slag, being environmentally acceptable as a recyclable by-product, is one of the more positive attributes of plasma arc gasification process for the management of MSW.

Another positive attribute of the plasma arc gasification process is that developments in the design of plasma arc gasification reactor have improved and lessened the need for pretreatment/preprocessing.[2–4,10]

WHAT IS MASS BURN (INCINERATION)?

Mass burn (Incineration) is a combustion process that uses an excess of oxygen and/or air to burn the SWs. The mass burn process operates with an "excess of oxygen" present and is therefore a "combustion" process as illustrated in Fig. 1.6. It is "NOT" a pyrolysis process.

Feedstock as MSW is preprocessed to remove saleable recyclables for the marketplace and remaining MSW may be shredded. MSW is fed into the fluid bed boiler with operating temperatures between 1,000 and 2,200°F. Excess air/O_2 is used for combustion of the combustibles in the MSW. High-pressure steam produced in the fluid bed boiler is sent to the power plant for energy generation. Hot exhaust gases from the fluid bed boiler are sent for gas cleanup and heat recovery sent to the power plant for generation of energy.

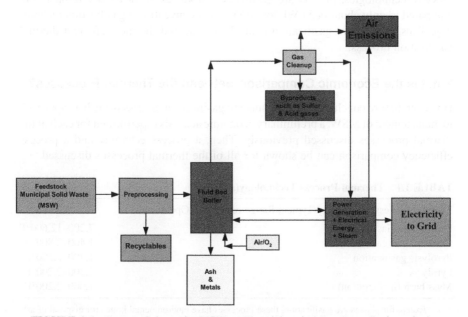

FIGURE 1.6 Process Schematic, MSW to Electricity via Mass Burn (Incineration).

The power plant produces electricity using steam turbines and saleable excess electricity to the grid.

One of the biggest drawbacks or negative environmental aspects of the mass burn process is the production of ash from the grate of the fluid bed boiler. This ash is typically sent to a landfill for disposal.

WHICH THERMAL PROCESS TECHNOLOGY IS THE MOST EFFICIENT AND ECONOMICAL?

Five Thermal Processes have been discussed so far but which process should be selected based upon the highest thermal efficiency and the best economics? To answer this question, the thermal efficiency and economics of the five technologies will be determined and compared.

Performance/Thermal Efficiency of Technologies

For the thermal process technologies discussed, the typical range of process operation is presented in Table 1.1 for comparison.[9]

The thermal efficiency of each thermal process technology previously discussed was determined by URS Corporation, which reported the net energy production of electricity to the Grid (area electrical distribution system) per ton of MSW, as shown in Fig. 1.7 and Table 1.2.[9,11]

On reviewing the net energy production to the grid for various types of thermal process technologies, plasma arc gasification produces about 816 kWh/ton MSW compared to only about 685 kWh/ton MSW for a conventional gasification technology. Thus, plasma arc gasification could be considered the most efficient thermal gasification process.

What is the Economic Comparison Between the Thermal Processes?

However, before concluding that plasma arc gasification process is the best approach to management of MSW, a preliminary economic analysis is performed for each of the thermal processes discussed previously. Then, a process economy and a process efficiency comparison can be shown for all of the thermal processes discussed.

TABLE 1.1   Thermal Process Technology(s)

Thermal Process Technology/Typical Range of Process Operation:	
Plasma Arc Gasification	7,200–12,600°F
Conventional gasification	1,400–2,800°F
Pyrolysis gasification	1,400–2,800°F
Pyrolysis	1,200–2,200°F
Mass burn (incineration)	1,000–2,200°F

Note: Except for plasma arc gasification, these processes have environmental issues for disposal of ash and slag.

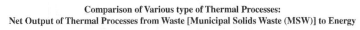

Comparison of Various type of Thermal Processes:
Net Output of Thermal Processes from Waste [Municipal Solids Waste (MSW)] to Energy

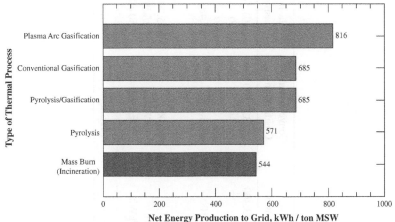

Note: Computations in this table were done (05-30-2007) by Dr. Gary C. Young from information
in the reference noted and Energy-from-waste, copyrighted 2007 Gary C. Young.
Reference: The Regional Municipality of Halton, Step 1B: EFW Technology Overview,
May 30, 2007 Submitted by Genivar, Ramboll, Jacques Whitford, Deloitte & URS,   Gary C. Young, Ph.D., 05-30-2007
Regional Municipality of Halton, 1151 Bronte Road, Oakville, ON L6M 3L1   VarThermalProcCompar 05 30 2007.pdw

FIGURE 1.7   Comparison of Various Types of Thermal Processes.

A preliminary economic analysis was completed on the five thermal processes. Parameters used in this economic evaluation are shown in Table 1.3 and were estimated from the available literature.[2–6,10]

Economic analysis with these parameters allows computation of the net revenue (before taxes) of each thermal process as shown in Fig. 1.8.

Mass burn shows negative net annual revenue (before taxes) while pyrolysis, pyrolysis/gasification, conventional gasification, and plasma arc gasification indicate positive net annual revenue (before taxes). Plasma arc gasification process has the highest net annual revenue. In addition, it should be pointed out that plasma arc gasification process produces vitrified slag that is an environmentally acceptable by-product with revenue as a road material at typically $15.00/ton.

On reviewing process characteristics of the thermal processes discussed, mass burn, pyrolysis, pyrolysis/gasification, and conventional gasification all typically produced ash as a by-product, which is not environmentally friendly since it must be disposed of in

TABLE 1.2   Thermal Process Technology and Net Energy to Grid

Type of Thermal Process Technology	Net Energy Production to Grid
Mass burn (incineration)	544 kWh/ton MSW
Pyrolysis	571 kWh/ton MSW
Pyrolysis/gasification	685 kWh/ton MSW
Conventional gasification	685 kWh/ton MSW
Plasma Arc Gasification	816 kWh/ton MSW

Source: Ref. (4).

TABLE 1.3 Parameters for the Economic Assessment of the Thermal Processes

Parameter	Mass Burn (Incineration)	Pyrolysis	Pyrolysis/ Gasification	Conventional Gasification	Plasma Arc Gasification
Capital investment at 6%, 20 years	$115,997,700	$86,936,900	$102,593,400	$80,337,800	$101,583,800
Plant capacity (tons MSW/day)	500	500	500	500	500
Energy production (kWh/ton MSW)	544	571	685	685	816
Operation and maintenance, capital budget, cost of ash disposal ($40/ton)[12] ($/year)	$8,216,600	$7,193,700	$7,711,100	6,871,800	$7,483,400
Tipping fee ($/ton MSW) (revenue)	$35.00	$35.00	$35.00	$35.00	$35.00
Green tags (revenue)	2 ¢/kWh	2 ¢/kWh	2 ¢/kWh	2 ¢/kWh	2 ¢/kWh
Production energy sales (revenue)	6.50 ¢/kWh	6.50 ¢/kWh	6.50 ¢/kWh	6.50 ¢/kWh	6.50 ¢/kWh
By-product	0.2	0.21	0.2	0.2	0.2
Residue (tons/ton MSW)	Ash	Ash/Char	Ash	Ash/Slag	Vitrified slag

Comparison of Various Types of Thermal Processes:
Net Annual Revenue (Before Taxes) From Waste [Municipal Solid Waste (MSW)] to Energy

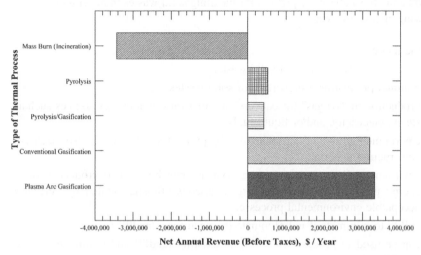

Dr. Gary C. Young, Ph.D.
GYCO, Inc.,   Nov. 18, 2008

ThermProcNetRevBarGdata 11 18 2008.pdw

FIGURE 1.8   Comparison of Various Types of Thermal Processes.

a landfill or other depository isolated from the environment. The plasma arc gasification process produces a "vitrified slag" as by-product. The vitrified slag is environmentally sound, since it is basically nonleaching and exceeds EPA leach test standards. Therefore, it can be used to produce other by-products such as rock wool, floor tiles, roof tiles, insulation, and landscaping blocks, or recycled as a road aggregate material.[2,4]

Toxicity leaching tests were conducted on the vitrified slag produced from MSW using a plasma arc gasification reactor.[4,10] Standard toxicity characteristics leaching procedure (TCLP) tests were conducted on vitrified sample materials from experiments. The results are shown in Table 1.4.

TABLE 1.4   Toxicity Leaching Test Results on Vitrified Slag

Heavy Metal	Permissible Concentration (mg/l)	Measured Concentration (mg/l)
Arsenic	5.0	<0.1
Barium	100.0	0.47
Cadmium	1.0	<0.1
Chromium	5.0	<0.1
Lead	5.0	<0.1
Mercury	0.2	<0.1
Selenium	1.0	<0.1
Silver	5.0	<0.1

Source: Ref. (4,10).

From the analysis above, it was concluded that plasma arc gasification process would be the most attractive process for handling solid wastes in general due to the following characteristics:

- thermal efficiency
- process variety of different solid wastes
- minimal pretreatment/presorting of solid wastes
- production of "syngas" for conversion into a variety of energy sources such as steam, electricity, and/or liquid fuels
- environmentally sound, since the solid by-product, vitrified slag, can be used as a construction material
- environmentally sound, since the "syngas" can be used to produce various energy products and any discharged gaseous effluents treated by currently acceptable environmental processes
- ability to minimize if not eliminate the need for a landfill
- can be used to process wastes in an existing landfill and eliminate the old landfill.

The plasma arc gasification process can be described as a "technologically advanced and environmentally friendly method of disposing of waste, converting it into commercially usable by-products. This process is a drastic nonincineration thermal process that uses extremely high temperatures in an oxygen-starved environment to completely decompose input waste material into very simple molecules. The intense and versatile heat generation capabilities of plasma technology enable a plasma gasification/vitrification facility to treat a large number of waste streams in a safe and reliable manner. The by-products of the process are a combustible gas and an inert slag. Plasma gasification consistently exhibits much lower environmental levels for both air emissions and slag leachate toxicity than other thermal technologies."[13]

Thus, the following Chapters 2, 3, 4, 5 and 7 are devoted to an estimate of the commercial economics and viability of plasma arc gasification. Chapter 6 presents the economic facts about cash flows from curbside pickup of garbage to landfill and the net cash revenues generated by the business segment of MSW management.

REFERENCES

1. Barbalace, K., The History of Waste. EnvironmentalChemistry.com. Aug. 2003. Accessed on-line:11/19/2008 http://EnvironmentalChemistry.com/yogi/environmental/wastehistory.html

2. Young, G.C., "Zapping MSW with Plasma Arc," Pollution Engineering, November 2006.

3. Young, G.C., "How Trash Can Power Ethanol Plants," Public Utilities Fortnightly, p. 72, February 2007.

4. Young, G.C., "From Waste Solids to Fuel," Pollution Engineering, February 2008.

5. "Summary Report: Evaluation of Alternative Solid Waste Processing Technologies," Prepared for: City of Los Angeles, Department of Public Works, Bureau of Sanitation, 419S. Spring Street, Suite 900, Los Angeles, CA 90013; Prepared by: URS Corporation, 915 Wilshire Boulevard, Suite 700, Los Angeles, CA 90017, September 2005.

6. "Conversion Technology Evaluation Report," Prepared for: The City of Los Angeles Department of Public Works and The Los Angeles County Solid Waste Management Committee/Integrated Waste Management Task Force's Alternative Technology Advisory Subcommittee; Prepared by: URS Corporation, 915 Wilshire Boulevard, Suite 700, Los Angeles, CA 90017, August 18, 2005.

7. U.S. Environmental Protection Agency, Landfill Methane Outreach Program (LMOP), www.epa.gov/lmop, Washington, D.C., 2008.

8. U.S. Environmental Protection Agency, An Overview of Landfill Gas Energy in the United States, Landfill Methane Outreach Program (LMOP), June 24, 2008.

9. Genivar, Ramboll, Jacques Whitford, Deloitte and URS, "The Regional Municipality of Halton, Step 1B: EFW Technology Overview," 30 May 2007, Oakville, Ontario, Canada.

10. Circeo, L.J., Engineering & Environmental Applications of Plasma Arc Technology, Technological Forum, Kirkwood Training and Outreach Services Center, Marion, Iowa, November 22, 2005.

11. Dodge, E., "Plasma-Gasification of Waste," Cornell University—Johnson Graduate School of Management, Queens University School of Business, July 2008.

12. Schneider, K., "Incinerator Operators Say Ruling Will Be Costly," The New York Times, May 3, 1994.

13. Moustakas, K., et al., "Demonstration plasma gasification/vitrification system for effective hazardous waste treatment," Journal of Hazardous Materials, vol. 123, pp. 120–126, 2005.

CHAPTER 2

How Can Plasma Arc Gasification Take Garbage To Electricity and a Case Study?

Subject: **Preliminary Economic Evaluation for a Municipal Solid Waste (MSW) Facility at 500 ton/day MSW Capacity using Plasma Arc Gasification Technology**

Background: **Many areas have a need for establishing and/or extending a landfill. As an alternative to the existing concept of a landfill, recently developed and early commercialization in Plasma Arc technology has been applied to the treatment of municipal solid waste (MSW). This recent development would eliminate or minimize the need for a landfill and this approach is known as Plasma Arc Gasification process for the treatment of MSW. A Plasma Arc Gasification facility with a 500 ton/day capacity has been stated as the size necessary for treating the Linn County MSW. Thus, this capacity was selected as reasonable with economy of scale for conducting a preliminary economic analysis.**

Plasma Arc Gasification process for the treatment of municipal solid waste (MSW) is a very high-temperature pyrolysis type of process (7,200–12,600°F) whereby the organics of waste solids (MSW) are converted to a synthesis gas (syngas) and the inorganics and minerals of the waste solids (MSW) produce a rocklike by-product.[1,2–5] The syngas is predominantly carbon monoxide (CO) and hydrogen (H_2).[3,6,5] The inorganics and minerals in the waste solids (MSW) are converted to a vitrified slag typically comprising metals and silica glass. This vitrified slag is basically nonleaching and exceeds EPA standards. Metals can be recovered from the slag and the slag can be used to produce other by-products such as rock wool, floor tiles, roof tiles, insulation, landscaping blocks to mention a few.[1,2–5] One of the simpler recyclable uses of the slag is as a road material but at a much lower economic value.[1] The synthesis type of gas can be used to produce electricity, and the rocklike by-product can be used as a material for road construction, since it is environmentally acceptable. The fuel/syngas produces by-products hydrochloric acid (HCl) and sulfur via the gas cleanup step.[4]

Municipal Solid Waste to Energy Conversion Processes: Economic, Technical, and Renewable Comparisons By Gary C. Young
Copyright © 2010 John Wiley & Sons, Inc.

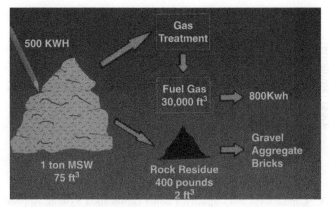

FIGURE 2.1   MSW to Energy, Plasma Arc Gasification.[1,2]

In this article, the economics are performed on a process for converting MSW by a Plasma Arc Gasification process (a pyrolysis process) to a syngas and a vitrified slag. The syngas is used to generate electricity and the slag as a road material. The bottom line is a process for treating MSW thereby eliminating the need for a landfill, and can be used to process existing landfill MSW sites.

MSW can be processed using the Plasma Arc Gasification process and simply represented as Fig. 2.1.[1,2]

One ton of MSW uses 500 kWh of the total electricity produced by the process but recent technological advances have reduced this usage to 200 kWh/ton of MSW.[7] The vitrified slag or rock residue produced is 400 pounds.

The above simplified representation of the MSW pyrolysis step (Fig. 2.1) can be shown as a simplified process flow diagram that shows that only the fuel/gas could be used for the generation of electricity and/or the synthesis of chemicals (Fig. 2.2).[4]

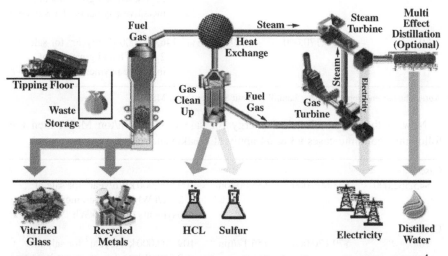

FIGURE 2.2   Process Flow Diagram, Plasma Gasification System, Applied to MSW.[4]

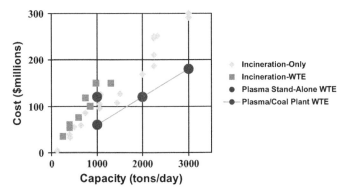

FIGURE 2.3 Capital Cost and Capacity, Plasma Gasification System, Applied to MSW.[1]

The capital cost for a Plasma Arc Gasification process to produce electricity and a vitrified slag can be deduced from Fig. 2.3 that depicts capital cost versus capacity.[1]

Thus, with the information presented so far, preliminary economics can be deduced for the needs of Linn County, Iowa at a plant capacity of 500 tons/day of MSW. As stated, the excess electrical energy produced for sale is 800 kWh/ton MSW. The following economic cases #1 and #2 apply at breakeven.

Capital Investment			
Government grant	Industry and Local Government (Financed)	Tipping Fee for MSW at Breakeven	Comments
Case 1			
$30,480,000	$48,690,000	$35.00/ton	146,000,000 kWh/year[2] for sale at 2.5 ¢/kWh and government incentive payments of 1.8 ¢/kWh
Case 2			
$0	$79,170,000	$49.20/ton	146,000,000 kWh/year[2] for sale at 2.5 ¢/kWh and government incentive payments of 1.8 ¢/kWh

Note: The current tipping fee at a landfill is about $35 per ton of MSW.

Now, if the excess electrical energy for sale is 600 kWh/ton MSW, then the following economic cases #3 & #4 apply at breakeven.

Case 3			
$45,562,000	$33,608,000	$35.00/ton	109,500,000 kWh/year[2] for sale at 2.5 ¢/kWh and government incentive payments of 1.8 ¢/kWh
Case 4			
$0	$79,170,000	$56.17/ton	109,500,000 kWh/year[2] for sale at 2.5 ¢/kWh and government incentive payments of 1.8 ¢/kWh

If the government incentive payment of 1.8 cents/kWh is replaced with a 1.8 cents/ kWh tax credit similar to that used for wind energy and a utility company sells the energy produced, a break-even case #5 applies.

Case 5			
$0	$79,170,000	$35.00/ton	109,500,000 kWh/year[2] for sale at 6.72 ¢/kWh and with government tax credits of 1.8 ¢/kWh

BASIS

A Plasma Arc Gasification facility operates at 500-tons/day capacity for the treatment of MSW with a "total" capital investment of $79,170,000.[1,8–10] Capital financed for the project would be at 5.75% interest for 20 years and making two payments per year. A government incentive of 1.8 cents/kWh revenue to produce a renewable energy up to maximum payment of $750,000.00 per year and a 1.0 cent/kWh revenue from green tags associated with renewable energy production were used in the financial analysis. In cases #1 & #2, the facility would generate 146,000,000 kWh/year of electrical energy (renewable energy) for sale to the grid system at 2.5 cents/kWh.[2] In cases #3 and #4, the facility would generate 109,500,000 kWh/year of electrical energy for sale to the grid system at 2.5 cents/kWh.[1] The by-product rock material would be sold as a road construction material at $15/ton.[1] Such cost considerations for Operations and Maintenance and a Capital Budget Reserve have been considered in the economic evaluation. Over 50 jobs would be created by this new plasma facility.

The basis for case 5 is similar to that for case −4, except that a government tax credit of 1.8 cents/kWh is used rather than a cash incentive payment. Also, the utility company sells the excess power generated at 6.72 cents/kWh for a break-even tipping fee of $35.00/ton MSW.

ECONOMIC CASES

From these "initial" economic analyses, one likely approach to a business plan involves a cooperative effort between a local utility and local governmental entities. Thus, the local government participates in the MSW treatment part of the Plasma Arc facility while the local utility participates in the electrical energy (renewable energy) producing part of the facility. Thus, the local government remains in the MSW business while the utility stays in the electrical energy business. This appears to be a win-win situation for the local government and the utility.

Case 1 presents the preliminary economics of a Plasma Arc Gasification facility for Linn County, Iowa for a facility producing 800 kWh/ton MSW for sale as excess energy. Private/Industry invests 61.5% of the total capital and a governmental grant for the remaining 38.5% of the capital requirements. The cooperative venture results in a "tipping fee" for the MSW of $35/ton at breakeven. A tipping fee of $35/ton of MSW is typical for Linn County, Iowa.[8,9]

Case 2 presents an economic evaluation similar to the previous case but private/industry finances 100% of the total capital requirements of $79,170,000. The net result at breakeven is for the "tipping fee" to be increased to $49.20/ton MSW. If, however, Case 2 had the electrical energy produced sold at 4.5 cents/kWh, the breakeven price for the tipping fee would be about $33/ton MSW.

Case 3 presents the preliminary economics of a Plasma Arc Gasification facility for Linn County, Iowa for a facility producing 600 kWh/ton MSW for sale as excess energy. Private/Industry invests 42.5% of the total capital and a governmental grant for the remaining 57.5% of the capital requirements. The cooperative venture results in a "tipping fee" for the MSW of $35/ton at breakeven. A tipping fee of $35/ton of MSW is typical for Linn County, Iowa.[8,9]

Case 4 presents an economic evaluation similar to the previous case but private/industry finances 100% of the total capital requirements of $79,170,000. The net result at breakeven is for the "tipping fee" to be increased to $56.17/ton MSW. If, however, case 4 had the electrical energy produced sold at 4.5 cents/kWh, the break-even price for the tipping-fee would be about $44/ton MSW.

Case 5 is similar to case 4 except that a government tax credit of 1.8 cents/kWh is used rather than a cash incentive payment. Also, the utility company sells the excess power generated at 6.72 cents/kWh for a break-even tipping fee of $35.00/ton MSW. Additionally, the utility company generates a tax credit of $1,971,000.00 per year. With the tax credit considered as a revenue source, the break-even selling price for energy produced becomes 4.92 ¢/kWh.

LOGICAL APPROACH FOR FUTURE PROGRESS

A logical approach is to take initial economics presented for a particular area and periodically "update" the analysis by a cooperative effort between governmental body(s) and industrial entity(s) so that both parties will have a fully transparent evolvement and trust in the final economic analysis. Thus, this factual, transparent, updated economic analysis will therefore determine the final approach taken by both government and industry for that particular area investigated to determine if Plasma Arc Gasification is an economically and environmentally attractive alternative to a landfill.[11]

As a final comment, the attractiveness of Plasma Arc Gasification technology is in the generation of renewable electrical energy with a useful by-product having environmentally acceptable properties. The bottom line is a process based upon "plasma technology" for treating MSW thereby eliminating the need for a landfill, and can be used to process existing landfill MSW sites.

In summation, a utility company in partnership with a local government would likely be the most economical combination and have the most positive benefit to the environment. In this article, the economics were stated as preliminary since such evaluations are site specific and should be done specifically for each case under study, and plasma processing of MSW is an emerging technology.

With some diligence, a viable business plan can be developed from the consideration of the many factors influencing the economics associated with a specific site selection and surrounding community.

Now that the key process concepts are presented in Fig. 2.2 and process economics, a discussion follows about more details of processing the crude syngas from the gasifier.

Raw syngas exiting the gasifier is typically cooled in a fire-tube boiler from high gasification temperature to about 700°F. Any char of fly ash produced from the final zone of the gasifier and entrained with the crude syngas stream is removed by a downstream cyclone and candle filter particulate control devices and returned to the gasifier. The candle filter is composed of ceramic candle filter elements and cleaned by the commonly accepted back-pulsing concept with syngas. The raw syngas exits at 700°F.

The raw syngas introduced into the fire-tube syngas cooler produces steam for use in a turbine that becomes part of the heat recovery system. Raw syngas then enters a scrubber at 700°F for removal of any remaining particulates and trace components. Next, the syngas stream is sent to a hydrolysis reactor whereby the carbonyl sulfide (COS) and hydrogen cyanide (HCN) are converted to hydrogen sulfide (H_2S).

The syngas stream is then processed by a mercury removal system such as by Eastman Chemical consisting of a packed bed of a special activated carbon that removes the mercury, arsenic, and trace materials. Following the mercury removal system, a Selexol unit can be used to remove H_2S and CO_2 from the cool syngas. A polishing unit of zinc oxide removes any residual H_2S and achieves a high-purity syngas stream.

As discussed, the process unit operations are presented to give a conceptual idea of the process equipment and technology currently available to clean up a raw syngas from a gasifier.

Many existing technologies and engineering experiences are available for design and further development for the cleanup of raw syngas from gasifiers. It is just important to become familiar with these technologies and to form the proper team whereby such important technologies are integrated properly into the waste-to-energy gasification process.[12] Final selection of the gas cleanup process parameters will depend on the wastes selected for feedstock to the gasification process.

REFERENCES

1. Circeo, L.J., Engineering & Environmental Applications of Plasma Arc Technology, Technological Forum, Kirkwood Training and Outreach Services Center, Marion, Iowa, November 22, 2005.

2. Circeo, L.J., "Engineering & Environmental Applications of Plasma Arc Technology," Presentation, Georgia Tech Research Institute, Atlanta, GA 2005.

3. Vera, R., "Organic Waste, Gasification and Cogeneration," Presentation, Trinity Plasma Technology, Technologies International Corporation, Trinity Consultants, Inc., 2005.

4. Recovered Energy, Inc., Pocatello, Idaho, www.recoveredenergy.com, "Process Flow Diagram," MSW into energy and useable by-products.

5. "Summary Report: Evaluation of Alternative Solid Waste Processing Technologies," Prepared for: City of Los Angeles, Department of Public Works, Bureau of Sanitation, 419 S. Spring Street, Suite 900, Los Angeles, CA 90013; Prepared by: URS Corporation, 915 Wilshire Boulevard, Suite 700, Los Angeles, CA 90017, September 2005.

6. Lee, C.C., "Plasma System," Standard Handbook of Hazardous Waste Treatment and Disposal, McGraw-Hill Book Company, New York, 1989, p. 8.169.

7. Circeo, L.J., Private Communication, April 4, 2006.

8. "Expert Touts Plasma Torch," The Gazette, November 22, 2005.

9. "Plasma Arc Technology may Help Linn Garbage Woes," The Gazette, November 20, 2005.

10. Circeo, L.J. and Smith, M.S., "Plasma Processing of MSW at Coal-Fired Power Plants," Presentation, Health and Environmental Systems Laboratory, Georgia Tech Research Institute, Georgia Institute of Technology, Atlanta, GA, 2005.

11. Young, G.C., "Zapping MSW with Plasma Arc," Pollution Engineering, November 2006.

12. DOE/NETL-2007/1260, "Baseline Technical and Economic Assessment of a Commercial Scale Fischer–Tropsch Liquids Facility," Final Report for Subtask 41817.401.01.08.001, National Energy Technology Laboratory (NETL), April 9, 2007.

CHAPTER 3

How Can Plasma Arc Gasification Take Garbage to Liquid Fuels and Case Studies?

Two approaches or synthesis routes will be presented in this chapter to produce a liquid fuel from MSW. Plasma Arc Gasification will be used to produce the Syngas. The Chemistry route or the Biochemistry route will then be used to convert the Syngas to a liquid fuel. The Chemistry route will use a Fischer-Tropsch catalyst method to convert the Syngas to a liquid fuel. The Biochemical route will use microbes to convert the Syngas to a liquid fuel.

MSW TO SYNGAS TO LIQUID FUELS VIA CHEMISTRY (FISCHER–TROPSCH SYNTHESIS) AND A CASE STUDY

Subject: Preliminary Economic Evaluation for a Municipal Solid Waste (MSW) Facility Using Plasma Arc Gasification Technology to Produce Liquid Fuels (Ethanol/Methanol) via a Chemistry Fischer-Tropsch synthesis route.

Background: Many areas in the United States have both carbonaceous resources such as coal deposits and renewable resources/wastes such as landfills, which are potential sources of energy. In the near future, more utilities will use these coal resources for gasification power plants utilizing Integrated Gasification Combined Cycle (IGCC) technology to generate electrical/steam energy. With new technology, the wastes in the landfills can be gasified and converted to electrical/steam energy. Much of this potential energy source in the landfill is in the form of municipal solid waste (MSW) and industrial waste (IW) produced daily and/or in an existing landfill. Whether coal or MSW is used as the raw material, new power plants will gasify these materials to predominantly carbon monoxide (CO) and hydrogen (H_2), known as syngas. This syngas can be converted to electrical/steam and/or liquid fuels/chemicals. Thus, a gasification power plant can be flexible with

Municipal Solid Waste to Energy Conversion Processes: Economic, Technical, and Renewable Comparisons By Gary C. Young
Copyright © 2010 John Wiley & Sons, Inc.

a syngas that can produce electricity on demand and simultaneously produce some liquid fuels/chemicals during off-peak periods. Such a synergy of syngas processes in producing electricity and fuels/chemicals should maximize the bottom line for a gasification facility. In reality, a gasification facility, due to syngas, places a utility into the "synthesis business," i.e., synthesizing electricity and/or fuels.

Both of these processes, coal gasification and MSW gasification to steam/electrical energy, have previously been economically evaluated commercially.[1,2] Thus, this article will evaluate the commercial gasification economics of MSW to syngas for the production of liquid fuels such as ethanol and methanol. Specifically, the economics of Plasma Arc Gasification technology will be evaluated as applied to the treatment of MSW, since this technology is considered the most efficient of the gasification processes.[3] For example, conventional gasification plant produces about 685 kWh/ton MSW (net energy to the grid compared to Plasma Arc Gasification plant with 816 kWh/ton MSW).

Plasma Arc Gasification process can be used to produce a syngas, which can be converted to liquid fuels via Fischer–Tropsch synthesis.[4] For example, consider an ethanol/methanol plant using MSW as a raw material with a capacity of 500 tons/day MSW. The MSW can be gasified to a syngas using Plasma Arc Gasification technology and then converted to liquid fuels (ethanol and methanol) via Fischer–Tropsch synthesis of the syngas. For this plant, a preliminary economic analysis was conducted. A simplified schematic is shown in Fig. 3.1 for such a process (combined Plasma Arc Gasification and ethanol/methanol Fischer-Tropsch synthesis plant).

Plasma Arc Gasification process for the treatment of MSW is a very high-temperature pyrolysis type of process (7,200–12,600°F) whereby the organics of waste solids (MSW) are converted to a synthesis gas (syngas) and the inorganics and

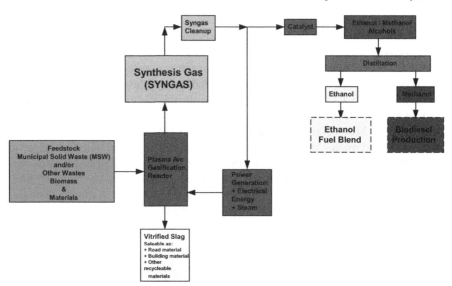

FIGURE 3.1   Selected Synthesis Gas Options for Syngas.

minerals of the waste solids (MSW) produce a rocklike by-product.[5,6] The syngas is predominantly CO and H_2.[6,7] The inorganics and minerals in the waste solids (MSW) are converted to a vitrified slag comprising typically metals and silica glass. This vitrified slag is basically nonleaching and exceeds EPA standards. Metals can be recovered from the slag and the slag can be used to produce other by-products such as rock wool, floor tiles, roof tiles, insulation, landscaping blocks to mention a few.[5,6]

One of the simpler recyclable uses of the slag is as a road material but at much lower economic value than the other uses mentioned.[5] The rocklike by-product is useful as a road construction material, since it is environmentally acceptable.[8,9] One primary requirement of the vitrified slag for use as a road material is to be highly nonleachable. A primary analysis used to show nonleachable characteristics is the Toxicity Characteristic Leaching Procedure (TCLP) as in the risk assessment portion of EPA's 1999 Report to Congress. TCLP tests conducted on the vitrified slag produced from MSW using Plasma Arc Gasification Reactor are shown in Table 3.1.[8]

Road material/stones derived from a slag, using a high-temperature melting process (vitrification process, 1,450°C), pass TCLP analysis and have been used as aggregates in asphalt and concrete.[9] When stones were used as a substitute for asphalt aggregate in constructing a public road in Kamagaya City, Japan, no difference was observed between the road construction area using these stones as substitute for coarse and fine aggregates compared to that using natural aggregates.[9] When these stones were used as concrete aggregates, the results were better than using natural aggregates. Concrete aggregate strength tests are shown in Table 3.2 when substituting 100% of the stones for coarse aggregates.

The synthesis type of gas can be used to produce electricity, steam, and/or liquid fuels. The syngas can be converted into liquid fuels such as ethanol and methanol via Fischer–Tropsch catalytic process. The fuel/syngas produces hydrochloric acid (HCl) and sulfur as by-products via the cleanup step.

The economics are performed on a process for converting MSW by a Plasma Arc Gasification process (a pyrolysis process) to a syngas and a vitrified slag. The syngas will be used to generate predominantly, ethanol and methanol, and the slag as a road material.

The ethanol product can be used in an ethanol blend motor fuel, and the methanol can be used in biodiesel manufacturing. The bottom line is a process for treating MSW

TABLE 3.1 TCLP Tests on Vitrified Slag

	TCLP	
Heavy Metal	Permissible Concentration (mg/l)	Measured Concentration (mg/l)
Arsenic	5.0	<0.1
Barium	100.0	0.47
Cadmium	1.0	<0.1
Chromium	5.0	<0.1
Lead	5.0	<0.1
Mercury	0.2	<0.1
Selenium	1.0	<0.1
Silver	5.0	<0.1

Source: Ref. (8).

TABLE 3.2   Concrete Aggregate Strength Test

	Result of Concrete Aggregate Strength Test		
	Compressive Strength (N/mm^2)	Flexural Strength (N/mm^2)	Tensile Strength (N/mm^2)
Base concrete	34.8	4.85	2.68
100% substitute for coarse aggregates	39.3	5.09	2.98

Source: Ref. (9).

TABLE 3.3   Economics of Plasma Arc Gasification Facility, Case Study

	Capital Investment Industry and Local Government	Tipping Fee for MSW	Ethanol Sales Price	Methanol Sales Price
Case				
Capital for plasma plant	$67,723,000	$35.00/ton	$2.0896/ gallon	$0.93/ gallon
Capital for ethanol/methanol plant[11]	$26,954,000			
Total capital investment	$94,677,000			

Net revenue before taxes from combined cacility operation = [revenue–expenditures] = $6,128,800.00 per year

Note: The current tipping fee at a landfill in the Cedar Rapids, Iowa area is about $35 per ton of MSW, which is typical for many parts of the country.

thereby eliminating the need for a landfill, and can be used to process existing landfill MSW sites and simultaneously produce valuable motor fuels.

The capital cost for the Plasma Arc Gasification process to produce electricity and/ or syngas and a vitrified slag can be found in a previous publication, and the capital cost for the ethanol/methanol synthesis process were determined elsewhere.[1,10] Thus, with the information presented so far, a preliminary evaluation and economics can be deduced for a Plasma Arc plant using MSW with sufficient capacity to supply both electricity and syngas for an adjacent process plant for producing ethanol and methanol.[1] Plant capacity was taken as 500 tons/day MSW for the economic case shown in Table 3.3.

Basis

A Plasma Arc Gasification facility operates at 500-tons/day capacity for treatment of MSW with a "total" Capital Investment of $67,723,000 for the Plasma

Arc processing plant and about $26,954,000 for the Fischer–Tropsch synthesis processing plant including the distillation unit for the separation of the two alcohols, ethanol and methanol.[5,12–14,10] The plasma plant supplies sufficient energy as electricity/steam to the overall plant for the production of about 10,406,400 gallons per year of alcohol mixture, which is about 60% ethanol and 40% methanol.[12] The alcohol mixture yield was taken at a conservative value of about 95 gallons per ton MSW (dry material). Capital cost of the overall plant is $94,677,000.[1,15,10] Capital financed for both plants would be at 6.00% interest for 20 years by making two payments per year. The by-product rock material would be sold as a road construction material at $15/ton.[5] Cost considerations for operations and maintenance, a capital budget reserve, process water, and sewer have been considered in the economic evaluation. Over 40+ jobs would be created by a combined plasma and liquid fuel (ethanol/methanol) facility. The combined facility is self-sufficient in electrical and steam energy requirements, since this energy is produced internally from a renewable raw material, MSW.

Economic Case

From these "initial" economic analyses, one likely approach to a business plan is for a cooperative effort between a local utility, local industry, and/or local governmental entities. Thus, the local government participates in the MSW treatment part of the Plasma Arc facility while the local utility participates in the electrical/steam energy/syngas and liquid fuels producing part of the facility. Alternately, a local industrial/business entity could own/operate the liquid fuel (ethanol/methanol) facility. Thus, the local government remains in the MSW business while the utility and/or industry stays in the electrical, steam, and liquid fuel energy businesses. This appears to be a win-win situation for the local government, utility, and local industry.

The preliminary economics of the combined Plasma Arc Gasification and Fischer–Tropsch Catalytic Synthesis facility was considered for a Linn County/Cedar Rapids, Iowa area facility processing about 500 tons/day of MSW. A positive cash flow is obtained for a cooperative venture with a "tipping fee" for the MSW of $35.00/ton and a selling price of ethanol at $2.0896 per gallon and methanol at $0.93 per gallon.[16,17] A tipping fee of $35/ton of MSW is typical for Linn County, Iowa.[12,13] Net revenue before taxes from combined facility operation = [revenue–expenditures] = $6,128,800.00 per year.

If the alcohol mixture yield was taken at a value of about 113 gallons per ton MSW (dry material) rather than the previous yield of 95 gallons/ton MSW (dry material), the net revenue before taxes from combined facility operation = [revenue–expenditures] = $8,661,800.00 per year.[10] The alcohol mixture produced would be 12,439,100 gallons per year.

As expected, this economic evaluation demonstrates the commercial importance of utilizing the Fischer–Tropsch catalyst with the highest yield for the products desired.[12] In today's energy market, Fischer–Tropsch catalysts with higher yields and selectivity for desired products are most likely to be just over the horizon.

Logical Approach for Future Progress

A logical approach is to take initial economics presented for a particular area and periodically "update" the analysis by a cooperative effort between governmental body(s) and industrial entity(s) so both parties will have a fully transparent evolvement and trust in the final economic analysis. Thus, this factual, transparent, updated economic analysis will therefore determine final approach taken by both government and industry for that particular area investigated to determine if Plasma Arc Gasification is an economically and environmentally attractive alternative to a landfill when integrated with a Fischer–Tropsch synthesis plant.

As a final comment, the attractiveness of Plasma Arc Gasification technology lies in the generation of renewable electrical/steam energy with useful by-products. The bottom line is a process based upon "plasma technology" for treating MSW thereby eliminating the need for a landfill, and can be used to process existing landfill MSW sites. The energy generated from the plasma plant can supply energy for itself and additional syngas for the production of saleable electricity and other products such as liquid fuels. Other fuels can be produced besides alcohols using Fischer–Tropsch synthesis such as illustrated in Fig. 3.2 for various options.[1,15,18,19]

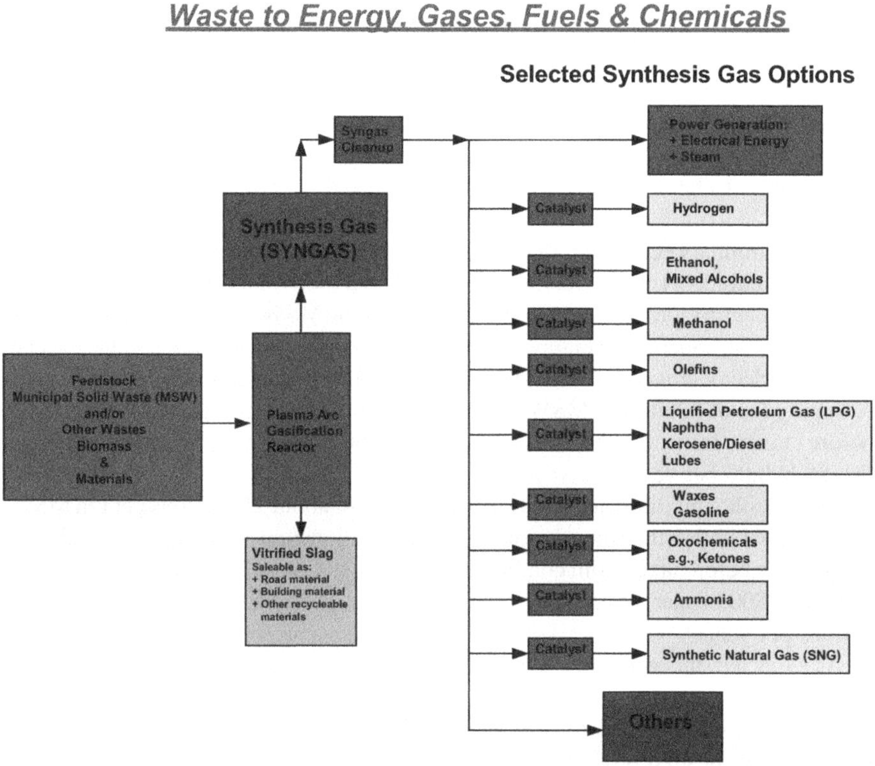

FIGURE 3.2   Selected Synthesis Gas Options for Gases, Fuels, and Chemicals.[4]

A utility and/or business company(s) in partnership with a local government would likely be the most economical combination and have the most positive benefit to the environment and financial reward to the local area. In this article, the economics were stated as preliminary, since such evaluations are site specific and should be done specifically for each case under study, and plasma processing of MSW is an emerging technology.

With some diligence, a viable business plan can be developed from the consideration of the many factors influencing the economics associated with a specific site selection, economy of scale, and the surrounding community.

MSW TO SYNGAS TO LIQUID FUEL VIA BIOCHEMISTRY AND A CASE STUDY

Subject: **Preliminary Economic Evaluation for a Municipal Solid Waste (MSW) Facility using Plasma Arc Gasification Technology to Produce Liquid Fuel (Ethanol) via a Biochemistry synthesis route.**

Background: **Areas in the United States typically have carbonaceous deposits such as wastes in landfills, which are potential sources of renewable energy. In the near future, more communities and/or utilities will use these resources for power plants with gasification technology to generate electrical/steam energy. With new technology, the wastes in the landfills can be gasified and converted to electrical/steam energy. Much of this potential energy source in the landfill is in the form of Municipal Solid Waste (MSW) and Industrial Waste (IW) produced daily and/or in an existing landfill. With MSW used as the raw material, new power plants will gasify this material to predominantly Carbon monoxide (CO) and Hydrogen (H_2), known as syngas. This Syngas can be converted to electricity/steam and/or liquid fuels/chemicals. Thus, a gasification power plant can be flexible with a syngas that can produce electricity on demand and simultaneously produce some liquid fuels/chemicals during off-peak periods. Such a synergy of syngas processes in producing electricity and fuels/chemicals should maximize the bottom line for a Gasification facility. In reality, a gasification facility, due to syngas, places a utility into the "synthesis business," i.e., synthesizing electricity and/or fuels.**

MSW gasification to steam/electrical energy has previously been economically evaluated commercially.[1,2] This article will evaluate the commercial gasification economics of MSW to syngas for the production of liquid fuel, such as ethanol.

Specifically, the economics of Plasma Arc Gasification technology will be evaluated as applied to the treatment of MSW, since this technology is considered the most efficient of the gasification processes.[3] Specifically, the syngas produced by the Plasma Arc Gasification step will be converted into ethanol via a biochemical process. This biochemical process is called the BRI process.[20,21,22,23]

Plasma Arc Gasification process can be used to produce a syngas, which can be converted to liquid fuel such as ethanol via biochemical technology. For example,

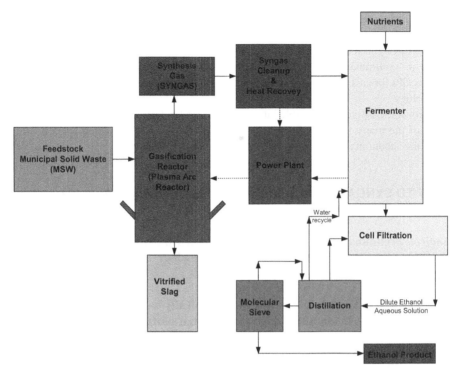

FIGURE 3.3   MSW to Plasma Arc Gasification—Syngas—Biochemical (BRI, Inc.)—ethanol.[4,23]

consider an ethanol plant using MSW as a raw material with a capacity of 500 tons/day MSW. The MSW can be gasified to a syngas using Plasma Arc Gasification technology and then converted to liquid fuel via biochemical fermentation process. For this plant, a preliminary economic analysis was conducted. A simplified schematic is shown for such a process (combined Plasma Arc Gasification and ethanol plant via BRI biochemical process) in Fig. 3.3.[4,23]

As illustrated in Fig. 3.3, MSW can be gasified using a Plasma Arc Gasification reactor producing a syngas (predominantly CO and H_2). The syngas is cleaned and cooled to 98°F prior to entering a fermenter (biocatalytic reactor) where ethanol is produced. A form of bacteria culture *Clostridium ljungdahlii* is used in the fermenter, and nutrients are added to provide cell growth and regeneration of the biocatalyst. The anaerobic fermentation takes place such that water and ethanol are the end products. The dilute ethanol aqueous solution from the fermenter is about 2–3% ethanol. Higher concentrations of ethanol inhibit bacteria metabolism. The dilute ethanol solution enters the distillation step and finished product is 99.5% pure industrial or fuel-grade ethanol.[20,22,23] The process recovers waste heat from syngas as it exits the gasifier and must be treated and cooled prior to the fermentation process. Overall, the process generates a surplus of energy that can be used to produce steam for export or produce electricity for sale to the grid.[20,21,22,23]

BASIS AND ECONOMICS

The preliminary economics of this process from MSW to finished ethanol product will be determined as illustrated in Fig. 3.3. Plasma Arc Gasification process for the treatment of MSW is a very high-temperature pyrolysis type of process (7,200–12,600°F) whereby the organics of waste solids (MSW) are converted to a syngas and the inorganics and minerals of the waste solids (MSW) produce a rocklike by-product.[5,6] The syngas is predominantly CO and H_2.[6,7] The inorganics and minerals in the waste solids (MSW) are converted to a vitrified slag typically comprising metals and silica glass. This vitrified slag is basically nonleaching and exceeds EPA standards. The vitrified slag will be sold as a road material. Syngas from the gasifier will be cleaned and cooled to 98°F prior to transport to the fermentation process whereby an aqueous stream of ethanol is produced and distilled into a finished fuel-grade ethanol.

A Plasma Arc Gasification facility, like a facility befitting Linn County of Iowa, would operate at 500-tons/day capacity for treatment of MSW at 23.2 wt.% moisture with a capital investment of $76,187,800 for the Plasma Arc processing plant and about $35,408,000 for the biochemical fermentation processing plant including the distillation unit for the separation of the ethanol from the aqueous stream. The plasma plant supplies sufficient energy as electricity/steam to the overall plant for the production of about 12,989,500 gallons/year saleable product. Capital cost of the overall plant is $111,595,800. Capital is financed at 6.00% interest for 20 years by making two payments per year. The by-product rock material (vitrified slag) is sold as a road construction material at $15/ton.[5] Cost considerations for operations and maintenance, a capital budget reserve, coke for use in the gasification reactor, process water, and sewer have been considered in the economic evaluation for a sum of $10,172,800 per year. Over 40 + jobs at an hourly wage of $28/hour would be created by a combined plasma and liquid fuel ethanol facility. The combined facility is self-sufficient in electrical and steam energy requirements, since this energy is produced internally from a renewable raw material, MSW. In fact, the overall process creates excess energy from sale to the grid of the amount of 0.0128 MW/(tons MSW/day) processed for syngas to the biochemical plant and sold at 4.5 ¢/kWh. A "tipping fee" of $35.00/ton MSW processed was considered as a revenue source, which is typical if this material was diverted instead to a landfill.

Selling price of the ethanol is $1.8966/gallon.[24] Net revenue before taxes from the combined facility operation = [revenue–expenditures] = $14,442,000/year. The manufacturing cost per gallon of ethanol produced was $1.5269/gallon.

This economic case is summarized in Table 3.4.

Another case was computed with a change in the capital cost for the facility as follows:

Capital	Gasification Plant—$67,722,500.00	Fermentation Plant—$30,789,500.00
	Total capital=$98,512,000.00 at 6.00% interest for 20 years at two payments/year	

TABLE 3.4 Economic Case, 500 ton/day MSW to Syngas—Fermentation—Ethanol

	Gasification Plant—$76,187,800.00	Fermentation Plant—$35,408,000.00
Capital	Total capital = $111,595,800.00 at 6.00% interest for 20 years at two payments/year	
MSW	500 tpd MSW, 182,500 ton/year at $35.00/ton (tipping fee) = $6,387,500.00/year	
Syngas	143,334 ton/year MSW gasified to syngas for use in fermentation	
Electricity (excess) energy (renewable energy) for sale	33,816,724 kWh/year at 4.5 ¢/kWh	
Green tags at 2.00 ¢/kWh for renewable energy sold to the grid		
State of Iowa tax credit at 1.5 ¢/kWh for renewable energy sold to grid		
Vitrified slag (0.2 ton vitrified slag/ton MSW): 36,500 tons/year sold at $15.00/ton		
Ethanol production	12,989,497 gallons/year at $1.8966/gallon	
Manufacturing cost per gallon of ethanol produced = $1.5269/gallon		

With all other parameters as the previous case, the net revenue before taxes from the combined facility operation = [revenue–expenditures] = $15,574,000/year. The manufacturing cost per gallon of ethanol produced was $1.4398/gallon.

Company(s) in partnership with a local government could be an economical combination and have the most positive benefit to the environment and financial reward for the local area. The economics were stated as preliminary, since such evaluations are site specific and should be done specifically for each case under study, and plasma processing of MSW as well as biochemical technology is an emerging technology.

With some diligence, a viable business plan can be developed from the consideration of the many factors influencing the economics associated with a specific site selection, economy of scale, and the surrounding community.

REFERENCES

1. Young, G.C., "Zapping MSW with Plasma Arc," Pollution Engineering, November 2006.
2. DOE/NETL-2007/1281, "Cost and Performance Baseline for Fossil Energy Plants, Vol. 1: Bituminous Coal and Natural Gas to Electricity, Final Report," May 2007.
3. Genivar, Ramboll, Jacques Whitford, Deloitte, and URS, "The Regional Municipality of Halton, Step 1B: EFW Technology Overview," 30 May 2007, Oakville, Ontario, Canada.
4. Young, G.C., "From Waste Solids to Fuel," Pollution Engineering, February 2008.
5. Circeo, L.J., Engineering and Environmental Applications of Plasma Arc Technology, Technological Forum, Kirkwood Training and Outreach Services Center, Marion, Iowa, November 22, 2005.
6. "Summary Report: Evaluation of Alternative Solid Waste Processing Technologies," Prepared for: City of Los Angeles, Department of Public Works, Bureau of Sanitation, 419 S. Spring Street, Suite 900, Los Angeles, CA 90013; Prepared by: URS Corporation, 915 Wilshire Boulevard, Suite 700, Los Angeles, CA 90017, September 2005.
7. Lee, C.C., "Plasma System," Standard Handbook of Hazardous Waste Treatment and Disposal, McGraw-Hill Book Company, New York, 1989, p. 8.169.
8. Circeo, L.J., "Evaluation of Plasma Arc Technology for the Treatment of Municipal Solid Wastes in Georgia," Georgia Tech Research Institute, Atlanta, GA, January 1997.
9. Nishida, K., et. al., "Melting and Stone Production Using MSW Incinerated Ash," Tsukishima Kikai Co., Ltd., Tokyo, Japan, 1999.
10. Syntec Biofuel Research, Inc., Private Communication, Vancouver, Canada, August 2007.
11. PacifiCorp, A MidAmerican Energy Holdings Company, Request for Proposal Technical Conference: Utah Docket 05-035-47, April 3, 2006; IGCC Cost of Energy Estimates; David Sokol, Chairman & CEO of MidAmerican Energy Holdings Company.
12. "Expert touts plasma torch," The Gazette, November 22, 2005.
13. "Plasma arc technology may help Linn garbage woes," The Gazette, November 20, 2005.
14. Circeo, L.J., "Engineering and Environmental Applications of Plasma Arc Technology," Presentation, Georgia Tech Research Institute, Atlanta, GA 2005.

15. Young, G.C., "How Trash Can Power Ethanol Plants," Public Utilities Fortnightly, p. 72, February 2007.

16. Methanex Company, www.methanex.com, August 1–August 31, 2007.

17. American Coalition of Ethanol, www.ethanol.org, State Average Ethanol Rack Prices, Iowa, August 16, 2007.

18. Sasol Chevron, www.sasolchevron.com, London, United Kingdom, 2007.

19. Jenkins, B.M. and Williams, R.B., "Thermal Technologies for Waste Management," University of California, Davis, April 2006.

20. Stewart, J.L., "The Co-Production of Ethanol and Electricity From Carbon-Based Wastes," BRI Energy, LLC, March 2006.

21. Bruce, W.F., "The Co-Production of Ethanol and Electricity From Carbon-Based Wastes," BRI Energy, LLC, 2006.

22. MacDonald, T., Schuetzie, D., Tamblyn, G., and Tornatore, F., "Assessment of Conversion Technologies for Bioalcohol Fuel Production," Western Governor's Association National Biomass State and Regional Partnership Report, Category XII—Fermentation of Syngas From Thermochemical Processes, Bioengineering Resources, Inc., Fayetteville, Arkansas, July 2006.

23. Vessia, O., Finden, P., and Skreiberg, O., Biofuels from lignocellulosic material—In the Norwegian context 2010—Technology, Potential and Cost, NTNU: Norwegian University of Science and Technology, December 20, 2005.

24. American Coalition for Ethanol, State Average Ethanol Rack Prices, $1.8966/gallon, Tuesday, November 18, 2008.

CHAPTER 4

Plasma Economics: Garbage/Wastes to Electricity, Case Study with Economy of Scale

Subject: "Preliminary" Economic Evaluation for the Management of Municipal Solid Waste (MSW), Commercial Waste (CW), and Industrial Waste (IW) Located in Marion, Cedar Rapids, and Linn County, Iowa Using Plasma Arc Gasification Technology

A preliminary economic analysis was conducted to evaluate the use of Plasma Arc Gasification technology for the management of municipal solid waste (MSW) for the Linn County, Cedar Rapids, and Marion, Iowa area. When MSW is gasified, the products are synthesis gas "syngas" and vitrified slag. For the purposes of this economic evaluation, the "syngas" is used to generate electricity and the slag is used as a road material. The basic Plasma Arc Gasification Process being evaluated is represented in Fig. 4.1.

For the economic analysis, the quantity of MSW for processing in the Plasma Arc Gasification facility was determined. The following quantities of waste to be treated are from the Cedar Rapids/Linn County Solid Waste Agency, Landfill Tonnage, Cedar Rapids/Linn County Solid Waste Agency, Fiscal Year-to-Date June 30 Comparison, November 12, 2007 and Bluestem Solid Waste Agency, Paper Mill Sludge Disposal Options Study, August 2003, Howard R. Green Company, Project No. 726830-J (Table 4.1).[1]

The characterization of the MSW can be found at Iowa Department of Natural Resources (IDNR).[2] Detailed information can be found in this reference.

As observed, the waste available on a daily basis for a Plasma Arc Gasification facility is 724 tons/day. Other parameters used in the economic analysis are shown in Table 4.2.

Municipal Solid Waste to Energy Conversion Processes: Economic, Technical, and Renewable Comparisons By Gary C. Young
Copyright © 2010 John Wiley & Sons, Inc.

Energy-from-Waste

Selected Synthesis Gas Option:
Electrical & Steam Energy

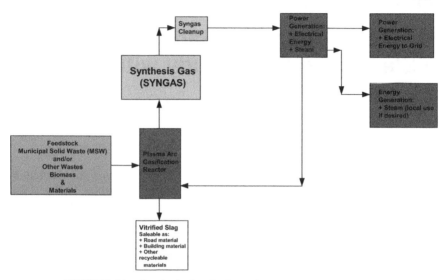

FIGURE 4.1 Synthesis Gas Option, Electrical, and Steam Energy.

With these economic factors, Fig. 4.2 was created.

With a feed rate of about 724 tons/day of waste, the Plasma Arc Gasification facility generates about $10 million annually, Net Annual Revenue Before Taxes (Total Annual Revenues–Total Annual Expenditures), if electricity is sold at 4.50 ¢/kWh. At a selling price to the grid of 5.50 and 6.50 ¢/kWh, the net annual revenue before taxes is about $13 million and $16 million per year, respectively. Capital investment would be about $130 million. In addition, a Plasma Arc Gasification facility would create about 50+ new jobs.

At a feed rate of 500 tons/day of waste, net annual revenue before taxes is about $5 million/year at 4.5 ¢/kWh, $7 million/year at 5.5 ¢/kWh, and $9 million/year at 6.5 ¢/kWh. Capital investment would be about $102 million.

It becomes clear from Fig. 4.2 that a Plasma Arc Gasification facility is near break-even point at a capacity of about 200–300 tons waste/day. The net annual revenue before taxes and the influence of plant capacity are collectively known as "economy of scale."

If the Cedar River Paper Company's waste at 233 tons/day were unavailable for the Plasma Arc Gasification facility, the feed to the plant would be about 490 tons/day of waste. The economic analysis at 500 tons/day reflects the net cash revenue values.

TABLE 4.1   Landfill Tonnage, Cedar Rapids/Linn County, Iowa

Waste Generator	Total (Tons/Day)	
City of Cedar Rapids, Iowa	55.26	
City of Marion, Iowa	14.40	
Cedar River paper	233.00	
Cedar Rapids, Iowa		
Hauler accounts		
Cash customers	52.52	
A-1 disposal	100.71	
ABC disposal	63.54	
Waste management	10.19	
Wilson Hauling	39.67	
BFI	36.02	
Rudd sanitation	17.03	
Banner valley	17.35	
Wapsi waste	3.83	
DW Zinser	11.61	
Johnson County refuse	4.03	
Absolute disposal	0.00	
Other charge accounts	64.78	
	723.94	Tons/Day
		Basis: 365 days per year

Source: Cedar Rapids/Linn County Solid Waste Agency, Fiscal Year—2007; Landfill Tonnage. Received by landfill—Site #2. Ref. (1).

It is clear from Fig. 4.2 that a Plasma Arc Gasification facility at a capacity of 1,000 tons/day has a net annual revenue before taxes of between $15 and $23 million per year depending on the selling price of electricity. Capital cost is approximately $154 million. Thus, the logical approach is a cooperative effort between one or more governmental bodies and industrial entities, so that the economy of scale is fully realized.

TABLE 4.2   Economic Parameters for Economic Analysis

Item/Parameter	Value
Capital	6.00% at 20 years
Tipping fee (revenue)	$35.00/ton waste
Vitrified slag selling price	$15.00/ton
Electricity selling price	4.50, 5.50, and 6.50 ¢/kWh
Green tags	2.00 ¢/kWh
State tax credit	1.50 ¢/kWh up to 20 MW
Labor	$28.00/hour

Other factors considered: Operation and Maintenance (O&M), Maintenance and Operational supplies, and Insurance and Capital Budget Reserve.
Note: Owner costs such as land acquisition and licensing fees were not considered, since these costs are site specific.

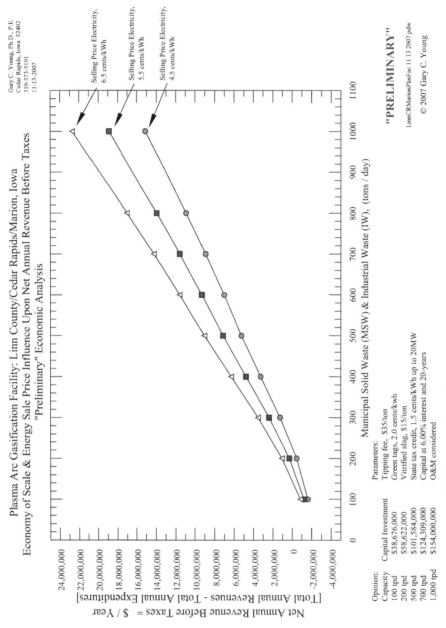

FIGURE 4.2 Plant Capacity and Net Annual Revenue, Economy of Scale.

38

As a final note, the net energy production from a Plasma Arc Gasification Facility is estimated at about 21, 30, and 43 MW for a capacity of 500, 700, and 1,000 tons/day, waste, respectively.

CONCLUSIONS AND RECOMMENDATIONS (OPINIONS)

Plasma Arc Gasification is an economically viable technology for the management of MSW in the Linn County, Cedar Rapids, and Marion, Iowa area. This technology can be used for the management of the following types of solid wastes: residential waste (RW), commercial waste (CW), industrial waste (IW), and MSW, which can be a mixture of these wastes. The Energy-from-Waste subcommittee determined that Plasma Arc Gasification process would be the most attractive and economically viable process for handling solid wastes in general. In addition, a Plasma Arc Gasification facility would create about 50 + new jobs.

Due to economy of scale, electrical energy selling price, and other economic factors, the Plasma Arc Gasification process and economics were developed to assist others in the community in evaluation of this relatively new technology for "all" of the citizens in Linn County, Cedar Rapids, and Marion area.

Environmentally, Plasma Arc Gasification technology exhibits much lower environmental levels of both air emissions and slag leachate toxicity than other potential thermal technologies.[3] In addition, this Plasma Arc Gasification technology can minimize, if not eliminate, the need for landfills and can be used to eliminate existing old landfills. The old wastes in existing landfills can be "mined/removed" and used as supplemental feed to a Plasma Arc Gasification facility. Thus, the use of the new Site #2 landfill can be minimized if not eventually eliminated and the old Site #1 landfill can be eliminated.

A logical approach is to take initial economics presented for a particular area and periodically update the analysis by a cooperative effort between one or more governmental bodies and industrial entities, so that both parties will have a fully transparent involvement and trust in the final economic analysis. This factual, transparent, updated economic analysis will therefore determine the final approach taken by both government and industry when determining if Plasma Arc Gasification is an economically and environmentally attractive alternative to a landfill.

The attractiveness of Plasma Arc Gasification technology is the generation of renewable electrical energy and the production of a useful by-product with environmentally acceptable properties. A utility company in partnership with a local government would likely be the most economical combination with a more positive benefit for the environment.

Keep in mind that the economics is site specific and thus should be repeated for each case. Yet this study suggests that with proper diligence, a viable business plan can be developed, an encouraging sign for the future of plasma arc technology. Due to economy of scale, the business plan should strive for a cooperative venture at a plant capacity of about 1,000 tons/day of wastes.

As a final comment, this study/report (preliminary economic evaluation) should be considered as Phase-I. Phase-II needs to generate economics of a site-specific case involving governmental agencies, industrial firms, and a utility company for the benefit of "all" citizens in the community.

REFERENCES

1. Harthun, B.K., "Bluestem Solid Waste Agency, Paper Mill Sludge Disposal Options Study," August 2003, Howard R. Green Company, Project No. 726830-J.
2. "Iowa Statewide Waste Characterization Study," Iowa Department of Natural Resources, R.W. Beck, Inc., January 2006.
3. Circeo, L.J., "Evaluation of Plasma Arc Technology for the Treatment of Municipal Solid Wastes in Georgia," Georgia Tech Research Institute, Atlanta, GA, January 1997.

CHAPTER 5

Plasma Economics: Garbage/Wastes to Power Ethanol Plants and a Case Study

Subject: Preliminary Economic Evaluation for a Municipal Solid Waste (MSW) Facility using Plasma Arc Gasification Technology to Power an Ethanol Plant using Corn

Background: Many areas in the Midwest have ethanol plants operating, under construction, or planning. These same areas also have municipal solid waste (MSW) produced daily and/or in an existing landfill. In addition, these areas have a need for establishing and/or extending a landfill. As an alternative to the existing concept of a landfill, recently developed and early commercialization in Plasma Arc technology has been applied to the treatment of MSW. This recent development would eliminate or minimize the need for a landfill and this approach is known as Plasma Arc Gasification process for the treatment of MSW. Plasma Arc Gasification process can generate an abundant amount of energy and electricity and/or steam. Nearby is typically an ethanol plant using corn and requiring electricity and a large amount of steam for the fermentation, distillation, and drying operations.

Consequently, a typical ethanol plant using corn with a capacity of 50 million gallons/year was considered with a Plasma Arc plant using MSW for conducting a preliminary economic analysis of such a joint process operation. A simplified schematic (Fig. 5.1), is shown for such a hybrid process (combined Plasma Arc plant and ethanol plant) below.

Plasma Arc Gasification process for the treatment of municipal solid waste (MSW) is a very high-temperature pyrolysis type of process (7200–12,600°F) whereby the organics of waste solids (MSW) are converted to a synthesis gas (syngas) and the inorganics and minerals of the waste solids (MSW) produce a rocklike by-product.[1–5] The syngas is predominantly carbon monoxide (CO) and hydrogen (H_2).[3,6,5] The inorganics and minerals in the waste solids (MSW) are converted to a vitrified slag

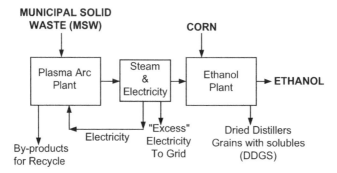

FIGURE 5.1   Hybrid Process, Combined Plasma Arc, and Ethanol Plant.

typically comprising of metals and silica glass. This vitrified slag is basically nonleaching and exceeds EPA standards. Metals can be recovered from the slag and the slag can be used to produce other by-products such as rock wool, floor tiles, roof tiles, insulation, landscaping blocks to mention a few.[1-5] One of the simpler recyclable uses of the slag is as a road material but at much lower economic value.[1] The synthesis type of gas can be used to produce electricity and the rocklike by-product can be used as a material for road construction since it is environmentally acceptable. The fuel/syngas produces by-products hydrochloric acid (HCl) and sulfur via the gas cleanup step.[4]

In this article, the economics are performed on a process for converting MSW by a Plasma Arc Gasification process (a pyrolysis process) to a syngas and a vitrified slag. The syngas will be used to generate electricity, steam and the slag as a road material. The bottom line is a process for treating MSW thereby eliminating the need for a landfill and can be used to process existing landfill MSW sites.

MSW can be processed using the Plasma Arc Gasification process and simply represented as Fig. 5.2.

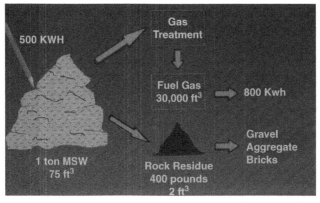

FIGURE 5.2   MSW to Energy, Plasma Arc Gasification.[1,2]

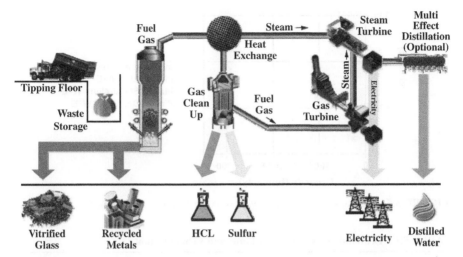

FIGURE 5.3   Process Flow Diagram, Plasma Gasification System, Applied to MSW.[4]

Note, 1 ton of MSW uses 500 kWh of the total electricity produced by the process but recent technological advances have reduced this usage to 200 kWh/ton of MSW.[7] The vitrified slag or rock residue produced is 400 pounds.

The simplified representation of the MSW pyrolysis step (Fig. 5.2) can be shown in a simplified process flow diagram (Fig. 5.3) which also illustrates that the fuel/gas could be used for generation of electricity, steam, and/or for synthesis of chemicals.[4]

A simplified process flow diagram (Fig. 5.4) is shown below for a typical ethanol plant converting the raw material, corn, into ethanol.

The capital cost for a Plasma Arc Gasification process to produce electricity and a vitrified slag can be deduced from Fig. 5.5 that shows capital cost versus capacity.[9]

Dry Grind Ethanol

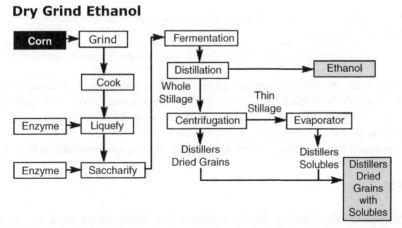

FIGURE 5.4   Process Flow Diagram, Corn-to-Ethanol Process, Dry Grind Process.[8]

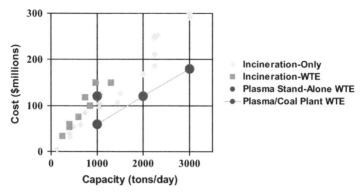

FIGURE 5.5   Capital Cost and Capacity, Plasma Gasification System, Applied to MSW.[9]

TABLE 5.1   Economic Cases, Plasma Plant and Ethanol Plant

	Capital Investment Industry & Local Government	Tipping Fee for MSW	Corn Purchase Price	Ethanol Sales Price
Case 1				
Capital for plasma plant	$97,344,000	$35.00/ton		
Capital for ethanol plant	$65,000,000		$3.75/bushel	$1.8000/gallon
Total Capital Investment	$162,344,000			

Net Revenue before taxes from combined facility operation = [revenue−expenditures] = $16,950,000.00 per year

Case 2				
Capital for plasma plant	$97,344,000	$35.00/ton		
Capital for ethanol plant	$65,000,000		$3.75/bushel	$2.5541/gallon[12]
Total Capital Investment	$162,344,000			

Net Revenue before taxes from combined facility operation = [revenue−expenditures] = $54,655,000.00 per year

Note: The current tipping fee at a landfill in the Cedar Rapids, Iowa area is about $35 per ton of MSW.

Thus, with the information presented so far, a preliminary evaluation and economics can be deduced for a Plasma Arc plant using MSW with sufficient capacity to supply both electricity and steam for an adjacent ethanol plant producing 50 million gallons of ethanol per year from corn.[10,11]

Consequently, economic cases #1 & #2 were evaluated and presented in Table 5.1.

BASIS

A Plasma Arc Gasification facility operates at a 706-tons/day capacity for the treatment of MSW with a "total" capital investment of $97,344,000.[1,13,14,9]

This capacity of plasma plant supplies sufficient energy as electricity and steam to an ethanol plant for the production of 50 million gallons of ethanol per year.[11] Capital cost of the ethanol plant is $65,000,000.[11] Capital financed for both plants would be at 5.75% interest for 20 years by making two payments per year. The Plasma Arc facility would generate 30,000,000 kWh/year of electrical energy (renewable energy) for sale to the ethanol plant at 4.5 cents/kWh.[2] A 2.0 cent/kWh revenue from green tags associated with renewable energy electrical production was used in the financial analysis for the Plasma Arc plant. In addition, the Plasma Arc plant would supply 1,421,405,000 lbs/year steam for sale to the ethanol plant at a sales price of $5.50/1000 lbs steam. The by-product rock material would be sold as a road construction material at $15/ton.[1] On the ethanol facility, 18,518,500 bushels/year of corn are purchased at $3.75/bushel.[11] A by-product, Dried Distillers Grain with Solubles (DDGS), is produced at 166,667 tons/year and sold as a cattle feed for $90/ton.[11] Cost considerations for operations and maintenance, a capital budget reserve, process water and sewer have been considered in the economic evaluation. Over 100 + jobs would be created by a combined plasma and ethanol facility. The combined facility is self-sufficient in electrical and steam energy requirements since this energy is produced internally from a renewable raw material, MSW.

ECONOMIC CASES

From these "initial" economic analyses, one likely approach to a business plan is for a cooperative effort between a local utility, local industry, and/or local governmental entities. Thus, the local government participates in the MSW treatment part of the Plasma Arc facility while the local utility participates in the electrical and steam energy (renewable energy) producing part of the facility. A local industrial/business entity could own/operate the ethanol facility. Thus, the local government remains in the MSW business while the utility and/or industry stay in the electrical, steam, and ethanol energy businesses. This appears to be a win-win situation for the local government, utility, and local industry.

Case 1 presents the preliminary economics of a Plasma Arc Gasification Facility combined with an ethanol plant such as for the Linn County/Cedar Rapids, Iowa area for a facility processing 706 tons/day of MSW. The cooperative venture results in a "tipping fee" for the MSW of $35.00/ton and a selling price of ethanol at $1.8000 per gallon. A tipping fee of $35/ton of MSW is typical for Linn County, Iowa.[13,14] Net revenue before taxes from combined facility operation = [revenue–expenditures] = $16,950,000.00 per year. Both the Plasma Arc plant and the ethanol plant contribute to positive net revenue.

Case 2 presents an economic evaluation similar to the previous case but the selling price of ethanol is $2.5541/gallon.[12] The net revenue before taxes from combined facility operation = [revenue-expenditures] = $54,655,000.00 per year. As before, both the Plasma Arc plant and the ethanol plant contribute to positive net revenue.

LOGICAL APPROACH FOR FUTURE PROGRESS

A logical approach is to take initial economics presented for a particular area and periodically "update" the analysis by a cooperative effort between governmental body (s) and industrial entity(s) so that both parties will have a fully transparent evolvement and trust in the final economic analysis. Thus, this factual, transparent, updated economic analysis will therefore determine the final approach taken by both government and industry for that particular area investigated to determine if Plasma Arc Gasification is an economically and environmentally attractive alternative to a landfill when integrated with an ethanol plant.[15]

As a final comment, the attractiveness of Plasma Arc Gasification technology is the generation of renewable electrical and steam energy with a useful by-product having environmentally acceptable properties. The bottom line is a process based upon "plasma technology" for treating MSW thereby eliminating the need for a landfill and can be used to process existing landfill MSW sites. The energy generated from the Plasma technology can supply necessary energy for other removable resources such as for ethanol plants.

A utility and/or business company(s) in partnership with a local government would likely be the most economical combination and have the most positive benefit to the environment and financial reward to the local area. In this article, the economics were stated as preliminary since such evaluations are site specific and should be done specifically for each case under study, and plasma processing of MSW is an emerging technology.

With some diligence, a viable business plan can be developed from the consideration of the many factors influencing the economics associated with a specific site selection and surrounding community.

As a final comment, the attractiveness of Plasma Arc Gasification technology is the generation of renewable electrical energy and with a useful by-product having environmentally acceptable properties. The bottom line is a process based upon "plasma technology" for treating MSW thereby eliminating the need for a landfill and can be used to process existing landfill MSW sites.

A utility company in partnership with a local government would likely be the most economical combination and have the most positive benefit to the environment. In this article, the economics were stated as preliminary since such evaluations are site specific, should be done specifically for each case under study, and plasma processing of MSW is an emerging technology.

With some diligence, a viable business plan can be developed from the consideration of the many factors influencing the economics associated with a specific site selection and surrounding community.

Now that the key process concepts and process economics are presented in Fig. 5.3, a discussion will follow about more details of processing the crude syngas from the gasifier.

Raw syngas exiting the gasifier is typically cooled in a fire-tube boiler from high gasification temperature to about 700°F. Any char of fly ash produced from the final zone of the gasifier and entrained with the crude syngas stream is removed by a downstream cyclone and candle filter particulate control devices and returned to the gasifier. The candle filter is composed of ceramic candle filter elements and cleaned by the commonly accepted back-pulsing concept with syngas. The raw syngas exits at 700°F.

The raw syngas introduced into the fire-tube syngas cooler produces steam for use in a turbine that becomes part of the heat recovery system. Raw syngas then enters a scrubber at 700°F for removing any remaining particulates and trace components. Next, the syngas stream is sent to a hydrolysis reactor whereby the carbonyl sulfide (COS) and hydrogen cyanide (HCN) are converted to hydrogen sulfide (H_2S).

The syngas stream then is processed by a mercury removal system such as by Eastman chemical consisting of a packed bed of a special activated carbon that removes the mercury, arsenic, and trace materials. Following the mercury removal system, a Selexol unit can be used to remove H_2S and CO_2 from the cool syngas. A polishing unit of zinc oxide removes any residual H_2S and achieves a high-purity syngas stream.

As discussed, the process unit operations are presented to give a conceptual idea of the process equipment and technology currently available to clean up a raw syngas from a gasifier. Many existing technologies and engineering experiences are available for design and further development for the cleanup of raw syngas from gasifiers. It is just important to become familiar with these technologies and to form the proper team whereby such important technologies are integrated properly into the waste-to-energy gasification process.[16] Final selection of the gas cleanup process parameters will depend on the wastes selected for feedstock to the gasification process.

REFERENCES

1. Circeo, L.J., Engineering and Environmental Applications of Plasma Arc Technology, Technological Forum, Kirkwood Training and Outreach Services Center, Marion, Iowa, November 22, 2005.

2. Circeo, L.J., "Engineering & Environmental Applications of Plasma Arc Technology," Presentation, Georgia Tech Research Institute, Atlanta, GA 2005.

3. Vera, R., "Organic Waste, Gasification and Cogeneration," Presentation, Trinity Plasma Technology, Technologies International Corporation, Trinity Consultants, Inc., 2005.

4. Recovered Energy, Inc., Pocatello, Idaho, www.recoveredenergy.com, "Process Flow Diagram," MSW into energy and useable by-products.

5. "Summary Report: Evaluation of Alternative Solid Waste Processing Technologies," Prepared for: City of Los Angeles, Department of Public Works, Bureau of Sanitation, 419 S. Spring Street, Suite 900, Los Angeles, CA 90013; Prepared by: URS Corporation, 915 Wilshire Boulevard, Suite 700, Los Angeles, CA 90017, September 2005.

6. Lee, C.C., "Plasma System," Standard Handbook of Hazardous Waste Treatment and Disposal, McGraw-Hill Book Company, New York, 1989, p. 8.169.

7. Circeo, L.J., Private Communication, April 4, 2006 and December 22, 2006.

8. Iowa Corn Organization, www.iowacorn.org and Iowa Corn Promotion Board/Iowa Corn Growers Association, December 2006.

9. Circeo, L.J. and Smith, M.S., "Plasma Processing of MSW at Coal-Fired Power Plants," Presentation, Health and Environmental Systems Laboratory, Georgia Tech Research Institute, Georgia Institute of Technology, Atlanta, GA, 2005.

10. Young, G.C., "Zapping MSW with Plasma Arc," Pollution Engineering, November 2006.

11. Ulrich, K., Private Communication, ICM, Inc., December 21, 2006.

12. American Coalition for Ethanol, www.ethanol.org, averages provided by Axis Petroleum, Sate Average Ethanol Rack Prices, Iowa, Tuesday, December 19, 2006.

13. "Expert Touts Plasma Torch," The Gazette, November 22, 2005.

14. "Plasma arc Technology may help Linn Garbage Woes," The Gazette, November 20, 2005.

15. Young, G.C., "How Trash Can Power Ethanol Plants," Public Utilities Fortnightly, p. 72, February 2007.

16. DOE/NETL-2007/1260, Baseline Technical and Economic Assessment of a Commercial Scale Fischer–Tropsch Liquids Facility, Final Report for Subtask 41817.401.01.08.001, National Energy Technology Laboratory (NETL), April 9, 2007.

CHAPTER 6

From Curbside to Landfill: Cash Flows as a Revenue Source for Waste Solids-to-Energy Management

Background: Nearly every community in the United States has municipal solid wastes (MSW) referred to by the public as "garbage or trash" generated by households. This MSW or garbage is typically picked up at the "curbside" by garbage trucks and usually hauled to local landfill(s) for disposal. This process of garbage collection from local residences and delivery to the front gate of the landfill for disposal will be referred to as "from curbside to landfill." This article presents the "preliminary cash flow" associated with "from curbside to landfill." The public should be aware of the "cash flow/economics" associated with garbage collection "from curbside to landfill" and how this money can benefit the community as a resource for further downstream processing of MSW/garbage and possible elimination of the downstream landfill with new technology.

The process of collecting "garbage or trash" (municipal solid waste, MSW) is straightforward and identified by the following steps referred to as "from curbside to landfill:"

As presented in Fig. 6.1, garbage/trash/MSW is placed at the curbside by residential customers for pickup by the garbage truck/trash hauler. A fee is charged from each customer for this curbside pickup. Next, once the garbage truck is filled, typically the truck delivers the trash to the landfill for disposal at a cost known as the "tipping fee." Finally, the empty truck returns and continues garbage collection by repeating these steps until the days scheduled route has been completed.

Our objective is to quantify the cash flows associated with the process of garbage collection from "curbside to landfill." "Predictive cash flows" can be quantified by the use of parameters for revenues and expenditures as shown in the following example for two typical towns in Iowa using a common landfill.

This predictive cash flow analysis from "curbside to landfill" can be understood by following an example completed for the cities of Marion, Iowa and Cedar Rapids,

Municipal Solid Waste to Energy Conversion Processes: Economic, Technical, and Renewable Comparisons By Gary C. Young
Copyright © 2010 John Wiley & Sons, Inc.

From Curbside Pickup (Garbage/Trash) to Landfill, Cash Flows

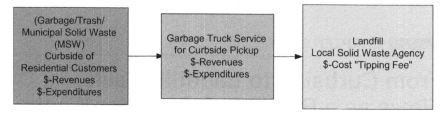

FIGURE 6.1 Curbside Pickup to Landfill, Garbage/Trash Flows.

Iowa associated with the landfill of the Cedar Rapids Linn County Solid Waste Agency in Marion, Iowa and with the parameters in Table 6.1.

Cash flows computed for this example with regard to the two cities in Iowa, Cedar Rapids and Marion, are

	Cedar Rapids, IA	Marion, IA
I. Residential curbside pickup cash flow	$2,981,070.06/year Total combined cash flow = $3,792,948.76/year	$811,878.70/year

Note: Total combined cash flow is "net annual revenue" = [total revenues–total expenditures]

II. Waste to landfill	78.64 tons/collection day Total combined waste, 85.41 tons/day waste to landfill (per collection day)	6.77 tons/collection day
III. "Net annual revenue" = [total revenues–total expenditures] = $170.81/ton waste, collected at curbside		

A preliminary cash flow analysis of "from curbside to landfill" has established a positive net annual revenue (cash flow) of $3,792,948.76/year. This cash flow is generated just by collecting and delivering residential solid waste to the landfill even after subtracting out the landfill tipping fee of $35/ton waste. Cash flow analysis from curbside to landfill was determined using the computer spreadsheet in Table 6.2 (Cash Flow Analysis from Residential Curbside (Waste Pickup) to Landfill).

If this cash flow was utilized toward financing a process for management of MSW to the landfill, what impact could it have? If a process such as Plasma Arc Gasification is considered, the economics for a Plasma Arc Gasification facility with a capacity of 500 tons/day MSW are as follows.[4–7]

Capital investment for a 500 tpd MSW Plasma Arc Gasification facility is around $101,584,000.00. Bond payment for the capital cost at 6% interest for 20-year financing is $8,789,511.29/year. If the Plasma Arc facility was used to produce electricity and sell it to the grid at 4.50¢/kWh, receive a revenue of $35.00/ton MSW processed (tipping fee), and receive $15.00/ton for the vitrified slag by-product as road material, the net revenue before taxes would be $5,235,041.27/year. Combined

TABLE 6.1   Economic Parameters as Input to Obtain Curbside to Landfill Cash Flows

Parameter	Cedar Rapids	Marion
Revenues		
a. Billings for curbside pickup[1,2]		
Number residences, solid waste service	45,000	9,475
Solid waste fee per residence	$0.3541/day	$5.50/month
Billings	365 days/year	12 months/year
Number residences, recycling service	45,000	9,475
Recycling fee	$0.1013/day	$5.50/month
Billings	365/year	12 months/year
b. Additional collection, garbage tags[1,2]		
Number tags sold	257,000/year	80,000/year
Selling price of tags	$1.25/tag	$1.25/tag
Expenditures		
Curbside pickup		
a. Time at curbside for solid waste pickup	0.509 minutes	0.208 minutes
b. Operating, hours per day for solid waste pickup	8.00 hours/day	8.00 hours/day
c. Operating, days per week for solids waste pickup	5.00 days/week	5.00 days/week
d. Capital cost of a new truck, rear loading refuse truck	$180,000.00	$180,000.00
e. Annual cost of capital for refuse truck at 5%, 6 years	$35,095.37/yr	$35,095.27/yr
f. Number of employees per refuse truck	Three persons/ truck	Three persons/ truck
g. Cost per employee per year, refuse truck operation[1]	$60,649.10	$60,649.10
h. Time at curbside for recycling pickup	0.408 minutes	0.163 minutes
i. Operating, hours per day for recycling pickup	8.00 hours/day	8.00 hours/day
j. Operating, days per week for recycling pickup	5.00 days/week	5.00 days/week
k. Capital cost of a new truck, recycling truck	$140,000.00	$140,000.00
l. Annual cost of capital for recycling truck at 5%, 6 years	$27,296.40/yr	$27,296.40/yr
m. Number of employees per recycling truck	Two persons/truck	Two persons/truck
n. Cost per employee per year, recycling truck operation[1]	$60,649.10	$60,649.10
o. Additional staff required for overall operations[1]		
Number of additional persons on payroll	8.00/year	0.00/year
Cost per employee per year, overall operations	$60,649.10	$0.00
p. Percentage of annual recycling revenues as a loss	10.00%	0.00%

(continued)

TABLE 6.1   (*Continued*)

Parameter	Cedar Rapids	Marion
Landfill disposal		
q. Tipping fee, $/ton waste solids[3]	$35.00/ton	$35.00/ton
r. Pounds waste per truck (refuse truck)[3]	16,480 lbs/truck	16,480 lbs/truck
s. Pounds recycling material per truck (recycle truck)[3]	9,420 lbs/truck	9,420 lbs/truck
t. Receiving trucks, operation	5.00 days/week	5.00 days/week
u. Percentage of recyclables picked up but landfilled	10.00%	0.00%
v. Drop-off fee to recycling center[2]	$0.00/ton	$10.00/ton
w. Additional fees[2]	$0.00/year	$179,122.00/year

net cash flow for the Plasma Arc facility and the curbside to landfill operations would be $9,027,990.03/year (positive cash flow).[7]

If clean renewable energy bonds (CREBs) financing was used, the bond payment for the Plasma Arc Gasification facility would be $7,256,509.97/year and a net revenue of $6,768,042.59/year for 14-year financing. Thus, the combined revenue for the Plasma Arc Gasification facility and the curbside to landfill operations would be $10,560,991.35/year.

It is clear that when the bond for the Plasma Arc Gasification facility could be paid off in about nine years, the positive cash flow becomes about $18,000,000/year (a "Cash Cow").

The economics here are shown for the two cities in Iowa, Cedar Rapids and Marion, working together with Linn County to provide a financially rewarding endeavor for all the people of both the cities and Linn County, Iowa. In addition, the Linn County landfill can be essentially eliminated/minimized whereby the negative liability impacts of a local landfill are mitigated.

The bottom line is that local solid waste/garbage collection and disposal, i.e., the refuse business, can create a positive cash flow for the local community. This positive cash flow can be used to finance new technology for converting the solid waste/garbage to energy. In addition, the energy generated could create additional revenue for the community. Also, the generated energy could be used to attract new industry to the area and thus result in the creation of new jobs.

Now that cash flows have been presented from curbside-to-landfill, fees (tipping fee), regulations, construction, operation, and environmental issues will be discussed associated with landfills. The presentation on landfills will be based upon the standards set forth in the Resource Conservation and Recovery Act (RCRA), subtitle D regulations as well as factual information accumulated over the years. A typical MSW landfill that is the depository for the MSW collected from the curbside is shown in Figs. 6.2 and 6.3.

In the previous cash flow analysis from curbside-to-landfill, a tipping fee of $35.00/ton was used for the MSW sent to and deposited in the local landfill. However, landfill fees (tipping fees) can vary considerably from one landfill to another. The varying landfill tipping fees are illustrated in Table 6.3 for average tipping fees for Wisconsin and neighboring states.[10]

TABLE 6.2 From Curbside to Landfill: Cash Flows, Linn County Solid Waste Agency Board, Site #2 Landfill, Iowa

Cedar Rapids/Linn County Solid Waste Agency Board: Site #2 Landfill		11/12/2008	GYCO, Inc.
Cedar Rapids, IA	Name I.: For Cash Flow for City or Company or Utility Being Computed		Cedar Rapids, IA
Marion, IA	Name II.: For Cash Flow for City or Company or Utility Being Computed		gycoinc@aol.com

NOTE: [ ] Denotes Input Data

Subject: Cash Flow Analysis from Residential Curbside (Waste Pickup) & Commercial Services & Industrial Waste Services Associated With Landfill, Site #2

"CONFIDENTIAL"
"PRELIMINARY"

I. RESIDENTIAL CURBSIDE PICKUP CASH FLOW ANALYSIS: Cedar Rapids, IA

1R.- REVENUES: City of Cedar Rapids, Iowa

Item	Description	Number of Residences	Waste Fee ($/day-residence)	Billings Days/year	Annual ($)	Comments
	Monthly Billing for Curbside pickup family residence Cedar Rapids, IA					
1Ra.	Solid Waste Service	45,000	$0.3624	365	$5,952,420.00	
1Rb.	Recycling Service	45,000	$0.1013	365	$1,663,852.50	
				Subtotal Revenue: Solid Waste Service & Recycle Service:		
	Cedar Rapids, IA		Annual Revenue for Curbside Pickup; Family Residence; Solid Waste & Recycling Service		$7,616,272.50	
1Rc.	Revenue from Selling Recyclable Waste from Curbside		Subtotal Revenue: Sale of Recyclables:		$2,000.00	
1RD.	Additional Garbage; Tags Sold: Revenue	257,000 Tags sold per year	$1.25 $/Tag		$321,250.00	
	Annual Revenues for Curbside Pickup; Cedar Rapids, Iowa; Family Residence; Solid Waste & Recycling Service; TOTAL REVENUES :				$7,939,522.50	

(continued)

TABLE 6.2 (Continued)

1E. - EXPENDITURES:

Cedar Rapids, IA

Item	No. Residences	

Monthly Billing for Curbside pickup
family residence
(by Solid Waste Back Compacted
Trucks hauling directly to landfill)
Cedar Rapids, IA

9.54 Number of trucks for Solid Waste Service (Calculated/Estimated)
0.509 minutes, time at curbside for each pickup, Solid Waste (back compacted truck)
8.00 hours per day operating   5.00 days/wk operating

Monthly Billing for Curbside pickup
family residence
(by Recyclables Truck
hauling to recycling center)
Cedar Rapids, IA

7.65 Number of trucks for Solid Waste Service (Calculated/Estimated)
0.408 minutes, time at curbside for each pickup, Solid Waste (back compacted truck)
8.00 hours per day operating   5.00 days/wk operating

1Ea. Solid Waste Service   **45,000** residences, NOTE: one pickup per week
9.54 Number of Garbage/Refuse Trucks Needed for Solid Waste Service

$180,000.00 Capital cost of a new truck, rear loading refuse truck
$35,095.37 /year/truck, Capital cost of a new truck at 5% interest for 6-years, 2-bond payment per year
$334,941.44 /year, annual Capital cost to purchase new trucks every 6-years
$106,890.00 /year, annual truck operating costs   3 number of empolyees per truck
$1,736,459.54 /year labor costs $60,649.10 annual cost per each employee
$2,178,290.98 /year, expenditures for Solid Waste Service $2,178,290.98   expenditures for Solid Waste Service

1Eb. Recycling Service   **45,000** residences, NOTE: one pickup per week
7.65 Number of Garbage/Refuse Trucks Needed for Recycling Service: (Calculated/Estimated)

$140,000.00 Capital cost of a new truck, Recycling truck
$27,296.40 /year/truck, Capital cost of a new truck at 5% interest for 6 years, two bond payment per year
$208,817.46 /year, annual Capital cost to purchase new trucks every 6-years
$106,890.00 /year, annual truck operating costs   2 number of empolyees per truck
$927,931.23 /year labor costs $60,649.10 annual cost per each employee
$1,243,638.69 /year, expenditures for Recycling Service $1,243,638.69   expenditures for Recycling Service for Pickup at Curbside of Residences

1Ec.   **10.00** % of Recycling Revenues per year that End Up as A Loss for the Department
beyond the cost to pickup the recyclables at the curbside of residences
$166,585.25 **per year, Cost for Operating the Recycling Center**
beyond the cost to pickup the recyclables at the curbside of residences $166,585.25   Loss Encured by Operating the Recycling Center

1Ed. **$485,192.80 /year, additional staff on payroll beyond calculated values.**
$60,649.10 annual cost per each employee
8.00 number of additional personnel on payroll
51.9 Total, number of personnel on payroll for Curbside Garbage and Recycle pickup $485,192.80
Subtotal Expenditures: Solid Waste Service & Recycling Service: $4,073,707.72   Expenditures for Solids Waste Servie & Recycling Service Plus Additional Cost of Recycle Center

1Ef. Expenditures for Tipping Fee at Landfill Site #2 in Marion, Iowa
$35.00 /ton Tipping fee at landfill   Subtotal Expenditures: Solid Waste Service & Recycling Service; Landfill Tipping Fee:
16,480 lbs waste/truck (back compacting loaded) of curbside Solid Waste to landfill
78.64 tons waste/day curbsite waste to landfill
9.420 lbs waste/truck (recycling truck) of curbsite Solid Waste
36.03 tons waste/day curbsite Recycleables picked up
5.00 days per week trucks operating
10.00 % of Recycleables that are Picked Up by Recycling trucks but end up in the landfill Site #2
82.24 tons waste (Solid Waste + Recyclables) / day [That go to Landfill Site #2]
21,383 tons waste (Solid Waste + Recyclables) / YEAR [That go to Landfill Site #2] $748,417.22   Expenditures for Solids Waste Serive & Recycling Service from Landfill Tipping Fees

1Eg. Fees Paid to Solid Waste Agency by   Cedar Rapids, IA   Additional Fees Paid by   Cedar Rapids, IA $0.00

Annual Expenditures for Curbside Pickup; Family Residence; Solid Waste & Recycling Service; **TOTAL EXPENDITURES:**
Cedar Rapids, IA   **$4,822,124.94**   TOTAL EXPENDITURES; SOLID WASTE & RECYCLE SERVICES

ANNUAL NET REVENUE = [TOTAL REVENUES - TOTAL EXPENDITURES] For Solid Waste & Recycle Services:
Cedar Rapids, IA   **$3,117,397.56**   **Annual Net Revenue for Solid Waste & Recycle Services Cedar Rapids, IA**

I RESIDENTIAL CURBSIDE PICKUP CASH FLOW ANALYSIS:

II. RESIDENTIAL CURBSIDE PICKUP CASH FLOW ANALYSIS:

Marion, IA

2R. - REVENUES:

Marion, IA

Item	Description	Number of Residences	Solid Waste Fee ($/month-residence)	No. Billings Months/year	Annual ($)	Comments
	Monthly Billing for Curbside pickup family residence Marion, IA					
2Ra.	Solid Waste Service	9,475	$5.50	12	$625,350.00	
2Rb.	Recycling Service	9,475	$5.50	12	$625,350.00	
	Subtotal Revenue: Solid Waste Service & Recycle Service:				$1,250,700.00	Revenues: Solid Waste Service & Recycle Service:
	Annual Revenue for Curbside Pickup; Family Residence; Solid Waste & Recycling Service Marion, IA					
2Rc.	Revenue from Selling Recyclable Waste from Curbside			Subtotal Revenue: Sale of Recyclables:	$2,000.00	Cardboard Sales
2Rd.	Additional Garbage; Tags Sold; Revenue		80,000 Tags sold per year	$1.25 $/Tag	$100,000.00	Additional Garbage; Tags Sold; Revenue;
	Annual Revenues for Curbside Pickup; Family Residence; Solid Waste & Recycling Service; TOTAL REVENUES :				$1,352,700.00	

(continued)

TABLE 6.2 (*Continued*)

2E.- EXPENDITURES:
Marion, IA

Marion, IA

Item		Number of Residences	

Monthly Billing for Curbside pickup
family residence
(by Solid Waste Back Compacted
Trucks hauling directly to landfill)
Marion, IA

0.82 Number of trucks for Solid Waste Service (Calculated/Estimated)

0.208 minutes, time at curbside for each pickup. Solid Waste (back compacted truck)
8.00 hours per day operating   5.00 days/wk operating

Monthly Billing for Curbside pickup
family residence
(by Recyclables Truck
hauling hauling to recycling center)
Marion, IA

0.64 No. of trucks for Solid Waste Service (Calculated/Estimated)

0.163 minutes, time at curbside for each pickup. Solid Waste (back compacted truck)
8.00 hours per day operating   5.00 days/wk operating

2Ea.   Solid Waste Service   9,475 residences, NOTE: one pickup per week
0.82 Number of Garbage/Refuse Trucks Needed for Solid Waste Service

$180,000.00 Capital cost of a new truck, rear loading Refuse truck
$35,095.37 /year/truck, Capital cost of a new truck at 5% interest for 6-years, 2-bond payment per year
$28,819.15 /year, annual Capital cost to purchase new trucks every 6-years
$9,197.07 /year, annual truck operating costs   3 number of empolyees per truck
$149,409.06 /year labor costs $60,649.10 annual cost per each employee
$187,425.27 /year, expenditures for Solid Waste Service

$187,425.27   expenditures for Solid Waste Service

2Eb.   Recycling Service   9,475 residences, NOTE: one pickup per week
0.64 Number of Garbage/Refuse Trucks Needed for Recycling Service: (Calculated/Estimated)

$140,000.00 Capital cost of a new truck, Recycling truck
$27,296.40 /year/truck, Capital cost of a new truck at 5% interest for 6-years, 2-bond payment per year
$17,565.52 /year, annual Capital cost to purchase new trucks every 6-years
$9,197.07 /year, annual truck operating costs   2 number of empolyees per truck
$78,056.66 /year labor costs $60,649.10 annual cost per each employee
$104,819.24 /year, expenditures for Recycling Service

$104,819.24   expenditures for Recycling Service for Pickup at Curbside of Residences

3.8 Total, number of personnel on payroll for Curbside Garbage and Recycle pickup

**0.00 % of Recycling Revenues per year that End Up as A Loss for the Department
beyond the cost to pickup the recycleables at the curbside of residences**
**$0.00 per year, Cost for Operating the Recycling Center
beyond the cost to pickup the recycleables at the curbside of residences**
Marion, IA

2Ec. $0.00   Loss Encured by Operating the Recycling Center

Subtotal Expenditures: Solid Waste Service & Recycling Service: $292,244.51   Expenditures for Solids Waste Servie & Recycling Service Plus
Additional Cost of Recycle Center

Expenditures for Tipping Fee at Landfill Site
Marion, IA

$35.00 /ton Tipping fee at landfill   Subtotal Expenditures: Solid Waste Service & Recycling Service: Landfill Tipping Fee: $69,454.79   Expenditures for Solids Waste Servie & Recycling Service
from Landfill Tipping Fees
16,480 lbs waste/truck (back compacting loaded) of curbsite Solid Waste to landfill
6.77 tons waste/day curbsite waste to landfill
9,420 lbs waste/truck (recycling truck) of curbsite Solid Waste
3.03 tons waste/day curbsite Recycleables picked up and sent to Recycling center at Site #2
5.00 days per week trucks operating
2Ed. $10.00 /ton, Drop Off Fee (Transportation) to Recycling Center at Site #2 Landfill for Recycling Picked Up at City of Marion,IA
0.00 % of Recycleables that are Picked Up by Recycling trucks but end up in the landfill Site #2
6.77 tons waste (Solid Waste + Recycleables) / day [That go to Landfill Site #2]
1,759 tons waste (Solid Waste + Recycleables) / YEAR [That go to Landfill Site #2]

2Ee.   Fees Paid to Solid Waste Agency by   Marion, IA $179,122.00   Additional Fees Paid by   Marion, IA

Annual Expenditures for Curbside Pickup; Family Residence, Solid Waste & Recycling Service; **TOTAL EXPENDITURES**:
Marion, IA
$540,821.30   TOTAL EXPENDITURES; SOLID WASTE & RECYCLE SERVICES

ANNUAL NET REVENUE = [TOTAL REVENUES - TOTAL EXPENDITURES] For Solid Waste & Recycle Services:
Marion, IA
$811,878.70   **Annual Net Revenue for Solid Waste & Recycle Services**
Marion, IA

II.RESIDENTIAL CURBSIDE PICKUP CASH FLOW ANALYSIS:

SUMMATION: ANNUAL NET REVENUE = [TOTAL REVENUES - TOTAL EXPENDITURES] For Solid Waste & Recycle Services:
I & II. RESIDENTIAL CURBSIDE PICKUP CASH FLOW ANALYSIS: Cedar Rapids, IA   &   Marion, IA

Annual Net Revenue for
Solid Waste & Recycle Services: $3,929,276.26
Cedar Rapids, IA   &   Marion, IA

SUMMARY: Solid Waste Landfill Site

I. RESIDENTIAL CURBSIDE PICKUP CASH FLOW   Cedar Rapids, IA
ANALYSIS:
ANNUAL NET REVENUE = [TOTAL REVENUES - TOTAL EXPENDITURES]
For Solid Waste & Recycle Services

$3,117,397.56   Annual Net Revenue for Solid Waste & Recycle Services   78.64 Tons / Day Waste
Cedar Rapids, IA   to Landfill Site

II. RESIDENTIAL CURBSIDE PICKUP CASH FLOW   Marion, IA
ANALYSIS:
ANNUAL NET REVENUE = [TOTAL REVENUES - TOTAL EXPENDITURES]
For Solid Waste & Recycle Services

$811,878.70   Annual Net Revenue for Solid Waste & Recycle Services   6.77 Tons / Day Waste
Marion, IA   to Landfill Site

COMBINED TOTAL CASH FLOW
ANALYSIS: RESIDENTIAL & COMMERCIAL & INDUSTRIAL: (FROM THIS STUDY)

$3,929,276.26   Annual Net Revenue for Services Associated with   TOTAL   85.41 Tons / Day Waste
                  Landfill Site (this study only)   to Landfill Site
ANNUAL NET REVENUE = [TOTAL REVENUES - TOTAL EXPENDITURES]   per Collection Day
For Solid Waste & Recycle Services   (Residential Curbside
                                                                               Garbage & Residential
                                                                               Curbside Recyclables)

Total Solid Waste from:
Cedar Rapids, IA
And
Marion, IA

	Estimated Collection Tons/Day
Cedar Rapids, IA	20,446.5 tons, collected / year
Marion, IA	1,769.3 tons, collected / year
Total	22,205.8 tons collected / year

COMBINED TOTAL CASH FLOW
ANALYSIS: RESIDENTIAL & COMMERCIAL & INDUSTRIAL: (FROM THIS STUDY)

$3,929,276.26   Annual Net Revenue for Services Associated with
                  Landfill Site (this study only)
ANNUAL NET REVENUE = [TOTAL REVENUES - TOTAL EXPENDITURES]
For Solid Waste & Recycle Services

Combined Total Cash Flow, Based Upon Net Revenue, from Curbside to Landfill (front gate) per Total Ton collected = $176.95   per ton waste solids collected
                                                                                                                                    (net cash flow from curbside to landfill per ton waste solids collected

FIGURE 6.2   Haulers Drop Off Trash at Landfill.[9]

FIGURE 6.3   Trash Loaded and Compacted into a Landfill.[9]

A study by an Iowa task force reported the landfill tipping fees in Iowa for the year 2002. The reported tipping fees for the Iowa landfills are shown in Table 6.4. Note, the range of landfill tipping fees varied from $15.00/ton to $60.00/ton for the year 2002.[11] As can be seen from Table 6.3, the average tipping fee for the state of Iowa is $37.17/ton for year 2004. Also shown in Table 6.4 are the Iowa tonnages placed into the respective Iowa landfills for fiscal year 2007, which is from July 1, 2006 to June 30, 2007. The U.S. Environmental Protection Agency considers MSW as characterized

TABLE 6.3   Landfill Average Tipping Fees for Wisconsin and Neighboring States, Year—2004

	Minnesota	Iowa	Wisconsin	Michigan	Illinois
Tipping fee, $/ton (average tipping fee)	$58.80/ton	$37.17/ton	$36.00/ton	$35.32/ton	$35.88/ton

Source: Ref. (10).

TABLE 6.4   Landfill Tipping Fees for Iowa, Year 2002[11] and Tonnages for Fiscal Year 2007

	Tipping Fee ($/ton)	Iowa Tonnages Fiscal Year 2007 (tons)
Adair—Adair County sanitary landfill	34.00	6,164.65
Audubon—Audubon County sanitary landfill	45.00	3,376.74
Black Hawk—Black Hawk County sanitary landfill	25.00	137,872.04
Boone—Boone County sanitary landfill	26.00	54,085.95
Bremer—Bremer County sanitary landfill	35.75	14,714.62
Buena Vista—Buena Vista County sanitary landfill	37.00	8,935.35
Carroll—Carroll County sanitary landfill/recycling	34.75	36,152.60
Cass—Cass County sanitary landfill	60.00	8,904.29
Cerro Gordo—Landfill of North Iowa	25.00	95,454.46
Cherokee—Cherokee County Sanitary Landfill	36.00	16,164.81
Clarke—Clarke County sanitary landfill—south side	35.00	10,351.85
Clinton—Clinton County sanitary landfill—East	44.00	16,453.47
Crawford—Crawford County sanitary landfill	35.00	12,144.86
Dallas—North Dallas County sanitary landfill	28.00	19,774.00
Dallas—South Dallas County sanitary landfill	36.00	23,328.91
Decatur—Wayne-Ringgold-Decatur County sanitary LF	25.50	9,648.29
Des Moines—Des Moines County sanitary landfill	30.00	59,024.81
Dickinson—Diskinson County sanitary landfill	37.65	56,894.54
Dubuque—Dubuque Metropolitan sanitary LF	29.73	98,406.23
Fremont—Fremont County sanitary landfill	35.00	7,560.81
Grundy—Grundy County sanitary landfill	45.00	6,707.00
Hamilton—Hamilton County sanitary landfill	45.00	14,292.91
Hardin—Rural Iowa sanitary landfill	35.00	23,362.39
Harrison—Harrison County sanitary landfill	35.00	9,982.98
IDA—IDA County sanitary landfill	30.00	3,641.54
Iowa—Iowa County sanitary landfill	40.00	10,932.62
Jasper—City of Newton sanitary landfill	34.00	24,028.69
Johnson—City of Iowa City sanitary landfill	38.50	124,093.81
Jones—Jones County sanitary landfill	40.00	12,003.09
Keokuk—Southeast multicounty sanitary landfill	25.00	26,814.24
Kossuth—Kossuth County sanitary landfill	25.00	16,505.00
Lee—Great River regional waste authority	30.00	57,420.34
Linn—Bluestem (Site #1-CR) solid waste agency	35.00	14,078.60
Linn—Bluestem (Site #2-Marion) solid waste agency	35.00	165,117.56
Madison—South Central Iowa sanitary landfill	22.50	32,607.02
Mahaska—Mahaska County sanitary landfill	16.25	75,746.77
Marion—South Central Iowa solid waste agency	22.00	61,812.38
Marshall—Marshall County sanitary landfill	52.00	33,482.41
Mills-Loess hills regional sanitary landfill	36.00	99,879.75
Mitchell—Floyd-Mitchell County sanitary landfill	28.00	28,879.75
Montgomery—Montgomery County sanitary landfill	48.70	16,455.47
Muscatine—Muscatine County sanitary landfill	38.00	35,564.00

(continued)

TABLE 6.4   (*Continued*)

	Tipping Fee ($/ton)	Iowa Tonnages Fiscal Year 2007 (tons)
Page—Page County sanitary landfill	50.00	11,089.12
Palo Alto—Northern Plains	17.27	34,187.47
Polk—Metro Park East sanitary landfill	31.00	489,588.61
Appanoose, Rathbun area solid waste comm	41.00	11,385.20
SAC—SAC County sanitary landfill	30.00	4,540.76
Scott—Scott area sanitary landfill	33.00	156,723.94
Sioux—Northwest Iowa area sanitary landfill	15.70	56,171.63
Tama—Tama County sanitary landfill	30.00	16,334.17
Union—Union County sanitary landfill	31.00	19,469.49
Wapello—Ottumwa-Wapello County sanitary LF	48.00	43,294.00
Webster—North Central Iowa regional sanitary LF	15.00	90,931.50
Winnebago—central disposal landfill	32.25	150,449.94
Winneshiek—Winneshiek County sanitary landfill	51.00	24,127.72
Woodbury—City of Sioux City sanitary landfill	24.67	64,174.59
Woodbury—Woodbury County sanitary landfill	20.00	1,256.48

Source: Ref. (8,11,13).

from sources such as residential, commercial, institutional, or industrial.[12] The tonnages reported are the wastes deposited into the respective landfills.[13]

In comparison to waste generated from the state of Iowa, the volume of New York City (NYC) waste (tons/day) alone is shown in Table 6.5 as 19,870 tons/day or 7,252,550 tons/year. The waste from NYC is transported to other destinations. The top destinations of NYC waste to landfills are shown in Table 6.5. The NYC volume is the tons/day from NYC and can be compared to the total volume (tons/day) deposited into the respective landfill as indicated.

As can be seen from Table 6.5, many waste-to-energy facilities could be potentially justified economically from the waste generated from NYC alone.[14]

Regulations for landfilling of MSW were promulgated in the Resource Conservation Recovery Act (RCRA) Subtitle D as mandated by the U.S. Congress. These regulations are based upon the "dry tomb" approach where MSW is landfilled into a "dry tomb" or say "entombed" in a plastic sheeting and compacted soil/clay bottom liner in the landfill and finally capped off the top of the landfill with a plastic sheeting and compacted soil/clay cover. A leachate collection system is required as well as a landfill gas (LFG)-collection system. A groundwater monitoring system is required. These systems are to be funded and maintained for a postclosure period of 30 years as specified by the Congress in the RCRA Subtitle D MSW landfilling regulations.[15]

Rather than discuss the RCRA Subtitle D regulations for landfills, a clear understanding of these regulations can be found by reviewing an example of a landfill design presented by the Iowa Department of Natural Resources (IDNR),

TABLE 6.5   The Top Destinations for New York City Waste (Volumes in Tons per Day)—2002

Facility/City/State	NYC volume	Total volume	Ownership/Owner Entity
GROWS Landfill/Morrisville/ PA	2,622	10,000	Private/Waste Management Inc./ Pennsylvania
Tullytown Landfill Tullytown/ PA	2,122	8,000	Private/Waste Management Inc./ Pennsylvania
Atlantic Waste Disposal Inc. Landfill/Waverly/VA	1,436	4,000	Private/Atlantic Waste Disposal Inc./(Waste Mgmt.)
American Ref-Fuel/Essex County (Newark) Facility/ Newark/NJ	1,388	2,800	Public/American Ref-Fuel Company
Conestoga/New Morgan Landfill/NewMorgan/PA	1,270	5,210	Private/Browning-Ferris Ind. Inc./ (AlliedWaste)
Superior Greentree Landfill/ Kersey/PA	1,225	4,240	Private/Superior Services, Inc. (Onyx)
Alliance Landfill/Taylor/PA	1,186	5,500	Private/Waste Management Inc
Modern Landfill and recycling/ York/PA	1,057	4,643	Private/Republic Services, Inc.
Greenridge Reclamation/ Scottdale/PA	642	2,000	Private/Republic Industries, Inc.
Shade Landfill/Cairnbrook/PA	640	2,100	Private/Waste Management Inc.
Hyland Landfill & Ash Monofill/Belmont/NY	638	850	Private/Casella Waste Systems, Inc.
Pottstown Landfill/Pottstown/ PA	623	3,500	Private/Waste Management Inc.
Keystone Sanitary Landfill/ Dunmore/PA	530	4,340	Private/Keystone Sanitary Landfill Inc.
Southern Alleghenies Landfill/ Davidsville/PA	477	1,500	Private/Waste Management Inc.
IESI Blue Ridge Landfill (R&A Bender)/Chambersburg/PA	391	700	Private/IESI Corp.
City of Leeper/Clarion County Landfill/Leeper/PA	388	1,000	Private/Allied Waste Industries Inc.
American Ref-Fuel/Delaware Co. (Chester) WTE/Chester/ PA	353	3,071	Public/Delaware County
IESI of Bethlehem Landfill/ Bethlehem/PA	346	635	Private/IESI Corp.
Warren County District Landfill/Oxford/NJ	252	300	Public/Warren County
Chrin Sanitary Landfill/Easton/ PA	211	1,200	Private/Chrin Sanitary Landfill Inc.
Grand Central Sanitary Landfill/PenArgyl/PA	191	2,750	Private/Waste Management, Inc.

(continued)

TABLE 6.5   *(Continued)*

Facility/City/State	NYC volume	Total volume	Ownership/Owner Entity
Cumberland County Landfill/ Shippensburg/PA	162	995	Private/Community Refuse Inc.
American Ref-Fuel/ Hempstead/WTE Facility/ Westbury/NY	154	2,505	Private/American Ref-fuel Company
Pine Grove Landfill/PineGrove/ PA	152	1,500	Private/Waste Management Inc.
Carbon Limestone Sanitary Landfill/Lowellville/OH	150	3,800	Private/Browning-Ferris Ind. Inc. (Allied Waste)
Laurel Highlands Landfill/ Vintondale/PA	149	1,475	Private/Waste Management Inc.
Lakeview Landfill/Erie/PA	147	2,500	Private/Waste Management Inc.
Clinton County Landfill (Wayne Township)/ McElhattan/PA	146	500	Public/Clinton County
Westmoreland County Sanitary Landfill/Monessen/PA	138	550	Public/Westmoreland County
Imperial Landfill/Imperial/PA	125	2,700	Private/Browning-Ferris Industries Inc.
McKean County/Kness Landfill/Mount Jewett/PA	107	400	Public/McKean County
Mountain View Reclamation Landfill/Greencastle/PA	90	1,453	Private/Waste Management Inc.
Westchester RESCO/Peekskill/ NY	83	2,200	Private/Waste Management Inc.
Kelly Run Sanitary Landfill/ Elizabeth/PA	73	700	Private/Kelly Run Sanitation Inc.
Charles City Landfill/Charles City/VA	71	3,000	Private/Waste Management Inc.
Pioneer Crossing Landfill/ Birdsboro/PA	52	700	Private/J P Mascaro & Sons Inc.
Arden Landfill/Washington/PA	37	2,400	Private/Waste Management Inc.
Commonwealth Environmental Systems andfill/Hegins/PA	15	2,000	Private/Commonwealth Environmental Systems LP
Maplewood Recycling and Disposal Facility/Amelia/VA	15	2,500	Private/Waste Management Inc.
Totals by Ownership			
Ownership	Count	Volume From NYC	Total Daily Volume
Private	36	17,486	98,996
Public	6	2,384	7,621

Source: Ref. (14).

which complies with RCRA Subtitle D regulations. Shaw Environmental, Inc. developed a cost model with these general landfill assumptions:[16]

- Total landfill footprint = 7.5 acres.
- Total landfill capacity = 429,590 cubic yards, inclusive of daily and intermediate cover. This corresponds to approximately 246,480 tons of capacity for waste, assuming a 15% daily/intermediate cover and a compacted waste density of 1,350 pounds per cubic yard.
- Daily waste acceptance of 29 tons per day. This corresponds with the current average waste acceptance for unlined facilities.
- Projected operating life of the landfill is 30 years.
- Cell construction size will be 0.65 acres, representing the smallest practical construction size. New cells will be constructed approximately every 2.6 years based on the assumed waste acceptance rate.
- Total cell capacity = 21,360 tons of waste, calculated by dividing the total landfill capacity by the relative cell area.
- Leachate is generated at a rate of 400 gallons per acre per day (per Bonaparte and Othman, Geotechnical News, March 1995).
- Landfills that construct composite liners must build composite final cover systems.
- 15% contingency costs.

A schematic representation of a single composite liner landfill containment system with the general landfill assumptions presented above is shown as Fig. 6.4.

In Fig. 6.4, the landfill is constructed from some parcel of land by excavating a subgrade. Then, a clay liner is placed over the subgrade and compacted. Over the clay liner is immediately laid a flexible membrane liner (FML) of high-density polyethylene (HDPE) plastic sheeting. A leachate collection pipe system consisting of

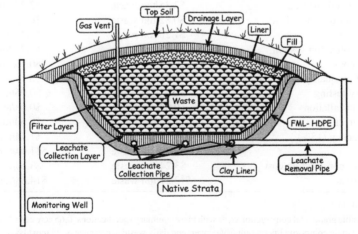

FIGURE 6.4   Single Composite Liner Landfill Containment System.[15]

leachate collection pipes, leachate collection layer, and filter layer is installed. A monitoring well is installed for monitoring and detection of any pollution of groundwater from the landfill. Waste is then placed into the constructed landfill until that portion of the constructed site is full. Then the landfill is capped, and a gas vent is installed for collection of LFG. Fill is placed over the waste and a liner is placed on top of the fill. A drainage layer is placed over the liner followed by top soil and vegetation.

The following discussion presents the construction costs for developing a landfill cell for the deposition of waste. The landfill cell comprises the single composite liner landfill containment system as shown in Fig. 6.4. For a single composite liner, the construction costs for a 0.65-acre cell are shown in Table 6.6. The composite cover construction costs are also shown in Table 6.6.[16]

By reviewing the possible total construction costs for the landfill at between $12.21/ton waste to 14.68/ton waste, these cost estimates indicate how subtitle D cell construction may impact current tipping fees. As shown in Table 6.3, the average tipping fee for the state of Iowa is $37.17/ton for year 2004.

Now that there is an understanding of a landfill construction meeting the regulations of the RCRA, Subtitle D for sanitary landfills, the technology, operation and details of Subtitle D landfilling of MSW will be presented.[15,17] The evolution of technology for landfilling of solid wastes was based upon a low-cost approach whereby urban areas deposited waste solids into nearby wetlands and low-value land for creation of a landfill. This approach to landfilling obviously had many flaws such as odor, attraction of vermin, rodents, flies, birds, etc. These shortcomings lead to having the daily deposited wastes being covered each day by a layer of soil. The concept that these poorly conceived waste depositories (landfills) could cause

TABLE 6.6 Single Composite Cell Costs and Final Cover Costs (0.65 acre Landfill Cell)

Construction Costs/ton Waste	Low Estimate	High Estimate
Excavation and subgrade	$2.35/ton waste	$3.77/ton waste
Groundwater diversion layer	$0.82/ton waste	$0.83/ton waste
Liner system	$1.26/ton waste	$1.42/ton waste
Leachate collection system	$2.15/ton waste	$2.22/ton waste
Miscellaneous	$0.64/ton waste	$0.76/ton waste
Construction quality assurance	$0.46/ton waste	$0.46/ton waste
Laboratory testing	$0.04/ton waste	$0.06/ton waste
Gas well installation	$0.08/ton waste	$0.08/ton waste
Total direct	$7.80/ton waste	$9.60/ton waste
Contingency (15%)	$1.17/ton waste	$1.44/ton waste
Total	$8.97/ton waste	$11.04/ton waste
Final cover construction cost	$3.24/ton waste	$3.64/ton waste
Grand total	$12.21/ton waste	$14.68/ton waste

Source: Ref. (16).

Note: Possible grand total construction costs with these summary specifications: 0.65 acre cell; 21,360 tons waste/cell; single composite liner; composite cover; and daily waste acceptance of 29 tons/day. Landfill to meet Resource Conservations and Recovery Act (RCRA), Subtitle D regulations.[16]

groundwater pollution or the gas generated by the landfills could be a threat to the public's health, where not considered seriously. Then in the 1980s, the U.S. EPA and state regulatory agencies adopted the "dry tomb" approach for the design and operation of landfills.[15,16]

The "dry tomb" approach to landfilling for the management of municipal and industrial solid waste is based upon the concept: isolate the waste in a landfill from water that generates the leachate (garbage juice), to prevent groundwater pollution. Some theorize that groundwater pollution is principally caused by the leachate formed from the solid waste and that constituents of the leachate (garbage juice) enter groundwater causing it to be polluted. Then, this can lead one to a conclusion that if water is prevented from contacting solid waste, leachate is not formed and pollution of groundwater is prevented. Thus, the design, construction, and operation of sanitary landfills as promulgated by governmental agencies adopted the "single composite liner landfill containment system" as illustrated in Fig. 6.4. This single composite liner approach to landfilling creates a situation whereby the wastes are isolated from the environment by compacted soil and a plastic sheeting creating the "lined tomb."[15–17]

The dry tomb landfilling approach is based upon using a thin plastic sheeting HDPE layer over a compacted soil/clay layer to form a "composite liner." Experience has shown that compacted clay layer alone will not prevent ground-water pollution. The plastic sheeting is thin and specified to be at least 60 mill thickness of HDPE. 60 mill thickness is less than $1/16^{th}$ of an inch, i.e., 0.96 (96%) of 1/16". The composite liner is created by placing the HDPE sheeting immediately upon the compacted soil/clay layer. Small holes in the HDPE plastic liner can lead to high leakage as stated. Note, the promulgated Subtitle D regulations are for a single composite liner and equivalent landfill cover as shown in Fig. 6.4. As mentioned, this approach was understood in the 1990s that it could only delay or postpone when groundwater pollution occurs by landfill leachate. Also, plastic sheeting of HDPE liner can allow dilute solutions of organic solvents to permeate through a liner with no holes.[15]

The next issue becomes the 30-year funding period for postclosure monitoring and maintenance as proposed by RCRA Subtitle D MSW landfilling regulations. This issue of postclosure of a landfill and the proper time, for monitoring and maintenance of the landfill to prevent pollution of groundwater and protect public safety, is not well understood. Is 30years enough? For example, it is reported that landfills developed in the Roman Empire about 2,000 years ago continue to produce leachate.[15] One could conclude that a better design for landfilling of MSW is needed and or a technology for eliminating landfills.

One key to the prevention of groundwater pollution is the collection of leachate from the landfill and removing it. Thus, a leachate collection and removal system is important for the prevention of groundwater pollution. Current leachate collection systems consist of a gravel or porous medium above the plastic liner to allow for the leachate to flow to the top of the HDPE liner. Once the leachate reaches the sloped liner, it is supposed to flow across the liner to collection pipes where the leachate is collected in a sump and pumped (removed) from the landfill. The issues here are minimization of the liquid (leachate) buildup on the liner due to many factors such as

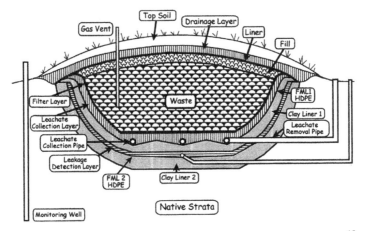

FIGURE 6.5 Double Composite Liner Landfill Containment System.[15]

plugging, biological growth, chemical precipitates, and solid particulate fines, which all contribute to buildup of leachate upon the top of the liner and potential leakage through the liner.

Currently, a number of states, such as Michigan, require double composite liners for MSW landfills. These liner systems are similar to those required for hazardous waste landfills. A schematic of a double composite liner landfill containment system is shown in Fig. 6.5.[15] Such containment systems with two composite liners are required to have a leak-detection system between the two liners so as to determine when the upper composite liner has failed.

Other issues on landfills are LFG and airborne emissions. Landfills contain organics that are converted to methane (CH_4) and carbon dioxide (CO_2) by bacteria. The CH_4 and CO_2 gaseous mixture produced is called LFG. Basically, landfill developers wish to control the concentration of CH_4 below 25% of the lower explosive limit (LEL) for CH_4 at landfill property boundary. The rate of LFG generation is dependent upon the moisture content of the wastes and prevention of water from entering the landfill from upper areas such as the cover. The more the moisture level, the greater the rate of LFG generation. LFG components can be malodorous compounds. MSW landfills are notorious for creating odor problems for miles around the landfill.

Conditions favorable for the production of LFG are: sufficient moisture content, proper nutrients, absence of oxygen and toxics, relatively neutral pH ranging from 6.7 to 7.2, alkalinity greater than 200 mg/L in the form of calcium carbonate, volatile acids less than 3000 mg/L in the form of acetic acid, and internal temperature between 86 and 131°F. The major constituents of LFG are: CH_4 (45–60% by volume), CO_2 (40–60% by volume), nitrogen (2–5% by volume), oxygen (0.1–1.0% by volume), ammonia (NH_3) (0.1–1.0% by volume), and hydrogen (0–0.2% by volume). LFG compositions of trace gases of less than 0.6% by volume are: odor-causing compounds, aromatic hydrocarbons, chlorinated solvents, aliphatic hydrocarbons, alcohols, and polyaromatic hydrocarbons. LFG yields can be 3–90 L/kg of dry solid waste (MSW). Estimates of LFG production rates are: (a) rapid degradation conditions, 3–7 years (4–10 L/kg/year); (b) moderate degradation conditions, 10–20 years (1.5–3 L/kg/year); and (c) slow degradation conditions, 20–40 years (0.7–1.5 L/kg/year), whereby kg is mass of MSW dry solids.[18]

Landfills have dust emissions and can impact adjacent properties. It is a good practice to prevent dust from a landfill from falling on adjacent properties. Dust can contain compounds that could affect the health of the public.

Stormwater runoff from a landfill can have adverse affects on the water quality of receiving waters of the runoff especially if any of the receiving waters are used for domestic water supplies and support aquatic life. Also, runoff from a landfill can contain pollutants that can be harmful to the public health. Stormwater runoff from landfills is polluted with wastes, chemicals, and pathogens, and thus the stormwater runoff must be controlled from the landfill, forever.[15]

To those who want to pursue many details about regulations on construction, operation, and managing a sanitary landfill for MSW based upon minimum standards pursuant to the RCRA for all municipal solid waste landfill (MSWLF) units under the Clean Water Act for MSWLFs that are used to dispose of sewage sludge, the following are excerpts from Chapter 113, Sanitary Landfills for MSW: Groundwater Protection Systems for the Disposal of Nonhazardous Wastes.[17]

567—113.7(455B) MSWLF unit design and construction standards. All MSWLF units shall be designed and constructed in accordance with this rule.

113.7(1) *Predesign meeting with the department.* A potential applicant for a new MSWLF unit may schedule a predesign meeting with the department's landfill permitting staff prior to beginning work on the plans and specifications of a modified or new MSWLF. The purpose of this meeting is to help minimize the need for revisions upon submittal of the official designs and specifications.

113.7(2) *Plans and specifications.*
a. Unless otherwise requested by the department, one copy of plans, specifications and supporting documents shall be sent to the department for review. Upon written department approval, the documents shall be submitted in triplicate to the department for proper distribution.
b. All new MSWLF units shall be constructed in compliance with the rules and regulations in effect at the time of construction. Previous department approval of plans and specifications for MSWLF units not yet constructed shall be superseded by the promulgation of new rules and regulations, after which plans and specifications shall be resubmitted to the department for approval prior to construction and operation.

113.7(3) *General site design and construction requirements.* An MSWLF shall have the following:
a. All-weather access roads to the facility.
b. A perimeter fence with a lockable gate(s) to help prevent unauthorized access.
c. A sign at the entrance to the facility specifying:
(1) Name and permit number of the facility.
(2) Days and hours that the facility is open to the public or a statement that the facility is not open to the public.

(3) A general list of materials that are not accepted.

(4) Telephone number of the official responsible for operation of the facility and the emergency contact person(s).

d. All-weather access roads within the facility.

e. Signs or pavement markings clearly indicating safe and proper on-site traffic patterns.

f. Adequate queuing distance for vehicles entering and exiting the property.

g. A scale certified by the Iowa department of agriculture and land stewardship.

113.7(4) *MSWLF unit subgrade.* The subgrade for a new MSWLF unit shall be constructed as follows:

a. All trees, stumps, roots, boulders, debris, and other material capable of deteriorating in situ material strength or of creating a preferential pathway for contaminants shall be completely removed or sealed off prior to construction of the MSWLF unit.

b. The material beneath the MSWLF unit shall have sufficient strength to support the weight of the unit during all phases of construction and operation. The loads and loading rate shall not cause or contribute to failure of the liner and leachate collection system.

IAC 7/4/07 Environmental Protection[567] Ch 113, p.17

c. The total settlement or swell of the MSWLF unit's subgrade shall not cause or contribute to failure of the liner and leachate collection system.

d. If the in situ material of the MSWLF unit's subgrade cannot meet the requirements of paragraphs 113.7(4)"b" and 113.7(4)"c," then such material shall be removed and replaced with material capable of compliance.

e. The subgrade of an MSWLF unit shall be constructed and graded to provide a smooth working surface on which to construct the liner.

f. The subgrade of an MSWLF unit shall not be constructed in or with frozen soil.

113.7(5) *MSWLF unit liners and leachate collection systems.* The liner and leachate collection system for a new MSWLF unit shall be constructed in accordance with the requirements of this subrule.

All active portions must have a composite liner or an alternative liner approved by the department.

An MSWLF unit must have a functioning leachate collection system during its active life.

a. *Liner systems.* An MSWLF unit shall have a liner system that complies with either the composite liner requirements of subparagraph 113.7(5)"a"(1) or an alternative liner system that complies with the requirements of subparagraph 113.7(5)"a"(2). Liners utilizing compacted soil must place the compacted soil in lifts no thicker than 8 inches after compaction.

(1) Composite liner systems.

1. A composite liner consists of two components, an upper flexible membrane liner (FML) and a lower compacted soil liner.

2. The upper component must consist of a minimum 30-mil flexible membrane liner (FML). FML components consisting of high-density polyethylene (HDPE) shall be at least 60 mil thick. The FML component must be installed in direct and uniform contact with the lower compacted soil component.

3. The lower component must consist of at least a 2-foot layer of compacted soil with a hydraulic conductivity of no more than 1×10^{-7} centimeters per second (cm/sec). The compacted soil must be placed in lifts no thicker than 8 inches after compaction.

4. The composite liner must be adequately sloped toward the leachate collection pipes to provide drainage of leachate. Unless alternative design requirements to this performance standard are approved as part of the permit under subrule 113.2(11) (relating to equivalency review procedure), the leachate collection system shall have a slope greater than or equal to 2 percent and not exceeding 33 percent.

(2) Alternative liner systems.

1. The design must ensure that the concentration values listed in Table I of rule 113.7(455B) will not be exceeded in the uppermost aquifer at the relevant point of compliance, as specified pursuant to numbered paragraph 113.7(5)"*a*"(2)"2." Alternative liners utilizing compacted soil must place the compacted soil in lifts no thicker than 8 inches.

2. The relevant point of compliance specified by the department must be within 50 feet of the planned liner or waste boundary, unless site conditions dictate otherwise, downgradient of the facility with respect to the hydrologic unit being monitored in accordance with subparagraph 113.10(2)"*a*"(2), and located on land owned by the owner of the MSWLF unit. The relevant point of compliance specified by the department shall be at least 50 feet from the property line of the facility.

3. When approving an alternative liner design, the department shall consider at least the following factors:

- The hydrogeologic characteristics of the facility and surrounding land.
- The climatic factors of the area.
- The volume and physical and chemical characteristics of the leachate.
- The sensitivities and limitations of the modeling demonstrating the applicable point of compliance.
- Practicable capability of the owner or operator.

Ch 113, p.18 Environmental Protection[567] IAC 7/4/07

4. The alternative liner must be adequately sloped toward the leachate collection pipes to provide drainage of leachate. Unless alternative design requirements to this performance standard are approved as part of the permit under subrule 113.2(11) (relating to equivalency review procedure), the leachate collection system shall have a slope greater than or equal to 2 percent and not exceeding 33 percent.

Table I

Chemical	MCL (mg/l)
Arsenic	0.01
Barium	1.0
Benzene	0.005
Cadmium	0.01
Carbon tetrachloride	0.005
Chromium (hexavalent)	0.05
2,4-Dichlorophenoxy acetic acid	0.1
1,4-Dichlorobenzene	0.075
1,2-Dichloroethane	0.005
1,1-Dichloroethylene	0.007
Endrin	0.0002
Fluoride	4.0
Lindane	0.004
Lead	0.05

(continued)

Table I (*Continued*)

Chemical	MCL (mg/l)
Mercury	0.002
Methoxychlor	0.1
Nitrate	10.0
Selenium	0.01
Silver	0.05
Toxaphene	0.005
1,1,1-Trichloromethane	0.2
Trichloroethylene	0.005
2,4,5-Trichlorophenoxy acetic acid	0.01
Vinyl chloride	0.002

b. Leachate collection system. All MSWLF units shall have a leachate collection system that complies with the following requirements:

(1) The leachate collection system shall be designed and constructed to function for the entire active life of the facility and the postclosure period.

(2) The leachate collection system shall be of a structural strength capable of supporting waste and equipment loads throughout the active life of the facility and the postclosure period.

(3) The leachate collection system shall be designed and constructed to minimize leachate head over the liner at all times. An MSWLF unit shall have a leachate collection system that maintains less than a 30-centimeter (i.e., 12-inch) depth of leachate over the liner. The leachate collection system shall have a method for accurately measuring the leachate head on the liner at the system's lowest point(s) within the MSWLF unit (e.g., sumps). Furthermore, an additional measuring device shall be installed to measure leachate directly on the liner in the least conductive drainage material outside of the sump and collection trench. Leachate head measurements from cleanout lines or manholes are not acceptable for the second measurement. All such measurement devices shall be in place before waste is placed in the MSWLF unit.

(4) If the leachate collection system is not designed and constructed factoring in leachate recirculation or bioreactor operations, the department may prohibit such activities within the MSWLF unit.

IAC 7/4/07 Environmental Protection[567] Ch 113, p.19

(5) The collection pipes shall be of a length and cross-sectional area that allow for cleaning and inspection through the entire length of all collection pipes at least once every three years. The collection pipes shall not be designed or constructed with sharp bends that prevent cleaning or inspection along any section of the collection pipe or that may cause the collection pipe to be damaged during cleaning or inspection.

(6) Leachate collection system designs shall attempt to minimize the potential for clogging due to mass loading.

(7) Unless alternative design requirements are approved as part of the permit under subrule 113.2(11) (relating to equivalency review procedure), the following design requirements shall apply:

1. A geotextile cushion over the flexible membrane liner (FML), if the liner utilizes an FML and granular drainage media. A geotextile cushion is not required if the granular drainage media is well rounded and less than 3/8 inch in diameter. The geotextile's mass shall be determined based on the allowable pressure on the geomembrane.

2. Collection pipe(s) at least 4 inches in diameter at the base of the liner slope(s), surrounded by the high hydraulic-conductivity material listed in numbered paragraph 113.7(5)"*b*"(7)"3" below. The collection pipe shall have slots or holes large enough to minimize the potential for clogging from fines conveyed by incoming leachate.

3. One of the following high hydraulic-conductivity materials:
- High hydraulic-conductivity material (e.g., gravel) of uniform size and a fines content of no more than 5 percent by weight passing a #200 sieve. The high hydraulic-conductivity material shall be at least 12 inches in depth and have a hydraulic conductivity of at least 1×10^{-2} cm/sec; or
- A geosynthetic drainage media (e.g., geonet). The transmissivity of geonets shall be tested with method ASTM D4716, or an equivalent test method, to demonstrate that the design transmissivity will be maintained for the design period of the facility. The testing for the geonet in the liner system shall be conducted using actual boundary material intended for the geonet at the maximum design normal load for the MSWLF unit, and at the design load expected from one lift of waste. At the maximum design normal load, testing shall be conducted for a minimum period of 100 hours unless data equivalent of the 100-hour period is provided, in which case the test shall be conducted for a minimum period of one hour. In the case of the design load from one lift of waste, the minimum period shall be one hour. For geonets used in final covers, only one test shall be conducted for a minimum period of one hour using the expected maximum design normal load from the cover soils and the actual boundary materials intended for the geonet. A granular layer at least 12 inches thick with a hydraulic conductivity of at least 1×10^{-3} cm/sec shall be placed above the geosynthetic drainage material that readily transmits leachate and provides separation between the waste and liner.

(8) Manholes within the MSWLF unit shall be designed to minimize the potential for stressing or penetrating the liner due to friction on the manhole exterior from waste settlement.

(9) The leachate drainage and collection system within the MSWLF unit shall not be used for the purpose of storing leachate. If leachate is to be stored, it shall be stored in designated storage structures outside of the MSWLF unit.

(10) All of the facility's leachate storage and management structures outside of the MSWLF unit (e.g., tanks, holding ponds, pipes, sumps, manholes, lift stations) and operations shall have containment structures or countermeasures adequate to prevent seepage to groundwater or surface water. The containment structures and countermeasures for leachate storage shall be at least as protective of groundwater at the liner of the MSWLF unit on a performance basis.

Ch 113, p.20 Environmental Protection[567] IAC 7/4/07

(11) Unless alternative design requirements are approved as part of the permit under subrule 113.2(11) (relating to equivalency review procedure), the leachate storage structures shall be able to store at least 7 days of accumulated leachate at the maximum generation rate used in designing the leachate collection system. Such minimum storage capacity may be constructed in phases over time so long as the 7-day accumulation capacity is maintained. The storage facility shall also have the ability to load tanker trucks in case sanitary sewer service is unavailable for longer than 7 days.

(12) The leachate collection system shall be equipped with valves or devices similar in effectiveness so that leachate can be controlled during maintenance.

(13) The leachate collection system shall be accessible for maintenance at all times and under all weather conditions.

(14) The permit holder shall annually submit a Leachate Control System Performance Evaluation (LCSPE) Report as a supplement to the facility Annual Water Quality Report, as

defined in subrule 113.10(10). The report shall include an evaluation of the effectiveness of the system in controlling the leachate, leachate head levels and elevations, the volume of leachate collected and transported to the treatment works or discharged under any NPDES permits, records of leachate contaminants testing required by the treatment works, proposed additional leachate control measures, and an implementation schedule in the event that the constructed system is not performing effectively.

113.7(6) *Quality control and assurance programs.* All MSWLF units shall be constructed under the supervision of a strict quality control and assurance (QC&A) program to ensure that MSWLF units are constructed in accordance with the requirements of rule 113.7(455B) and the approved plans and specifications. At a minimum, such a QC&A program shall consist of the following.

a. The owner or operator shall designate a quality control and assurance (QC&A) officer. The QC&A officer shall be a professional engineer (P.E.) registered in Iowa. The QC&A officer shall not be an employee of the facility, the construction company or construction contractor. The owner or operator shall notify the department of the designated QC&A officer and provide the department with that person's contact information. The QC&A officer may delegate another person or persons who are not employees of the facility to supervise or implement an aspect of the QC&A program.

b. The QC&A officer shall document compliance with rule 113.7(455B), and the approved plans and specifications, for the following aspects of construction:

(1) The MSWLF unit's subgrade.

(2) The liner system, as applicable, below:

1. The flexible membrane liner (FML). Destructive testing of the FML shall be kept to side slopes when continuous seams are utilized. Patches over FML destructive testing areas shall be checked with nondestructive methods.

2. The compacted clay component of the liner system. A minimum of five field moisture density tests per 8-inch lift per acre shall be performed to verify that the correct density, as correlated to permeability by a laboratory analysis, has been achieved. Laboratory hydraulic conductivity testing of Shelby tube samples from the constructed soil liner or test pad, or field hydraulic conductivity testing of the constructed soil liner or test pad, or other methods approved by the department, shall be utilized as a QC&A test.

(3) The leachate collection, conveyance and storage systems.

(4) Any other aspect of construction as required by the department.

c. A sampling and testing program shall be implemented by the QC&A officer as part of the QC&A program. The sampling and testing program shall:

(1) Verify full compliance with the requirements of rule 113.7(455B), and the approved plans and specifications.

(2) Be approved by the department prior to construction of the MSWLF unit.

(3) Detail how each stage of construction will be verified for full compliance with the requirements of rule 113.7(455B), and the approved plans and specifications.

IAC 7/4/07 Environmental Protection[567] Ch 113, p.21

(4) Be based on statistically significant sampling techniques and establish criteria for the acceptance or rejection of materials and constructed components of the MSWLF unit.

(5) Detail what actions will take place to remedy and verify any material or constructed component that is not in compliance with the requirements of rule 113.7(455B), and the approved plans and specifications.

d. The QC&A officer shall document the QC&A program. Upon completion of the MSWLF unit construction, the QC&A officer shall submit a final report to the department that verifies compliance with the requirements of rule 113.7(455B), and the approved plans and specifications. A copy of the final report shall also be maintained by the facility in the operating record. At a minimum, the final report shall include the following.

(1) A title page and index.

(2) The name and permit number of the facility.

(3) Contact information for the QC&A officer and persons delegated by the QC&A officer to supervise or implement an aspect of the QC&A program.

(4) Contact information for all construction contractors.

(5) Copies of daily reports containing the following information.

1. The date.

2. Summary of weather conditions.

3. Summary of locations on the facility where construction was occurring.

4. Summary of equipment, materials and personnel utilized in construction.

5. Summary of meetings held regarding the construction of the MSWLF unit.

6. Summary of construction progress.

7. Photographs of the construction progress, with descriptions of the time, subject matter and location of each photograph.

8. Details of sampling and testing program for that day. At a minimum, this report shall include details of where sampling and testing occurred, the methods utilized, personnel involved and test results.

9. Details of how any material or constructed component that was found not to be in compliance via the sampling and testing program was remedied.

(6) A copy of detailed as-built drawings with supporting documentation and photographic evidence.

This copy shall also include a narrative explanation of changes from the original departmentapproved plans and specifications.

(7) A signed and sealed statement by the QC&A officer that the MSWLF unit was constructed in accordance with the requirements of rule 113.7(455B), and the approved plans and specifications.

113.7(7) *Vertical and horizontal expansions of MSWLF units.* All vertical and horizontal expansions of disposal airspace over existing and new MSWLF units shall comply with the following requirements.

a. Horizontal expansions shall, at a minimum, comply with the following requirements:

(1) Horizontal expansions are new MSWLF units and, at a minimum, shall be designed and constructed in accordance with subrules 113.7(4), 113.7(5) and 113.7(6).

(2) The slope stability of the horizontal expansion between the existing unit and new MSWLF unit shall be analyzed. The interface between two MSWLF units shall not cause a slope failure of either of the MSWLF units.

(3) A horizontal expansion may include a vertical elevation increase of an existing MSWLF unit, pursuant to paragraph 113.7(7)"*b*," if approved by the department.

b. Vertical expansions shall, at a minimum, comply with the following requirements:

(1) A vertical expansion of an MSWLF unit shall not be allowed if the MSWLF unit does not have an approved leachate collection system and a composite liner or a leachate collection system and an alternative liner modeled at an approved point of compliance.

Ch 113, p.22 Environmental Protection[567] IAC 7/4/07

(2) An analysis of the structural impacts of the proposed vertical expansion on the liner and leachate collection system shall be completed. The vertical expansion shall not contribute to the structural failure of the liner and leachate collection system.

(3) An analysis of the impact of the proposed vertical expansion on leachate generation shall be completed. The vertical expansion shall not overload the leachate collection system or contribute to excess head on the liner.

(4) An analysis of the effect of the proposed vertical expansion on run-on, runoff and discharges into waters of the state shall be completed. The vertical expansion shall not cause a violation of subrule 113.7(8).

(5) The proposed vertical expansion shall be in compliance with the final slopes required at closure pursuant to paragraph 113.12(1)"*e.*"

(6) An analysis of the potential impact of the proposed vertical expansion on litter generation shall be completed. Landfill management strategies may need to be amended to help prevent increased litter.

(7) An analysis of the impact of the proposed vertical expansion on lines-of-sight and any visual buffering utilized by the landfill shall be completed.

113.7(8) *Run-on and runoff control systems.*
a. Owners or operators of all MSWLF units must design, construct, and maintain the following:
(1) A run-on control system to prevent flow onto the active portion of the landfill during the peak discharge from a 25-year storm;
(2) A runoff control system from the active portion of the landfill to collect and control at least the water volume resulting from a 24-hour, 25-year storm.
b. Runoff from the active portion of the MSWLF unit must be handled in accordance with paragraph 113.10(1)"*a.*"

567—113.8(455B) Operating requirements. The requirements of this rule shall be consolidated in a development and operations plan (DOPs) pursuant to subrule 113.8(4) and the emergency response and remedial action plan (ERRAP) pursuant to subrule 113.8(5), as applicable.

113.8(1) *Prohibited operations and activities.* For the purposes of this subrule, "regulated hazardous waste" means a solid waste that is a hazardous waste, as defined in Iowa Code section 455B.411.

a. Waste screening for prohibited materials. Owners or operators of all MSWLF units must implement a program at the facility for detecting and preventing the disposal of regulated hazardous wastes, polychlorinated biphenyls (PCB) wastes and other prohibited wastes listed in paragraph 113.8(1)"*b.*" This program must include, at a minimum:
(1) Random inspections of incoming loads unless the owner or operator takes other steps to ensure that incoming loads do not contain regulated hazardous wastes, PCB wastes or other prohibited wastes listed in paragraph 113.8(1)"*b*";
(2) Records of any inspections;
(3) Training of facility personnel to recognize regulated hazardous wastes, PCB wastes and other prohibited wastes listed in paragraph 113.8(1)"*b*"; and
(4) Notification of the EPA regional administrator if regulated hazardous wastes or PCB wastes are discovered at the facility.
b. Materials prohibited from disposal. The following wastes shall not be accepted for disposal by an MSWLF. Some wastes may be banned from disposal via the multiple categories listed below.
(1) Hazardous waste, whether it is a chemical compound specifically listed by EPA as a regulated hazardous waste or a characteristic hazardous waste pursuant to the characteristics below:
1. Ignitable in that the waste has a flash point (i.e., it will ignite) at a temperature of less than 140 degrees Fahrenheit.
2. Corrosive in that the waste has a pH less than 2 or greater than 12.5

IAC 7/4/07 Environmental Protection[567] Ch 113, p.23

3. Reactive in that the waste is normally unstable; reacts violently with water; forms an explosive mixture with water; contains quantities of cyanide or sulfur that could be released into the air in sufficient quantity to be a danger to human health; or can easily be detonated or exploded.

4. Toxicity characteristic leaching procedure (TCLP) (EPA Method 1311) toxic, in that a TCLP listed chemical constituent exceeds the EPA assigned concentration standard in 40 CFR Part 261 or the department assigned concentration standard in Table I of rule 113.7(455B). Waste from a residential building that is contaminated by lead-based paint (i.e., the waste fails the TCLP test for lead only) may be disposed of in an MSWLF unit. The purpose of this exclusion is to help prevent the exposure of children to lead-based paint. Therefore, the meaning of "residential building" in regard to this TCLP exclusion shall be interpreted broadly and include any building which children or parents may utilize as a residence (temporarily or permanently). Such residential buildings include, but are not limited to, single-family homes, apartment buildings, townhomes, condominiums, public housing, military barracks, nursing homes, hotels, motels, bunkhouses, and campground cabins.

(2) Polychlorinated biphenyl (PCB) wastes with a concentration equal to or greater than 50 parts per million (ppm).

(3) Free liquids, liquid waste and containerized liquids. For purposes of this subparagraph, "liquid waste" means any waste material that is determined to contain "free liquids" as defined by Method 9095B (Paint Filter Liquids Test), as described in Test Methods for Evaluating Solid Wastes, Physical/Chemical Methods (EPA Pub. No. SW-846). For the purposes of this subparagraph, "gas condensate" means the liquid generated as a result of the gas recovery process(es) at the MSWLF unit. However, free liquids and containerized liquids may be placed in MSWLF units if:

1. The containerized liquid is household waste other than septic waste. The container must be a small container similar in size to that normally found in household waste;

2. The waste is leachate or gas condensate derived from the MSWLF unit, whether it is a new or existing MSWLF unit or lateral expansion, and is designed with a composite liner and leachate collection system as described in paragraph 113.7(5)"*a.*" The owner or operator must demonstrate compliance with this subparagraph and place the demonstration in the operating record; or

3. The MSWLF unit is a research, development and demonstration (RD&D) project in which the department has authorized the addition of liquids and meets the applicable requirements of subrule 113.4(10).

(4) Septage, which is the raw material, liquids and pumpings from a septic system, unless treated pursuant to 567—Chapter 68.

(5) Appliances as defined pursuant to 567—Chapter 118, unless there is documentation that the appliance has been demanufactured pursuant to 567—Chapter 118.

(6) Radioactive waste, excluding luminous timepieces and other items using very small amounts of tritium.

(7) Infectious waste, unless managed and disposed of pursuant to 567—Chapter 109.

(8) Hot loads, meaning solid waste that is smoking, smoldering, emitting flames or hot gases, or otherwise indicating that the solid waste is in the process of combustion or close to igniting. Ash that has not been fully quenched or cooled is considered a hot load. Such wastes may be accepted at the gate, but shall be segregated and completely extinguished and cooled in a manner as safe and responsible as practical before disposal.

(9) Asbestos-containing material (ACM) waste with greater than 1 percent asbestos, unless managed and disposed of pursuant to 567—Chapter 109.

(10) Petroleum-contaminated soil, unless managed and remediated pursuant to 567—Chapter 120.

(11) Grit and bar screenings, and grease skimmings, unless managed and disposed of pursuant to 567—Chapter 109.

(12) Waste tires, unless each tire is processed into pieces no longer than 18 inches on any side. The department encourages the recycling of all waste tires, even if processed to disposal standards.

Ch 113, p.24 Environmental Protection[567] IAC 7/4/07

(13) Yard waste.

(14) Lead-acid batteries.

(15) Waste oil and materials containing free-flowing waste oil. Materials contaminated with waste oil may be disposed of if no free-flowing oil is retained in the material, and the material is not a hazardous waste.

(16) Baled solid waste, unless the waste is baled on site after the waste has been visually inspected for prohibited materials.

c. Open burning and fire hazards. No open burning of any type shall be allowed within the permitted boundary of an MSWLF facility. The fueling of vehicles and equipment, and any other activity that may produce sparks or flame, shall be conducted at least 50 feet away from the working face.

d. Scavenging and salvaging. Scavenging shall not be allowed at the MSWLF facility. However, salvaging by MSWLF operators may be allowed.

e. Animal feeding and grazing. Feeding animals MSW shall not be allowed at an MSWLF facility.

The grazing of domestic animals on fully vegetated areas of the MSWLF facility not used for disposal, including closed MSWLF units, may be allowed by the department so long as the animals do not cause damage or interfere with operations, inspections, environmental monitoring and other required activities. Large, hoofed animals (including but not limited to buffalo, cattle, llamas, pigs, and horses) shall not be allowed on closed MSWLF units.

113.8(2) *Disposal operations and activities.* All MSWLFs shall comply with the following requirements.

a. Survey controls and monuments. Survey controls and monuments shall be maintained as follows.

(1) The property boundary, the permitted boundary and the boundaries of all MSWLF units shall be surveyed and marked by a professional land surveyor at least once prior to closure.

(2) Prior to waste placement, all new MSWLF unit boundaries shall be surveyed and marked by a professional engineer.

(3) Survey monuments shall be established to check vertical elevations and the progression of fill sequencing. The survey monuments shall be established and maintained by a professional land surveyor.

(4) All survey stakes and monuments shall be clearly marked.

(5) A professional engineer shall biennially inspect all survey monuments and replace missing or damaged survey monuments.

b. First lift. The first lift and initial placement of MSW over a new MSWLF unit liner and leachate collection system shall comply with the following requirements.

(1) Waste shall not be placed in the new MSWLF unit until the QC&A officer has submitted a signed and sealed final report to the department pursuant to paragraph 113.7(6)"*d*" and that report has been approved by the department.

(2) Construction and earth-moving equipment shall not operate directly on the liner and leachate management system. Waste disposal operations shall begin at the edge of the new MSWLF unit by pushing MSW out over the liner and leachate collection system. Compactors and other similarly heavy equipment shall not operate directly on the leachate collection system until a minimum of 4 feet of waste has been mounded over the top of the leachate collection system.

(3) Construction and demolition debris and materials clearly capable of spearing through the leachate collection system and liner shall not be placed in the first 4 feet of waste over the top of the leachate collection system. The first 4 feet of waste shall consist of select waste that is unlikely to damage the liner and performance of the leachate collection system.

(4) The owner or operator must place documentation in the operating record and submit a copy to the department that adequate cover material was placed over the top of the leachate collection system in the MSWLF unit or that freeze/thaw effects had no adverse impact on the compacted clay component of the liner.

IAC 7/4/07 Environmental Protection[567] Ch 113, p.25

c. *Fill sequencing.* The rate and phasing of disposal operations shall comply with the following requirements.

(1) The fill sequencing shall be planned and conducted in a manner and at a rate that do not cause a slope failure, lead to extreme differential settlement, or damage the liner and leachate collection system.

(2) The fill sequencing shall be planned and conducted in a manner compliant with the run-on and runoff requirements of subrule 113.7(8) and surface water requirements of rule 113.10 (455B).

d. *Working face.* The working face shall comply with the following requirements.

(1) The working face shall be no larger than necessary to accommodate the rate of disposal in a safe and efficient manner.

(2) The working face shall not be so steep as to cause heavy equipment and solid waste collection vehicles to roll over or otherwise lose control.

(3) Litter control devices of sufficient size to help prevent blowing litter shall be utilized at the working face. The operation of the working face shall attempt to minimize blowing litter.

(4) The operation of the working face shall prevent the harborage of vectors and attempt to minimize the attraction of vectors.

(5) Employees at the working face shall be trained to visually recognize universal symbols, markings and indications of prohibited wastes pursuant to paragraph 113.8(1)"*b.*"

e. *Special wastes.* Special wastes shall be managed and disposed of pursuant to 567—Chapter 109.

f. *Cover material and alternative cover material.* Pursuant to 567—Chapter 108, alternative cover material of an alternative thickness (e.g., tarps, spray covers) may be authorized if the owner or operator demonstrates to the approval of the department that the alternative material and thickness control vectors, fires, odors, blowing litter, and scavenging without presenting a threat to human health and the environment. Cover material or alternative cover material shall be available for use during all seasons in all types of weather. Cover material and alternative cover material shall be utilized as follows unless otherwise approved by the department pursuant to 567—Chapter 108:

(1) Daily cover. Six inches of cover material or an approved depth or application of alternative cover material shall be placed and maintained over waste in the active portion at the end of each operating day, or at more frequent intervals if necessary, to control vectors, fires, odors, blowing litter, and scavenging.

(2) Intermediate cover. At least 1 foot of compacted cover material or an approved depth or application of alternative cover material shall be placed and maintained over waste in the active portion that has not or will not receive more waste for at least 30 days. At least 2 feet of compacted cover material or alternative cover material shall be placed and maintained over waste in the active portion that has not or will not receive waste for at least 180 days. Such active portions shall be graded to manage runon and runoff pursuant to subrule 113.7(8). Such active portions shall be seeded if they will not receive waste for a full growing season.

(3) Scarification of cover. To help prevent leachate seeps by aiding the downward flow of leachate, cover material or alternative cover material, which prevents the downward flow of leachate and is at least 5 feet from the outer edge of the MSWLF unit, shall be scarified prior to use of that area as a working face. Cover material or alternative cover material that does not impede the downward flow of leachate, as approved by the department, does not require scarification. Scarification may be as simple as the spearing or breaking up of a small area of the cover. Areas of intermediate cover may require removal of some of the cover material or alternative cover material to aid the downward flow of leachate.

(4) Final cover. Final cover over an MSWLF unit that is to be closed shall be constructed and maintained according to the closure and postclosure requirements of rules 113.12(455B) and 113.13(455B).

Ch 113, p.26 Environmental Protection[567] IAC 7/4/07

g. Leachate seeps. Leachate seeps shall be contained and plugged upon being identified. Leachate seeps shall not be allowed to reach waters of the state. Soils outside of the MSWLF unit that are contaminated by a leachate seep shall be excavated and then disposed of within the MSWLF unit. Such soils may be used for daily cover material.

h. Leachate recirculation. The department must approve an MSWLF unit for leachate recirculation.

The primary goal of the leachate recirculation system is to help stabilize the waste in a more rapid, but controlled, manner. The leachate recirculation system shall not contaminate waters of the state, contribute to erosion, damage cover material, harm vegetation, or spray persons at the MSWLF facility.

Leachate recirculation shall be limited to MSWLF units constructed with a composite liner.

i. Differential settlement. Areas of differential settlement sufficient to interfere with runoff and run-on shall be brought back up to the contours of the surrounding active portion. Differential settlement shall not be allowed to cause ponding of water on the active portion.

113.8(3) *Facility operations and activities.* All MSWLFs shall comply with the following requirements.

a. Controlled access. Owners or operators of all MSWLF units must control public access and prevent unauthorized vehicular traffic and illegal dumping of wastes by using artificial barriers, natural barriers, or both, as appropriate to protect human health and the environment.

b. Scales and weights. A scale certified by the Iowa department of agriculture and land stewardship shall weigh all solid waste collection vehicles and solid waste transport vehicles. The owner or operator shall maintain a record of the weight of waste disposed of.

c. All-weather access to disposal. A disposal area shall be accessible during all weather conditions.

d. Salvaged and processed materials. Salvaged and processed materials (e.g., scrap metal, compost, mulch, aggregate, tire chips) shall be managed and stored in an orderly manner that does not create a nuisance or encourage the attraction or harborage of vectors.

e. Vector control. Owners or operators must prevent or control the on-site populations of vectors using techniques appropriate for the protection of human health and the environment.

f. Litter control. The operator shall take steps to minimize the production of litter and the release of windblown litter off site of the facility. All windblown litter off site of the facility shall be collected daily unless prevented by unsafe working conditions. On-site litter shall be collected daily unless prevented by working conditions. A dated record of unsafe conditions that prevented litter collection activities shall be maintained by the facility.

g. Dust. The operator shall take steps to minimize the production of dust so that unsafe or nuisance conditions are prevented. Leachate shall not be used for dust control purposes.

h. Mud. The operator shall take steps to minimize the tracking of mud by vehicles exiting the facility so that slick or unsafe conditions are prevented.

i. Leachate and wastewater treatment. The leachate management system shall be managed and maintained pursuant to the requirements of paragraph 113.7(5)"*b.*" Leachate collection pipes shall be cleaned and inspected as necessary, but not less than once every three years. Leachate and wastewater shall be treated as necessary to meet the pretreatment limits, if any, imposed by an agreement between the MSWLF and a publicly owned wastewater treatment works (POTW) or by the effluent discharge limits established by an NPDES permit. Documentation of the POTW agreement or NPDES permit must be submitted to the department. All leachate and wastewater treatment systems shall conform to department wastewater design standards.

j. Financial assurance. Financial assurance shall be maintained pursuant to rule 113.14(455B).
IAC 7/4/07
IAC 7/4/07 Environmental Protection[567] Ch 113, p.27

113.8(4) *Development and operations plan (DOPs).* An MSWLF unit shall maintain a development and operations plan (DOPs). At a minimum, the DOPs shall detail how the facility will operate and how compliance with the requirements of rule 113.8(455B) will be maintained. The DOPs shall contain at least the following components.

a. A title page and table of contents.

b. Telephone number of the official responsible for the operation of the facility and an emergency contact person if different.

c. Service area of the facility and political jurisdictions included in that area.

d. Days and hours of operation of the facility.

e. Details of how the site will comply with the prohibited operations and activity requirements of subrule 113.8(1) and any related permit conditions.

f. Details of how the site will comply with the disposal operation and activity requirements of subrule 113.8(2) and any related permit conditions.

g. Details of how the site will comply with the facility operations and activity requirements of subrule 113.8(3), any related permit conditions, and any leachate and wastewater treatment requirements.

113.8(5) *Emergency response and remedial action plan (ERRAP).* All MSWLFs shall develop, submit to the department for approval, and maintain on site an ERRAP.

a. ERRAP submittal requirements. An updated ERRAP shall be submitted to the department with any permit modification or renewal request that incorporates facility changes that impact the ERRAP.

b. Content. The ERRAP is intended to be a quick reference during an emergency. The content of the ERRAP shall be concise and readily usable as a reference manual by facility managers and operators during emergency conditions. The ERRAP shall contain and address at least the following components, unless facility conditions render the specific issue as not applicable. To facilitate department review, the rationale for exclusion of any issues that are not applicable must be provided either in the body of the plan or as a supplement. Additional ERRAP requirements unique to the facility shall be addressed as applicable.

(1) Facility information.

1. Permitted agency.

2. DNR permit number.

3. Responsible official and contact information.

4. Certified operator and contact information.

5. Facility description.

6. Site and environs map.

(2) Regulatory requirements.

1. Iowa Code section 455B.306(6)"*d*" criteria citation.

2. Reference to provisions of the permit.

(3) Emergency conditions, response activities and remedial action.

1. Failure of utilities.
   • Short-term (48 hours or less).
   • Long-term (over 48 hours).

2. Evacuation procedures during emergency conditions.

3. Weather-related events.
   • Tornado and wind events.
   • Snow and ice.
   • Intense rainstorms, mud, and erosion.
   • Lightning strikes.
   • Flooding.
   • Event and postevent conditions.

Ch 113, p.28 Environmental Protection[567] IAC 7/4/07

4. Fire and explosions.
   • Waste materials.
   • Buildings and site.
   • Equipment.
   • Fuels.
   • Utilities.
   • Facilities.
   • Working area.
   • Hot loads.
   • Waste gases.
   • Explosive devices.

5. Regulated waste spills and releases.
   • Waste materials.
   • Leachate.
   • Waste gases.
   • Waste stockpiles and storage facilities.
   • Waste transport systems.
   • Litter and airborne particulate.
   • Site drainage system.
   • Off-site releases.

6. Hazardous material spills and releases.
   • Load-check control points.
   • Mixed waste deliveries.

- Fuels.
- Waste gases.
- Site drainage systems.
- Off-site releases.

7. Mass movement of land and waste.
   - Earthquakes.
   - Slope failure.
   - Waste shifts.
   - Waste subsidence.

8. Emergency and release notification and reporting.
   - Federal agencies.
   - State agencies.
   - County and city agencies including emergency management services.
   - News media.
   - Public and private facilities with special populations within five miles.
   - Reporting requirements and forms.

9. Emergency waste management procedures.
   - Communications.
   - Temporary discontinuation of services—short-term and long-term.
   - Facilities access and rerouting.
   - Waste acceptance.
   - Wastes in process.

10. Primary emergency equipment inventory.
    - Major equipment.
    - Fire hydrants and water sources.
    - Off-site equipment resources.

IAC 7/4/07 Environmental Protection[567] Ch 113, p.29

11. Emergency aid.
    - Responder contacts.
    - Medical services.
    - Contracts and agreements.

12. ERRAP training requirements.
    - Training providers.
    - Employee orientation.
    - Annual training updates.
    - Training completion and record keeping.

13. Reference tables, figures and maps.

113.8(6) *MSWLF operator certification.* Sanitary landfill operators shall be trained, tested, and certified by a department-approved certification program.

a. A sanitary landfill operator shall be on duty during all hours of operation of a sanitary landfill, consistent with the respective certification.

b. To become a certified operator, an individual shall complete a basic operator training course that has been approved by the department or an alternative, equivalent training approved by the department and shall pass a departmental examination as specified by this subrule. An operator certified by another state may have reciprocity subject to approval by the department.

c. A sanitary landfill operator certification is valid until June 30 of the following even-numbered year.

d. The required basic operator training course for a certified sanitary landfill operator shall have at least 25 contact hours and shall address the following areas, at a minimum:

(1) Description of types of wastes.

(2) Interpreting and using engineering plans.

(3) Construction surveying techniques.

(4) Waste decomposition processes.

(5) Geology and hydrology.

(6) Landfill design.

(7) Landfill operation.

(8) Environmental monitoring.

(9) Applicable laws and regulations.

(10) Permitting processes.

(11) Leachate control and treatment.

e. Alternate basic operator training must be approved by the department. The applicant shall be responsible for submitting any documentation the department may require to evaluate the equivalency of alternate training.

f. Fees.

(1) The examination fee for each examination is $20.

(2) The initial certification fee is $8 for each one-half year of a two-year period from the date of issuance to June 30 of the next even-numbered year.

(3) The certification renewal is $24.

(4) The penalty fee is $12.

g. Examinations.

(1) The operator certification examinations shall be based on the basic operator training course curriculum.

Ch 113, p.30 Environmental Protection[567] IAC 7/4/07

(2) All individuals wishing to take the examination required to become a certified operator of a sanitary landfill shall complete the Operator Certification Examination Application, Form 542-1354.

A listing of dates and locations of examinations is available from the department upon request. The application form requires the applicant to indicate the basic operator training course taken. Evidence of training course completion must be submitted with the application for certification. The completed application and the application fee shall be sent to the department and addressed to the central office in Des Moines. Application for examination must be received by the department at least 30 days prior to the date of examination.

(3) A properly completed application for examination shall be valid for one year from the date the application is approved by the department.

(4) Upon failure of the first examination, the applicant may be reexamined at the next scheduled examination. Upon failure of the second examination, the applicant shall be required to wait a period of 180 days between each subsequent examination.

(5) Upon each reexamination when a valid application is on file, the applicant shall submit to the department the examination fee at least ten days prior to the date of examination.

(6) Failure to successfully complete the examination within one year from the date of approval of the application shall invalidate the application.

(7) Completed examinations will be retained by the department for a period of one year after which they will be destroyed.

(8) Oral examinations may be given at the discretion of the department.

h. Certification.

(1) All operators who passed the operator certification examination by July 1, 1991, are exempt from taking the required operator training course. Beginning July 1, 1991, all operators are required to take the basic operator training course and pass the examination in order to become certified.

(2) Application for certification must be received by the department within 30 days of the date the applicant receives notification of successful completion of the examination. All applications for certification shall be made on a form provided by the department and shall be accompanied by the certification fee.

(3) Applications for certification by examination which are received more than 30 days but less than 60 days after notification of successful completion of the examination shall be accompanied by the certification fee and the penalty fee. Applicants who do not apply for certification within 60 days of notice of successful completion of the examination will not be certified on the basis of that examination.

(4) For applicants who have been certified under other state mandatory certification programs, the equivalency of which has been previously reviewed and accepted by the department, certification without examination will be recommended.

(5) For applicants who have been certified under voluntary certification programs in other states, certification will be considered. The applicant must have successfully completed a basic operator training course and an examination generally equivalent to the Iowa examination. The department may require the applicant to successfully complete the Iowa examination.

(6) Applicants who seek Iowa certification pursuant to subparagraphs 113.8(6)"*h*"(4) and (5) shall submit an application for examination accompanied by a letter requesting certification pursuant to those subparagraphs. Application for certification pursuant to those subparagraphs shall be received by the department in accordance with subparagraphs 113.8(6)"*h*"(2) and (3).

IAC 7/4/07 Environmental Protection[567] Ch 113, p.31

i. Renewals. All certificates shall expire every two years, on even-numbered years, and must be renewed every two years to maintain certification. Application and fee are due prior to expiration of certification.

(1) Late application for renewal of a certificate may be made, provided that such late application shall be received by the department or postmarked within 30 days of the expiration of the certificate.

Such late application shall be on forms provided by the department and accompanied by the penalty fee and the certification renewal fee.

(2) If a certificate holder fails to apply for renewal within 30 days following expiration of the certificate, the right to renew the certificate automatically terminates. Certification may be allowed at any time following such termination, provided that the applicant successfully completes an examination.

The applicant must then apply for certification in accordance with paragraph 113.8(6)"*h.*"

(3) An operator shall not continue to operate a sanitary landfill after expiration of a certificate without renewal thereof.

(4) Continuing education must be earned during the two-year certification period. All certified operators must earn ten contact hours per certificate during each two-year period. The two-year period will begin upon issuance of certification.

(5) Only those operators fulfilling the continuing education requirements before the end of each two-year period will be allowed to renew their certificates. The certificates of operators not fulfilling the continuing education requirements shall be void upon expiration, unless an extension is granted.

(6) All activities for which continuing education credit will be granted must be related to the subject matter of the particular certificate to which the credit is being applied.

(7) The department may, in individual cases involving hardship or extenuating circumstances, grant an extension of time of up to three months within which the applicant may fulfill the minimum continuing education requirements. Hardship or extenuating circumstances include documented health-related confinement or other circumstances beyond the control of the certified operator which prevent attendance at the required activities. All requests for extensions must be made 60 days prior to expiration of certification.

(8) The certified operator is responsible for notifying the department of the continuing education credits earned during the period. The continuing education credits earned during the period shall be shown on the application for renewal.

(9) A certified operator shall be deemed to have complied with the continuing education requirements of this subrule during periods that the operator serves honorably on active duty in the military service; or for periods that the operator is a resident of another state or district having a continuing education requirement for operators and meets all the requirements of that state or district for practice there; or for periods that the person is a government employee working as an operator and is assigned to duty outside the United States; or for other periods of active practice and absence from the state approved by the department.

j. Discipline of certified operators.

(1) Disciplinary action may be taken on any of the following grounds:

1. Failure to use reasonable care or judgment or to apply knowledge or ability in performing the duties of a certified operator. Duties of certified operators include compliance with rules and permit conditions applicable to landfill operation.

2. Failure to submit required records of operation or other reports required under applicable permits or rules of the department, including failure to submit complete records or reports.

3. Knowingly making any false statement, representation, or certification on any application, record, report or document required to be maintained or submitted under any applicable permit or rule of the department.

Ch 113, p.32 Environmental Protection[567] IAC 7/4/07

(2) Disciplinary sanctions allowable are:

1. Revocation of a certificate.

2. Probation under specified conditions relevant to the specific grounds for disciplinary action. Additional education or training or reexamination may be required as a condition of probation.

(3) The procedure for discipline is as follows:

1. The department shall initiate disciplinary action. The commission may direct that the department investigate any alleged factual situation that may be grounds for disciplinary action under subparagraph 113.8(6)"*j*"(1) and report the results of the investigation to the commission.

2. A disciplinary action may be prosecuted by the department.

3. Written notice shall be given to an operator against whom disciplinary action is being considered.

The notice shall state the informal and formal procedures available for determining the matter.

The operator shall be given 20 days to present any relevant facts and indicate the operator's position in the matter and to indicate whether informal resolution of the matter may be reached.

4. An operator who receives notice shall communicate verbally, in writing, or in person with the department, and efforts shall be made to clarify the respective positions of the operator and department.

5. The applicant's failure to communicate facts and positions relevant to the matter by the required date may be considered when determining appropriate disciplinary action.

6. If agreement as to appropriate disciplinary sanction, if any, can be reached with the operator and the commission concurs, a written stipulation and settlement between the department and the operator shall be entered into. The stipulation and settlement shall recite the basic facts and violations alleged, any facts brought forth by the operator, and the reasons for the particular sanctions imposed.

7. If an agreement as to appropriate disciplinary action, if any, cannot be reached, the department may initiate formal hearing procedures. Notice and formal hearing shall be in accordance with 567—Chapter 7 related to contested and certain other cases pertaining to license discipline.

k. Revocation of certificates. Upon revocation of a certificate, application for certification may be allowed after two years from the date of revocation. Any such applicant must successfully complete an examination and be certified in the same manner as a new applicant.

l. Temporary certification. A temporary operator of a sanitary landfill may be designated for a period of six months when an existing certified operator is no longer available to the facility. The facility must make application to the department, explain why a temporary certification is needed, identify the temporary operator, and identify the efforts which will be made to obtain a certified operator. A temporary operator designation shall not be approved for greater than a six-month period except for extenuating circumstances. In any event, not more than one six-month extension to the temporary operator designation may be granted. Approval of a temporary operator designation may be rescinded for cause as set forth in paragraph 113.8(6)"*j.*" All MSWLFs shall have at least one MSWLF operator trained, tested and certified by a department-approved program.

567—113.9(455B) Environmental monitoring and corrective action requirements for air quality and landfill gas. All MSWLFs shall comply with the following environmental monitoring and corrective action requirements for air quality and landfill gas.

113.9(1) *Air criteria.* Owners or operators of all MSWLFs must ensure that the units do not violate any applicable requirements developed under a state implementation plan (SIP) approved or promulgated by the department pursuant to Section 110 of the Clean Air Act.

113.9(2) *Landfill gas.* All MSWLFs shall comply with the following requirements for landfill gas.

For purposes of this subrule, "lower explosive limit" means the lowest percent by volume of a mixture of explosive gases in air that will propagate a flame at 25°C and atmospheric pressure.

IAC 7/4/07 Environmental Protection[567] Ch 113, p.33

a. Owners or operators of all MSWLF units must ensure that:

(1) The concentration of methane gas generated by the facility does not exceed 25 percent of the lower explosive limit for methane in facility structures (excluding gas pipeline, control or recovery system components);

(2) The concentration of methane gas does not exceed the lower explosive limit for methane at the facility property boundary; and

b. Owners or operators of all MSWLF units must implement a routine methane-monitoring program to ensure that the standards of paragraph 113.9(2)"*a*" are met. Such a program shall include routine subsurface methane monitoring (e.g., at select groundwater wells, at gas monitoring wells).

(1) The type and frequency of monitoring must be determined based on the following factors:
1. Soil conditions;
2. The hydrogeologic conditions surrounding the facility;
3. The hydraulic conditions surrounding the facility;
4. The location of facility structures (including potential subsurface preferential pathways such as, but not limited to, pipes, utility conduits, drain tiles and sewers) and property boundaries; and
5. The locations of structures near the outside of the facility to which or along which subsurface migration of methane gas may occur. Examples of such structures include, but are not limited to, houses, buildings, basements, crawl spaces, pipes, utility conduits, drain tiles and sewers.

(2) The minimum frequency of monitoring shall be quarterly.

c. If methane gas levels exceeding the limits specified in paragraph 113.9(2)"*a*" are detected, the owner or operator must:

(1) Immediately take all necessary steps to ensure protection of human health and notify the department and department field office with jurisdiction over the MSWLF;

(2) Within 7 days of detection, place in the operating record and notify the department and department field office with jurisdiction over the MSWLF of the methane gas levels detected and a description of the steps taken to protect human health; and

(3) Within 60 days of detection, implement a remediation plan for the methane gas releases, place a copy of the plan in the operating record, and notify the department and department field office with jurisdiction over the MSWLF that the plan has been implemented. The plan shall describe the nature and extent of the problem and the proposed remedy.

d. The owner or operator shall submit an annual report to the department detailing the gas monitoring sampling locations and results, any action taken, and the results of steps taken to address gas levels exceeding the limits of paragraph 113.9(2)"*a*" during the previous year. This report shall include a site map that delineates all structures, perimeter boundary locations, and other monitoring points where gas readings were taken. The site map shall also delineate areas of landfill gas migration outside the MSWLF units, if any. The report shall contain a narrative explaining and interpreting all of the data collected during the previous year. The report shall be due each year at a date specified by the department in the facility's permit.

567—113.10(455B) Environmental monitoring and corrective action requirements for groundwater and surface water. All MSWLFs shall comply with the following environmental monitoring and corrective action requirements for groundwater and surface water.

113.10(1) *General requirements for environmental monitoring and corrective action for groundwater and surface water.* The following general requirements apply to all provisions of this rule.

a. Surface water requirements. MSWLF units shall not:

(1) Cause a discharge of pollutants into waters of the United States, including wetlands, that violates any requirements of the Clean Water Act, including, but not limited to, the National Pollutant Discharge Elimination System (NPDES) requirements, pursuant to Section 402 of the Clean Water Act.

Ch 113, p.34 Environmental Protection[567] IAC 7/4/07

(2) Cause the discharge of a nonpoint source of pollution into waters of the United States, including wetlands, that violates any requirement of an areawide or statewide water quality management plan that has been approved under Section 208 or 319 of the Clean Water Act.

b. A new MSWLF unit must be in compliance with the groundwater monitoring requirements specified in subrules 113.10(2), 113.10(4), 113.10(5) and 113.10(6) before waste can be placed in the unit.

c. Once established at an MSWLF unit, groundwater monitoring shall be conducted throughout the active life and postclosure care period of that MSWLF unit as specified in rule 113.13 (455B).

d. For the purposes of this rule, a "qualified groundwater scientist" means a scientist or an engineer who has received a baccalaureate or postgraduate degree in the natural sciences or engineering and has sufficient training and experience in groundwater hydrology and related fields demonstrated by state registration, professional certifications, or completion of accredited university programs that enable that individual to make sound professional judgments regarding groundwater monitoring, contaminant fate and transport, and corrective action.

e. The department may establish alternative schedules for demonstrating compliance with:

(1) Subparagraph 113.10(2)"*e*"(3), pertaining to notification of placement of certification in operating record;

(2) Subparagraph 113.10(5)"*c*"(1), pertaining to notification that statistically significant increase (SSI) notice is in operating record;

(3) Subparagraphs 113.10(5)"*c*"(2) and (3), pertaining to an assessment monitoring program;

(4) Paragraph 113.10(6)"b," pertaining to sampling and analyzing Appendix II constituents;

(5) Subparagraph 113.10(6)"*d*"(1), pertaining to placement of notice (Appendix II constituents detected) in record and notification of placement of notice in record;

(6) Subparagraph 113.10(6)"*d*"(2), pertaining to sampling for Appendices I and II;

(7) Paragraph 113.10(6)"g," pertaining to notification (and placement of notice in record) of SSI above groundwater protection standard;

(8) Numbered paragraph 113.10(6)"g"(1)"4" and paragraph 113.10(7)"a," pertaining to assessment of corrective measures;

(9) Paragraph 113.10(8)"a," pertaining to selection of remedy and notification of placement in record;

(10) Paragraph 113.10(9)"f," pertaining to notification of placement in record (certification of remedy completed).

113.10(2) *Groundwater monitoring systems.* All MSWLFs shall have a groundwater monitoring system that complies with the following requirements:

a. A groundwater monitoring system must be installed that meets the following objectives:

(1) Yields groundwater samples from the uppermost aquifer that represent the quality of background groundwater that has not been affected by leakage from a unit. A determination of background quality may include sampling of wells that are not hydraulically upgradient of the waste management area where either:

1. Hydrogeologic conditions do not allow the owner or operator to determine which wells are hydraulically upgradient; or

2. Sampling at other wells will provide an indication of background groundwater quality that is as representative as or more representative than that provided by the upgradient wells.

IAC 7/4/07 Environmental Protection[567] Ch 113, p.35

(2) Yields groundwater samples from the uppermost aquifer that represent the quality of groundwater passing the relevant point of compliance specified by the department under numbered paragraph 113.7(5)"a"(2)"2." The downgradient monitoring system must be installed at the relevant point of compliance specified by the department under numbered

paragraph 113.7(5)"a"(2)"2" that ensures detection of groundwater contamination in the uppermost aquifer. When physical obstacles preclude installation of groundwater monitoring wells at the relevant point of compliance at existing units, the downgradient monitoring system may be installed at the closest practicable distance, hydraulically downgradient from the relevant point of compliance specified by the department under numbered paragraph 113.7(5)"a"(2)"2," that ensures detection of groundwater contamination in the uppermost aquifer.

(3) Provides a high level of certainty that releases of contaminants from the site can be promptly detected. Downgradient monitoring wells shall be placed along the site perimeter, within 50 feet of the planned liner or waste boundary unless site conditions dictate otherwise, downgradient of the facility with respect to the hydrologic unit being monitored. Each groundwater underdrain system shall be included in the groundwater detection monitoring program under subrule 113.10(5). The maximum drainage area routed through each outfall shall not exceed 10 acres unless it can be demonstrated that site-specific factors such as drain flow capacity or site development sequencing require an alternative drainage area. If contamination is identified in the groundwater underdrain system pursuant to subrule 113.10(5), the owner or operator shall manage the underdrain discharge as leachate in lieu of assessment monitoring and corrective action.

(4) Be designed and constructed with the theoretical release evaluation pursuant to subparagraph 113.6(3)"e"(6) taken into consideration.

b. For those facilities which are long-term, multiphase operations, the department may establish temporary waste boundaries in order to define locations for monitoring wells. The convergence of groundwater paths to minimize the overall length of the downgradient dimension may be taken into consideration in the placement of downgradient monitoring wells provided that the multiphase unit groundwater monitoring system meets the requirements of paragraphs 113.10 (2)"a," 113.10(2)"c,"

113.10(2)"d" and 113.10(2)"e" and will be as protective of human health and the environment as the individual monitoring systems for each MSWLF unit, based on the following factors:

(1) Number, spacing, and orientation of the MSWLF units;

(2) Hydrogeologic setting;

(3) Site history;

(4) Engineering design of the MSWLF units; and

(5) Type of waste accepted at the MSWLF units.

c. Monitoring wells must be constructed and cased by a well contractor certified pursuant to 567—Chapter 82 in a manner that maintains the integrity of the monitoring well borehole. This casing must be screened or perforated and packed with gravel or sand, where necessary, to enable collection of groundwater samples. The annular space (i.e., the space between the borehole and well casing) above the sampling depth must be sealed to prevent contamination of samples and the groundwater. Monitoring wells constructed in accordance with the rules in effect at the time of construction shall not be required to be abandoned and reconstructed as a result of subsequent amendments to these rules unless the department finds that the well is no longer providing representative groundwater samples. See Figure 1 for a general diagram of a properly constructed monitoring well.

(1) The owner or operator must notify the department that the design, installation, development, and decommission of any monitoring wells, piezometers and other measurement, sampling, and analytical devices documentation has been placed in the operating record.

(2) The monitoring wells, piezometers, and other measurement, sampling, and analytical devices must be operated and maintained so that they perform to design specifications throughout the life of the monitoring program.

Ch 113, p.36 Environmental Protection[567] IAC 7/4/07

(3) Each groundwater monitoring point must have a unique and permanent number, and that number must never change or be used again at the MSWLF. The types of groundwater monitoring points shall be identified as follows:

1. Monitoring wells by "MW# (Insert unique and permanent number)".

2. Piezometers by "PZ# (Insert unique and permanent number)".

3. Groundwater underdrain systems by "GU# (Insert unique and permanent number)".

(4) Monitoring well construction shall be performed by a certified well contractor (pursuant to 567—Chapter 82) and shall comply with the following requirements:

1. In all phases of drilling, well installation and completion, the methods and materials used shall not introduce substances or contaminants that may alter the results of water quality analyses.

2. Drilling equipment that comes into contact with contaminants in the borehole or above-ground shall be thoroughly cleaned to avoid spreading contamination to other depths or locations. Contaminated materials or leachate from wells must not be discharged onto the ground surface or into waters of the state so as to cause harm in the process of drilling or well development.

3. The owner or operator must ensure that, at a minimum, the well design and construction log information is maintained in the facility's permanent record using DNR Form 542-1277 and that a copy is sent to the department.

(5) Monitoring well casings shall comply with the following requirements:

1. The diameter of the inner well casing (see Figure 1) of a monitoring well shall be at least 2 inches.

2. Plastic-cased wells shall be constructed of materials with threaded and nonglued joints that do not allow water infiltration under the local subsurface pressure conditions and when the well is evacuated for sampling.

3. Well casing shall provide sufficient structural stability so that a borehole or well collapse does not occur. Flush joint casing is required for small diameter wells installed through hollow stem augers.

(6) Monitoring well screens shall comply with the following requirements:

1. Slot size shall be based on sieve analysis of the sand and gravel stratum or filter pack. The slot size must keep out at least 90 percent of the filter pack.

2. Slot configuration and open area must permit effective development of the well.

3. The screen shall be no longer than 10 feet in length, except for water table wells, in which case the screen shall be of sufficient length to accommodate normal seasonal fluctuations of the water table.

The screen shall be placed 5 feet above and below the observed water table, unless local conditions are known to produce greater fluctuations. Screen length for piezometers shall be 2 feet or less. Multiplescreened, single-cased wells are prohibited.

(7) Monitoring well filter packs shall comply with the following requirements:

1. The filter pack shall extend at least 18 inches above and 12 inches below the well screen.

2. The size of the filter pack material shall be based on sieve analysis when sand and gravel are screened. The filter pack material must be 2.5 to 3 times larger than the 50 percent grain size of the zone being monitored.

3. In stratum that is neither sand nor gravel, the size of the filter pack material shall be selected based on the particle size of the zone being monitored.

(8) Monitoring well annular space shall comply with the following requirements:

1. Grouting materials must be installed from the top of the filter pack up in one continuous operation with a tremie tube.

2. The annular space between the filter pack and the frostline must be backfilled with bentonite grout.

3. The remaining annular space between the protective casing and the monitoring well casing must be sealed with bentonite grout from the frostline to the ground surface.

IAC 7/4/07 Environmental Protection[567] Ch 113, p.37

(9) Monitoring well heads shall be protected as follows:

1. Monitoring wells shall have a protective metal casing installed around the upper portion of the monitoring well casing as follows:

- The inside diameter of the protective metal casing shall be at least 2 inches larger than the outer diameter of the monitoring well casing.
- The protective metal casing shall extend from a minimum of 1 foot below the frostline to slightly above the well casing top; however, the protective casing shall be shortened if such a depth would cover a portion of the well screen.
- The protective casing shall be sealed and immobilized with a concrete plug around the outside. The bottom of the concrete plug must extend at least 1 foot below the frostline; however, the concrete plug shall be shortened if such a depth would cover a portion of the well screen. The top of the concrete plug shall extend at least 3 inches above the ground surface and slope away from the well. Soil may be placed above the plug and shall be at least 6 inches below the cap to improve runoff.
- The inside of the protective casing shall be sealed with bentonite grout from the frostline to the ground surface.
- A vented cap shall be placed on the monitoring well casing.
- A vented, locking cap shall be placed on the protective metal casing. The cap must be kept locked when the well is not being sampled.

2. All monitoring wells shall have a ring of brightly colored protective posts or other protective barriers to help prevent accidental damage.

3. All monitoring wells shall have a sign or permanent marking clearly identifying the permanent monitoring well number (MW#).

4. Run-on shall be directed away from all monitoring wells.

(10) Well development is required prior to the use of the monitoring well for water quality monitoring purposes. Well development must loosen and remove fines from the well screen and gravel pack.

Any water utilized to stimulate well development must be of sufficient quality that future samples are not contaminated. Any gases utilized in well development must be inert gases that will not contaminate future samples. Following development, the well shall be pumped until the water does not contain significant amounts of suspended solids.

d. Groundwater monitoring points that are no longer functional must be sealed. Groundwater monitoring points that are to be sealed and are in a future waste disposal area shall be reviewed to determine if the method utilized to seal the monitoring point needs to be more protective than the following requirements. All abandoned groundwater-monitoring points (e.g., boreholes, monitoring wells, and piezometers) shall be sealed by a well contractor certified pursuant to 567—Chapter 82 and in accordance with the following requirements.

(1) The following information shall be placed in the operating record and a copy sent to the department:

1. The unique, permanent monitoring point number.

2. The reasons for abandoning the monitoring point.

3. The date and time the monitoring point was sealed.

4. The method utilized to remove monitoring point materials.

5. The method utilized to seal the monitoring point.

6. Department Form 542-1226 for Water Well Abandonment Plugging Record.

(2) The monitoring point materials (e.g., protective casing, casing, screen) shall be removed. If drilling is utilized to remove the materials, then the drilling shall be to the maximum depth of the previously drilled monitoring point. All drilling debris shall be cleaned from the interior of the borehole.

Ch 113, p.38 Environmental Protection[567] IAC 7/4/07

(3) The cleared borehole shall be sealed with impermeable bentonite grout via a tremie tube. The end of the tremie tube shall be submerged in the grout while filling from the bottom of the borehole to the top of the ground surface. Uncontaminated water shall be added from the surface as needed to aid grout expansion.

(4) After 24 hours, the bentonite grout shall be retopped if it has settled below the ground surface.

e. Hydrologic monitoring system plan (HMSP). Unless otherwise approved by the department in writing, the number, spacing, and depth of groundwater monitoring points shall be:

(1) Determined based upon site-specific technical information, including but not limited to the soil and hydrogeologic investigation pursuant to subrule 113.6(3) and the site exploration and characterization report pursuant to subrule 113.6(4), that must include thorough characterization of:

1. Aquifer thickness, groundwater flow rate, and groundwater flow direction including seasonal and temporal fluctuations in groundwater flow; and

2. Saturated and unsaturated geologic units and fill materials overlying the uppermost aquifer, materials comprising the uppermost aquifer, and materials comprising the confining unit defining the lower boundary of the uppermost aquifer, including, but not limited to: thicknesses, stratigraphy, lithology, hydraulic conductivities, porosities and effective porosities; and

3. Projected paths and rates of movement of contaminants found in leachate pursuant to subparagraph 113.6(3)"*e*"(6).

(2) Designed and constructed with a maximum of 300 feet between downgradient groundwater monitoring wells, unless it is demonstrated by site-specific analysis or modeling that an alternative well spacing is justified. The convergence of groundwater paths to minimize the overall length of the downgradient dimension may be taken into consideration in the placement of downgradient monitoring wells provided that the groundwater monitoring system meets the requirements of paragraphs 113.10(2)"*a*," 113.10(2)"*c*," 113.10(2)"*d*," and 113.10(2)"*e*."

(3) Certified by a qualified groundwater scientist, as defined in paragraph 113.10(1)"*d*," and approved by the department. Within 14 days of this certification and approval by the department, the owner or operator must notify the department that the certification has been placed in the operating record.

IAC 7/4/07 Environmental Protection[567] Ch 113, p.39

Ch 113, p.40 Environmental Protection[567] IAC 7/4/07

f. Monitoring well maintenance and performance reevaluation plan. A monitoring well maintenance and performance reevaluation plan shall be included as part of the hydrologic monitoring system plan. The plan shall ensure that all monitoring points remain reliable. The plan shall provide for the following:

(1) A biennial examination of high and low water levels accompanied by a discussion of the acceptability of well location (vertically and horizontally) and exposure of the screened interval to the atmosphere.

(2) A biennial evaluation of water level conditions in the monitoring wells to ensure that the effects of waste disposal or well operation have not resulted in changes in the hydrologic setting and resultant flow paths.

(3) Measurements of well depths to ensure that wells are physically intact and not filling with sediment.

Measurements shall be taken annually in wells which do not contain dedicated sampling pumps and every five years in wells containing dedicated sampling pumps.

(4) A biennial evaluation of well recharge rates and chemistry to determine if well deterioration is occurring.

113.10(3) *Surface water monitoring systems.* The department may require an MSWLF facility to implement a surface water monitoring program if there is reason to believe that a surface water of the state has been impacted as a result of facility operations (i.e., leachate seeps, sediment pond discharge) or a groundwater SSI over background has occurred.

a. A surface water monitoring program must be developed that consists of a sufficient number of monitoring points, designated at appropriate locations, to yield surface water samples that:

(1) Provide a representative sample of the upstream quality of a surface water of the state if the surface water being monitored is a flowing body of water.

(2) Provide a representative sample of the downstream quality of a surface water of the state if the surface water being monitored is a flowing body of water.

b. Surface water levels must be measured at a frequency specified in the facility's permit, within 1/10 of a foot at each surface water monitoring point immediately prior to sampling, each time surface water is sampled. The owner or operator must determine the rate and direction of surface water flow, if any, each time surface water is sampled. Surface water level and flow measurements for the same surface water of the state must be measured on the same day to avoid temporal variations that could preclude accurate determination of surface water flow and direction.

c. The owner or operator must notify and receive approval from the department for the designation or decommission of any surface water monitoring point, and must place that approval in the operating record.

d. The surface water monitoring points shall be designated to maintain sampling at that monitoring point throughout the life of the surface water monitoring program.

e. Each surface water monitoring point must have a unique and permanent number, and that number must never change or be used again at the MSWLF. Surface water monitoring points shall be identified by "SW# (Insert unique and permanent number)".

f. The number, spacing, and location of the surface water monitoring points shall be determined based upon site-specific technical information, including:

(1) Water level, including seasonal and temporal fluctuations in water level; and

(2) Flow rate and flow direction, including seasonal and temporal fluctuations in flow.

g. The MSWLF may discontinue the surface water monitoring program if monitoring data indicates that facility operations are not impacting surface water.

IAC 7/4/07 Environmental Protection[567] Ch 113, p.41

113.10(4) *Groundwater sampling and analysis requirements.*

a. The groundwater monitoring program must include consistent sampling and analysis procedures that are designed to ensure monitoring results that provide an accurate representation of groundwater quality at the background and downgradient wells installed in compliance with subrule 113.10(2). The groundwater monitoring program shall utilize a laboratory certified by the department.

The owner or operator must notify the department that the sampling and analysis program documentation has been placed in the operating record, and the program must include procedures and techniques for:

(1) Sample collection;

(2) Sample preservation and shipment;

(3) Analytical procedures;

(4) Chain of custody control; and

(5) Quality assurance and quality control.

b. The groundwater monitoring programs must include sampling and analytical methods that are appropriate for groundwater sampling and that accurately measure hazardous constituents and other monitoring parameters in groundwater samples. Groundwater samples shall not be field-filtered prior to laboratory analysis.

c. The sampling procedures and frequency must be protective of human health and the environment, and consistent with subrule 113.10(5).

d. Groundwater elevations must be measured at a frequency specified in the facility's permit, within 1/100 of a foot in each well immediately prior to purging, each time groundwater is sampled.

The owner or operator must determine the rate and direction of groundwater flow each time groundwater is sampled. Groundwater elevations in wells which monitor the same waste management area must be measured within a period of time short enough to avoid temporal variations in groundwater flow which could preclude accurate determination of groundwater flow rate and direction.

e. The owner or operator must establish background groundwater quality in a hydraulically upgradient or background well(s) for each of the monitoring parameters or constituents required in the particular groundwater monitoring program that applies to the MSWLF unit, as determined under paragraph 113.10(5)"*a*" or 113.10(6)"*a*." Background groundwater quality may be established at wells that are not located hydraulically upgradient from the MSWLF unit if the wells meet the requirements of subparagraph 113.10(2)"*a*"(1).

f. The number of samples collected to establish groundwater quality data must be consistent with the appropriate statistical procedures determined pursuant to paragraph 113.10(4)"*g*." The sampling procedures shall be those specified under paragraphs 113.10(5)"*b*" for detection monitoring, 113.10(6)"*b*" and 113.10(6)"*d*" for assessment monitoring, and 113.10(7)"*b*" for corrective action.

g. The owner or operator must specify in the operating record which of the following statistical methods will be used in evaluating groundwater monitoring data for each hazardous constituent. The statistical test chosen shall be conducted separately for each hazardous constituent in each well.

(1) A parametric analysis of variance (ANOVA) followed by multiple comparisons procedures to identify statistically significant evidence of contamination. The method must include estimation and testing of the contrasts between each compliance well's mean and the background mean levels for each constituent.

(2) An analysis of variance (ANOVA) based on ranks followed by multiple comparisons procedures to identify statistically significant evidence of contamination. The method must include estimation and testing of the contrasts between each compliance well's median and the background median levels for each constituent.

(3) A tolerance or prediction interval procedure in which an interval for each constituent is established from the distribution of the background data, and the level of each constituent in each compliance well is compared to the upper tolerance or prediction limit.

Ch 113, p.42 Environmental Protection[567] IAC 7/4/07

(4) A control chart approach that gives control limits for each constituent.

(5) Another statistical test method that meets the performance standards of paragraph 113.10 (4)"*h.*" The owner or operator must place a justification for this alternative in the operating record and notify the department of the use of this alternative test. The justification must demonstrate that the alternative method meets the performance standards of paragraph 113.10 (4)"*h.*"

h. The statistical method required pursuant to paragraph 113.10(4)"*g*" shall comply with the following performance standards:

(1) The statistical method used to evaluate groundwater monitoring data shall be appropriate for the distribution of chemical parameters or hazardous constituents. If the distribution of the chemical parameters or hazardous constituents is shown by the owner or operator to be inappropriate for a normal theory test, then the data shall be transformed or a distribution-free theory test shall be used. If the distributions for the constituents differ, more than one statistical method may be needed.

(2) If an individual well comparison procedure is used to compare an individual compliance well constituent concentration with background constituent concentrations or a groundwater protection standard, the test shall be done at a Type I error level not less than 0.01 for each testing period. If a multiple comparisons procedure is used, the Type I experimentwise error rate for each testing period shall be not less than 0.05; however, the Type I error level of not less than 0.01 for individual well comparisons must be maintained.

(3) If a control chart approach is used to evaluate groundwater monitoring data, the specific type of control chart and its associated parameter values shall be protective of human health and the environment.

The parameters shall be determined after the number of samples in the background data base, the data distribution, and the range of the concentration values for each constituent of concern have been considered.

(4) If a tolerance interval or a predictional interval is used to evaluate groundwater monitoring data, the levels of confidence and, for tolerance intervals, the percentage of the population that the interval must contain, shall be protective of human health and the environment. These parameters shall be determined after the number of samples in the background data base, the data distribution, and the range of the concentration values for each constituent of concern have been considered.

(5) The statistical method shall account for data below the limit of detection (LD) by recording such data at one-half the limit of detection (i.e., LD/2) or as prescribed by the statistical method. Any practical quantitation limit (pql) that is used in the statistical method shall be the lowest concentration level that can be reliably achieved within specified limits of precision and accuracy during routine laboratory operating conditions that are available to the facility.

(6) If necessary, the statistical method shall include procedures to control or correct for seasonal and spatial variability as well as temporal correlation in the data.

i. The owner or operator must determine whether or not there is an SSI over background values for each parameter or constituent required in the particular groundwater monitoring program that applies to the MSWLF unit, as determined under paragraph 113.10(5)"*a*" or 113.10(6)"*a.*"

(1) In determining whether an SSI has occurred, the owner or operator must compare the groundwater quality of each parameter or constituent at each monitoring well designated pursuant to subrule 113.10(2) to the background value of that constituent, according to the statistical procedures and performance standards specified under paragraphs 113.10(4)"*g*" and 113.10(4)"*h*."

(2) Within 45 days after completing sampling and analysis, the owner or operator must determine whether there has been an SSI over background at each monitoring well.

IAC 7/4/07 Environmental Protection[567] Ch 113, p.43

113.10(5) *Detection monitoring program.*
a. Detection monitoring is required at MSWLF units at all groundwater monitoring wells defined under subrule 113.10(2). At a minimum, a detection monitoring program must include the monitoring for the constituents listed in Appendix I and any additional parameters required by the department on a site-specific basis. An alternative list of constituents may be used if it can be demonstrated that the constituents removed are not reasonably expected to be in or derived from the waste contained in the unit and if the alternative list of constituents is expected to provide a reliable indication of leachate leakage or gas impact from the MSWLF unit.

(1) The department may establish an alternative list of inorganic indicator parameters for an MSWLF unit within Appendix I, in lieu of some or all of the heavy metals (constituents 1 to 15 in Appendix I), if the alternative parameters provide a reliable indication of inorganic releases from the MSWLF unit to the groundwater. In determining alternative parameters, the department shall consider the following factors:
1. The types, quantities and concentrations of constituents in wastes managed at the MSWLF unit;
2. The mobility, stability and persistence of waste constituents or their reaction products in the unsaturated zone beneath the MSWLF unit;
3. The detectability of indicator parameters, waste constituents and reaction products in the groundwater; and
4. The concentration or values and coefficients of variation of monitoring parameters or constituents in the groundwater background.

(2) Reserved.
b. The monitoring frequency for all constituents listed in Appendix I or in the alternative list approved in accordance with subparagraph 113.10(5)"*a*"(1) shall be at least semiannual (i.e., every six months) during the active life of the facility (including closure) and the postclosure period. Where insufficient background data exist, a minimum of five independent samples from each well, collected at intervals to account for seasonal and temporal variation, must be analyzed for the constituents in Appendix I or in the alternative list approved in accordance with subparagraph 113.10(5)"*a*"(1) during the first year. At least one sample from each well must be collected and analyzed during subsequent semiannual sampling events. The department may specify an appropriate alternative frequency for repeated sampling and analysis for constituents in Appendix I or in the alternative list approved in accordance with subparagraph 113.10(5)"*a*"(1) during the active life (including closure) and the postclosure care period. The alternative frequency during the active life (including closure) shall be not less than annually. The alternative frequency shall be based on consideration of the following factors:
(1) Lithology of the aquifer and unsaturated zone;
(2) Hydraulic conductivity of the aquifer and unsaturated zone;

(3) Groundwater flow rates;

(4) Minimum distance between upgradient edge of the MSWLF unit and downgradient monitoring well screen (minimum distance of travel); and

(5) Resource value of the aquifer.

c. If the owner or operator determines, pursuant to paragraph 113.10(4)"*i*," that there is an SSI over background for one or more of the constituents listed in Appendix I or in the alternative list approved in accordance with subparagraph 113.10(5)"*a*"(1) at any monitoring well specified under subrule 113.10(2), then the owner or operator:

(1) Must, within 14 days of this finding, place a notice in the operating record indicating which constituents have shown statistically significant changes from background levels, and notify the department that this notice was placed in the operating record.

(2) Must establish within 90 days an assessment monitoring program meeting the requirements of subrule 113.10(6) except as provided in subparagraph 113.10(5)"*c*"(3).

Ch 113, p.44 Environmental Protection[567] IAC 7/4/07

(3) The owner or operator may demonstrate that a source other than an MSWLF unit caused the contamination or that the SSI resulted from error in sampling, analysis, statistical evaluation, or natural variation in groundwater quality. A report documenting this demonstration must be certified by a qualified groundwater scientist, approved by the department, and placed in the operating record. If resampling is a part of the demonstration, resampling procedures shall be specified prior to initial sampling.

If a successful demonstration to the department is made and documented, the owner or operator may continue detection monitoring as specified in subrule 113.10(5). If, after 90 days, a successful demonstration is not made, the owner or operator must initiate an assessment monitoring program as required in subrule 113.10(6).

113.10(6) *Assessment monitoring program.*

a. Assessment monitoring is required whenever an SSI over background has been confirmed pursuant to paragraph 113.10(5)"*c*" to be the result of a release from the facility.

b. Within 90 days of triggering an assessment monitoring program, and annually thereafter, the owner or operator must sample and analyze the groundwater for all constituents identified in Appendix II. A minimum of one sample from each downgradient well shall be collected and analyzed during each sampling event. For any constituent detected in the downgradient wells as a result of the complete Appendix II analysis, a minimum of four independent samples from each well must be collected and analyzed to establish background for the constituents. The department may specify an appropriate subset of wells to be sampled and analyzed for Appendix II constituents during assessment monitoring.

The department may delete any of the Appendix II monitoring parameters for an MSWLF unit if it can be shown that the removed constituents are not reasonably expected to be in or derived from the waste contained in the unit.

c. The department may specify an appropriate alternate frequency for repeated sampling and analysis for the full set of Appendix II constituents required by paragraph 113.10(6)"*b*" during the active life (including closure) and postclosure care period of the unit. The following factors shall be considered:

(1) Lithology of the aquifer and unsaturated zone;

(2) Hydraulic conductivity of the aquifer and unsaturated zone;

(3) Groundwater flow rates;

(4) Minimum distance between upgradient edge of the MSWLF unit and downgradient monitoring well screen (minimum distance of travel);

(5) Resource value of the aquifer; and

(6) Nature (fate and transport) of any constituents detected in response to this paragraph.

d. After obtaining the results from the initial or subsequent sampling events required in paragraph

113.10(6)"*b*," the owner or operator must:

(1) Within 14 days, place a notice in the operating record identifying the Appendix II constituents that have been detected and notify the department that this notice has been placed in the operating record;

(2) Within 90 days, and on at least a semiannual basis thereafter, resample all wells specified by subrule 113.10(2) and conduct analyses for all constituents in Appendix I or in the alternative list approved in accordance with subparagraph 113.10(5)"*a*"(1), and for those constituents in Appendix II that are detected in response to the requirements of paragraph 113.10(6)"*b*." Concentrations shall be recorded in the facility operating record. At least one sample from each well must be collected and analyzed during these sampling events. The department may specify an alternative monitoring frequency during the active life (including closure) and the post-closure period for the constituents referred to in this subparagraph. The alternative frequency for constituents in Appendix I or in the alternative list approved in accordance with subparagraph 113.10(5)"*a*"(1) during the active life (including closure) shall be no less than annual. The alternative frequency shall be based on consideration of the factors specified in paragraph 113.10(6)"*c*";

IAC 7/4/07 Environmental Protection[567] Ch 113, p.45

(3) Establish background concentrations for any constituents detected pursuant to paragraph 113.10(6)"*b*" or subparagraph 113.10(6)"*d*"(2); and

(4) Establish groundwater protection standards for all constituents detected pursuant to paragraph 113.10(6)"*b*" or 113.10(6)"*d*." The groundwater protection standards shall be established in accordance with paragraph 113.10(6)"*h*" or 113.10(6)"*i*."

e. If the concentrations of all Appendix II constituents are shown to be at or below background values, using the statistical procedures in paragraph 113.10(4)"*g*" for two consecutive sampling events, the owner or operator must notify the department of this finding and may return to detection monitoring.

f. If the concentrations of any Appendix II constituents are above background values, but all concentrations are below the groundwater protection standard established under paragraph 113.10(6)"*h*" or 113.10(6)"*i*," using the statistical procedures in paragraph 113.10(4)"*g*," the owner or operator must continue assessment monitoring in accordance with this subrule.

g. If one or more Appendix II constituents are detected at statistically significant levels above the groundwater protection standard established under paragraph 113.10(6)"*h*" or 113.10(6)"*i*" in any sampling event, the owner or operator must, within 14 days of this finding, place a notice in the operating record identifying the Appendix II constituents that have exceeded the groundwater protection standard and notify the department and all other appropriate local government officials that the notice has been placed in the operating record. The owner or operator also:

(1) Must, within 90 days of this finding, comply with the following requirements or the requirements in subparagraph 113.10(6)"*g*"(2):

1. Characterize the nature and extent of the release by installing additional monitoring wells as necessary until the horizontal and vertical dimensions of the plume have been defined to background concentrations;

2. Install at least one additional monitoring well at the facility boundary in the direction of contaminant migration and sample this well in accordance with subparagraph 113.10(6)"g"(2);

3. Notify all persons who own the land or reside on the land that directly overlies any part of the plume of contamination if contaminants have migrated off site when indicated by sampling of wells in accordance with subparagraph 113.10(6)"g"(1); and

4. Initiate an assessment of corrective measures as required by subrule 113.10(7).

(2) May demonstrate that a source other than an MSWLF unit caused the contamination, or that the SSI resulted from error in sampling, analysis, statistical evaluation, or natural variation in groundwater quality. A report documenting this demonstration must be certified by a qualified groundwater scientist, approved by the department, and placed in the operating record. If a successful demonstration is made, the owner or operator must continue monitoring in accordance with the assessment monitoring program pursuant to subrule 113.10(6), and may return to detection monitoring if the Appendix II constituents are at or below background as specified in paragraph 113.10(6)"e." Until a successful demonstration is made, the owner or operator must comply with paragraph 113.10(6)"g" including initiating an assessment of corrective measures.

h. The owner or operator must establish a groundwater protection standard for each Appendix II constituent detected in the groundwater. The groundwater protection standard shall be:

(1) For constituents for which a maximum contaminant level (MCL) has been promulgated under Section 1412 of the Safe Drinking Water Act (codified) under 40 CFR Part 141, the MCL for that constituent;

(2) For constituents for which MCLs have not been promulgated, the background concentration for the constituent established from wells in accordance with subrule 113.10(2); or

(3) For constituents for which the background concentration is higher than the MCL identified under subparagraph 113.10(6)"h"(1) or health-based concentrations identified under paragraph 113.10(6)"i," the background concentration.

Ch 113, p.46 Environmental Protection[567] IAC 7/4/07

i. The department may establish an alternative groundwater protection standard for constituents for which MCLs have not been established. These groundwater protection standards shall be appropriate health-based concentrations that comply with the statewide standards for groundwater established pursuant to 567—Chapter 137.

j. In establishing alternative groundwater protection standards under paragraph 113.10(6)"i," the department may consider the following:

(1) The policies set forth by the Groundwater Protection Act;

(2) Multiple contaminants in the groundwater with the assumption that the effects are additive regarding detrimental effects to human health and the environment;

(3) Exposure threats to sensitive environmental receptors; and

(4) Other site-specific exposure or potential exposure to groundwater.

113.10(7) *Assessment of corrective measures.*

a. Within 90 days of finding that any of the constituents listed in Appendix II have been detected at a statistically significant level exceeding the groundwater protection standards defined under paragraph 113.10(6)"h" or 113.10(6)"i," the owner or operator must initiate an assessment of corrective measures. Such an assessment must be completed and submitted to the department for review and approval within 180 days of the initial finding unless otherwise authorized or required by the department.

b. The owner or operator must continue to monitor in accordance with the assessment monitoring program as specified in subrule 113.10(6).

c. The assessment shall include an analysis of the effectiveness of potential corrective measures in meeting all of the requirements and objectives of the remedy as described under subrule 113.10(8), addressing at least the following:

(1) The performance, reliability, ease of implementation, and potential impacts of appropriate potential remedies, including safety impacts, cross-media impacts, and control of exposure to any residual contamination;

(2) The time required to begin and complete the remedy;

(3) The costs of remedy implementation; and

(4) The institutional requirements such as state or local permit requirements or other environmental or public health requirements that may substantially affect implementation of the remedy(ies).

d. Within 60 days of approval from the department of the assessment of corrective measures, the owner or operator must discuss the results of the corrective measures assessment, prior to the selection of a remedy, in a public meeting with interested and affected parties. The department may establish an alternative schedule for completing the public meeting requirement. Notice of public meeting shall be sent to all owners and occupiers of property adjacent to the permitted boundary of the facility, the department, and the department field office with jurisdiction over the facility. A copy of the minutes of this public meeting and the list of community concerns must be placed in the operating record and submitted to the department.

113.10(8) *Selection of remedy.*

a. Based on the results of the corrective measures assessment conducted under subrule 113.10 (7), the owner or operator must select a remedy within 60 days of holding the public meeting that, at a minimum, meets the standards listed in paragraph 113.10(8)"*b.*" The department may establish an alternative schedule for selecting a remedy after holding the public meeting. The owner or operator must submit a report to the department, within 14 days of selecting a remedy, describing the selected remedy, stating that the report has been placed in the operating record, and explaining how the selected remedy meets the standards in paragraph 113.10(8)"*b.*"

b. Remedies must:

(1) Be protective of human health and the environment;

IAC 7/4/07 Environmental Protection[567] Ch 113, p.47

(2) Attain the groundwater protection standards specified pursuant to paragraph 113.10(6)"*h*" or 113.10(6)"*i*";

(3) Control the source(s) of releases so as to reduce or eliminate, to the maximum extent practicable, further releases of Appendix II constituents into the environment that may pose a threat to human health or the environment; and

(4) Comply with standards for management of wastes as specified in paragraph 113.10(9)"*d.*"

c. In selecting a remedy that meets the standards of paragraph 113.10(8)"*b*," the owner or operator shall consider the following evaluation factors:

(1) The long-term and short-term effectiveness and protectiveness of the potential remedy(ies), along with the degree of certainty that the remedy will prove successful based on consideration of the following:

1. Magnitude of reduction of existing risks;

2. Magnitude of residual risks in terms of likelihood of further releases due to waste remaining following implementation of a remedy;

3. The type and degree of long-term management required, including monitoring, operation, and maintenance;

4. Short-term risks that might be posed to the community, workers, or the environment during implementation of such a remedy, including potential threats to human health and the environment associated with excavation, transportation, redisposal, or containment;
5. Time period until full protection is achieved;
6. Potential for exposure of humans and environmental receptors to remaining wastes, considering the potential threat to human health and the environment associated with excavation, transportation, redisposal, or containment;
7. Long-term reliability of the engineering and institutional controls; and
8. Potential need for replacement of the remedy.
(2) The effectiveness of the remedy in controlling the source to reduce further releases based on consideration of the following factors:
1. The extent to which containment practices will reduce further releases; and
2. The extent to which treatment technologies may be used.
(3) The ease or difficulty of implementing a potential remedy(ies) based on consideration of the following factors:
1. Degree of difficulty associated with constructing the technology;
2. Expected operational reliability of the technology;
3. Need to coordinate with and obtain necessary approvals and permits from other agencies;
4. Availability of necessary equipment and specialists; and
5. Available capacity and location of needed treatment, storage, and disposal services.
(4) Practicable capability of the owner or operator, including a consideration of technical and economic capabilities.
(5) The degree to which community concerns, including but not limited to the concerns identified at the public meeting required pursuant to paragraph 113.10(7)"*d*," are addressed by a potential remedy(ies).
d. The owner or operator shall specify as part of the selected remedy a schedule(s) for initiating and completing remedial activities. Such a schedule must require the initiation of remedial activities within a reasonable period of time taking into consideration the factors set forth in subparagraphs 113.10(8)"*d*"(1) to (8). The owner or operator must consider the following factors in determining the schedule of remedial activities:
(1) Extent and nature of contamination;

Ch 113, p.48 Environmental Protection[567] IAC 7/4/07
(2) Practical capabilities of remedial technologies in achieving compliance with groundwater protection standards established under paragraph 113.10(6)"*h*" or 113.10(6)"*i*" and other objectives of the remedy;
(3) Availability of treatment or disposal capacity for wastes managed during implementation of the remedy;
(4) Desirability of utilizing alternative or experimental technologies that are not widely available, but which may offer significant advantages over already available technologies in terms of effectiveness, reliability, safety, or ability to achieve remedial objectives;
(5) Potential risks to human health and the environment from exposure to contamination prior to completion of the remedy;
(6) Resource value of the aquifer including:
1. Current and future uses;
2. Proximity and withdrawal rate of users;

3. Groundwater quantity and quality;
4. The potential damage to wildlife, crops, vegetation, and physical structures caused by exposure to waste constituents;
5. The hydrogeologic characteristics of the facility and surrounding land;
6. Groundwater removal and treatment costs; and
7. The cost and availability of alternative water supplies;
(7) Practicable capability of the owner or operator; and
(8) Other relevant factors.

113.10(9) *Implementation of the corrective action plan.*
a. Based on the schedule established under paragraph 113.10(8)"*d*" for initiation and completion of remedial activities, the owner or operator must:
(1) Establish and implement a corrective action groundwater monitoring program that:
1. At a minimum, meets the requirements of an assessment monitoring program under subrule 113.10(6);
2. Indicates the effectiveness of the corrective action remedy; and
3. Demonstrates compliance with groundwater protection standards pursuant to paragraph 113.10(9)"*e*";
(2) Implement the corrective action remedy selected under subrule 113.10(8); and
(3) Take any interim measures necessary to ensure the protection of human health and the environment.
Interim measures should, to the greatest extent practicable, be consistent with the objectives of and contribute to the performance of any remedy that may be required pursuant to subrule 113.10(8). The following factors must be considered by an owner or operator in determining whether interim measures are necessary:
1. Time period required to develop and implement a final remedy;
2. Actual or potential exposure of nearby populations or environmental receptors to hazardous constituents;
3. Actual or potential contamination of drinking water supplies or sensitive ecosystems;
4. Further degradation of the groundwater that may occur if remedial action is not initiated expeditiously;
5. Weather conditions that may cause hazardous constituents to migrate or be released;
6. Risk of fire or explosion, or potential for exposure to hazardous constituents as a result of an accident or the failure of a container or handling system; and
7. Other factors that may pose threats to human health and the environment.

IAC 7/4/07 Environmental Protection[567] Ch 113, p.49
b. An owner or operator may determine, based on information developed after implementation of the remedy has begun or other information, that compliance with the requirements of paragraph 113.10(8)"*b*" is not being achieved through the remedy selected. In such cases, the owner or operator must notify the department and implement other methods or techniques that could practicably achieve compliance with the requirements, unless the owner or operator makes the determination under paragraph 113.10(9)"*c*." The notification shall explain how the proposed alternative methods or techniques will meet the standards in paragraph 113.10(8)"*b*," or the notification shall indicate that the determination was made pursuant to paragraph 113.10(9)"*c*." The notification shall also specify a schedule(s) for implementing and completing the remedial activities to comply with paragraph 113.10(8)"*b*" or the alternative measures to comply with paragraph 113.10(9)"*c*." Within 90 days of approval by the department for the proposed alternative methods or techniques or the

determination of impracticability, the owner or operator shall implement the proposed alternative methods or techniques meeting the standards of paragraph 113.10(8)"*b*" or implement alternative measures meeting the requirements of subparagraphs 113.10(9)"*c*" (2) and (3).

c. If the owner or operator determines that compliance with requirements under paragraph 113.10(8)"*b*" cannot be practicably achieved with any currently available methods, the owner or operator must:

(1) Obtain certification of a qualified groundwater scientist and approval by the department that compliance with requirements under paragraph 113.10(8)"*b*" cannot be practicably achieved with any currently available methods;

(2) Implement alternate measures to control exposure of humans or the environment to residual contamination, as necessary to protect human health and the environment;

(3) Implement alternate measures for control of the sources of contamination, or for removal or decontamination of equipment, units, devices, or structures that are:

1. Technically practicable; and

2. Consistent with the overall objective of the remedy; and

(4) Notify the department within 14 days that a report justifying the alternate measures prior to implementation has been placed in the operating record.

d. All solid wastes that are managed pursuant to a remedy required under subrule 113.10(8), or an interim measure required under subparagraph 113.10(9)"*a*"(3), shall be managed in a manner:

(1) That is protective of human health and the environment; and

(2) That complies with applicable RCRA, state and local requirements.

e. Remedies selected pursuant to subrule 113.10(8) shall be considered complete when:

(1) The owner or operator complies with the groundwater protection standards established under paragraph 113.10(6)"*h*" or 113.10(6)"*i*" at all points within the plume of contamination that lie beyond the groundwater monitoring well system established under subrule 113.10(2).

(2) Compliance with the groundwater protection standards established under paragraph 113.10 (6)"*h*" or 113.10(6)"*i*" has been achieved by demonstrating that concentrations of Appendix II constituents have not exceeded the groundwater protection standard(s) for a period of three consecutive years using the statistical procedures and performance standards in paragraphs 113.10(4)"*g*" and 113.10(4)"*h*." The department may specify an alternative length of time during which the owner or operator must demonstrate that concentrations of Appendix II constituents have not exceeded the groundwater protection standard(s), taking into consideration:

1. The extent and concentration of the release(s);

2. The behavior characteristics of the hazardous constituents in the groundwater;

3. The accuracy of monitoring or modeling techniques, including any seasonal, meteorological, or other environmental variables that may affect accuracy; and

4. The characteristics of the groundwater.

(3) All actions required by the department to complete the remedy have been satisfied.

Ch 113, p.50 Environmental Protection[567] IAC 7/4/07

f. Upon completion of the remedy, the owner or operator must notify the department within 14 days that a certification has been placed in the operating record verifying that the remedy has been completed in compliance with the requirements of paragraph 113.10(9)"*e*." The certification must be signed by the owner or operator and by a qualified groundwater scientist and approved by the department.

g. When, upon completion of the certification, the owner or operator determines that the corrective action remedy has been completed in accordance with the requirements under paragraph 113.10(9)"*e*," the owner or operator shall be released from the requirements for financial assurance for corrective action pursuant to subrule 113.14(5).

113.10(10) *Annual water quality reports.* The owner or operator shall submit an annual report to the department detailing the water quality monitoring sampling locations and results, assessments, selection of remedies, implementation of corrective action, and the results of corrective action remedies to address SSIs, if any, during the previous year. This report shall include a site map that delineates all monitoring points where water quality samples were taken, and plumes of contamination, if any.

The report shall contain a narrative explaining and interpreting all of the data collected during the previous year. The report shall be due each year on a date set by the department in the facility's permit.

567—113.11(455B,455D) Record-keeping and reporting requirements. The primary purpose of the record-keeping and reporting activities is to verify compliance with this chapter and to document the construction and operations of the facility. The department can set alternative schedules for recordkeeping and notification requirements as specified in subrules 113.11(1) and 113.11(2), except for the notification requirements in paragraph 113.6(2)"*a*" and numbered paragraph 113.10(6)"*g*"(1)"3."

All MSWLFs shall comply with the following record-keeping and reporting requirements.

113.11(1) *Record keeping.* The owner or operator of an MSWLF unit must record and retain near the facility in an operating record or in an alternative location approved by the department the following information as it becomes available:

a. Permit application, permit renewal and permit modification application materials pursuant to rule 113.5(455B);

b. The site exploration and characterization reports pursuant to subrule 113.6(4);

c. Design and construction plans and specifications, and related analyses and documents, pursuant to rule 113.7(455B). The QC&A final reports, and related analyses and documents, pursuant to paragraph 113.7(6)"*d*";

d. Inspection records, training procedures, and notification procedures required in rule 113.8 (455B);

e. Any MSWLF unit design documentation for placement of leachate or gas condensate in an MSWLF unit as required under numbered paragraphs 113.8(1)"*b*"(3)"2" and "3";

f. Gas monitoring results from monitoring and any remediation plans required by rule 113.9 (455B);

g. Any demonstration, certification, finding, monitoring, testing, or analytical data required by rule 113.10(455B);

h. Closure and postclosure care plans and any monitoring, testing, or analytical data as required by rules 113.12(455B) and 113.13(455B); and

i. Any cost estimates and financial assurance documentation required by this chapter.

113.11(2) *Reporting requirements.* The owner or operator must notify the department when the documents required in subrule 113.11(1) have been placed in the operating record. All information contained in the operating record must be furnished upon request to the department and be made available at all reasonable times for inspection by the department.

IAC 7/4/07 Environmental Protection[567] Ch 113, p.51

567—113.12(455B) Closure criteria. All MSWLFs shall comply with the following closure requirements.

113.12(1) Owners or operators of all MSWLF units must install a final cover system that is designed to minimize infiltration and erosion. The final cover system must be designed and constructed to:
a. Have a permeability less than or equal to the permeability of any bottom liner system (for MSWLFs with some type of liner) or have a permeability no greater than 1×10^{-7} cm/sec, whichever is less;
b. Minimize infiltration through the closed MSWLF by the use of an infiltration layer that contains a minimum of 18 inches of compacted earthen material;
c. Minimize erosion of the final cover by the use of an erosion layer that contains a minimum of 24 inches of earthen material that is capable of sustaining native plant growth;
d. Have an infiltration layer and erosion layer that are a combined minimum of 42 inches of earthen material at all locations over the closed MSWLF unit; and
e. Have a slope between 5 percent and 25 percent. Steeper slopes may be used if it is demonstrated that a steeper slope is unlikely to adversely affect final cover system integrity.

113.12(2) The department may approve an alternative final cover design that includes:
a. An infiltration layer that achieves reduction in infiltration equivalent to the infiltration layer specified in paragraphs 113.12(1)"*a*" and 113.12(1)"*b*"; and
b. An erosion layer that provides protection from wind and water erosion equivalent to the erosion layer specified in paragraphs 113.12(1)"*c*" and 113.12(1)"*d.*"

113.12(3) The owner or operator must prepare a written closure plan that describes the steps necessary to close all MSWLF units at any point during the active life in accordance with the cover design requirements in subrule 113.12(1) or 113.12(2), as applicable. The closure plan, at a minimum, must include the following information:
a. A description of the final cover including source, volume, and characteristics of cover material, designed in accordance with subrule 113.12(1) or 113.12(2) and the methods and procedures to be used to install the cover;
b. An estimate of the largest area of the MSWLF unit requiring a final cover, as required under subrule 113.12(1) or 113.12(2), at any time during the active life;
c. An estimate of the maximum inventory of wastes on site over the active life of the landfill facility; and
d. A schedule for completing all activities necessary to satisfy the closure criteria in rule 113.12 (455B).

113.12(4) The owner or operator must notify the department that the closure plan has been placed in the operating record no later than the initial receipt of waste in a new MSWLF unit.

113.12(5) At least 180 days prior to beginning closure of each MSWLF unit as specified in subrule 113.12(6), an owner or operator must notify the department of the intent to close the MSWLF unit, and that a notice of the intent to close the unit has been placed in the operating record. If the MSWLF facility will no longer be accepting MSW for disposal, then the owner or operator must also notify all local governments utilizing the facility and post a public notice of the intent to close and no longer to accept MSW.

113.12(6) The owner or operator must begin closure activities of each MSWLF unit:
a. No later than 30 days after the date on which the MSWLF unit receives the known final receipt of wastes; or

Ch 113, p.52 Environmental Protection[567] IAC 7/4/07
b. If the MSWLF unit has remaining capacity and there is a reasonable likelihood that the MSWLF unit will receive additional wastes, no later than one year after the most recent receipt of wastes. Extensions beyond the one-year deadline for beginning closure may be granted by the department if the owner or operator demonstrates that the MSWLF unit has the capacity to receive additional wastes and the owner or operator has taken and will continue to take all steps necessary to prevent threats to human health and the environment from the unclosed MSWLF unit.

113.12(7) The owner or operator of all MSWLF units must complete closure activities of each MSWLF unit in accordance with the closure plan within 180 days following the beginning of closure as specified in subrule 113.12(6). Extensions of the closure period may be granted by the department if the owner or operator demonstrates that closure will, of necessity, take longer than 180 days and that the owner or operator has taken and will continue to take all steps to prevent threats to human health and the environment from the unclosed MSWLF unit.

113.12(8) Following closure of each MSWLF unit, the owner or operator must submit to the department certification, signed by an independent professional engineer (P.E.) registered in Iowa, verifying that closure has been completed in accordance with the closure plan. Upon approval by the department, the certification shall be placed in the operating record.

113.12(9) Following closure of all MSWLF units, the owner or operator must record a notation on the deed to the landfill facility property, or some other instrument that is normally examined during title search in lieu of a deed notification, and notify the department that the notation has been recorded and a copy has been placed in the operating record. The notation on the deed must in perpetuity notify any potential purchaser of the property that:
a. The land has been used as a landfill facility; and
b. Its use is restricted under paragraph 113.13(3)*"c."*

113.12(10) The owner or operator may request permission from the department to remove the notation from the deed if all wastes are removed from the facility.

567—113.13(455B) Postclosure care requirements. All MSWLFs shall comply with the following postclosure care requirements.

113.13(1) Following closure of each MSWLF unit, the owner or operator must conduct postclosure care. Postclosure care must be conducted for 30 years, except as provided under subrule 113.13(2), and consist of at least the following:
a. Maintaining the integrity and effectiveness of any final cover, including making repairs to the cover as necessary to correct the effects of settlement, subsidence, erosion, or other events, and preventing run-on and runoff from eroding or otherwise damaging the final cover;
b. Maintaining and operating the leachate collection system in accordance with the requirements in paragraphs 113.7(5)*"b"* and 113.8(3)*"i,"* if applicable. The department may allow the owner or operator to stop managing leachate if the owner or operator demonstrates that leachate no longer poses a threat to human health and the environment;
c. Monitoring the groundwater in accordance with the requirements of rule 113.10(455B) and maintaining the groundwater monitoring system; and

d. Maintaining and operating the gas monitoring system in accordance with the requirements of rule 113.9(455B).

113.13(2) The length of the postclosure care period may be:

a. Decreased by the department if the owner or operator demonstrates that the reduced period is sufficient to protect human health and the environment and this demonstration is approved by the department; or

b. Increased by the department if the department determines that the lengthened period is necessary to protect human health and the environment.

IAC 7/4/07 Environmental Protection[567] Ch 113, p.53

113.13(3) The owner or operator of all MSWLF units must prepare a written postclosure plan that includes, at a minimum, the following information:

a. A description of the monitoring and maintenance activities required in subrule 113.13(1) for each MSWLF unit, and the frequency at which these activities will be performed;

b. Name, address, and telephone number of the person or office to contact about the facility during the postclosure period; and

c. A description of the planned uses of the property during the postclosure period. Postclosure use of the property shall not disturb the integrity of the final cover, liner(s), or any other components of the containment system, or the function of the monitoring systems unless necessary to comply with the requirements in this chapter. The department may approve any other disturbance if the owner or operator demonstrates that disturbance of the final cover, liner or other component of the containment system, including any removal of waste, will not increase the potential threat to human health or the environment.

113.13(4) The owner or operator must notify the department that a postclosure plan has been prepared and placed in the operating record by the date of initial receipt of waste.

113.13(5) Following completion of the postclosure care period for each MSWLF unit, the owner or operator must submit to the department a certification, signed by an independent professional engineer (P.E.) registered in Iowa, verifying that postclosure care has been completed in accordance with the postclosure plan. Upon department approval, the certification shall be placed in the operating record.

113.14(9) *Amount of required financial assurance.* A financial assurance mechanism established pursuant to subrule 113.14(6) shall be in the amount of the third-party cost estimates required by subrules 113.14(3), 113.14(4), and 113.14(5) except that the amount of the financial assurance may be reduced by the sum of the cash balance in a trust fund or local government dedicated fund established to comply with subrule 113.14(8) plus the current value of investments held by said trust fund or local government dedicated fund if invested in one or more of the investments listed in Iowa Code section 12B.10(5).

567—113.15(455B,455D) Variances. A request for a variance to this chapter shall be submitted in writing pursuant to 561—Chapter 10. Some provisions of this chapter are minimum standards required by federal law (see 40 CFR 258), and variances to such provisions shall not be granted unless they are as protective as the applicable minimum federal standards.

Appendix I   Constituents for Detection Monitoring[1]

Inorganic Constituents:

    (1) Antimony (Total)

    (2) Arsenic (Total)

    (3) Barium (Total)

    (4) Beryllium (Total)

    (5) Cadmium (Total)

    (6) Chromium (Total)

    (7) Cobalt (Total)

    (8) Copper (Total)

    (9) Lead (Total)

  (10) Nickel (Total)

  (11) Selenium (Total)

  (12) Silver (Total)

  (13) Thallium (Total)

  (14) Vanadium (Total)

  (15) Zinc (Total)

Organic Constituents:

  (16) Acetone 67-64-1

  (17) Acrylonitrile 107-13-1

  (18) Benzene 71-43-2

  (19) Bromochloromethane 74-97-5

  (20) Bromodichloromethane 75-27-4

  (21) Bromoform; Tribromomethane 75-25-2

  (22) Carbon disulfide 75-15-0

  (23) Carbon tetrachloride 56-23-5

  (24) Chlorobenzene 108-90-7

  (25) Chloroethane; Ethyl chloride 75-00-3

  (26) Chloroform; Trichloromethane 67-66-3

  (27) Dibromochloromethane; Chlorodibromomethane 124-48-1

  (28) 1,2-Dibromo-3-chloropropane; DBCP 96-12-8IAC 7/4/07Ch 113, p.70 Environmental Protection[567] IAC 7/4/07

  (29) 1,2-Dibromoethane; Ethylene dibromide; EDB 106-93-4

  (30) o-Dichlorobenzene; 1,2-Dichlorobenzene 95-50-1

  (31) p-Dichlorobenzene; 1,4-Dichlorobenzene 106-46-7

  (32) trans-1,4-Dichloro-2-butene 110-57-6

  (33) 1,1-Dichloroethane; Ethylidene chloride 75-34-3

  (34) 1,2-Dichloroethane; Ethylene dichloride 107-06-2

(35) 1,1-Dichloroethylene; 1,1-Dichloroethene; Vinylidene chloride 75-35-4

(36) cis-1,2-Dichloroethylene; cis-1,2-Dichloroethene 156-59-2

(37) trans-1,2-Dichloroethylene; trans-1,2-Dichloroethene 156-60-5

(38) 1,2-Dichloropropane; Propylene dichloride 78-87-5

(39) cis-1,3-Dichloropropene 10061-01-5

(40) trans-1,3-Dichloropropene 10061-02-6

(41) Ethylbenzene 100-41-4

(42) (42) 2-Hexanone; Methyl butyl ketone 591-78-6

(43) Methyl bromide; Bromomethane 74-83-9

(44) Methyl chloride; Chloromethane 74-87-3

(45) Methylene bromide; Dibromomethane 74-95-3

(46) Methylene chloride; Dichloromethane 75-09-2

(47) Methyl ethyl ketone; MEK; 2-Butanone 78-93-3

(48) Methyl iodide; Iodomethane 74-88-4

(49) (49) 4-Methyl-2-pentanone; Methyl isobutyl ketone 108-10-1

(50) Styrene 100-42-5

(51) 1,1,1,2-Tetrachloroethane 630-20-6

(52) 1,1,2,2-Tetrachloroethane 79-34-5

(53) Tetrachloroethylene; Tetrachloroethene; Perchloroethylene 127-18-4

(54) Toluene 108-88-3

(55) 1,1,1-Trichloroethane; Methylchloroform 71-55-6

(56) 1,1,2-Trichloroethane 79-00-5

(57) Trichloroethylene; Trichloroethene 79-01-6

(58) Trichlorofluoromethane; CFC-11 75-69-4

(59) 1,2,3-Trichloropropane 96-18-4

(60) Vinyl acetate 108-05-4

(61) Vinyl chloride 75-01-4

(62) Xylenes 1330-20-7

Notes:

[1]This list contains 47 volatile organics for which possible analytical procedures provided in EPA Report SW-846 "Test Methods for Evaluating Solid Waste," third edition, November 1986, as revised December 1987, includes Method 8260; and 15 metals for which SW-846 provides either Method 6010 or a method from the 7000 series of methods.

[2]Common names are those widely used in government regulations, scientific publications, and commerce; synonyms exist for many chemicals.

[3]Chemical Abstracts Service registry number. Where "Total" is entered, all species in the groundwater that contain this element are included.

IAC 7/4/07 Environmental Protection[567] Ch 113, p.71

Appendix II List of Hazardous Inorganic and Organic Constituents[1]

Common Name[2]	CAS RN[3]	Chemical abstracts index name[4]	Suggested Method[5]	PQL, $(\mu g/L)$[6]
Acenaphthene	83-32-9	Acenaphthylene, 1,2-dihydro-	8100, 8270	200, 10
Acenaphthylene	208-96-8	Acenaphthylene	8100, 8270	200, 10
Acetone	67-64-1	2-Propanone	8260	100
Acetonitrile; Methyl cyanide	75-05-8	Acetonitrile	8015	100
Acetophenone	98-86-2	Ethanone, 1-phenyl-	8270	10
2-Acetylaminofluorene; 2-AAF	53-96-3	Acetamide, N-9H-fluoren-2-yl-	8270	20
Acrolein	107-02-8	2-Propenal	8030, 8260	5, 100
Acrylonitrile	107-13-1	2-Propenenitrile	8030, 8260	5, 200
Aldrin	309-00-2	1,4:5,8-Dimethanonaphthalene, 1,2,3,4, 10,10-hexachloro-1,4,4a,5,8,8ahexahydro-(1I,4I,4aJ,5I,8I,8aJ)-	8080, 8270	0.05, 10
Allyl chloride	107-05-1	1-Propene, 3-chloro-	8010, 8260	5, 10
4-Aminobiphenyl	92-67-1	[1,11-Biphenyl]-4-amine	8270	20
Anthracene	120-12-7	Anthracene	8100, 8270	200, 10
Antimony	(Total)	Antimony	6010, 7040, 7041	300, 2000, 30
Arsenic	(Total)	Arsenic	6010, 7060, 7061	500, 10, 20
Barium	(Total)	Barium	6010, 7080	20, 1000
Benzene	71-43-2	Benzene	8020, 8021, 8260	2, 0.1, 5
Benzo[a]anthracene;Benzanthracene	56-55-3	Benz[a]anthracene	8100, 8270	200, 10
Benzo[b]fluoranthene	205-99-2	Benz[e]acephenanthrylene	8100, 8270	200, 10
Benzo[k]fluoranthene	207-08-9	Benzo[k]fluoranthene	8100, 8270	200, 10
Benzo[ghi]perylene	191-24-2	Benzo[ghi]perylene	8100, 8270	200, 10
Benzo[a]pyrene	50-32-8	Benzo[a]pyrene	8100, 8270	200, 10
Benzyl alcohol	100-51-6	Benzenemethanol	8270	20

(continued)

109

Appendix II (*Continued*)

Common Name[2]	CAS RN[3]	Chemical abstracts index name[4]	Suggested Method[5]	PQL, (μg/L)[6]
Beryllium	(Total)	Beryllium	6010, 7090, 7091	3, 50, 2
alpha-BHC	319-84-6	Cyclohexane, 1,2,3,4,5,6-hexachloro-,(1I,2I,3J,4I,5I,6J)-	8080, 8270	0.05, 10
beta-BHC	319-85-7	Cyclohexane, 1,2,3,4,5,6-hexachloro-,(1I,2I,3I,4J,5I,6J)-	8080, 8270	0.05, 20
delta-BHC	319-86-8	Cyclohexane, 1,2,3,4,5,6-hexachloro-,(1I,2I,3I,4J,5I,6J)-	8080, 8270	0.1, 20
gamma-BHC; Lindane	58-89-9	Cyclohexane, 1,2,3,4,5,6-hexachloro-,(1I,2I,3J,4I,5I,6J)-	8080, 8270	0.05, 20
Bis(2-chloroethoxy)methane	111-91-1	Ethane, 1,11-[methylene-bis(oxy)] bis[2-chloro-	8110, 8270	5, 10
Bis(2-chloroethyl) ether; Dichloroethyl ether	111-44-4	Ethane, 1,11-oxybis[2-chloro-	8110, 8270	3, 10
Bis-(2-chloro-1-methylethyl) ether; 2,21-Dichlorodiisopropyl ether; DCIP, see Note 7	108-60-1	Propane, 2,21-oxybis[1-chloro-	8110, 8270	10, 10
Bis(2-ethylhexyl) phthalate	117-81-7	1,2-Benzenedicarboxylic acid, bis(2-ethylhexyl) ester	8060	20
Bromochloromethane; Chlorobromomethane	74-97-5	Methane, bromochloro-	8021, 8260	0.1, 5
Bromodichloromethane; Dibromochloromethane	75-27-4	Methane, bromodichloro-	8010, 8021, 8260	1, 0.2, 5
Bromoform; Tribromomethane	75-25-2	Methane, tribromo-	8010, 8021, 8260	2, 15, 5
4-Bromophenyl phenyl ether	101-55-3	Benzene, 1-bromo-4-phenoxy-	8110, 8270	25, 10
Butyl benzyl phthalate; Benzyl butyl phthalate	85-68-7	1,2-Benzenedicarboxylic acid,	8060, 8270	5, 10

	butylphenylmethyl ester			
Cadmium	Cadmium	(Total)	6010, 7130, 7131	40, 50, 1
Carbon disulfide	Carbon disulfide	75-15-0	8260	100
Carbon tetrachloride	Methane, tetrachloro-	56-23-5	8010, 8021, 8260	1, 0.1, 10
Chlordane See Note 8		4,7-Methano-1Hindene, 1,2,4,5,6,7,8,		
p-Chloroaniline	8-octachloro-2,3,3a,4,7,7ahexahydro- Benzenamine, 4-chloro-	106-47-8	8080, 8270 8270	0.1, 50
Chlorobenzene	Benzene, chloro-	108-90-7	8010, 8020, 8021, 8260 22, 0.1, 5	20
Chlorobenzilate	Benzeneacetic acid, 4-chlor-1-(4- chlorophenyl)-1-hydroxy-, ethyl ester	510-15-6	8270	10
p-Chloro-m-cresol; 4-Chloro-3-methylphenol	Phenol, 4-chloro-3-methyl-	59-50-7	8040, 8270	5, 20
Chloroethane; Ethyl chloride	Ethane, chloro-	75-00-3	8010, 8021, 8260	51, 10
Chloroform; Trichloromethane	Methane, trichloro-	67-66-3	8010, 8021, 8260	0.5, 0.2, 5
2-Chloronaphthalene	Naphthalene, 2-chloro-	91-58-7	8120, 8270	10, 10
2-Chlorophenol	Phenol, 2-chloro-	95-57-8	8040, 8270	5, 10
4-Chlorophenyl phenyl ether	Benzene, 1-chloro-4-phenoxy-	7005-72-3	8110, 8270	40, 10
Chloroprene	1,3-Butadiene, 2-chloro-	126-99-8	8010, 8260	50, 20
Chromium	Chromium	(Total)	6010, 7190, 7191	70, 500, 10
Chrysene	Chrysene	218-01-9	8100, 8270	200, 10
Cobalt	Cobalt	(Total)	6010, 7200, 7201	70, 500, 10
Copper	Copper	(Total)	6010, 7210, 7211	60, 200, 10
m-Cresol; 3-methylphenol	Phenol, 3-methyl-	108-39-4	8270	10
o-Cresol; 2-methylphenol	Phenol, 2-methyl-	95-48-7	8270	10
p-Cresol; 4-methylphenol	Phenol, 4-methyl-	106-44-5	8270	10
Cyanide	Cyanide	57-12-5	9010	200

(continued)

111

Appendix II (*Continued*)

Common Name[2]	CAS RN[3]	Chemical abstracts index name[4]	Suggested Method[5]	PQL. (μg/L)[6]
2,4-D; 2,4-Dichlorophenoxyacetic Acid	94-75-7	Acetic acid, (2,4-dichlorophenoxy)-	8150	10
4,41-DDD	72-54-8	Benzene 1,11-(2,2-dichloroethyl-idene) bis[4-chloro-	8080, 8270	0.1, 10
4,41-DDE	72-55-9	Benzene, 1,11-(dichloroethyenyl-idene) bis[4-chloro-	8080, 8270	0.05, 10
4,41-DDT	50-29-3	Benzene, 1,11-(2,2,2-trichloroethylidene)bis[4-chloro-	8080, 8270	0.1, 10
Diallate	2303-16-4	Carbamothioic acid, bis(1-methyl-ethyl)-, S-(2,3-dichloro-2-propenyl) ester	8270	10
Dibenz[a,h]anthracene	53-70-3	Dibenz[a,h]anthracene	8100, 8270	200, 10
Dibenzofuran	132-64-9	Dibenzofuran	8270	10
Dibromochloromethane; Chlorodibromomethane	124-48-1	Methane, dibromochloro-	8010, 8021, 8260	1, 0.3, 5
1,2-Dibromo-3-chloropropane; DBCP	96-12-8	Propane, 1,2-dibrome-3-chloro-	8011, 8021, 8260	0.1, 30, 25
1,2-Dibromoethane; Ethylene dibromide; EDB	106-93-4	Ethane, 1,2-dibromo-	8011, 8021, 8260	0.1, 10, 5
Di-n-butyl phthalate	84-74-2	1,2-Benzenedicarboxylic acid, dibutyl ester	8060, 8270	5, 10
o-Dichlorobenzene; 1,2-Dichlorobenzene	95-50-1	Benzene, 1,2-dichloro-	8010, 8020, 8021 8120, 8260, 8270	25, 0.5, 10 5, 10
m-Dichlorobenzene; 1,3-Dichlorobenzene	541-73-1	Benzene, 1,3-dichloro-	8010, 8020, 8021	55, 0.2, 10

	CAS			
p-Dichlorobenzene; 1,4-Dichlorobenzene	106-46-7	Benzene, 1,4-dichloro-	8120, 8260, 8270	5, 10
3,31-Dichlorobenzidine	91-94-1	[1,11-Biphenyl]-4,41-diamine, 3,31-dichloro-	8010, 8020, 8021 / 8120, 8260, 8270 / 8270	25, 0.1, 15 / 5, 10 / 20
trans-1,4-Dichloro-2-butene	110-57-6	2-Butene, 1,4-dichloro-, (E)-	8260	100
Dichlorodifluoromethane; CFC 12	75-71-8	Methane, dichlorodifluoro-	8021,8260	0.5, 5
1,1-Dichloroethane; Ethyldidene chloride	75-34-3	Ethane, 1,1-dichloro-	8010, 8021,8260	1, 0.5, 5
1,2-Dichloroethane; Ethylene dichloride	107-06-2	Ethane, 1,1-dichloro-	8010,8021, 8260	0.5, 0.3, 5
1,1-Dichloroethylene; 1,1-Dichloroethene; Vinylidene chloride	75-35-4	Ethene, 1,1-dichloro-	8010, 8021, 8260	1, 0.5, 5
cis-1,2-Dichloroethylene; cis-1,2-Dichloroethene	156-59-2	Ethene, 1,2-dichloro-, (Z)-	8021, 8260	0.2, 5
trans-1,2-Dichloroethylene; trans-1,2-Dichloroethene	156-60-5	Ethene, 1,2-dichloro-, (E)-	8010, 8021, 8260	1, 0.5, 5
2,4-Dichlorophenol	120-83-2	Phenol, 2,4-dichloro-	8040, 8270	5, 10
2,6-Dichlorophenol	87-65-0	Phenol, 2,6-dichloro-	8270	10
1,2-Dichloropropane; Propylene dichloride	78-87-5	Propane, 1,2-dichloro-	8010, 8021, 8260	0.5, 0.05, 5
1,3-Dichloropropane; Trimethylene dichloride	142-28-9	Propane, 1,3-dichloro-	8021, 8260	0.3, 5
2,2-Dichloropropane; Isopropylidene chloride	594-20-7	Propane, 2,2-dichloro-	8021, 8260	0.5, 15
1,1-Dichloropropene	563-58-6	1-Propene, 1,1-dichloro-	8021, 8260	0.2, 5

(continued)

Appendix II (*Continued*)

Common Name[2]	CAS RN[3]	Chemical abstracts index name[4]	Suggested Method[5]	PQL, (μg/L)[6]
cis-1,3-Dichloropropene	10061-01-5	1-Propene, 1,3-dichloro-, (Z)-	8010, 8260	20, 10
trans-1,3-Dichloropropene	10061-02-6	1-Propene, 1,3-dichloro-, (E)-	8010, 8260	5, 10
Dieldrin 60-57-1 2,7:3,6-Dimethanonaphth[2,3-b]oxirene, 3,4,5,6,9,9-hexa,chloro-1a, 2,2a,3,6,6a,7, 7a-octahydro-,		(1a1,2J,2a1,3J,6J,6a1,7J, 7a1)-	8080, 8270	0.05, 10
Diethyl phthalate	84-66-2	1,2-Benzenedicarboxylic acid, diethyl ester	8060, 8270	5, 10
0,0-Diethyl 0-2-pyrazinyl phosphorothioate; Thionazin	297-97-2	Phosphorothioic acid, 0,0-diethyl 0-pyrazinyl ester	8141, 8270	5, 20
Dimethoate	60-51-5	Phosphorodithioic acid, 0,0-dimethyl S- [2-(methylamino)-2-oxoethyl] ester	8141, 8270	3, 20
p-(Dimethylamino)azobenzene 60-11-7 Benzenamine, N,N-dimethyl-4-(phenylazo)-				
7,12-Dimethylbenz[a]anthracene	57-97-6	Benz[a]anthracene, 7,12-dimethyl-	8270	10
3,31-Dimethylbenzidine	119-93-7	[1,11-Biphenyl]-4,41-diamine, 3,31-dimethyl-	8270	10
2,4-Dimethylphenol;m-Xylenol	105-67-9	Phenol, 2,4-dimethyl-	8040, 8270	5, 10
Dimethyl phthalate	131-11-3	1,2-Benzenedicarboxylic acid, dimethyl ester	8060, 8270	5, 10
m-Dinitrobenzene	99-65-0	Benzene, 1,3-dinitro-	8270	20
4,6-Dinitro-o-cresol 4,6-Dinitro-2-methylphenol	534-52-1	Phenol, 2-methyl-4,6-dinitro-	8040, 8270	150, 50
2,4-Dinitrophenol	51-28-5	Phenol, 2,4-dinitro-	8040, 8270	150, 50

2,4-Dinitrotoluene	Benzene, 1-methyl-2,4-dinitro-	121-14-2	8090, 8270	0.2, 10
2,6-Dinitrotoluene	Benzene, 2-methyl-1,3-dinitro-	606-20-2	8090, 8270	0.1, 10
Dinoseb; DNBP; 2-sec-Butyl-4,6-dinitrophenol	Phenol, 2-(1-methylpropyl)-4,6-dinitro-	88-85-7	8150, 8270	1, 20
Di-n-octyl phthalate	1,2-Benzenedicarboxylic acid, dioctyl ester	117-84-0	8060, 8270	30, 10
Diphenylamine	Benzenamine, N-phenyl-	122-39-4	8270	10
Disulfoton	Phosphorodithioic acid, 0,0-diethyl S-[2-(ethylthio)ethyl]ester	298-04-4	8140, 8141, 8270	2, 0.5, 10
Endosulfan I	6,9-Methano-2,4,3-benzodioxathiepin, 6,7,8,9,10,10-hexachloro-1,5,5a,6,9,9a-hexahydro-, 3-oxide	959-98-8	8080, 8270	0.1, 20
Endosulfan II	6,9-Methano-2,4,3-benzodioxathiepin, 6,7,8,9,10,10-hexachloro-1,5,5a,6,9,9a-hexahydro-, 3-oxide,(3I,5aI,6J,9J,9aI)-	33213-65-9	8080, 8270	0.05, 20
Endosulfan sulfate	6,9-Methano-2,4,3-benzodioxathiepin, 6,7,8,9,10,10-hexachloro-1,5,5a,6,9,9a-hexahydro-, 3-3-Dioxide	1031-07-8	8080, 8270	0.5, 10
Endrin	2,7:3,6-Dimethanonaphth[2,3-b]oxirene, 3,4,5,6,9,9-hexachloro-1a,2,2a,3,6,6a, 7,7a-octahydro-,(1aI,2J,2aJ,3I,6I,6aJ,7J,7aI)-	72-20-8	8080, 8270	0.1, 20
Endrin aldehyde	1,2,4-Methenocyclopenta[cd]pentalen	7421-93-4	8080, 8270	0.2, 10

(continued)

115

Appendix II (*Continued*)

Common Name[2]	CAS RN[3]	Chemical abstracts index name[4]	Suggested Method[5]	PQL, (μg/L)[6]
		e- 5-carboxaldehyde,2,2a,3, 3,4,7-hexachlorodecahydro-, (1I,2I, 2aJ,4J,4aJ,5J,6aJ,6bJ,7R*)-		
Ethylbenzene	100-41-4	Benzene, ethyl-	8020, 8221,8260	2, 0.05, 5
Ethyl methacrylate	97-63-2	2-Propenoic acid, 2-methyl-, ethyl ester	8015, 8260, 8270	5, 10, 10
Ethyl methanesulfonate	62-50-0	Methanesulfonic acid, ethyl ester	8270	20
Famphur	52-85-7	Phosphorothioic acid, 0-[4-[(dimethyl-amino)sulfonyl]phenyl] 0,0-dimethyl ester	8270	20
Fluoranthene	206-44-0	Fluoranthene	8100, 8270	200, 10
Fluorene	86-73-7	9H-Fluorene	8100,8270	200, 10
Heptachlor	76-44-8	4.7-Methano-1H-indene, 1,4,5,6,7, 8,8- heptachloro-3a,4,7,7a-tetrahydro-	8080, 8270	0.05, 10
Heptachlor epoxide	1024-57-3	2,5-Methano-2H-indeno[1,2-b]oxirene, 2,3,4,5,6,7,7-heptachloro-1a,1b,5,5a,6,6a-hexahydro-(1aI,1bJ, 2I, 5I,5aI, 6J, 6aI)	8080, 8270	1, 10
Hexachlorobenzene	118-74-1	Benzene, hexachloro-	8120, 8270	0.5, 10
Hexachlorobutadiene	87-68-3	1,3-Butadiene, 1,1,2,3,4, 4-hexachloro-	8021, 8120, 8260, 8270	0.5, 5, 10, 10
Hexachlorocyclopentadiene	77-47-4	1,3-Cyclopentadiene,	8120, 8270	5, 10

Hexachloroethane	67-72-1	1,2,3,4,5,5-hexachloro-Ethane, hexachloro-	8120, 8260, 8270	0.5, 10,10
Hexachloropropene	1888-71-7	1-Propene, 1,1,2,3,3,3-hexachloro-	8270	10
2-Hexanone; Methyl butyl Ketone	591-78-6	2-Hexanone	8260	50
Indeno(1,2,3-cd)pyrene	193-39-5	Indeno(1,2,3-cd)pyrene	8100, 8270	200, 10
Isobutyl alcohol	78-83-1	1-Propanol, 2-methyl-	8015, 8240	50, 100
Isodrin	465-73-6	1,4,5,8-Dimethanonaphthalene,1,2, 3,4, 10,10- hexachloro-1,4,4a,5,8, 8ahexahydro-(1I,4I,4aJ,5J,8J,8aJ)-	8270, 8260	20, 10
Isophorone	78-59-1	2-Cyclohexen-1-one, 3,5,5-trimethyl-	8090, 8270	60, 10
Isosafrole	120-58-1	1,3-Benzodioxole, 5-(1-propenyl)-	8270	10
Kepone	143-50-0	1,3,4-Metheno-2Hcyclobuta[cd]pentalen-2-one,1,1a,3,3a,4,5,5,5a,5b,6-decachlorooctahydro-	8270	20
Lead	(Total)		6010, 7420, 7421	400, 1000,10
Mercury	(Total)		7470	2
Methacrylonitrile	126-98-7	2-Propenenitrile, 2-methyl-	8015, 8260	5, 100
Methapyrilene	91-80-5	1,2-Ethanediamine, N.Ndimethyl-N1-2-pyridinyl-N1/2-thienylmethyl)-	8270	100
Methoxychlor	72-43-5	Benzene,1,11-(2,2,2, trichloroethylidene)bis[4-methoxy-	8080, 8270	2, 10
Methyl bromide; Bromomethane; Methyl chloride;	74-83-9	Methane, bromo-	8010,8021	20, 10

(continued)

Appendix II (*Continued*)

Common Name[2]	CAS RN[3]	Chemical abstracts index name[4]	Suggested Method[5]	PQL, (μg/L)[6]
Chloromethane	74-87-3	Methane, chloro-	8010, 8021	1, 0.3
3-Methylcholanthrene	56-49-5	Benz[j]aceanthrylene, 1,2-dihydro- 3-methyl-	8270	10
Methyl ethyl ketone; MEK;2-Butanone	78-93-3	2-Butanone	8015, 8260	10, 100
Methyl iodide; Iodomethane	74-88-4	Methane, iodo-	8010, 8260	40, 10
Methyl methacrylate	80-62-6	2-Propenoic acid, 2-methyl-, methyl ester	8015, 8260	2, 30
Methyl methanesulfonate	66-27-3	Methanesulfonic acid, methyl ester	8270	10
2-Methylnaphthalene	91-57-6	Naphthalene, 2-methyl-	8270	10
Methyl parathion; Parathion methyl	298-00-0	Phosphorothioic acid, 0,0-dimethyl	8140, 8141, 8270	0.5, 1, 10
4-Methyl-2-pentanone; Methyl isobutyl ketone	108-10-1	2-Pentanone, 4-methyl-	8015, 8260	5, 100
Methylene bromide; Dibromomethane	74-95-3	Methane, dibromo-	8010, 8021, 8260	15, 20,10
Methylene chloride; Dichloromethane	75-09-2	Methane, dichloro-	8010, 8021, 8260	5, 0.2, 10
Naphthalene	91-20-3	Naphthalene	8021, 8100, 8260, 8270	0.5,200, 5, 10
1,4-Naphthoquinone	130-15-4	1,4-Naphthalenedione	8270	10
1-Naphthylamine	134-32-7	1-Naphthalenamine	8270	10

Common name	CAS number	Chemical name	Method	Detection limit
2-Naphthylamine	91-59-8	2-Naphthalenamine	8270	10
Nickel	(Total)	Nickel	6010, 7520	150, 400
o-Nitroaniline; 2-Nitroaniline	88-74-4	Benzenamine, 2-nitro-	8270	50
m-Nitroaniline; 3-Nitroanile	99-09-2	Benzenamine, 3-nitro-	8270	50
p-Nitroaniline; 4-Nitroaniline	100-01-6	Benzenamine, 4-nitro-	8270	20
Nitrobenzene	98-95-3	Benzene, nitro-	8090, 8270	40, 10
o-Nitrophenol; 2-Nitrophenol	88-75-5	Phenol, 2-nitro-	8040, 8270	5, 10
p-Nitrophenol; 4-Nitrophenol	100-02-7	Phenol, 4-nitro-	8040, 8270	10, 50
N-Nitrosodi-n-butylamine	924-16-3	1-Butanamine, N-butyl-Nnitroso-	8270	10
N-Nitrosodiethylamine	55-18-5	Ethanamine, N-ethyl-Nnitroso-	8270	20
N-Nitrosodimethylamine	62-75-9	Methanamine, N-methyl-Nnitroso-	8070	2
N-Nitrosodiphenylamine	86-30-6	Benzenamine, N-nitroso-Nphenyl-	8070	5
N-Nitrosodipropylamine; N-Nitroso-N-dipropylamine; Di-n-propylnitrosamine	621-64-7	1-Propanamine, N-nitroso-N-propyl-	8070	10
N-Nitrosomethylethalamine	10595-95-6	Ethanamine, N-methyl-Nnitroso-	8270	10
N-Nitrosopiperidine	100-75-4	Piperidine, 1-nitroso-	8270	20
N-Nitrosopyrrolidine	930-55-2	Pyrrolidine, 1-nitroso-	8270	40
5-Nitro-o-toluidine	99-55-8	Benzenamine, 2-methyl-5-nitro-	8270	10
Parathion	56-38-2	Phosphorothioic acid, 0,0-diethyl 0-(4-nitrophenyl) ester	8141, 8270	0.5, 10
Pentachlorobenzene	608-93-5	Benzene, pentachloro-	8270	10
Pentachloronitrobenzene	82-68-8	Benzene, pentachloronitro-	8270	20

(continued)

119

Appendix II (*Continued*)

Common Name[2]	CAS RN[3]	Chemical abstracts index name[4]	Suggested Method[5]	PQL, (μg/L)[6]
Pentachlorophenol	87-86-5	Phenol, pentachloro-	8040, 8270	5, 50
Phenacetin	62-44-2	Acetamide, N-(4-ethoxyphenl)	8270	20
Phenanthrene	85-01-8	Phenanthrene	8100, 8270	200, 10
Phenol	108-95-2	Phenol	8040	1
p-Phenylenediamine	106-50-3	1,4-Benzenediamine	8270	10
Phorate	298-02-2	Phosphorodithioic acid, 0,0-diethyl S-[(ethylthio)methyl]ester	8140, 8141, 8270	2, 0.5, 10
Polychlorinated biphenyls; PCBs; Aroclors	See Note 9	1,1[prime]-Biphenyl, Chloroderivatives	8080, 8270	50, 200
Pronamide	23950-58-5	Benzamide, 3,5-dichloro-N-(1,1-dimethyl-2-propynyl)-	8270	10
Propionitrile; Ethyl cyanide	107-12-0	Propanenitrile	8015, 8260	60, 150
Pyrene	129-00-0	Pyrene	8100, 8270	200, 10
Safrole	94-59-7	1,3-Benzodioxole, 5-(2-propenyl)-	8270	10
Selenium	(Total)	Selenium	6010, 7740, 7741	750, 20, 20
Silver	(Total)	Silver	6010, 7760, 7761	70, 100, 10
Silvex; 2,4,5-TP	93-72-1	Propanoic acid, 2-(2,4,5-trichlorophenoxy)-	8150	2
Styrene	100-42-5	Benzene, ethenyl-	8020, 8021, 8260	1, 0.1, 10
Sulfide	18496-25-8	Sulfide	9030, 4000	
2,4,5-T; 2,4,5-Trichlorophenoxyacetic Acid	93-76-5	Acetic acid, (2,4,5-trichlorophenoxy)-	8150	2
1,2,4,5-Tetrachlorobenzene	95-94-3	Benzene, 1,2,4,5-tetrachloro-	8270	10

1,1,1,2-Tetrachloroethane	Ethane, 1,1,1,2-tetrachloro-	630-20-6	8010, 8021, 8260	5, 0.05, 5
1,1,2,2-Tetrachloroethane	Ethane, 1,1,2,2-tetrachloro-	79-34-5	8010, 8021, 8260	0.5, 0.1, 5
Tetrachloroethylene; Tetrachloroethene; Perchloroethylene	Ethene, tetrachloro-	127-18-4	8010, 8021, 8260	0.5, 0.5, 5
2,3,4,6-Tetrachlorophenol	Phenol, 2,3,4,6-tetrachloro-	58-90-2	8270	10
Thallium	Thallium	(Total)	6010, 7840,7841	400, 1000, 10
Tin	Tin	(Total)	6010	40
Toluene	Benzene, methyl-	108-88-3	8020, 8021, 8260	2, 0.1, 5
o-Toluidine	Benzenamine, 2-methyl-	95-53-4	8270	10
Toxaphene	Toxaphene	See Note 10	8080	2
1,2,4-Trichlorobenzene	Benzene, 1,2,4-trichloro-	120-82-1	8021, 8120, 8260, 8270	0.3, 0.5, 10, 10
1,1,1-Trichloroethane; Methylchloroform	Ethane, 1,1,1-trichloro-	71-55-6	8010, 8021, 8260	0.3, 0.3, 5
1,1,2-Trichloroethane	Ethane, 1,1,2-trichloro-	79-00-5	8010, 8260	0.2, 5
Trichloroethylene; Trichloroethene;	Ethene, trichloro-	79-01-6	8010, 8021, 8260	1, 0.2, 5
Trichlorofluoromethane; CFC-11	Methane, trichlorofluoro-	75-69-4	8010, 8021, 8260	10, 0.3, 5
2,4,5-Trichlorophenol	Phenol, 2,4,5-trichloro-	95-95-4	8270	10
2,4,6-Trichlorophenol	Phenol, 2,4,6-trichloro-	88-06-2	8040, 8270	5, 10
1,2,3-Trichloropropane	Propane, 1,2,3-trichloro-	96-18-4	8010, 8021, 8260	10, 5, 15
0,0,0-Triethyl phosphorothioate	Phosphorothioic acid, 0,0,0-triethyl ester	126-68-1	8270	10
sym-Trinitrobenzene	Benzene, 1,3,5-trinitro-	99-35-4	8270	10
Vanadium	Vanadium	(Total)	6010, 7910,7911	80, 2000, 40
Vinyl acetate	Acetic acid, ethenyl ester	108-05-4	8260	50
Vinyl chloride; Chloroethene	Ethene, chloro-	75-01-4	8010, 8021, 8260	2, 0.4, 10

(continued)

Appendix II (*Continued*)

Common Name[2]	CAS RN[3]	Chemical abstracts index name[4]	Suggested Method[5]	PQL, (μg/L)[6]
Xylene	(total) See Note 11	Benzene, dimethyl-	8020, 8021, 8260	5, 0.2, 5
Zinc	(Total)	Zinc	6010, 7950, 7951	20, 50, 0.5

Notes:

[1]The regulatory requirements pertain only to the list of substances; the right-hand columns (Methods and PQL) are given for informational purposes only. See also footnotes 5 and 6.

[2]Common names are those widely used in government regulations, scientific publications, and commerce; synonyms exist for many chemicals.

[3]Chemical Abstracts Service registry number. Where "Total" is entered, all species in the groundwater that contain this element are included.

[4]CAS index names are those used in the 9th Collective Index.

IAC 7/4/07

Ch 113, p.86 Environmental Protection[567] IAC 7/4/07, 10/10/07

[5]Suggested Methods refer to analytical procedure numbers used in EPA Report SW-846 "Test Methods for Evaluating Solid Waste," third edition, November 1986, as revised, December 1987. Analytical details can be found in SW-846 and in documentation on file at the agency. CAUTION: The methods listed are representative SW-846 procedures and may not always be the most suitable method(s) for monitoring an analyte under the regulations.

[6]Practical Quantitation Limits (PQLs) are the lowest concentrations of analytes in groundwaters that can be reliably determined within specified limits of precision and accuracy by the indicated methods under routine laboratory operating conditions. The PQLs listed are generally stated to one significant figure. PQLs are based on 5 mL samples for volatile organics and 1 L samples for semivolatile organics. CAUTION: The PQL values in many cases are based only or a general estimate for the method and not on a determination for individual compounds; PQLs are not a part of the regulation.

[7]This substance is often called Bis(2-chloroisopropyl) ether, the name Chemical Abstracts Service applies to its noncommercial isomer, Propane, 2,2[sec]-oxybis[2-chloro- (CAS RN 39638-32-9).

[8]Chlordane: This entry includes alpha-chlordane (CAS RN 5103-71-9), beta-chlordane (CAS RN 5103-74-2), gamma-chlordane (CAS RN 5566-34-7), and constituents of chlordane (CAS RN 57-74-9 and CAS RN 12789-03-6). PQL shown is for technical chlordane. PQLs of specific isomers are about 20 μg/L by method 8270.

[9]Polychlorinated biphenyls (CAS RN 1336-36-3); this category contains congener chemicals, including constituents of Aroclor 1016 (CAS RN 12674-11-2), Aroclor 1221 (CAS RN 11104-28-2), Aroclor 1232 (CAS RN 11141-16-5), Aroclor 1242 (CAS RN 53469-21-9), Aroclor 1248 (CAS RN 12672-29-6), Aroclor 1254 (CAS RN 11097-69-1), and Aroclor 1260 (CAS RN 11096-82-5). The PQL shown is an average value for PCB congeners.

[10]Toxaphene: This entry includes congener chemicals contained in technical toxaphene (CAS RN 8001-35-2), i.e., chlorinated camphene.

[11]Xylene (total): This entry includes o-xylene (CAS RN 96-47-6), m-xylene (CAS RN 108-38-3), p-xylene (CAS RN 106-42-3), and unspecified xylenes (dimethylbenzenes) (CAS RN 1330-20-7). PQLs for method 8021 are 0.2 μg/L for o-xylene and 0.1 for m- or p-xylene. The PQL for m-xylene is 2.0 μg/L by method 8020 or 8260.

These rules are intended to implement Iowa Code section 455B.304.

[Filed 11/21/02, Notice 9/18/02—published 12/11/02, effective 1/15/03]

[Filed 6/14/07, Notice 12/6/06—published 7/4/07, effective 10/1/07*]

REFERENCES

1. City of Cedar Rapids, Iowa, October, 2008.
2. City of Marion, Iowa, October, 2008.
3. Cedar Rapids Linn County Solid Waste Agency, 2008.
4. Young, G.C., "Zapping MSW with Plasma Arc," Pollution Engineering, November 2006.
5. Young, G.C., "How Trash Can Power Ethanol Plants," Public Utilities Fortnightly, p. 72, February 2007.
6. Young, G.C., "From Waste Solids to Fuel," Pollution Engineering, February 2008.
7. Young, G.C. and Lumsden, K. J., "From Curbside to Landfill," Pollution Engineering, April 2009.
8. "Iowa Statewide Waste Characterization Study," Report: Iowa Department of Natural Resources, by R-W-Beck, January 2006.
9. How Landfills Work, http://science.howstuffworks.com/landfill3.htm, 09/16/2009.
10. Iowa Department of Natural Resources (IDNR), Compiled from state agency websites, DNR survey and personal communications with state agency staff, October 19, 2005.
11. Demolition Site Waste Task Force, Senate File 2325, Report prepared by: Demolition Site Waste Material Task Force, December 2002, www.iowadnr.wmad.org.
12. "Renewable Energy Annual 1996," Energy Information Administration, Office of Coal, Nuclear, Electric and Alternate Fuels, U.S. Department of Energy, Washington, DC, April 1997.
13. Iowa Department of Natural Resources (IDNR), Landfills, Iowa Tonnages for Fiscal Year 2007, July 1, 2006–June 30, 2007.
14. Chartwell Information and the New York City Department of Sanitation, reproduced from Solid Waste Digest 12:7 (July, 2002), p. 6, http://www.wasteinfo.com
15. Lee, F.G. and Jones-Lee, A., "Flawed Technology of Subtitle D Landfilling of Municipal Solid Waste," G. Fred Lee & Associates, September 2009.
16. Iowa Department of Natural Resources (IDNR), Iowa DNR Workshop #1 Handout Subtitle D Economics, Shaw Environmental, Inc., May 2005.
17. Chapter 113, Sanitary Landfills for Municipal Solid Waste: Groundwater Protection Systems for the Disposal of Nonhazardous Wastes, IAC 7/4/2007, 10/10/07, Environmental Protection [567], Ch 113, pp. 1–87.
18. Thorneloe, Susan, US EPA, Research Triangle Park, NC, 2007.

CHAPTER 7

Plasma Economics: Garbage/Wastes to Power, Case Study with Economics of a 94 ton/day Facility

Subject: **Preliminary Economic Evaluation for a Municipal Solid Waste (MSW) Facility at 94 ton/day MSW Capacity using Plasma Arc Gasification Technology.**

Background: **Plasco Energy Group, Inc. of Ottawa, Canada built a demonstration plant to process 94 tons/day of municipal solid waste (MSW) to energy/ electricity using plasma arc technology. The facility is located across the road from Trail Road landfill southwest of Ottawa. The following preliminary analysis will determine economics of such a small-scale facility and the parameters in Canada that make it a viable demonstration facility.**

The gasification test facility will process 94 tons/day for the city of Ottawa, Canada from the Trail Road landfill. The plasma gasification process breaks down the garbage and produces a syngas for the generation of electricity. The by-product is vitrified slag and it is used as a road material.[1]

A process flow diagram illustrating the Plasma Gasification Process is shown in Fig. 7.1.

It can be seen from Fig. 7.1 that feedstock to the Plasma Arc Gasification reactor is municipal solid waste (MSW). Prior to entering the gasification reactor, feedstock is pretreated using methods such as shredding and metal recovery. Then the MSW is fed into the Plasma Arc Gasification reactor and heated to about 700°C before entering the upper part near 1,000°C. The raw synthesis gas (syngas, predominantly CO and H_2) is withdrawn overhead. The bottoms from the reactor are cooled and collected as vitrified slag. The raw syngas is cleaned and cooled, and impurities such as sulfur and acid gases are removed. Cleaned syngas is used as a fuel and converted to energy in the power plant. Air emission control equipment, using the best-available technology, treats gaseous exhaust from the syngas cleanup and power plant prior to release to the

Municipal Solid Waste to Energy Conversion Processes: Economic, Technical, and Renewable Comparisons By Gary C. Young
Copyright © 2010 John Wiley & Sons, Inc.

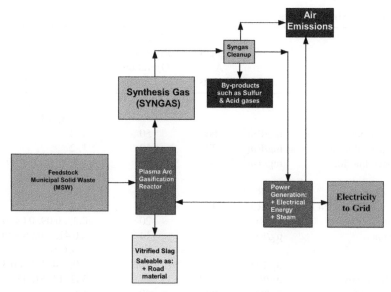

FIGURE 7.1   Process Drawing, MSW to Electricity via Plasma Arc Gasification.

atmosphere. The power plant produces energy for the facility and exports energy as electricity to the grid.[2,3]

Using the process drawing of Fig. 7.1, a preliminary economic analysis of the Plasco Trail Road (PTR) facility was conducted. Economic analysis parameters for this facility are according to Table 7.1 and monetary values from Plasco Energy Group.[4]

The economic analysis computes a net revenue [total revenues–total expenditures] of $118,022.05/year (before taxes). Two key economic parameters that determine the positive net revenue are grant of $5,950,300.00 and a selling price for electricity of 9.91 ¢/kWh to the grid.

As reported by the Plasco Energy Group, Inc. at a recent public meeting in Ottawa, Canada, air emissions from the Plasco Trail Road are shown in Table 7.2.[1]

TABLE 7.1   Parameters for Analysis of Plasco Trail Road Plasma Gasification Facility, Ottawa, Canada: (Parameters—U.S. Currency & *t*-ton = 2,000 pounds)

Capital: $22,878,600.00 at 6.00% for 20 years and grant for $5,950,300.00
(Total Capital = $28,828,900.00)
Energy value: 1,277 kWh/ton MSW
Energy to grid: 1,021 kWh/ton MSW at selling price of 9.91 cents/kWh
Plant capacity: 94 tons/day MSW
Personnel to operate facility: 15 at $28.00/hour
Maintenance and operation supplies: $1,078,600.00 per year
Capital budget reserve: $328,300 per year
Vitrified slag: 0.20 tons of vitrified slag/ton MSW at selling price of $15.00/ton
Tipping fee (revenue): $32.68/ton MSW

TABLE 7.2   Air Emissions

Parameter	Units	Ontario A-7 Limits	Plasco Trail Road (PTR) Operational Limits	Current Data
Particulate matter	mg/Rm^3	17	12	10.9, 9.9, 20.2, 3.9, 3.5, 6.8
Organic mater	mg/Rm^3	66	50	4, 4
Hydrogen chloride	mg/Rm^3	27	19	1, 2
Hydrogen fluoride	mg/Rm^3	—	—	<0.03, 0.11, 0.27, 0.05, 0.03
Sulfur dioxide	mg/Rm^3	56	37	28, 38
NOx expressed as NO$_2$	mg/Rm^3	207	207	112, 114
Mercury	μg/Rm^3	20	20	0.25, 0.09, 0.04, 0.06
Cadmium	μg/Rm^3	14	14	<0.41, <0.06, <0.07, <0.14
Lead	μg/Rm^3	142	142	3.9, 7.4, 1.2, <0.3
Class III metals	μg/Rm^3	—	—	342, 14, 51, 45
Dioxins and furans	pg/Rm^3	80	40	10.8

Source: Ref. (1).
Note: All values are expressed at 11% O$_2$ and reference conditions (101.3 kPa, 25°C)

As shown in Table 7.2, air emissions from the Plasma Arc Gasification facility processing MSW meet the air requirements as stipulated by the controlling authorities. Plasco Energy Group, Inc. states for air: "Emissions are meeting and we are improving upon the regulated limits."[1]

Recently, Plasco Energy Group, Inc. mentioned, "that the Ottawa City Council unanimously agreed to issue a letter of intent to PlascoEnergy to build, own and operate a 400 tonnes-per-day, [441 tons/day], waste conversion facility that will process residual household waste that would otherwise be sent to landfill."[5]

MORE RECENT EVENTS ABOUT THE PROJECT

Ottawa, Canada generates about 1,017,000 tonnes annually of solid waste for approximately 330,000 residential and 697,000 commercial and industrial (C&I). Unless the solid waste can be recycled, it continues to be dumped into a landfill.

The city of Ottawa and Plasco Energy Group formed a joint venture for the city to provide the waste and Plasco to operate a state-of-the-art facility to evaluate and convert 85 tonnes per day of waste into electricity at the PTR Gasification Process Demonstration facility. Household wastes from the city will be diverted to the facility including about 8% of nonrecyclable plastics that would otherwise be diverted to a landfill.

The Plasco gasification facility converts the waste into syngas (CO and H$_2$) for further conversion into electricity for sale to Hydro Ottawa. The system is sealed so

TABLE 7.3   Recent Air Emissions, Engine Data

Parameter	Units	Ontario Limits	Plasco Trail Road (PTR) Operational Limits	Current Data (07-13-2009 to 08-16-2009)
Hydrogen chloride	ppmv	18	13	<8
SO_x	ppmv	21	14	<12
Organic matter	ppmv	225	200	<135

Source: Ref. (6). Plasco Energy Group, Inc., Plasco Conversion Process, August 2009.

that no emissions go directly to the atmosphere. Power is generated by an internal combustion engine with the syngas as fuel. The exhaust from the engines is closely monitored so as to contain emissions within regulatory standards.

Recent data from the engine exhausts are shown in Table 7.3.[6]

The Plasco conversion process for conversion of MSW to electricity is shown in Fig. 7.2. First, any waste materials of high reclamation value are removed from the waste stream and sent to recycling. Then, the MSW is shredded and any remaining high-value materials are again sent to recycling. The MSW enters into a conversion chamber where the waste is converted to crude syngas. The crude syngas is processed in a refinement chamber where plasma torches refine the crude syngas into a cleaner syngas. Now, the syngas is sent to a gas quality control unit to remove and recover sulfur, acid gases, and segregate heavy metals from the syngas. The process creates a clean syngas from the waste with no air emissions. The clean syngas is used to fuel internal combustion engines that generate electricity. The waste heat recovered from the engines and from cooling the syngas is used to produce steam in a heat recovery steam generation (HRSG) unit. The steam produced can be used to generate

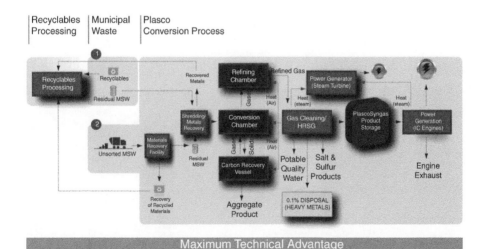

FIGURE 7.2   Process Drawing, Plasco Conversion Process, MSW to Electricity.[8]

electricity from a steam turbine or can provide steam heating for local industries and/ or for local district heating.

The solid residue from the conversion chamber is sent to a high-temperature carbon recovery vessel with plasma torches for conversion into additional crude syngas. The additional crude syngas is recycled back into the main process crude syngas conversion chamber. Any remaining solids are stabilized into a melted slag and cooled into pellets.[7] The slag pellets are a vitrified solid that is inert and sold as construction aggregate.[7,8]

REFERENCES

1. Plasco Energy Group, Inc., Plasco Trail Road, Public Meeting, Ottawa, Canada, July 24, 2008.
2. Ebert, J., "Landfill Eliminators, The Process is Called Plasma Gasification and the Technology for Creating and Harnessing Plasma has Been Around for Decades. However, Plasma Gasification Technology is Now Being Used for a New Purpose—The Conversion of Municipal Solid Waste-To-Energy," Biomass Magazine, October 2007.
3. Plasco Energy Group, Inc., Plasco Conversion Process, November 2008.
4. Plasco Energy Group, Inc., Private Communication, May 3, 2007.
5. Plasco Energy Group, Inc., News release, June 25, 2008.
6. Plasco Energy Group, Inc., Plasco Conversion Process, August 2009.
7. Plasco Energy Group, Inc., Monthly Engineer's Report, Plasco Trail Road, Gasification Process Demonstration Project, Review period: June 1, 2009 to June 30, 2009.
8. Plasco Energy Group, Inc., www.plascoenergygroup.com, 09/02/2009.

CHAPTER 8

Plant Operations: Eco-Valley Plant in Utashinai, Japan: An Independent Case Study

Subject: **Evaluation of the Commercial Plant Operations at the Eco-Valley Plant in Utashinai, Japan Using Westinghouse Plasma Corporation Plasma Gasification Technology Processing Municipal Solid Waste (MSW) and Auto Shredder Residue**.

Background: **Municipal solid waste (MSW) and auto shredder residue (ASR) are commercially processed via a process known as Plasma Arc Gasification. This commercial operation forms the basis of the Eco-Valley plant in Utashinai, Japan, which Hitachi Metals, Ltd. partially owns and operates. The Eco-Valley plant uses the Plasma Arc Gasification technology of Westinghouse Plasma Corporation, a division of Alter Nrg. The plant has been operating since April 2003. Plant capacity is about 180 tonnes/day (200 ton/day) of MSW and ASR converting these solid wastes into energy, steam, and electricity**

This commercial plant gasifies the organic wastes by a pyrolysis type of process (7,200–12,600°F) where the wastes are converted to a synthesis gas composed primarily of carbon monoxide (CO) and hydrogen (H_2). The inorganics of the solid wastes produce a rocklike material as a by-product. This by-product is a vitrified slag, typically composed of silica glass and metals.[1–7]

The Eco-Valley plant at Utashinai, Japan is illustrated in Fig. 8.1 whereby the feedstock to the plasma arc reactor is municipal solid waste (MSW) and auto shredder residue (ASR). Coke and limestone are added as process aides.

The organic materials are gasified in the reactor to syngas and exit the top for cleanup and heat recovery prior to entering the power plant. In the power plant, the syngas supplies the energy for the boiler producing steam for the steam turbine, which produces the electricity. Some energy from the power plant is used by the plant

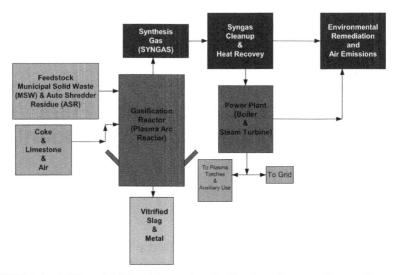

FIGURE 8.1   MSW and ASR to Plasma Arc Gasification—Syngas—Energy (Steam and Electricity).

(plasma torches and auxiliary use) and the remainder sold to the grid. The vitrified slag and metal are by-products recycled for commercial use.[1-7] It has been reported that for this plant, with feedstock of MSW at 23 wt.% H_2O and 23 wt.% ash, high heating value (HHV) of 5,660 Btu/lb, and coke (for generation of internal heat in the gasifier) at about 4 wt.% of the MSW feed, on a N_2-free basis the syngas would have a component composition as 40.37 mole% CO, 15.88 mole% H_2, 37.33 mole% H_2O, and 3.55 mole% CO_2 for an HHV of 276 Btu/ft^3.[8]

The Plasma Arc Gasification system consists basically of the following:

Westinghouse plasma torches, Marc 3a, power: 80–300 kW for each torch, four torches per reactor, two reactors, boiler, and steam turbine.[9-12,1]

In addition, the plant design criteria are as follows:

- 200 tons/day feedstock of MSW and auto shredder waste (ASR);
- energy production from the plant is 7.9 MW of electricity, of which 3.6 MW is consumed by the plant (torches and auxiliaries) and 4.3 MW is sent to the grid.[9-12,1]
- energy production from the plant at 300 tons/day MSW will produce an equivalent amount of total energy, i.e., 7.9 MW.[12]

With the previous outlined process and equipment described for the Eco-Valley plant at Utashinai, Japan, the plant operations can be discussed and determined by technical analysis. First, the operation of the plasma arc torches as supplied by Westinghouse Plasma Corporation (WPC) can be discussed as to their operability in a commercial plant setting. As mentioned previously, the plasma torches are Marc 3a with the capability of operating a power supplied between 80 and 300 kW per torch. With a quantity of four torches per reactor, this design permits a reactor to operate from 80 to 1,200 kW. In process operations, this represents a turndown ratio of 15/1, i. e., maximum to minimum reactor operation. When considering a waste solid like MSW, this would indicate that one reactor could operate with a feedstock ranging from about 16 to 238 tons/day MSW. With two reactors in parallel, a feedstock ranging from about 16 to 470 tons/day of MSW could be processed. These factors indicate that the Plasma Arc Gasification reactors are designed to be flexible in regard to feedstock variations and should be reliable for a commercial plant operation. As quoted in an article, "The facilities also suffer operational problems, though not with the plasma torch itself,…."[10] Thus, one can readily conclude from the torch and reactor design that much flexibility exists for the plasma arc gasifiers, verified by comments from the Utashinai facility.

Second, a discussion about the output of energy from the Utashinai plant is of keen interest to many since comments have been published as quoted, "… Tashinai's plant pumps out … megawatts of power per year, all of which is used to run the plant."[10] By considering a mass and energy balance on the plant, an understanding of this comment can be substantiated. Consider the following mass and energy computational procedure as:

a. From the plant design criteria, a feedstock of 200 tons/day of MSW and ASR will produce 7.9 MW of energy. If 300 tons/day of MSW is used as feedstock, the plant will produce 7.9 kW.

b. Now, with a mass and energy balance using the result of (a), the computations will produce the following results at plant design conditions:

   MSW feed is 110 tons/day.

   ASR feed is 90 tons/day.

   Total feed is 200.0 tons/day.

These feed materials produce 3.6 MW for plant use (torches and auxiliaries) and 4.3 MW for the grid, a total of 7.9 MW per the plant design specifications.

Now consider the quotation from an article, "On average, the plant only processes 60% of trash volume that the company expected."[10] By using the previous results determined for the feed of MSW and ASR to the plant but at 60% of design feed rate, the mass and energy balance yields the following:

MSW feed is 66 tons/day.

ASR feed is 54 tons/day.

Total feed is 120.0 tons/day.

Thus, the new feed rates produce 1.7 MW energy from MSW and 3.0 MW from ASR for a total of 4.7 MW. At feedstock at 50% of design, 1.4 MW energy is generated from MSW and 2.5 MW from ASR **for a total of 3.9 MW. Note, the plant design** specifications state that 3.6 MW is consumed by the plant (torches and auxiliaries). The conclusion is obvious that at 50–60% of the design feedstock, energy is produced only to operate the plant, with no excess energy produced for the grid. The results are summarized in Table 8.1.

The bottom line is that comments reported in articles on the Eco-Valley plant in Utashinai, Japan have been substantiated by these mass and energy balances. With proper quantity of feedstock for the Utashinai plant, one would expect the plant to perform according to design, i.e., the Utashinai Plasma Arc Gasification plant on MSW and ASR will produce an excess of energy for sale to the grid. In summary, from this analysis, the Utashinai plant is producing according to plant design and technology.

With recent interest in the ever-increasing cost for the disposal of, MSW, auto fluff, hazardous materials, and medical wastes, the Eco-Valley plant in Utashinai, Japan continues to draw the public attention. The existing facilities in Japan and other countries typically own the rights to testing information and thus are not required to publically disclose process and emission technology or data. Thus, public information is limited at best and therefore results that appear in public documents are, and can be, confusing and potentially misleading. The evaluation of the Utashinai facility presented earlier in this chapter was done principally to dispel myths generated by any confusion or misunderstandings based upon sound scientific analysis. In addition, the following comments are made regarding the Utashinai facility, depicted in Fig. 8.2, to bring additional attention to this important facility upon which the advancement of Plasma Arc Gasification technology currently depends.

The plasma arc facility at Utashinai uses the Westinghouse (Alter Nrg) plasma arc torches. Alter Nrg is the parent company of WPC technology. With experience exceeding 40 years, Alter Nrg has designed and engineered many applications. The WPC technology was originally developed with NASA for use in the Apollo space program to test high-temperature re-entry conditions of over 9,932°F or 5,500°C. With this experience with WPC technology since the 1980s, WPC has proven itself in

TABLE 8.1   Summaries of Utashinai Plant Design and Operations

Material	Design Feed	Design Energy Generated	60% of Design Feed	60% of Design Energy Generated
MSW	110 tons/day	3.6 MW	66 tons/day	1.7 MW
ASR	90 tons/day	4.3 MW	54 tons/day	3.0 MW
Total	200 tons/day	7.9 MW	120.0 tons/day	4.7 MW

Material	Design Feed	Design Energy Generated	50% of Design Feed	50% of Design Energy Generated
MSW	110 tons/day	3.6 MW	55 tons/day	1.4 MW
ASR	90 tons/day	4.3 MW	45 tons/day	2.5 MW
Total	200 tons/day	7.9 MW	100.0 tons/day	3.9 MW

Note: Design specification for energy consumed by the plant operation (torches and auxiliaries) is 3.6 MW. When plant is at 50–60% of design feed rate, the energy generated from the plant is about 3.6 MW, which is just sufficient for plant operations.

FIGURE 8.2   Eco-Valley Waste-to-Energy Facility, Processing MSW, Utashinai, Japan.[13]

industrial applications such as General Motors and Hitachi Metals. Such industrial applications have made the WPC torches robust and reliable under strenuous environments.[13,14]

The Utashinai, Japan facility (Fig. 8.2) has a capacity of between 200 and 300 tons/day of MSW and ASR. This plant was designed for 200 tons/day of auto fluff or 300 tons/day MSW. Combination of the two feedstocks can also be processed. Once again, public information on the research conducted by this facility is limited. The evaluation presented earlier in this chapter is presented to contribute to a better understanding of the current operation and design of this facility.

REFERENCES

1. Circeo, L.J., Engineering and Environmental Applications of Plasma Arc Technology, Technological Forum, Kirkwood Training and Outreach Services Center, Marion, Iowa, November 22, 2005.
2. Circeo, L.J., "Engineering & Environmental Applications of Plasma Arc Technology," Presentation, Georgia Tech Research Institute, Atlanta, GA 2005.
3. Recovered Energy, Inc., Pocatello, Idaho, www.recoveredenergy.com, "Process Flow Diagram," MSW into energy and useable by-products.
4. "Summary Report: Evaluation of Alternative Solid Waste Processing Technologies," Prepared for: City of Los Angeles, Department of Public Works, Bureau of Sanitation, 419 S. Spring Street, Suite 900, Los Angeles, CA 90013; Prepared by: URS Corporation, 915 Wilshire Boulevard, Suite 700, Los Angeles, CA 90017, September 2005.
5. Young, G.C., "Zapping MSW with Plasma Arc," Pollution Engineering, November 2006.
6. Young, G.C., "From Waste Solids to Fuel," Pollution Engineering, February 2008.
7. Young, G.C., "How Trash Can Power Ethanol Plants," Public Utilities Fortnightly, February 2007.
8. Cobb, J.T., Jr., "Production of Synthesis Gas by Biomass Gasification—A Tutorial," 2007 AIChE Spring National Meeting, April 22–26, 2007.

9. Westinghouse Plasma Corporation, a division of Alter Nrg, www.westinghouse-plasma. com, Commercially Proven, December 2008.

10. Cyranoski, D., "One Man's Trash...," Nature, November 16, 2006, p. 262.

11. Bodurow, C.(USEPA/OPPTS/OPPT/RAD), Circeo, L.J., Caravati, K.C., Martin, R.C., and Smith, M.S.(Georgia Tech Research Institute), "Plasma Processing of MSW for Energy Recovery," EPA Science Forum, 2005.

12. Circeo, L.J., Private Communication, December 23, 2008.

13. Alter NRG, Eco-Valley Waste to Energy, www.alternrg.ca, September 2009.

14. Aitkin County Plasma Gasification Study, Prepared by: Northspan Group, Inc., Duluth, Minnesota, April 2008.

CHAPTER 9

Municipal Solid Waste and Properties

WHAT IS MUNICIPAL SOLID WASTE (MSW) AND HOW MUCH IS GENERATED IN THE UNITED STATES?

Municipal solid waste (MSW) is typically known by the public as "trash" or "garbage." Basically MSW consists of product packaging, grass clippings, furniture, clothing, bottles, food scraps, newspapers, appliances, and batteries. MSW is disposed in landfills as illustrated in Fig. 9.1.

Some of the items in MSW are recycled wherever economical and/or practical. In addition, materials, not generally included as MSW but sent to landfills, are construction and demolition debris, municipal wastewater treatment sludges, and nonhazardous industrial wastes.

In other words, MSW does "not" include wastes of other types or from other sources such as automobile bodies, municipal sludges, combustion ash, and industrial process wastes that might also be disposed in municipal waste landfills.[2]

Sources of MSW are both residential and commercial. Residential waste sources (including waste from multifamily dwellings) were estimated to be 55–65% of total MSW generation. Commercial waste sources (including waste from schools, some industrial sites where packaging is generated, and businesses) constitutes between 35 and 45% of MSW generation. Final breakdown of quantities involved with residential waste and commercial waste depend on location.[2]

MSWs can be characterized as coming from residential, commercial, institutional, or industrial sources. Some examples of the types of MSW that come from each of the broad categories of sources are listed in Table 9.1.[2]

"MSW" does "not" include everything that is landfilled in Subtitle D landfills (Subtitle D of the Resource Conservation and Recovery Act deals with wastes other than the hazardous wastes that are covered under Subtitle C.) As shown in Table 9.2, however, RCRA Subtitle D includes many kinds of wastes. It has been a common practice to landfill wastes such as municipal sludges, nonhazardous industrial wastes, residue from automobile salvage operations, and construction and demolition debris

Municipal Solid Waste to Energy Conversion Processes: Economic, Technical, and Renewable Comparisons By Gary C. Young
Copyright © 2010 John Wiley & Sons, Inc.

FIGURE 9.1   MSW Disposed in Landfills.[1]

TABLE 9.1   Sources of MSW, Year 2005

Sources and Examples	Example Products
Residential: single- and multifamily homes	Newspapers, clothing, disposable tableware, food packaging, cans and bottles, food scraps, yard trimmings
Commercial: office buildings, retail and papers, wholesale establishments, restaurants	Corrugated boxes, food scraps, office disposable tableware, paper napkins, yard trimmings
Institutional: schools, libraries, hospitals, prisons	Cafeteria and restroom trash can wastes, office papers, classroom wastes, yard trimmings
Industrial: packaging and administrative; not process wastes	Corrugated boxes, plastic film, wood pallets, lunchroom wastes, office papers

TABLE 9.2   MSW and RCRA Subtitle D Wastes

The Subtitle D wastes included here in MSW are:
    Containers and packaging such as soft drink bottles and corrugated boxes
    Durable goods such as furniture and appliances
    Nondurable goods such as newspapers, trash bags, and clothing
    Other wastes such as food scraps and yard trimmings
The Subtitle D wastes not included here in MSW, are:
    Municipal sludges
    Agricultural wastes
    Industrial nonhazardous wastes
    Oil and gas wastes
    Construction and demolition debris
    Mining wastes

Source: Ref. (2).

along with MSW, but these other kinds of wastes are "not" included as MSW in the estimates presented.

MSW generation in the United States has increased over the years to 245.7 million tons of MSW in 2005. This amounts to 4.5 pounds of MSW per person per day in 2005. MSW consists of: paper and paperboard, yard trimmings, glass, metals, plastics, wood, and food scraps. In addition, rubber, leather, and textiles are combined as a group and a final group is listed as miscellaneous. These constituents in MSW are broken down by percent weight and illustrated in Fig. 9.2.

The generation and recovery (recycle) of materials in MSW for 2005 are presented in Table 9.3.[2]

Latest generation of MSW in the United States for the year 2007 are summarized in Table 9.4.

The MSW industry can be pictorially shown in Fig. 9.3 with four basic components: recycling, composting, landfilling, and waste-to-energy. As discussed, the U.S. Environmental Protection Agency defines MSW as wastes from residential, commercial, institutional, or industrial sources.[4]

A statewide waste characterization study for Iowa was conducted in the year 2005 for Iowa Department of Natural Resources (IDNR). The results of this study are summarized in Table 9.5 for materials, composition, and quantities.[5]

The estimated generation of MSW for the 38 largest metropolitan areas in the United States is presented in Table 9.6. A total metropolitan population of about 124 million has an estimated generation of MSW of about 57,893,000 tons/year.[6]

MSW PROPERTIES

The design of a process for the management of MSW and the resultant economic evaluation and development of a viable business plan require the properties of MSW. Hence, the MSW properties are presented to assist those performing such design and economic evaluations.

In the presentation of these properties, one should keep in mind the material for which these properties are determined. A picture of the material, MSW, is vividly displayed in Fig. 9.4 for a typical landfill.[5]

Some typical properties of MSW of interest are the density of various components as shown in Table 9.7. Typical density for various uncompacted wastes in MSW are listed such as food wastes, paper, plastics, garden trimmings, glass, and ferrous metals.[7]

The table also lists typical moisture content and range of moisture content for various components of MSW such as residential, commercial, construction and demolition, industrial, and agricultural. A typical proximate analysis is also shown for wastes such as mixed food, mixed paper, mixed plastics, yard wastes, glass, and residential MSW. The typical proximate analysis values for moisture, volatiles, carbon and ash are shown in weight percent. In addition, an elemental analysis is shown for the wastes of mixed food, mixed paper, mixed plastics, yard wastes, and

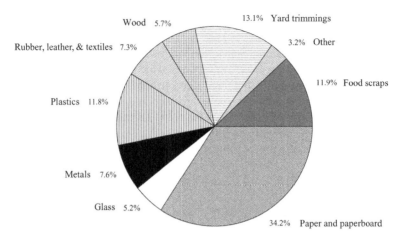

FIGURE 9.2 2005—Total MSW Generation in United States—246 Million Tons (Before Recycling).

TABLE 9.3 Generation and Recovery (Recycle) of Materials in the United States of MSW (Year 2005)

Material	Weight Generated (Millions Tons)	Weight Recovered (Millions Tons)	Percent Recovered (%)
Paper and paperboard	84.0	42.0	50.0
Glass	12.8	2.76	21.6
Metals			
Steel	13.8	4.93	35.7
Aluminum	3.21	0.69	21.5
Other nonferrous metals[a]	1.74	1.26	72.4
Total metals	18.75	6.88	36.7
Plastics	28.9	1.65	5.7
Rubber and leather	6.70	0.96	14.3
Textiles	11.1	1.70	15.3
Wood	13.9	1.31	9.4
Other materials	4.57	1.17	25.6
Total materials in products	180.72	58.43	32.3
Other wastes			
Food, other[b]	29.2	0.69	2.4
Yard trimmings	32.1	19.9	62.0
Miscellaneous inorganic wastes	3.69	Neg.	Neg.
Total other wastes	64.99	20.59	31.7
Total municipal solid waste (MSW)	245.7	79.0	32.1

Source: Ref. (2).
Notes: The table includes waste from residential, commercial, and institutional sources. Neg = Less than 5,000 tons or 0.05%.
[a] Includes lead from lead-acid batteries.
[b] Includes recovery of other MSW organics for composting.

TABLE 9.4   MSW in the United States in 2007 (Facts)

Generated
  254.1 million tons
  1,686.3 pounds per person per year
  4.62 pounds per person per day

  Yard trimmings, food wastes, corrugated boxes, glass bottles, and newspapers are the largest items in MSW before recycling

Recycled
  63.3 million tons, a 24.9% rate
  1.15 pounds per person per day
  419.75 pounds per person per year
  Corrugated boxes, newspapers, office paper, and glass bottles are the most recycled by weight. Lead-acid batteries, newspapers, corrugated boxes, and office papers have the highest recycling rates

Composted
  21.7 million tons of yard and food waste
  8.5% composting rate for all MSW
  64.1% composting rate for yard waste
  2.6% rate for food waste
  0.39 pounds per person per day
  142.35 pounds per person per year

Landfill density (1997 data)
  323,812,000 cubic yards of MSW was landfilled

  Corrugated boxes, clothing and footwear, yard waste, and food waste occupy the most space in landfills
  Aluminum cans and plastic bottles have the lowest landfill density
  Glass bottles and food waste have the highest density
  An average pound of trash has a landfill density of 739 pounds per cubic yard

Source: Ref. (3).

refuse derived fuel (RDF), and elements contained in these wastes such as carbon (C), hydrogen (H), oxygen (O), nitrogen (N), sulfur (S), and ash on a % by weight basis.[7]

Some materials of interest appear in MSW and respective properties need identification for moisture, ash, and calorific values (dry basis). Some typical materials that appear in most areas are shown in Table 9.8.[8]

An elemental analysis was performed for the MSW and coal used at the Ames Municipal Power Plant in Ames, Iowa. The results of this analysis are shown in Table 9.9.[9]

A detailed chemical analysis for an average MSW, presented in the *Brazilian Journal of Physics*, is given in Table 9.10.[10]

An elemental analysis of MSW and heating values in comparison to bituminous coal are presented in Table 9.11.[11]

Heating values are presented in the literature in units of kJ/kg, BTU/lb, kWh/lb, kWh/ton, and kWh/tonne. A ton is 2,000 pounds and a tonne (metric ton) is 2,205 pounds. To help understand the magnitude of these heating values, presented in any of these units, Table 9.12 was generated.

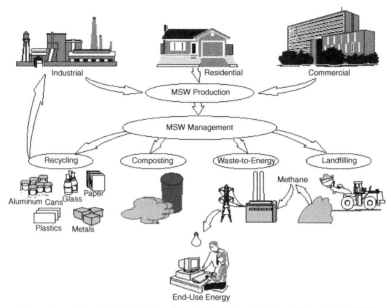

Source: Energy Information Administration, Office of Coal, Nuclear, Electric and Alternate Fuels (1996)

FIGURE 9.3 Components of MSW: Recycling, Composting, Landfilling, and Waste-to-Energy.[4]

A study was conducted for New York City solid wastes to determine the energy related to the management of MSW. This study examined the energy of the combustible components of MSW and their use as fuel to generate process steam and electricity. Even though MSW has considerable heterogeneity of materials, the mean hydrocarbon structure was found to be approximated by $C_6H_{10}O_4$. The comparison of the experimental heating values of the various waste materials is shown in Fig. 9.5.[12]

Other data for heating values, moisture, proximate analysis, and elemental analysis of wastes are shown in Table 9.13.

The name "refuse" is given to a solid waste in studies for gasification processes. In Nordrhein-Westfalen a "standard refuse" was developed with composition presented in Table 9.14.[15]

Some caution should be used about compositions presented in any literature because the refuse and/or MSW presented is site-specific and depends on local recycling of household wastes and can have a big influence upon the final heating value.

Additional waste material discussed routinely is given the term "biomass." Biomass covers a variety of materials that can be used as raw materials for gasification. Biomass is generally considered as a material derived from living organisms but "excluding" fossil fuels such as coal, oil, and gas that are derived from ancient life forms. Biomass can be generally considered agricultural and forestry wastes and wastes from processes such as biochemical processes using renewable raw materials. Some biomass materials are listed in Table 9.15.[16]

The properties of Illinois No. 6 coal are presented in Table 9.16, as some energy studies on gasification processes use this particular coal as a raw material.[17]

TABLE 9.5   MSW Materials, Iowa Statewide—2005

Materials	Composition (Weight%)	Quantities (Tons)
Compostable paper	6.5	173,764
High grade office	2.5	65,743
Magazines	1.8	48,784
Mixed recyclable paper	6.9	186,153
Newsprint	4.0	107,465
Non-recyclable paper	2.8	73,844
OCC and draft paper	2.8	227,093
Total paper	32.9	882,846
#1 PET deposit beverage containers	0.2	5,243
#1 PET beverage containers	0.4	11,472
#2 HDPE containers	1.0	27,033
Film/wrap/bags	6.5	175,191
Other #1 PET containers	0.3	7,243
Other plastic containers	0.4	10,447
Other plastic products	6.0	160,666
Total plastics	14.8	397,294
Aluminum beverage containers	0.1	3,773
Aluminum deposit beverage containers	0.2	4,306
Ferrous food and beverage containers	1.0	26,288
Other aluminum containers	0.1	2,841
Other ferrous metals	2.8	75,770
Other non-ferrous scrap	0.5	13,314
Total metals	4.7	126,293
Blue glass	0.0	919
Brown glass	0.0	1,079
Clear glass	0.7	19,205
Glass deposit containers	0.3	7,240
Green glass	0.1	2,913
Other mixed cullet	0.5	14,645
Total glass	1.6	46,000
Pumpkins	0.3	7,307
Yard waste	1.3	36,163
Total yard waste	1.6	43,470
Food waste	10.8	289,268
Non-treated	3.4	90,449
Treated	4.6	122,276
Total wood	8.0	213,225
Demolition/renovation/construction debris	5.5	146,823
Cell phones and charges	0.0	112
Central processing units/peripherals	0.2	5,191
Computer monitors/TV's	0.1	1,429

(continued)

TABLE 9.5 (*Contineud*)

Materials	Composition (Weight%)	Quantities (Tons)
Electrical and household appliances	2.1	57,248
Other durables	2.7	71,571
Total durables	5.1	135,550
Textiles and leathers	4.9	131,328
Diapers	2.4	63,786
Rubber	0.4	11,089
Automotive products	0.0	768
Household cleaners	0.0	253
Lead acid batteries	0.0	—
Mercury containing products	0.0	23
Other batteries	0.2	5,500
Other HHM (household hazardous material)	0.1	2,995
Paints and solvent	0.1	3,013
Pesticides, herbicides, fungicides	0.0	204
Total HHMs	0.5	12,756
Sharps	0.0	252
Other organic	1.5	39,818
Other inorganic	2.4	63,569
Fines/super mix	2.4	63,948
Other	0.5	12,384
Grand total	100.0	2,679,700

Source: Ref. (5).

The heating value is determined by combustion from a sample analyzed in a calorimeter. However, if a heating value is not available, then the high heating value (HHV) can be determined from the ultimate or proximate analyses by the following Equation (9.1):[18]

$$HHV(MJ/kg) = (34.91 \times C) + (117.83 \times H) - (10.34 \times O) - (1.51 \times N)$$
$$+ (10.05 \times S) - (2.11 \times Ash) \tag{9.1}$$

where C, H, O, N, S, and ash are the mass fractions of the elements from the ultimate and proximate analyses.

Example: An MSW has an ultimate analysis (mass fractions) as follows:[11]

$$C = 0.37418$$
$$H = 0.05138$$
$$O = 0.29908$$
$$N = 0.01186$$
$$S = 0.00132$$
$$Ash = 0.2556$$

TABLE 9.6   Estimated Generation of MSW for the 38 Largest U.S. Metropolitan Areas

City	Population (millions)	Municipal Solid Waste (MSW) (1000 tons/year)
New York, NY	15.000	7005
Los Angeles, CA	13.000	6071
Chicago, IL	8.008	3740
Philadelphia, PA	4.950	2312
Dallas-Ft. Worth, TX	4.910	2293
Washington, DC	4.740	2214
Detroit, MI	4.475	2090
San Francisco-Oakland, CA	4.035	1884
Houston, TX	4.011	1873
Atlanta, GA	3.857	1801
Miami-Ft. Lauderdale, FL	3.711	1733
Boston, MA	3.297	1540
Seattle-Tacoma, WA	3.260	1522
Phoenix-Mesa, AZ	3.014	1408
Minneapolis-St. Paul, MN	2.872	1341
San Diego, CA	2.821	1317
St. Louis, MO	2.569	1200
Baltimore, MD	2.491	1163
Pittsburgh, PA	2.331	1089
Tampa-St. Petersburg, FL	2.278	1064
Cleveland, OH	2.221	1037
Denver, CO	1.979	924
Portland, OR–Vancouver, WA	1.846	862
Kansas City, MO	1.756	820
San Jose, CA	1.647	769
Cincinnati, OH	1.628	760
Sacramento, CA	1.585	740
San Antonio, TX	1.565	731
Norfolk-Virginia Beach, VA	1.563	730
Indianapolis, IN	1.537	718
Orlando, FL	1.535	717
Columbus, OH	1.489	695
Milwaukee, WI	1.462	683
Charlotte-Gastonia, NC	1.417	662
Las Vegas, NV	1.381	645
New Orleans, LA	1.305	609
Salt Lake-Ogden, UT	1.275	595
Hartford, CT	1.147	536
Total Metropolitan United States	123.968	57,893

Source: Ref. (6).

FIGURE 9.4   MSW Landfill.[5]

which computes an HHV of 15.48 MJ/kg dry from Equation (9.1) for the MSW, or 6,661 BTU/pound. This calculated HHV is in agreement with the value reported for the ultimate analysis.[11]

The heating value can be estimated using the Dulong formula as Equation (9.2) based upon the ultimate analysis:

$$HHV(MJ/kg) = (33.86 \times C) + [144.4 \times (H-(O/8)] + (9.428 \times S) \qquad (9.2)$$

where C, H, O, and S are the mass fractions of the elements from the ultimate analysis.[19]

Example: A coal has the following proximate analysis expressed in mass fractions:[17]

C = 0.7172
H = 0.0506
O = 0.0775
S = 0.0282

which computes an HHV of 30.48 MJ/kg dry from Equation (9.2) for Illinois No. 6 coal, or 13,115 BTU/lb.

The computed heating value (HHV) of 30.48 MJ/kg from Equation (9.2) is in close agreement with the HHV value of 30.51 MJ/kg (13,126 BTU/lb) reported for this Illinois No. 6 coal with the proximate analysis.[17]

As a final comment, caution should be used with some due diligence when properties are selected for a particular waste material, as such properties can vary widely for different site-specific locations.

The process of gasification of materials can be understood by looking at the basic chemical equations involved but knowing that it is a much more complex process. In

TABLE 9.7 Physical Properties of MSW

<table>
<tr><td colspan="2">Typical Properties of Uncompacted Wastes (USA Data)—Density</td></tr>
<tr><td></td><td>Density (kg/m^3)</td></tr>
<tr><td>Food wastes</td><td>288</td></tr>
<tr><td>Paper</td><td>81.7</td></tr>
<tr><td>Plastics</td><td>64</td></tr>
<tr><td>Garden trimmings</td><td>104</td></tr>
<tr><td>Glass</td><td>194</td></tr>
<tr><td>Ferrous metals</td><td>320</td></tr>
</table>

Typical moisture Contents of Wastes

	Moisture content (wt.%)	
	Range	Typical
Residential		
Food wastes (mixed)	50–80	70
Paper	4–10	6
Plastics	1–4	2
Yard wastes	30–80	60
Glass	1–4	2
Commercial		
Food wastes	50–80	70
Rubbish (mixed)	10–25	15
Construction and demolition		
Mixed demolition Combustibles	4–15	8
Mixed construction combustibles	4–15	8
Industrial		
Chemical sludge (wet)	75–99	80
Sawdust	10–40	20
Wood (mixed)	30–60	35
Agricultural		
Mixed agricultural waste	40–80	50
Manure (wet)	75–96	94

Typical Proximate Analysis Values (% by weight)

Type of waste	Moisture	Volatiles	Carbon	Ash
Mixed food	70.0	21.4	3.6	5.0
Mixed paper	10.2	75.9	8.4	5.4
Mixed plastics	0.2	95.8	2.0	2.0
Yard wastes	60.0	30.0	9.5	0.5
Glass	2.0	—	—	96-99
Residential MSW	21.0	52.0	7.0	20.0

Typical Elemental Analysis (% by weight):

Type of waste	C	H	O	N	S	Ash
Mixed food	73.0	11.5	14.8	0.4	0.1	0.2
Mixed paper	43.3	5.8	44.3	0.3	0.2	6.0
Mixed plastics	60.0	7.2	22.8	—	—	10.0
Yard wastes	46.0	6.0	38.0	3.4	0.3	6.3
Refuse derived fuel	44.7	6.2	38.4	0.7	<0.1	9.9

Source: Ref. (7).

TABLE 9.8 Calorific Values of Various Materials

Material	Calorific Value (BTU/lb)	Ash Content (wt.%)	Moisture Content (wt.%)
Soft wood	6,330	0.1	19
Fiber board 90% paper	7,600	4.6	7.5
Damp wood	5,690	1.2	27.5
Leather trimmings	7,670	5.2	10.4
Cotton seed hulls	10,600	2.47	8.9
Sludge material (steel mill)	9,150	24.5	1.9
Nitrile rubber	15,240	3.4	
Cardboard granulated	8,592	12.3	6.4
Carbon residue	13,681	8.7	Nil
Wood waste, sawdust	7,500	0.8	14
Nut shells	7,980	1.75	11.85

this regard, the chemical equations with the heat of reactions are presented for understanding.

Basic chemical reactions in the gasification process for producing syngas are:

$$C + O_2 \rightarrow CO_2 \quad -394 \, MJ/kmol \quad (\text{exothermic}) \quad (9.3)$$

$$C + H_2O \rightleftharpoons CO + H_2 \quad +131 \, MJ/kmol \quad (\text{endothermic}) \quad (9.4)$$

$$C + 2H_2 \rightleftharpoons CH_4 \quad -75 \, MJ/kmol \quad (\text{exothermic}) \quad (9.5)$$

$$C + CO_2 \rightleftharpoons 2CO \quad +172 \, MJ/kmol \quad (\text{endothermic}) \quad (9.6)$$

$$CO + H_2O \rightleftharpoons CO_2 + H_2 \quad -41 \, MJ/kmol \quad (\text{exothermic}) \quad (9.7)$$

$$C_nH_m + nH_2O \rightleftharpoons nCO + (n + \tfrac{1}{2}m)H_2 \quad (\text{endothermic}) \quad (9.8)$$

TABLE 9.9 Elemental Analysis of MSW and Coal

Constituent	MSW (wt.%)	Coal (wt.%)
C	35.5	51.7
H	4.8	3.5
O	25.1	12.6
N	0.40	0.73
S	0.15	0.22
Cl	0.59 (max)	0.02
Moisture	26.5	27.0
Ash	7.6	4.38

Source: Ref. (9).
Note: Elemental analysis for MSW and Coal used at Ames Municipal Power Plant, Ames, Iowa. MSW burned at the Ames facility. Coal used: Rochelle North Antelope western coal.

TABLE 9.10   Chemical Composition of Average MSW

Component	Composition (mg/kg)	Weight %
C	196,548.69	1.9655E + 01
H	16,787.29	1.6787E + 00
O	79,525.46	7.9525E + 00
N	3,239.47	3.2395E − 01
S	3,351.38	3.3514E − 01
Cd	0.6436	6.4360E − 05
Cr	6.9193	6.9193E − 04
Cu	67.2243	6.7224E − 03
Ni	4.3536	4.3536E − 04
Pb	28.2185	2.8219E − 03
Zn	93.9698	9.3970E − 03
Sn	1.1246	1.1246E − 04
Cl	1,774.97	1.7750E − 01
F	21.2468	2.1247E − 03
Al	2,864.85	2.8649E − 01
Fe	11,473.79	1.1474E + 00
Hg	0.0179	1.7900E − 06
As	0.4451	4.4510E − 05
Sb	0.00056	5.6000E − 08
Tl	8.4870	8.4870E − 04
Ag	3.0185	3.0185E − 04
Br	1.1253	1.1253E − 04
Ba	4.2113	4.2113E − 04
Se	4.0474	4.0474E − 04
Si	4.0747	4.0747E − 04
V	49.5798	4.9580E − 03
Na_2CO_3	96,805.03	9.6805E + 00
$CaCO_3$		
SiO_3		
(Glass)		
H_2O	548,575.67	5.4858E + 01
Ash/inert	31,603.09	3.1603E + 00

Source: Ref. (10).

For the steam methane reforming reaction, where $n = 1$ and $m = 4$, Equation (9.8) becomes:

$$CH_4 + H_2O \rightleftharpoons CO + 3H_2 \quad + 204 \, MJ/kmol \quad (endothermic) \qquad (9.9)$$

Note that an exothermic reaction produces heat (output) whereas an endothermic reaction requires heat (input). Also note that Equation (9.3) is a combustion reaction that is sometimes used in the gasification process to provide some internal (*in situ*) process heat for the endothermic syngas reactions but only a (limited) amount of

TABLE 9.11 Elemental Analysis of MSW and Heating Values

Constituent	MSW	Bituminous Coal
C, Carbon (kg/kg dry)	0.37418	0.73798
H, Hydrogen (kg/kg dry)	0.05138	0.04896
O, Oxygen (kg/kg dry)	0.29908	0.09133
N, Nitrogen (kg/kg dry)	0.01186	0.01405
S, Sulfur (kg/kg dry)	0.00132	0.00801
Cl, Chlorine (kg/kg dry)	0.00659	0.00199
Ash, inert (kg/kg dry)	0.2556	0.09769
Water, wet basis (kg/kg wet)	0.241	0.089
Lower heating value (MJ/kg dry) (calculated)	14.36	29.37
Upper heating value (MJ/kg dry) (calculated)	15.48	30.44

Source: Ref. (11).

Note: In Table 9.11, both lower heating and upper heating values are presented. Heating values express the amount of energy released on combustion of a given quantity of fuel. The upper heating value includes the heat obtained by condensing the water vapor produced by combustion. The lower heating value (LHV) does not include the water vapor condensed from combustion. Typically, if a process exhausts the water vapor produced by combustion, then the LHV may be used. If the process condenses the water vapor produced by combustion, then the upper heating value may be used.

oxygen is supplied so the gasification reactor environment remains conducive for the production of syngas. The remaining process heat is supplied externally to the gasification reactor, for example through the heat transfer surfaces of the reactor heating system.

In a study of biofuel feedstock with forest and wood residues, properties of these cellulose materials were presented for those located in Norway. Densities were presented for some hardwoods and softwoods, for example: Scandinavian spruce

TABLE 9.12 Energy Conversions for Heating Values

Heating Value (kJ/kg)		Heating Value (BTU/lb)		Heating Value (kWh/lb)		Heating Value (kWh/ton)		Heating Value (kWh/tonne)
2,000	=	860.6	=	0.2520	=	504.0	=	555.7
3,000	=	1,290.8	=	0.3780	=	756.1	=	833.6
5,000	=	2,151.4	=	0.6300	=	1,260.1	=	1,389.3
10,000	=	4,302.8	=	1.2601	=	2,520.2	=	2,778.5
12,500	=	5,378.5	=	1.5751	=	3,150.2	=	3,473.1
15,000	=	6,454.2	=	1.8901	=	3,780.3	=	4,167.8
17,500	=	7,529.9	=	2.2052	=	4,410.3	=	4,862.4
20,000	=	8,605.6	=	2.5202	=	5,040.4	=	5,557.0
25,000	=	10,757.1	=	3.1502	=	6,300.5	=	6,946.3
30,000	=	12,908.5	=	3.7803	=	7,560.6	=	8,335.5
35,000	=	15,059.9	=	4.4103	=	8,820.6	=	9,724.8

Note: ton = 2000 lbs; tonne = 2205 lbs; New York City (NYC) MSW is about 12,500 kJ/kg heating value.

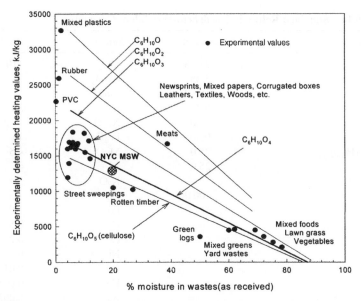

FIGURE 9.5 Comparison of Experimental Heating Values of Various Waste Materials (Hollander, Tchobanoglous, 1980). Lines Show Respective Thermochemical Values for $C_6H_{10}O_x$ Materials.[12]

TABLE 9.13   Heating Value, Moisture, Proximate Analysis, and Elemental Analysis of Wastes

Heating Value of MSW and Moisture Content

Type of waste	Heating value (dry basis)[13] [kJ/kg]	Moisture content[14] [Wt.%]
Food waste	4,600	70
Paper	16,700	6
Plastics	32,500	2
Textiles	17,500	10
Rubber	23,300	2
Wood	18,600	60
Garden trimmings	6,500	60

Composition of MSW by Proximate Analysis

Type of waste	Moisture (wt.%)	Ash (wt.%)	Volatile matter (wt.%)	Fixed carbon (wt.%)
Mixed paper	10.2	5.4	75.9	8.5
Yard waste	75.2	1.6	18.6	4.6
Food waste	78.3	1.0	17.0	3.7
Plastics (PE)	0.2	1.2	98.5	0.1

Composition of MSW by Elemental Analysis

Element	Wt. %
Carbon, C	51.9
Hydrogen, H	7.0
Oxygen, O	39.6
Nitrogen, N	1.1
Sulfur, S	0.4

TABLE 9.14   Standard Refuse in Nordrhein-Westfalen (Germany)

Composition	Mass (wt.%)	Ash Composition	kg/1000 kg
C	27.16	SiO_2	110
H	3.45	Al_2O_3	34
O	18.39	CaO	31
N	0.30	Fe	30
S	0.20	Na_2O	15.2
Cl	0.50	Fe_2O_3	15
Moisture (H_2O)	25.0	MgO	4.5
Ash	25.0	Al	4
		K_2O	3
Total	100.00	Zn	1.5
		Pb	1.0
		Cu	0.5
		Cr	0.2
		Ni	0.075
		Cd	0.01
		Hg	0.005
		As	0.005
Lower heating value, 10 MJ/kg			

Source: Ref. (15).

(softwood), density of 380 kg dry wood/m^3; Scandinavian pine (softwood), density of 440 kg dry wood/m^3; and Scandinavian birch (hardwood), density of 500 kg dry wood/m^3. Bark was considered to have a density of 500 kg dry wood/m^3. Another property of importance was the influence of moisture content on the heating value of these wood cellulose products. Water in these cellulose products takes considerable energy to evaporate, such as 2,444 kJ/kg (1,052 BTU/lb) water at 25°C. It was noted that at a moisture content of about 87%, the energy content of the wood is equivalent to

TABLE 9.15   Properties of Some Biomasses

Biomass	High Heating Value (MJ/kg)	Moisture (wt.%)	Ash (wt.%)	Sulfur (wt.% dry)	Chlorine (wt.% dry)
Charcoal	25–32	1–10	0.5–6		
Wood	10–20	10–60	0.25–1.7	0.01	0.01
Straw	14–16	10	4–5	0.07	0.49
Cotton residues (stalks)	16	10–20	0.1		
Rice husk	13–14	9–15	15–20		
Soya straw	15–16	8–9	5–6		
Cotton residue (gin trash)	14	9	12		
Maize (stalk)	13–15	10–20	2 (3–7)	0.05	1.48
Sawdust	11	35	2		
Bagasse	8–10	40–60	1–4		

Source: Refs. (15, 16).

TABLE 9.16   Characteristics of Illinois No. 6 Coal

Rank	High Volatile Bituminous	
Seam	Illinois #6 (Herrin)	
Source	Old Ben mine	

	Proximate Analysis (weight %)[a]	
	As Received	Dry
Moisture	11.12	0.00
Ash	9.70	10.91
Volatile Matter	34.99	39.37
Fixed Carbon	44.19	49.72
HHV (Btu/lb)	11,666	13,126

	Ultimate Analysis (weight %)	
	As Received	Dry
Carbon	63.75	71.72
Moisture	11.12	0.00
Hydrogen	4.50	5.06
Nitrogen	1.25	1.41
Chlorine	0.29	0.33
Sulfur	2.51	2.82
Ash	9.70	10.91
Oxygen[b]	6.88	7.75
Total	100.00	100.00

Source: Ref. (17).
[a]The above proximate analysis assumes that sulfur is volatile matter.
[b]By difference.

the energy required to evaporate the moisture. At around 50–55% moisture, any further increase in moisture level reduces the energy value of the wood significantly. LHV is 2,120 kWh/m^3 for dry spruce, 2,230 kWh/m^3 for dry pine, 2,650 kWh/m^3 for dry birch, and 2,050 kWh/m^3 for dry bark.[20]

TABLE 9.17   Composition of a Hardwood, Typical Yellow Poplar

Component	Dry Basis (% wt.)
Cellulose	42.67
Xylan	19.05
Arabinan	0.79
Mannan	3.93
Galactan	0.24
Acetate[a]	4.64
Lignin	27.68
Ash	1.00
Total	100.00
Moisture	47.90

Source: Ref. (21).
Acetate refers to the acetate groups present in the hemicellulose polymer. They are generally converted to acetic acid in the prehydrolysis reactor.

TABLE 9.18 Wastes: MSW, Medical Waste, and Used Tires

MSW (3,100 kWh/ton)	wt.%	Medical Waste (4,800 kWh/ton)	wt.%	Used Tires (8,900 kWh/ton)	wt.%
Paper	25.97	Cellulose	18.40	Carbon black	30.50
Yard	8.25	Glass	11.04	Rubber	46.00
Food	7.43	Polypropylene	11.04	Sulfur	2.50
Plastics	7.16	Polyethylene	11.04	Nitrogen	1.00
Metals	5.32	Polystyrene	7.36	Steel	15.00
Rubber, leather, textile	4.50	Organic tissue	3.50	Ash	5.00
Glass	3.75	Stainless steel	3.68		
Wood	3.61	PVC	3.68		
Others	2.18				
Moisture	31.83	Moisture	30.26		
Total	100.00		100.00		100.00

Source: Ref. (22).

Much experimental work has been conducted using a hardwood, typically yellow poplar, as a typical feedstock. Therefore, the composition of yellow poplar is presented in Table 9.17.[21]

Properties of wastes as resources are reported with regard to energy content and constituent composition in Table 9.18 for MSW, medical waste, and used tires.[22]

TABLE 9.19 Fuel-Grade Ethanol Specifications (Denatured)

Test	Non-Detergent Grade
Apparent proof, 60°F	200 (minimum)–203 (maximum)
Specific gravity, 60/60°F	0.7870–0.7950
Water, mass %	0.50 (nominal) –0.82 (maximum)
Ethanol content (vol. %)	92.1 minimum
Methanol (vol. %)	0.50 maximum
Sulfur, mass %	0.0010 maximum
Benzene (vol. %)	0.06 maximum
Olefins (vol. %)	0.50 maximum
Aromatics (vol. %)	1.70 maximum
Chloride ion content (mg/l)	32 maximum
Copper content (mg/kg)	0.08 maximum
Acidity (as acetic acid, CH_3COOH), mass %	0.0070 maximum, 0.0042 maximum (shipments to Canada)
Appearance	Clear and bright, visibly free of suspended and/or settled contaminants
Hydrocarbon denaturant	4.76 (maximum)–1.96 (minimum)
pH	6.5 (minimum)–9.0 (maximum)
Sulfate (mg/kg)	4.0 maximum
Solvent-washed gum	5.0 mg/100 ml max

Source: Ref. (23).

TABLE 9.20   Properties of Some Common Types of Asbestos

	Asbestos Type		
	Chrysotile (Hydrated Mg silicate)	Crocidolite (Fe and Na silicate)	Amosite (Fe and Mg silicate)
Chemical composition (essential elements) (%)			
SiO_2	38–42	50–56	≈50
FeO	0–2	4–20	≈40
Fe_2O_3	0.9–1.5	13–18	—
MgO	38–43	1–13	≈6.5
Na_2O	—	6–7	—
H_2O	11–14	2–3	≈2
Physical characteristic	0.1–1	1–2	1–2
Diameter of industrial fiber, μm			
Chemical resistance	Resistance to alkalis	Resistance to acids	Resistance to acids
Fusion point (°C)	1500	1000	1100
Mechanical property (kg/mm^2)	50–200	75–225	10–60

Source: Ref. (24).

The property of fuel-grade ethanol is of interest when reviewing the processes for the manufacture of ethanol from waste materials. The specifications for denatured fuel-grade ethanol are shown in Table 9.19.[23]

High-temperature gasification processes have been developed for the destruction of asbestos waste. Asbestos is a natural mineral material that is fibrous and crystallized. Properties of some common asbestos materials are shown in Table 9.20.[24]

REFERENCES

1. Westinghouse Plasma Corp, website, www.westinghouse-plasma.com, 04/01/2009.

2. United States Environmental Protection Agency, Office of Solid Waste (5306P), EPA530-R-06-001, "Municipal Solid Waste in The United States: 2005 Facts and Figures", October 2006.

3. Miller, C., "MSW 2007," Waste Age, December 1, 2008.

4. "Renewable Energy Annual 1996," Energy Information Administration, Office of Coal, Nuclear, Electric and Alternate Fuels, U.S. Department of Energy, Washington, DC, April 1997.

5. Iowa Department of Natural Resources (IDNR), Report: Iowa Statewide Waste Characterization Study, by R.W. Beck, January 2006.

6. Feed System Innovation for Gasification of Locally Economical Alternative Fuels (FIGLEAF), Final Report: DOE Cooperative Agreement No. DE-FC26-OONT40904,

U.S. DOE & NETL, Prepared by Michael L. Swanson, Mark A. Musich, Darren D. Schmidt and Joseph K. Schultz, Energy & Environmental Research Center, University of North Dakota, Grand Forks, ND, February 2003.

7. Tchobanoglous, G., et. al., "Integrated Solid Waste Management: Engineering Principles and Management Issues," McGraw-Hill Science Engineering, January 1993.

8. Heat Treatment International (HTI), Dutton, Michigan, 2008.

9. Report prepared for City of Ames, Iowa, Electric Department, by Dr. Robert C. Brown, Director and Professor, Center for Sustainable Environmental Technologies, Iowa State University, Ames, Iowa, June 26, 2001.

10. Leal-Quiros, E. "Plasma Processing of Municipal Solid Waste," Brazilian Journal of Physics, vol. 34 (4B), 2004.

11. Wochele, J., CAWAF, Composition and Analysis of Waste, Raw Material and Fuels, A data compilation of elemental analysis, PSI 2003.

12. Themelis, N.J., Kim, Y.H. and Brady, Mark H., "Energy recovery from New York City solid wastes," ISWA Journal: Waste Management and Research vol. 20, 223–233, 2002.

13. Schwartz, S.C. and Brunner, C.R., "Energy and Resource Recovery Wastes," Noyes Data Corp. , 1983.

14. Vesilind, P.A., Worrell, W.A. and Reinhart, D.R., "Solid Waste Engineering," Nelson Engineering, 2001.

15. Higman, C. and van der Burgt, M., "Gasification," 2nd edition, Elsevier, 2008.

16. Quaak, P., Knoef, H. and Stassen, H., "Energy from Biomass: A Review of Combustion and Gasification Technologies," World Bank, Washington, DC, 1999.

17. DOE/NETL-2007/1260, Final Report, Baseline Technical and Economic Assessment of a Commercial Scale Fischer-Tropsch Liquids Facility, April 9, 2007.

18. Channiwala, S.A. and Parikh, P.P., "A unified correlation for estimating HHV of solid, liquid and gaseous fuels," Fuel, vol. 81 (8), pp. 1051–1063, (2002).

19. Perry, R.H. and Chilton, C.H., "Chemical Engineer's Handbook," 5th edition, McGraw-Hill-Kogakusha, Tokyo, 1973.

20. Vessia, O., Finden, P. and Skreiberg, O., "Biofuels From Lignocellulosic Material, In The Norwegian Context 2010, Technology, Potential and Costs," Norwegian University of Science and Technology (NTNU), December 20, 2005.

21. Wooley, R., Ruth, M., Sheehan, J., Ibsen, K., Majdeski, H. and Galvez, A., "Lignocellulosic Biomass to Ethanol Process Design and Economics Utilizing Co-Current Dilute Acid Prehydrolysis and Enzymatic Hydrolysis Current and Futuristic Scenarios," National Renewable Energy Laboratory (NREL), NREL/TP-580-26157, July 1999.

22. Startech Environmental Corp., "Thermal Conversion of Wastes to Hydrogen for Fuel Cell Applications," March 31, 2003, www.Startech.net.

23. "Fuel-Grade Ethanol Specifications (Denatured)," Aventine Renewable Energy, Inc., Pekin, IL, March 14, 2007, www.aventinerei.com.

24. Blary, F. and Rollin, M., "Vitrification of Asbestos Waste," INERTAM, circa 1995.

CHAPTER 10

MSW Processes to Energy with High-Value Products and Specialty By-Products

Gasification is an old technology used for the production of energy from solid materials. Even today, gasification is an important technology and is becoming recognized as a technology for the management of municipal solid waste (MSW). In the general sense, gasification is the conversion of a carbonaceous material into a gaseous product for the production of energy products and by-products in an oxygen-starved environment. The carbonaceous or "organic" material in MSW is processed in an oxygen-deficient environment with heat to produce a syngas comprising mostly of carbon monoxide (CO) and hydrogen (H_2). Chemical bonds of the carbonaceous materials (complex chemical compounds) break down with heat to produce the more simple and thermodynamically stable gaseous molecules of CO and H_2. The inorganic or "mineral" materials are converted to a solid rocklike material called slag or vitrified slag or ash.[1,2,3]

This chapter will discuss the various products from the gasification of MSW with particular emphasis upon the high-value products and specialty by-products that can be produced or have great potential to be produced in the near future. The key factor to remember is that syngas contains the "basic building blocks, (CO and H_2)" for the production of other chemicals, fuels, and materials through chemistry. The focus will be on the processes with potential to convert these basic building blocks into high-value products and specialty by-products from the gasification of MSW. In addition, some focus will be given to what high-value products can be made from the slag or vitrified slag produced from the inorganic or "mineral" components of MSW.

A typical process for the gasification of MSW to syngas and slag or vitrified slag is illustrated in Fig. 10.1. As mentioned, the organic components of MSW are converted into syngas, the primary product, and the mineral components are converted into slag or vitrified slag or ash, a by-product. For some gasification processes, some (limited) amount of oxygen is supplied to the gasification reactor to provide heat in the form of

Municipal Solid Waste to Energy Conversion Processes: Economic, Technical, and Renewable Comparisons By Gary C. Young
Copyright © 2010 John Wiley & Sons, Inc.

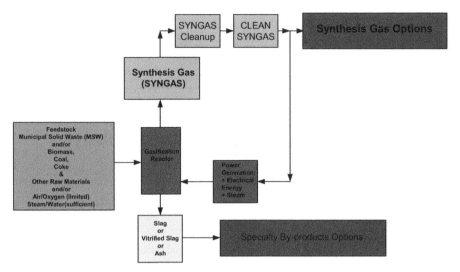

FIGURE 10.1 Gasification of MSW to Syngas and Vitrified Slag or Slag or Ash.

combustion heat necessary for the oxygen-deficient reactions to produce syngas, which are endothermic reactions. Sufficient steam/water is supplied to the gasification reactor to promote some syngas reactions.

Basic chemical reactions in the gasification process for producing syngas are:

$$C + O_2 \rightarrow CO_2 \quad \text{(exothermic)} \tag{10.1}$$

$$C + H_2O \rightleftharpoons CO + H_2 \quad \text{(endothermic)} \tag{10.2}$$

$$C + 2H_2 \rightleftharpoons CH_4 \quad \text{(exothermic)} \tag{10.3}$$

$$C + CO_2 \rightleftharpoons 2CO \quad \text{(endothermic)} \tag{10.4}$$

$$CO + H_2O \rightleftharpoons CO_2 + H_2 \quad \text{(exothermic)} \tag{10.5}$$

$$C_nH_m + nH_2O \rightleftharpoons nCO + (n + \tfrac{1}{2}m) H_2 \quad \text{(endothermic)} \tag{10.6}$$

where for the steam methane (CH_4) reforming reaction (where $n = 1$ and $m = 4$), Equation (10.6) becomes:

$$CH_4 + H_2O \rightleftharpoons CO + 3H_2 \quad \text{(endothermic)} \tag{10.7}$$

Note that an exothermic reaction produces heat (output) whereas an endothermic reaction requires heat (input). Also note that Equation (10.1) is a combustion reaction that is sometimes used in the gasification process to provide some internal (*in situ*) process heat for the endothermic syngas reactions but only a (limited) amount of oxygen is supplied so the gasification reactor environment remains conducive for the production of syngas. The remaining process heat is supplied externally to the gasification reactor, for example through the heat transfer surfaces of the reactor heating system.

In the following text, the clean syngas will be presented as the starting point while discussing what uses or potential uses (synthesis gas options) can be derived from these simple chemical building blocks of chemistry, i.e., CO and H_2.

PRODUCTION OF AMMONIA (NH₃) FROM SYNGAS VIA CHEMICAL SYNTHESIS ROUTE

Ammonia (NH_3) is the second largest synthetic chemical product. Most of the NH_3 produced is used to make fertilizers in the form of urea or ammonium salts (nitrate, phosphate, and sulfate). A small fraction of NH_3 is used for the manufacture of organic chemical feedstocks for the plastics industry, such as polyamides, caprolactam, and others, and for the production of explosives (hydrazine, nitriles, etc.). NH_3 is also converted to nitric acid and cyanides. The basic process steps for the industrial production of NH_3 are synthesis gas production, gas conditioning, compressions, and NH_3 synthesis. The purpose of the synthesis gas production and gas conditioning steps is to provide a clean H_2 stream for input to the NH_3 converter/synthesis reactor. About 50% of the H_2 produced from syngas processes is used for NH_3 production.[4]

NH_3 can be produced from syngas via processes in existence. As an example, a somewhat simplified process flow diagram for NH_3 production is illustrated in Fig. 10.2.[5]

Consider the feed stream to the process in Fig. 10.2 contains a clean syngas, nitrogen, and steam in proper stochiometric quantities for the NH_3 synthesis. Next, a water–gas shift conversion reaction converts CO into H_2 with a two-stage high-temperature shift reaction followed by a low-temperature shift reaction as shown in Equation (10.5). CO_2 is removed from the gaseous stream in an absorber column, such as by using a hot potassium carbonate solution to absorb the carbon dioxide (CO_2). Then, the CO_2 is stripped off in a stripper with steam. CO and CO_2 can poison the NH_3 synthesis catalyst downstream; therefore removal is important, which explains the reason for conversion of CO into CO_2 by Equation (10.5) and impurities of CO and CO_2 into CH_4 in the methanator by Equations (10.8 and 10.9).

$$CO + 3H_2 \rightleftharpoons CH_4 + H_2O \tag{10.8}$$

$$CO_2 + 4H_2 \rightleftharpoons CH_4 + 2H_2O \tag{10.9}$$

After removal of CO and CO_2 via the methanator, the gaseous stream of H_2 and N_2 is cooled and excess steam is removed by condensation prior to becoming feed for the NH_3 synthesis system. The gaseous stream is compressed and preheated prior to entering the NH_3 synthesis reactor by exothermic reaction (Eq. (10.10). The basic design of the NH_3 synthesis reactor is a pressure vessel with sections for catalyst beds and heat exchangers. The main design challenge is to remove the heat generated by the

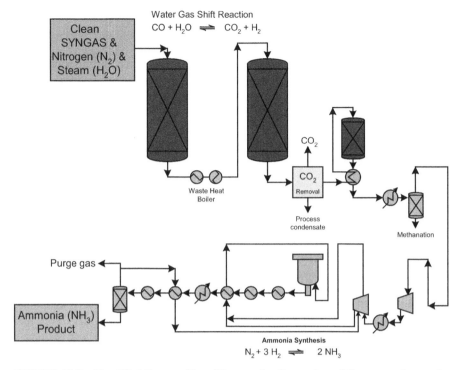

FIGURE 10.2 Simplified Process Flow Diagram for Conversion of Syngas to Ammonia (NH_3).[5]

exothermic synthesis reaction to control reactor temperature. Commercial NH_3 synthesis catalysts are basically the same Fe-promoted catalysts developed many years ago. Recent developments in NH_3 synthesis catalysts include Ru-based formulations that show promise. With effective cleanup and conditioning of the synthesis gas, catalyst lifetimes of 5–8 years are reported.[4]

$$N_2 + 3H_2 \rightleftharpoons 2NH_3 \qquad (10.10)$$

Since the reaction is highly exothermic, maximum conversion at equilibrium occurs at high pressures and low temperatures.

The final NH_3 is condensed, removed by a separator, and sent to storage.

PRODUCTION OF GAS TO LIQUIDS FROM SYNGAS VIA CHEMICAL SYNTHESIS ROUTE

A feasibility study was conducted that incorporates coal gasification for the production of commercial-grade diesel and naphtha liquids by Fischer–Tropsch (F-T) technology. Summary of the plant's performance, 50,000 barrel per day (bbl/day)

TABLE 10.1   Plant Performance Summary, Coal Gasification to Diesel and Naphtha via F-T

Parameter	Value
Coal feed flow rate (tons/day)	24,533
Diesel production (bbl/day)	27,819
Naphtha production (bbl/day)	22,173
Net plant power (MW$_e$)	124.3
Elemental sulfur production (tons/day)	612
Slag (tons/day)	2,470
CO_2 capture, tons/day	32,481

Source: Ref. (6).

Note: Net plant power is the excess power generated and available for export to the grid. CTL facility uses F-T technology using iron-based catalyst. Coal used in the plant is Illinois No. 6 bituminous. Gasification uses ConocoPhillips' E-Gas™ gasification technology.

coal-to-liquids (CTL) facility, is shown in Table 10.1.[6] Plant site size is approximately 300 acres.

Although the process is about coal gasification using ConocoPhillips' E-Gas™ gasification technology, the process technology used for conversion of the syngas from gasifier into finished products via F-T synthesis utilizes much of the same technologies if other feedstocks were used. In addition, much of the gasifier designs/technologies developed or under development are specialized or custom forms of the gasifiers originally used to commercialize coal gasification. Thus, much can be learned and utilized from the conversion technology of coal gasification, especially from syngas to energy/liquid-fuels. Thus, a process from CTL will be presented for understanding the downstream processing from syngas to liquid fuels. The liquid fuels produced from this CTL process are a commercial-grade diesel and naphtha liquids.

The basic block flow diagram used in the gasification CTL study is shown in Fig. 10.3.[6]

The process is based upon the ConocoPhillips' E-Gas™ gasification technology and will be discussed with reference to Fig. 10.3. Coal, Illinois No. 6 bituminous coal is fed as a slurry to the gasifier. The gasifier is a two-stage, oxygen-blown, entrained-flow, refractory-lined gasifier with continuous slag removal. An air separation unit (ASU) supplies 95 mole% pure oxygen to the gasifiers. The coal slurry reacts with oxygen in the primary stage at about 2,500°F and 500 psia. The oxygen is consumed in the first stage undergoing partial combustion thereby producing heat that causes the gasification reactions (oxygen starved) to proceed rapidly. The second stage of the gasifier enhances the endothermic gasification reactions and the exit temperature of around 1,900°F. Raw syngas exiting the gasifier is cooled to about 700°F in a fire-tube boiler. Any char and fly ash produced in the second gasification zone is entrained in the syngas stream and removed by the downstream cyclone and candle filter particulate control devices for return to the gasifier. The filter is composed of ceramic candle elements and cleaned by back-pulsing with syngas. The raw syngas exists at about 700°F.

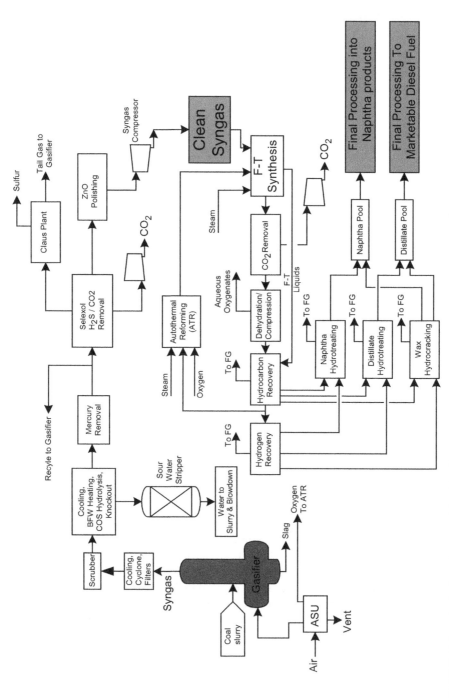

FIGURE 10.3 Block Flow Diagram of Coal Gasification to Syngas to Diesel and Naphtha Production.[6] Some Acronyms and Abbreviations Used in the Figure are: ASU, Air Separation Unit; ATR, Autothermal Reformer; BFW, Boiler Feed Water; FG, Fuel Gas; F-T, Fischer–Tropsch; AGR, Acid Gas Removal; COS, Carbonyl Sulfide; HCN, Hydrogen Cyanide; H$_2$S, Hydrogen Sulfide; ZnO, Zinc Oxide; and CO$_2$, Carbon Dioxide.

TABLE 10.2   Properties of Illinois No. 6 Coal

Rank	High Volatile Bituminous	
Seam	Illinois #6 (Herrin)	
Source	Old Ben mine	
	Proximate Analysis (wt.%)	
	As Received	Dry
Moisture	11.12	0.00
Ash	9.70	10.91
Volatile matter	34.99	39.37
Fixed carbon	44.19	49.72
HHV, Btu/lb	11,666	13,126
	Ultimate Analysis (wt.%)	
	As Received	Dry
Carbon	63.75	71.72
Moisture	11.12	0.00
Hydrogen	4.50	5.06
Nitrogen	1.25	1.41
Chlorine	0.29	0.33
Sulfur	2.51	2.82
Ash	9.70	10.91
Oxygen[b]	6.88	7.75
Total	100.00	100.00

Source: Ref. (6).
[a]The above proximate analysis assumes that sulfur is volatile matter.
[b]By difference.

Ash fuses at these conditions, and thus molten ash flows out from the bottom of the gasifier. The molten ash is then water quenched forming an inert vitreous slag.

Coal fed to the gasifiers is an Illinois No. 6 bituminous coal with properties shown in Table 10.2.[6]

Raw syngas to the fire-tube syngas cooler produces high-pressure steam for the steam turbine, which is part of the heat recovery system. Then the raw syngas enters a scrubber at 700°F for removing particulates and trace components. Next, the syngas stream is reheated and sent to a hydrolysis reactor where carbonyl sulfide (COS) and hydrogen cyanide (HCN) are converted to hydrogen sulfide (H_2S).

Then, the syngas enters a mercury removal system, Eastman Chemical, consisting of a packed bed of sulfur-impregnated activated carbon that removes mercury, arsenic, and trace materials. Following the mercury removal system, a selexol unit removes H_2S and CO_2 from the cool syngas. The polishing unit of zinc oxide (ZnO) effectively removes any residual H_2S and achieves high purity of the syngas stream. This high-purity gaseous stream is shown as clean syngas at this stage of the process prior to entering the F-T synthesis unit.

The clean syngas is sent to the F-T reactors to produce hydrocarbon liquids. The F-T systems distillation unit separates the liquid hydrocarbon products into basically naphtha, distillate, and wax fractions with the light components, fuel gas (FG), being

TABLE 10.3 Compositions of Raw Syngas Stream and Clean Syngas Stream

Component	Raw Syngas (Mole Fraction)	Clean Syngas (Mole Fraction)
Ar	0.0100	0.0136
CH$_4$	0.0279	0.0380
CO	0.4040	0.5495
CO$_2$	0.1387	0.0073
COS	0.0005	0.0000
H$_2$	0.2773	0.3771
H$_2$O	0.1193	0.0001
H$_2$S	0.0080	0.0000
N$_2$	0.0120	0.0144
NH$_3$	0.0023	0.0000
O$_2$	0.0000	0.0000
SO$_2$	0.0000	0.0000
Total	1.0000	1.0000

Source: Ref. (6).

used to produce electricity in gas turbines, GE 6FA units. Energy produced in the F-T system is effectively recovered and utilized in the process with a net excess of 124.3 MW$_e$ for export to the grid.

Compositions of raw syngas downstream of the scrubber and clean syngas, as illustrated in Fig. 10.3, are shown in Table 10.3. The clean syngas feed to the F-T system mass rate is 2,510,720 lb/hour.[6] The F-T process converts the clean syngas to hydrocarbon liquids using an iron-based F-T catalyst that also promotes the water–gas shift reaction (Eq. (10.5)) for H$_2$ required in the F-T synthesis reaction (Eq. (10.11)).

$$nCO + 2nH_2 = C_nH_{2n} + nH_2O \qquad (10.11)$$

The F-T synthesis is carried out in F-T slurry-bed synthesis reactors. The slurry-bed F-T synthesis reactors convert the CO and H$_2$ into hydrocarbons, CO$_2$, and water.

Naphtha fraction from the distillation unit is further processed in a catalytic hydrotreating unit for final conversion to naphtha products. The distillate and wax fractions are converted in the catalytic hydrotreating and catalytic hydrocracking units, respectively. Then, these two streams are combined and further processed into a marketable diesel fuel. The overall compounds in the naphtha and distillate pools shown in Fig. 10.3 are shown in Tables 10.4 and 10.5, respectively.

The F-T diesel is a high-value product, since it is free of sulfur, nitrogen, and aromatics.

As a final comment on this process for CTL, the environmental control equipment was based upon best available control technology (BACT) guidelines. The 2006 New Source Performance Standards (NSPS) regulations are shown in Table 10.6.[6] BACT

TABLE 10.4   Naphtha Components

Naphtha Products	Product Distribution (Liquid Volume)
C5–C6 (paraffins)	38%
C7 + to 300°F boiling point	48%
300 to 350°F boiling point	14%

Source: Ref. (6).
Note: Production: 22,173 Bbl/day Naphtha Production.

TABLE 10.5   Diesel Components

Diesel Products	Product Distribution (Liquid Volume)
300–500°F boiling point	42%
500 + °F boiling point	58%

Source: Ref. (6).
Note: Production: 27,819 Bbl/day Diesel Production.

TABLE 10.6   Current Regulations of 2006 New Source Performance Standards (NSPS)

	Emission Limit	% Reduction
PM	$0.015 \, lb/10^6 \, Btu$	99.9%
SO_2	1.4 lb/MWh	95%
NO_x	1.0 lb/MWh	N/A

Source: Ref. (6).

TABLE 10.7   Gasification Technologies and BACT Guidelines

Pollutant	Control Technology	Limit
Sulfur	Selexol/Econamine Plus/ Sulfinol-M + Claus plant	99 + % or less than or equal to $0.050 \, lb/10^6 \, Btu$
NO_x	Low-NO_x burners and dilution	15 ppmvd (@ 15% O_2)
PM	Cyclone/barrier filter/wet scrubber/AGR absorber	$0.006 \, lb/10^6 \, Btu$
Hg	Activated carbon bed	95% removal

Source: Ref. (6).

technologies have specific emission limits and corresponding control equipment as shown in Table 10.7.[6]

The BACT technologies of this process study meet the emission requirement of the 2006 New Source Performance Standards (NSPS).

PRODUCTION OF METHANOL (CH₃OH) FROM SYNGAS VIA CHEMICAL SYNTHESIS ROUTE

Methanol is a clear biodegradable liquid. Methanol is used in many industrial and consumer products such as synthetic textiles, recyclable plastics, household paints, adhesives, foam cushions, and pillows and medicines. The market is expanding for the use of methanol in biodiesel production. Other uses of methanol are in the manufacture of: silicones, refrigerants, specialty plastics and coatings, methanol-based windshield antifreeze, formaldehyde, and acrylic plastic. Global methanol market is sometimes tight between supply and demand. Methanol is a commodity chemical and one of the top ten chemicals produced globally. Consumption for year 2008 was 44,763,000 metric tonne, and supply was 45,433,000 metric tonne.[7,8,9] The majority of methanol is currently synthesized from syngas. Catalytic methanol synthesis from syngas is produced via a high-temperature, high-pressure, exothermic, and equilibrium-limited reaction for conversion efficiency of 99%. Removing the excess heat of reaction and thermodynamic limitations are challenges in the commercial synthesis of methanol.[4]

The production of methanol will be discussed here by starting with the necessary syngas (CO and H_2) without elaborating on the technology of generating the syngas. Methanol is produced by the chemical reactions in Equations (10.12 and 10.13).[10]

$$CO + 2H_2 \rightleftharpoons CH_3OH \qquad (10.12)$$

$$CO_2 + 3H_2 \rightleftharpoons CH_3OH + H_2O \qquad (10.13)$$

Syngas is desirable at a pressure of 50–100 bar over a copper catalyst and with a composition of 3 mole% CO_2, a stoichiometric ratio (SR) of $2.03 = (H_2 - CO_2)/(CO - CO_2)$, H_2S at <0.1 ppm, and inerts plus CH_4 at a minimum.

By starting with the clean syngas, the process for the manufacture of methanol (CH_3OH) can be basically understood by the block flow diagram in Fig. 10.4. The proper SR of gaseous feed to the methanol synthesis systems is created by mixing a bypass of syngas with a portion of syngas modified by the water–gas shift reaction (Eq. (10.5)). After the water–gas shift conversion, CO_2 is first removed by the CO_2

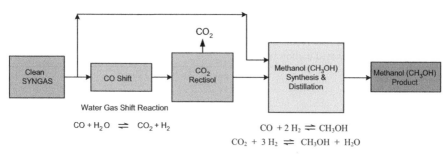

FIGURE 10.4   Simplified Block Flow Diagram (BFD) of Syngas to Methanol (CH_3OH).[11]

TABLE 10.8   Gaseous Feedstock for Methanol Synthesis System

Feedstock Component	Mole%[11]	Mole%[10]
CO_2	3.52	3.50
CO	27.86	27.59
H_2	67.97	67.24
CH_4	0.21	0.86
Inerts	0.44	—
N_2	—	0.33
Ar	—	0.48
$(H_2-CO_2)/(CO+CO_2)$	2.05	2.05

Source: Refs. (10,11).

Rectisol process. The end result is a mixture of bypassed syngas with water–gas shift conversion with CO_2 removal gas which has a SR of about 2.03. This feed gaseous mixture is then fed to the methanol synthesis system for the production of methanol via Equations (10.12 and 10.13). The final product methanol (CH_3OH) is recovered by distillation.[11]

An example of gaseous feedstock for the methanol synthesis system is shown in Table 10.8.[10]

Further details for the methanol synthesis process are illustrated in the simplified process flow diagram for conversion of clean syngas to methanol (Fig. 10.5). A sidestream of clean syngas is converted to proper H_2 and CO by a Lurgi CO shift process followed by CO_2 removal using a CO_2 Rectisol process. A bypass stream of clean syngas is then mixed with the gaseous stream exiting the Rectisol process to form a gaseous feedstock mixture with the proper composition, i.e., SR. The feedstock is converted into methanol (CH_3OH) in the methanol synthesis reactor. The methanol synthesis process is followed by distillation step to produce a methanol product for the marketplace.[11]

The CO shift conversion unit reacts CO with steam (H_2O) to form H_2 and CO_2 by water–gas shift reaction (Eq. (10.5)). The water–gas shift reaction is exothermic and supplies an appreciable amount of the thermodynamic steam requirements for the reaction. A two-stage reactor system, Lurgi CO Shift Process, is shown using a conventional shift catalyst. A widely used commercial isothermal methanol converter is the Lurgi methanol converter. It is a shell and tube design and the tubes contain a proprietary Lurgi methanol catalyst ($Cu/ZnO/Cr_2O_3$ + promoters) surrounded by boiling water. Varying the pressure of the boiling water controls the reactor temperature. By-product steam is produced at 40–50 bar and can be used to run the compressor or provide heat for the distillation process. These units can operate at 50–100 bar and 230–265 °C. Commercial methanol synthesis catalysts have lifetimes of about 3–5 years under normal operating conditions.[4]

The CO_2 rectisol removal system is a physical washing unit consisting of a CO_2 absorber. The CO_2 removal section operates at temperatures around −60 °C.

After mixing the slip-stream clean syngas bypass with the syngas of the water–gas shift conversion followed by CO_2 removal, the feedstock with proper SR is sent to

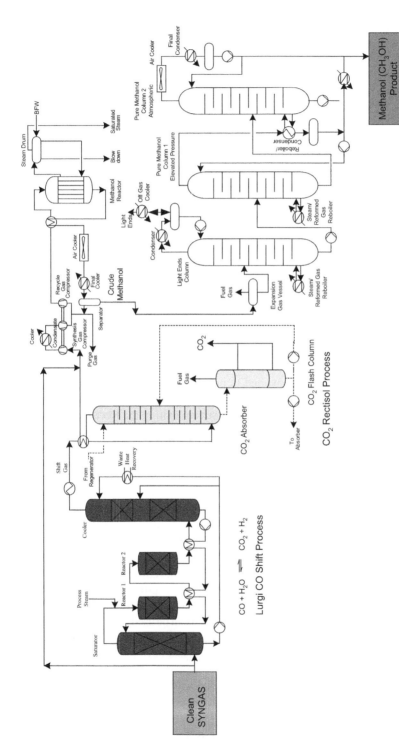

FIGURE 10.5 Simplified Process Flow Diagram for Conversion of Syngas to Methanol (CH₃OH).[11]

a Lurgi low-pressure methanol process. The SR is $(H_2 - CO_2)/(CO + CO_2)$ and lies in the practical process range for the specific plant design. Operating pressure of the methanol synthesis can be between 50 bar and somewhat higher. A tubular reactor can be used for conversion of the carbon oxides, CO and CO_2, and H_2 into methanol via reactions (Eqs. (10.12) and (10.13)).

A steam pressure of around 40 bar corresponds to an exit gas temperature of 250 °C. The unconverted gas with methanol leaves the reactor system and is cooled whereby the methanol (CH_3OH) is condensed as crude methanol liquid. The crude methanol is separated out using a separator and fed to the distillation section. Some gas from the separator is purged and the remainder is fed back into the methanol reactor system.

Crude methanol is processed into finished product in the three-column distillation system. The crude methanol is sent to an expansion gas vessel where a small amount of light components and dissolved gases are flashed off and used as a FG. Low-boiling impurities are removed in the light crude column and methanol purified in the elevated-pressure pure methanol column-1. The final pure methanol column-2 is at atmospheric pressure and purifies the methanol by distillation to the final desired methanol product specifications. Specifications for U.S. Federal Grade AA methanol are methanol purity of >99.85 wt.% and water of <0.10 wt.%.

PRODUCTION OF SYNTHETIC NATURAL GAS (SNG) FROM SYNGAS VIA CHEMICAL SYNTHESIS ROUTE

Periodic or perceived shortages of natural gas (CH_4) have created much interest in gasification technology for use in the production of SNG. One project that is operational is in Beulah, North Dakota and was started by the Dakota Gasification Company. The facility, Great Plains Synfuels Plant, was created from the abundant lignite resources underlying the North Dakota plains. The facility operates and produces more than 54 billion standard cubic feet of natural gas annually. Coal consumption is 6 + million tons per year. SNG leaves the plant through a 2-foot-diameter pipeline to a major pipeline to supply thousands of homes and businesses. Other synfuels are also produced at this plant.[12–16]

SNG consists predominately of CH_4. The SNG is synthesized by reacting syngas (CO and H_2) over a nickel catalyst via Equations (10.8 and 10.9), i.e., $CO + 3H_2 = CH_4 + H_2O$, $-206 MJ/kmol$ and $CO_2 + 4H_2 = CH_4 + 2H_2O$, $-165 MJ/kmol$, respectively. Both reactions are exothermic. Reaction chemistry requires an $SR = (H_2/(3CO + 4CO_2))$ of between 0.98 to 1.03.[10,17]

In Fig. 10.6, the gasification process for synthesis of SNG begins with crude syngas (CO and H_2).[12] Crude syngas stream discharges from the gasifier into a scrubber that scrubs and partially cools the crude syngas. Condensables from the gaseous stream are also removed. The crude gas stream is then divided whereby about half or more goes to the gas cooling unit. The remaining portion of the crude gas stream goes to the shift conversion units where steam is added for the waster gas shift reaction (Eq. (10.5)).

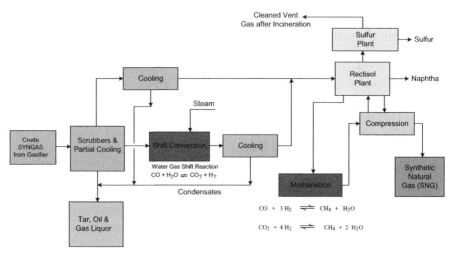

FIGURE 10.6 Simplified Process Block Flow Diagram for Conversion of Syngas (CO and H_2) to Synthetic Natural Gas (SNG).[12]

The waster gas shift reaction is catalytic, exothermic, and sulfur resistant. Some catalyst deterioration occurs but carbon deposition requires a periodic reactivation of the catalyst. The shifted crude gas enters a cooling unit and then is combined with the bypassed crude gas prior to entering the rectisol unit. The combined gaseous stream (crude synthesis gas) has a H_2 to CO ratio of about 3.5, which is acceptable for methanation.

The crude synthesis gas contains sulfur compounds, such as H_2S, COS, and CS_2. It also contains amounts of vapor and gaseous components such as naphtha, NH_3, and HCN. The gas enters the rectisol unit and is very important for preparing the crude synthesis gas for proper purity requirements for the methanation unit. The rectisol unit must remove all sulfur compounds and provide a clean synthesis gas. The rectisol unit is capable of removing impurities in the crude synthesis gas and some CO_2 as required for the nickel catalyst. Crude synthesis gas entering the rectisol unit passes to a cold methanol wash where residual water and naphtha are removed. The naphtha-free gas enters the absorber where H_2S and COS are removed. The heat of absorption is removed by refrigeration and the temperature of the cold methanol wastes can be between $-20°F$ and $-50°F$. Some of the absorbed gases are removed from methanol by multistage flashes in the regenerator and the remainder stripped in the hot regenerator. The off-gas streams are collected and sent to the sulfur recovery unit. Gas leaving the rectisol unit is clean synthesis gas.

Naphtha is recovered from the methanol and water by naphtha extraction and an azeotropic column. Methanol is recovered in a methanol–water distillation column.

From the rectisol unit, clean synthesis gas is fed to the methanator reactor unit. The methanator converts low-BTU synthesis gas (375–430 BTU/cu.ft.) to a CH_4 rich, high-BTU gas of about 980 BTU/cu.ft. by the exothermic catalytic reactions (Eqs. 10.8 and 10.9).

TABLE 10.9   Product Composition and Heating Value of Synthetic Natural Gas (SNG)

Component	Mole% (% by Volume)
H_2	3.00
CH_4	95.95
CO	0.05
CO_2	0.40
N_2 + Ar	0.60
Total	100.00
	Higher Heating Value of the gas: 1027 BTU/SCF

Source: Ref. (12)

The feed gas entering the methanation reactors is passed through a fixed bed containing a pelleted reduced-type nickel catalyst. The temperature rise of the reactors is controlled and maintained by recycling of methanated effluent gas by mixing with feed gas. The reaction heat is removed by waste heat exchangers at the outlet of each reactor. The gas leaving the synthesis loop is passed through a cleanup reactor to completely convert any remaining CO.

Finally, the gas leaving the methanator unit is cooled to condense out water and sent to the gas compression unit. The final moisture is removed from the gas with a glycol dehydration unit following compression. The gas is then SNG and is ready for distribution to a pipeline and final distribution to customers. Composition of the SNG product is shown in Table 10.9.

PRODUCTION OF HYDROGEN (H₂) FROM SYNGAS VIA CHEMICAL SYNTHESIS ROUTE(S)

H_2 is a large market and covers a wide range of industries such as the petroleum, food, chemical, metals, refining, and electronics. About half of the H_2 produced today is from steam reforming of natural gas. Annual production of H_2 in year 2006 by source was: natural gas, 24 million tonnes/year; oil, 15 million tonnes/year; coal, 9 million tonnes/year; and electrolysis, 2 million tonnes/year for a total of 50 million tonnes/year. Nearly half of the H_2 is produced in the United States, China, and Europe, each producing 18%, 16%, and 14%, respectively.[18,19,20]

Current technologies produce H_2 from coal, H_2 from natural gas, and H_2 from photovoltaic (PV)/water electrolysis. Methods considered here would be those starting with syngas (CO and H_2) and presenting a technology either current or advanced for the production of a H_2 product. H_2 production is the largest use of syngas. H_2 is mainly consumed for NH_3 production followed by refining and methanol production.[4]

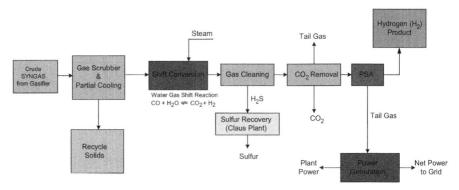

FIGURE 10.7 Simplified Process Block Flow Diagram for Conversion of Syngas (CO and H_2) to H_2.[18,21,22]

In Fig. 10.7, a Texaco quench gasification system produces a crude syngas with conventional scrubbing, cooling, shift conversion, acid gas removal, CO_2 removal, and pressure swing adsorption (PSA) for the H_2 recovery. Crude syngas is produced from gasification of coke in the gasifier. The crude syngas exits the gasifier and proceeds to a gas scrubber and partial cooling. Syngas then proceeds to a shift conversion that is supplied with process steam for the water–gas shift reaction. After conversion, the gaseous stream enters the gas cleaning unit for removal of sulfur compounds for final production of high-purity sulfur in a Claus plant. Next, approximately half of the CO_2 is removed from the gaseous stream in the CO_2 removal unit prior to the PSA unit that generates a tail gas fuel. The PSA system produces a H_2 product with purity exceeding 95%. The PSA unit can obtain H_2 purity greater than 99.99%.[4] In this manner, a commercially proven technology is used effectively to convert coke into H_2, power, and steam.[18,21,22]

Another approach to the production of H_2 is shown in Fig. 10.8. Advanced E-Gas™ gasification technology of ConocoPhillips is used in combination with a ceramic membrane system operating at 600 °C. This membrane technology is capable of shifting and separating H_2 from the syngas. Assumptions have been made that 90 mole% of the syngas is converted to H_2 in the membrane system. The remaining syngas contains mostly CO_2 and some CO and H_2. This remaining syngas is sent to a combustor and combusted with Oxygen (O_2) to form a rich stream of CO_2 that passes to a gas turbine for generation of power for the plant. The hot exhaust gases from the gas turbine proceed to a heat recovery steam generator (HRSG) that generates process steam for the plant. The process generates excess power for sale to the grid.[18,23]

The hot raw syngas from the gasifier is first cleaned of larger particulates in a cyclone and then cooled in a fire-tube boiler to 1,100°F. Then, the syngas is sent to the hot gas sulfur removal unit that removes sulfur compounds from the syngas. After exiting the candle filter, the clean syngas is mixed with steam/water and proceeds to the tube-side H_2 separation device where high-purity H_2 is produced from water–gas shift reaction. Final H_2 product is produced with a purity of >99.5%. A pressure swing adsorption (PSA) unit can obtain H_2 purity greater than 99.99%.[4]

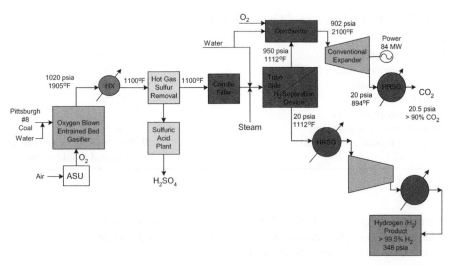

FIGURE 10.8 Simplified Process Block Flow Diagram for Conversion of Coal to Syngas (CO and H_2) to H_2.[18,23]

A more detailed process description is discussed now for more insight into this process to produce a high-purity H_2 product from the syngas produced by a gasifier.

The hot raw syngas leaves the gasifier and is cooled from 1,905°F to 1,100°F by first being cleaned of larger particulates in a cyclone and then cooled in a fire-tube boiler to 1,100°F before entering the hot gas desulfurization (hot gas sulfur removal) and particulate removal (candle filter). The hot gas sulfur removed is converted to sulfuric acid (H_2SO_4).

In the next process step, the desulfurized and particulate removal of syngas at 605°F is carried out by a hydrogen separator device (HSD), which is a high-temperature membrane device in a shell and tube configuration. The membrane is composed of inorganic membrane tubes with pore sizes of specific diameters. The inorganic membrane can be made of Al_2O_3 or other ceramic materials. The membrane acts similar to a molecular sieve that excludes molecules of larger diameters. The syngas with added steam enters the tube side of the HSD, which has gas contact catalytic properties for promotion of the water–gas shift reaction. With excess steam present, the water–gas shift reaction is toward conversion of CO and H_2O to CO_2 and H_2. The H_2 passes through the membrane, which also promotes the water–gas shift reaction to replenish the H_2 passed through the membrane. The H_2 rich gas exits the HSD at 1,112°F and 20 psia. Several stages are arranged to the tube side exit will be a rich stream of CO_2 (retentate) and the shell side H_2 rich gas will be collected together as H_2-rich (permeate).

The CO_2-rich gas (retentate) leaving the HSD is an FG containing about 5% of the fuel value of the inlet syngas stream. The FG goes to a gas turbine combustor and is reacted with oxygen to convert CO and H_2 to CO_2 and H_2O. The hot gas is expanded in a conventional gas turbine expander to produce electric power. The gas exiting the gas

TABLE 10.10   Performance of Coal to Hydrogen Plant

Coal feed	221,631 lb/hr
Oxygen feed (95%)	224,519 lb/hr
Hydrogen product stream	35,903 lb/hr
CO_2 product stream	582,566 lb/hr
Sulfuric acid product	19,482 lb/hr
Gross power production	84 MW
Auxiliary power requirement	77 MW
Net power production (to grid)	7 MW (Note: 8.3% of power generated is sold to grid.)

Source: Ref. (23).

turbine expander is cooled in an HRSG to produce steam for use in the plant as process steam. The CO_2-rich stream is cooled to 100°F, dewatered, and sent off-site.

The CO_2-rich stream is >90% CO_2.

The H_2-rich (retentate) stream leaving the HSD is at 1,112°F and 20 psia. Note that the gas enters the HSD at 605°F and exits at 1,112°F as a result of the exothermic water–gas shift reaction. The hot H_2-rich gas is cooled by an HRSG and compressed to 346 psia for a H_2 product of high purity (>99.5%). Steam is produced in the HRSG and is used as process steam in the plant.

Performance of this process as evaluated is shown in Table 10.10.

The process units are described in detail to get a better understanding of the process equipment involved in this new technology.[23]

Gasifier

The gasifier used in this study was a Destec high-pressure entrained flow gasifier that consists of two stages to gasify a coal–water slurry feed with oxygen. The slurry is prepared by grinding coal to about 200 mesh and mixing it with water to about 65 wt. % solids to 35 wt.% water. Operation in two stages permits an adjustment of flows to the gasifier to achieve a desired outlet temperature of the crude syngas. Typical operating temperature of the Destec gasifier is 1,900°F, which represents a 78/22 flow split between the first and second stages. Slag is formed in the first stage and flows to the bottom where it falls into a water bath and is cooled into a shattered inert material. Gas leaving the gasifier at 1,905°F enters a cyclone that separates entrained particulates from the gaseous stream for recycle to the gasifier. A fire-tube boiler cools the gaseous stream leaving the cyclone to about 1,100°F.

Air Separation Unit (ASU)

The ASU is a conventional cryogenic ASU. The ASU is designed to produce 95% pure O_2. The high-pressure plant is designed with liquefaction and liquid oxygen storage to supply the plant.

Hot Gas Cleanup System

The hot gas cleanup system consists of a desulfurizer reactor with a riser tube, a disengager, and a standpipe for both the absorber section and regeneration section. A sorbent from the absorber passes through the regenerator riser, disengages, and transfers back to the absorber through the standpipe. Regeneration is done with neat air to minimize heat release and limit temperature. The regeneration heat has negligible effect on the sorbent temperature in the absorber. The regeneration off-gas contains predominantly SO_2 and is sent to the sulfuric acid plant. Any particles are separated by high-efficiency cyclones at the top of the absorber. A candle filter downstream removes any remaining particulates.

Sulfuric Acid Plant

This sulfuric acid plant is a double-absorption contact process using an intermediate absorber in a four-pass converter developed by Monsanto. The conversion of SO_2 to SO_3 is an exothermic reaction. The contact plant uses a vanadium catalyst and takes advantage of mass transfer considerations by first allowing the gases to pass over a part of the catalyst and at 800°F and then allowing the temperature to increase as the reaction proceeds. The reaction essentially stops from mass transfer considerations when the temperature reaches 1,100°F. The gas is cooled in a waste heat boiler and then passed over subsequent stages of the catalyst until the gas temperature does not exceed 800°F. The gases pass through about three layers of catalyst and are cooled before passing through an intermediate absorber tower where some SO_3 is removed with 98% H_2SO_4. Gases leaving the absorber are then heated and passed through the remaining layers of catalyst in the converter. The gases are then cooled and pass through the final absorber tower where more than 99.7% of the SO_2 is converted into SO_3 and finally sulfuric acid (H_2SO_4) product.

CO₂-Rich Separated Gas Stream/Conventional Turbine Expander

The FG leaving the HSD (tube side H_2 separation device) has a fuel value of about 15 BTU/scf. A conventional expansion turbine extracts the energy from the gaseous stream and produces power and steam. The FG stream (CO and H_2) is combusted with oxygen (O_2) forming CO_2 and water (H_2O) vapor. Water is injected into the combustor to moderate the stream temperature to about 2,100°F to be compatible with the conventional turbine expander. The turbine expander reduces the gas pressure to 20 psia and temperature to 905°F. Power of 84 MW is produced by the generator. In-plant power requirements require 77 MW, leaving 7 MW for sale to the grid, which amounts to selling about 8.3% of the electricity generated by the plant as saleable electrical energy to the grid. The gas leaving the conventional expander enters a HRSG where it is cooled to 250°F in the process of producing steam for process plant use. The CO_2-rich product leaving the HRSG is cooled to 100°F and dried before sending it off-site.

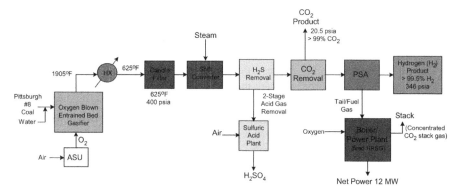

FIGURE 10.9 Block Flow Diagram for Conventional Process from Coal to Syngas to Hydrogen.[23]

In summation, 221,631 lb/hr of coal is gasified to a syngas (CO and H_2) and further processed into marketable products: H_2, sulfuric acid, and net power for export to the grid. H_2 production is 35,903 lb/hr at >99.5% purity. Sulfuric acid production is 19,482 lb/hr. Net power production for export sale to the grid is 7 MW of electricity, which is 8.3% of the total electricity generated by the plant, i.e., a total of 84 MW.

Another conceptual design was done to convert coal to H_2 with CO_2 removal using conventional gas stream processing. This process is illustrated in Fig. 10.9. With E-Gas™ technology using Destec gasifier, coal is gasified and the high-pressure syngas is produced, which exits the gasifier at 1,905°F. The exit temperature from the gasifier is controlled by adjustments in the second stage of the gasifier. The syngas proceeds to the fire-tube heat exchanger and is cooled to 625°F. Next, the ceramic candle filter removes particulates. The following step used a sulfur-tolerant catalyst in the shift converter to promote the water–gas shift reaction, i.e., $CO + H_2O \rightleftharpoons CO_2 + H_2$. The gas is then cleaned of sulfur compounds and CO_2 in a double-stage selexol unit. H_2S from the acid gas removal system is converted into sulfuric acid in the sulfuric acid plant. H_2 is purified by the PSA unit producing a H_2 product with purity of >99.5% and pressure of 346 psia. Tail/FG from the PSA unit is combusted with oxygen in a boiler (fired HRSG) power plant to produce power for the plant and net power for sale to the grid. The stack will also contain a concentrated CO_2 stream as a result of using the tail/FG from the PSA unit as fuel with oxygen in the boiler/power plant. Any steam produced from hot gas cooling and energy recovered by any HRSG is used to produce steam and power for the plant and any excess will be exported for sale.

In summation, this conventional H_2 plant with CO_2 removal produces a high-purity CO_2 stream and a high-purity H_2 stream as shown in Table 10.11.

These processes for the production of H_2 from the gasification of coal and other carbonaceous materials as feedstocks will continue to be developed and refined. The benefits of these processes are the direct result of producing a high-purity product of H_2 and CO_2. Both of these simple materials are some of the key building blocks or raw

TABLE 10.11   Performance of Conventional Process Plant for Coal to Hydrogen

Coal feed (dry basis)	2500 tpd
Plant size	317.8 tons H$_2$/day, 114 MMscfd @346 psia
CO$_2$ product stream	6233 tpd
Net power production (to grid)	12 MW

Source: Ref. (23).

materials in chemistry for the production of industrial and consumer products by various chemical and/or biochemical processes.

PRODUCTION OF ETHANOL (CH$_3$CH$_2$OH) FROM SYNGAS VIA CHEMICAL SYNTHESIS ROUTE

Production of Ethanol and Methanol from Syngas using Fischer–Tropsch Synthesis Process

Waste solids can be gasified to syngas (CO & H$_2$) and then converted to fuels such as ethanol and methanol mixture by F-T synthesis. A process for this option to produce ethanol for use as a fuel is shown in Fig. 10.10. MSW was gasified using a Plasma Arc gasifier to produce a syngas.

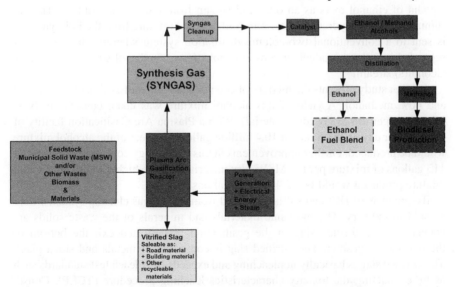

FIGURE 10.10   Selected Synthesis Gas Options for Syngas.[3]

The clean syngas can be used to produce power generation for use in the plant such as electricity and steam while the remainder can be used to produce liquid fuels. These liquid fuels can be ethanol or methanol or a mixture of these alcohols via F-T synthesis.

From a mixture of ethanol and methanol, a distillation system can separate the ethanol for blending with gasoline for use as a motor fuel and the methanol for manufacturing biodiesel.[3]

A similar approach was presented but with water being adsorbed by molecular sieves before the distillation step involves separation of the alcohol mixture into three streams: methanol, ethanol, and a mixture of higher-molecular-weight alcohols. A thermochemical process such as the F-T process determined in a study that first drying the entire mixed alcohol stream before any separation would be preferable. The water was adsorbed by molecular sieves and then desorbed by depressurization and flushing with methanol. The resulting methanol/water mixture from flushing of the molecular sieves is recycled back into the process. The dry mixture of alcohols stream exiting the molecular sieve dehydrator is fed to the first column of a two-column distillation system. The distillation system consists of typical columns using trays, overhead condensers, and reboilers. The dry feed alcohol mixture is separated in the first column, with 99% of the incoming ethanol recovered in the overhead with nearly all the methanol. The bottom stream contains 99% of the incoming propanol, 1% of incoming ethanol, and all of the butanol and pentanol. This bottom stream from the first column is considered a by-product of the plant. The methanol/ethanol overhead stream from the first column is sent to the second distillation column. The ethanol recovered in the second column as bottom stream consists of 99% of the incoming ethanol with a 0.5 mole% methanol concentration (maximum), and therefore the bottom stream meets specifications as an ethanol fuel. The methanol and small amount of ethanol exits as an overhead stream from the second column. Thus in summation, the dehydrated (water removed) alcohol mixture from the F-T synthesis is sent to a conventional two-column distillation system whereby the mixture is separated into three product streams: ethanol stream, methanol stream, and higher alcohol(s) stream.[24,25]

Another study presents the mixture of ethanol and methanol as about 60% ethanol and 40% methanol. A yield of this alcohol mixture was taken conservatively at 95 gallons per ton MSW (dry material). With a Plasma Arc Gasification facility of 500 tons per day of MSW, about 10.4 million gallons per year of the alcohol mixture would be produced. Further improvements in this technology could permit a yield of 113 gallons of mixture per ton MSW (dry material). At the higher yield, the alcohol mixture produced would be 12.4 million gallons per year.[3]

By-products of HCl and sulfur produced from the syngas cleanup system would be sold to industry. The inorganic materials and minerals of the waste solids are converted to a vitrified slag in the gasification process and exit the bottom of the plasma arc gasifier. The vitrified slag is composed of metals and silica glass. The vitrified slag is basically nonleaching and exceeds EPA leach test standards such as by conducting the toxicity characteristics leaching procedure (TCLP). Consequently, the vitrified slag can be sold as a road construction material.[3]

TABLE 10.12   Process Conditions for Mixed Alcohol Synthesis via F-T catalysis

Parameter	Process Condition
Temperature	$\sim 300\,^\circ C$
Pressure	1500–2000 psia
H_2/CO ratio	1.0–1.2
CO_2 concentration	0.0%–7% (mole%)
Sulfur concentration	50–100 ppmv

Source: Ref. (27).

On the subject of producing an alcohol mixture using syngas via the F-T catalysts, the technology of a modified F-T catalyst will be discussed at this time, specifically a molybdenum-disulfide-based (MoS_2) catalyst. This catalyst is by Dow/UCC and is interesting due to its relatively high ethanol selectivity and with some production of linear alcohols. The catalyst uses a high surface area of MoS_2 promoted by alkali metal sales (potassium carbonate) and cobalt (CoS). These aspects of the catalyst promote the shift from hydrocarbons to alcohols. Also, the catalyst can be supported on alumina or activated carbon or be used unsupported. Parameters of the catalyst for the use of the synthesis process are shown in Table 10.12. Maintaining a reactor temperature is important, since the reactions are highly exothermic. The temperature has an impact on the selectivity of the alcohol and product distribution. High pressures typically favor alcohol production. MoS_2 catalysts are efficient F-T catalysts at ambient or low pressures. But, raising the pressure and pH helps to shift the pathway from hydrocarbons towards alcohol production. It has been reported that CO_2 concentration elevated to 30 vol.% can decrease the conversion of CO but not alter the alcohol/hydrocarbon ratio. Thus, recycling unconverted syngas should be viewed with caution so as to avoid a buildup in CO_2 concentrations accumulating to unacceptable value, thereby negatively impacting the yield for the desired alcohol. CO_2 concentrations up to 6.7 vol.% appear acceptable but can decrease somewhat the chain alcohol yield relative to the methanol yield. A CO_2 concentration of 5 mole% is recommended when conditioning the syngas prior to the F-T synthesis reactor. One of the major benefits of the catalyst being discussed is the sulfur tolerance.[25,26]

The overall reaction for alcohol synthesis for the F-T synthesis can be presented as Equation (10.14).

$$nCO + 2nH_2 \rightleftharpoons C_nH_{2n+1}OH + (n-1)H_2O \tag{10.14}$$

Equation (10.14) suggests an optimum H_2/CO ratio of 2.0 but the catalyst has significant water–gas shift activity that generates some H_2 from CO and H_2O by the water–gas shift reaction (Eq. (10.5)).

$$CO + H_2O \rightleftharpoons CO_2 + H_2 \quad \text{(exothermic)} \tag{10.5}$$

This water–gas shift reaction shifts the optimal ratio near 1.0 and shifts more by-product production of CO_2. The typical ratio of H_2/CO used is between 1.0 and 1.2.

TABLE 10.13   Alcohol Distribution of Mixed Alcohol Product

Alcohol	Dow (wt.%)
Methanol	30–70%
Ethanol	34.5%
Propanol	7.7%
Butanol	1.4%
Pentanol +	1.5%
Acetates (C_1 and C_2)	2.5%
Water	2.4%

Source: Ref. (28).

As a final comment, a lifetime for the catalyst is selected as 5 years in some technical process evaluations.[25,27]

A range of product distribution for alcohol mixtures is shown in Table 10.13 produced by F-T catalysis. As a final comment, a lifetime for the catalyst was selected as 5 years in some technical process evaluations and simulations.[25,28]

PRODUCTION OF ETHANOL FROM SYNGAS VIA A BIO-CHEMICAL SYNTHESIS ROUTE

The current route for production of ethanol (CH_3CH_2OH) is via fermentation of sugar and starch crops, such as the corn-to-ethanol fermentation process where the ethanol product is blended as a motor fuel. However, these crops used to produce ethanol generally have high food value and competition between motor fuels and food can eventually drive up the price of food unless other market forces prevail. The process approach discussed here is to start with syngas (CO and H_2) that can be produced from a variety of sources (besides food sources) and produce ethanol from a biochemical synthesis route. The advantage of this process route is that the gasification step to produce syngas can be performed using all kinds of biomass/organic feedstocks. Thus, a synthesis route to ethanol using syngas as a feedstock for a biochemical process route will be discussed as illustrated in Fig. 10.11.

As shown in Fig. 10.11, the clean syngas is sent to a biochemical (fermentation) process for the production of ethanol. One such fermentation process is based upon the bacteria *Clostridium ljungdahlii* where the reactants are CO, H_2, and CO_2 in an aqueous phase. The bacteria culture is anaerobic and dies in the presence of air. Typical pressure ranges from 0.8 to 2 bar and temperature is 35–37 °C (95–98.6°F). The fermentation follows the given reactions (Eqs. 10.15 and 10.16).

$$6CO + 3H_2O \rightleftharpoons C_2H_5OH + 4CO_2 \quad \Delta G = -48.7 \, \text{kcal/mol} \qquad (10.15)$$

$$6H_2 + 2CO_2 \rightleftharpoons C_2H_5OH + 3H_2O \quad \Delta G = +28.7 \, \text{kcal/mol} \qquad (10.16)$$

Selected Synthesis Gas Option: Biochemical

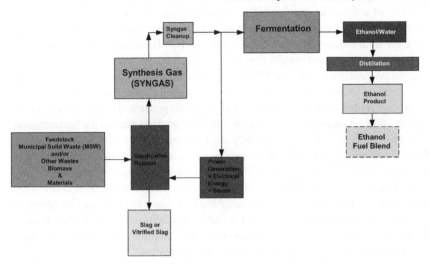

FIGURE 10.11   Process for Production of Ethanol from Syngas via a Biochemical Synthesis Route.

Some production of acetic acid is possible from Equations (10.17 and 10.18).

$$4CO + 2H_2O \rightleftharpoons CH_3COOH + 2CO_2 \qquad \Delta G = -39.2 \, kcal/mol \qquad (10.17)$$

$$4H_2 + 2CO_2 \rightleftharpoons CH_3COOH + 2H_2O \qquad \Delta G = -25.8 \, kcal/mol \qquad (10.18)$$

where ΔG is the free energy.

The free energy, ΔG, indicates whether a reaction goes spontaneously at constant temperature and pressure. A ΔG (negative) indicates a spontaneous reaction is most likely to occur whereas a (positive) ΔG indicates less likelihood. Thus, one can conclude by the free energy value (ΔG) from Equation (10.15) that the most likely preference is for CO as a reactant. A lesser preference is for H_2 as indicated by Equation (10.16). The reactions described by Equations (10.17 and 10.18) indicate that some acetic acid production will likely occur. A 90% conversion into ethanol is reported for CO and 70% for H_2.

The production of ethanol over acetic acid/acetate ions depends on fermentation conditions. The effect of pH on the fermentation process is important and is typically controlled near pH 4.5. The influence of the CO/H_2 ratio is not as important, as the bacteria have a preference for CO. Even though the bacteria prefer CO to H_2, both CO and H_2/CO_2 mixtures can be simultaneously converted to ethanol.

High concentrations of ethanol can be toxic to the bacteria, so the ethanol concentration in the fermenter is kept below 3%. It is reported that the fermentation reactions are very selective with relatively high yields, the products being only ethanol (C_2H_5OH), CO_2, and water. However, acetic acid/acetate ions will also be produced as noted previously.[29,30,31,4,32]

The conversion of syngas to ethanol by this biochemical synthesis route has been proven in a pilot plant in Arkansas for several years on various feedstocks by

Bioengineering Resources, Inc. (BRI, Inc.). This pilot plant operation shows the practicality of combining the flexibility of gasification to produce syngas and then coupling it with a biochemical synthesis route, using fermentation/bacteria, to produce the final ethanol product. Some of the carbon-based feedstocks to the pilot plant for gasification and production of ethanol via the final fermentation and distillation process are as follows: MSW, biosolids, corn stover and other agricultural residues, timber and wood waste forestry materials, used tires or plastics, coal, natural gas, and other hydrocarbons and other waste materials generated in the United States. For gasification, a Consutech gasifier was added to supply syngas to the fermentation reactor. This process technology is ready for commercialization.

The Consutech gasifier comprises two stages where the syngas temperature is elevated to 1,288–1,370 °C (2,350–2,500°F) in the second stage for cracking of any heavy hydrocarbons so that CO and H_2 are mainly produced. The thermal gasification process occurs in an oxygen-starved atmosphere to produce the syngas. The syngas is cooled to 37 °C (99°F) before being sent to the fermentation reactors. The bacteria in the fermenters/reactors are a modified culture of "*C. ljungdahlii.*" Nutrients are added to promote cell growth and regeneration of the biocatalyst.[29,33]

In the process of cooling the hot syngas from the gasifier from about 2,500°F to about 97°F, large quantities of waste heat are recovered as high-temperature steam for the generation of electricity from steam turbines.

The reactions for the biochemical anaerobic fermentation process are Equations (10.17 and 10.18):

$$4CO + 2H_2O \rightleftharpoons CH_3COOH + 2CO_2 \qquad \Delta G = -39.2\,kcal/mol \qquad (10.17)$$

$$4H_2 + 2CO_2 \rightleftharpoons CH_3COOH + 2H_2O \qquad \Delta G = -25.8\,kcal/mol \qquad (10.18)$$

It is stated that this process produces ethanol, water, and H_2. Ethanol concentration in the process is kept to 3% or less, since higher concentration of ethanol inhibits bacteria metabolism. A slip stream of aqueous ethanol is continuously withdrawn from a membrane unit that retains cells for recycle. The aqueous slip stream of ethanol is sent to a conventional distillation system with molecular sieve unit for the production of final ethanol product that can be pure industrial- or fuel-grade ethanol.[29,34] The BRI process is shown in Fig. 10.12.

Water and nutrients are recycled from the distillation system to the fermenter. Fermentation times of a few minutes have been achieved at ambient temperature and pressure. BRI states that the entire process, from the time the waste material is fed into the gasifier to the creation of ethanol, takes less than 7 minutes. BRI reports that one of the great strengths of the process is the rapid biochemical conversion to ethanol.[34]

The BRI process is stated to utilize an enzyme from patented bacteria that consume synthesis gas from the gasifier to discharge ethanol at a yield of 75 gallons/dry ton of biomass and can yield 150 gallons or more per ton from used tires or hydrocarbons.

A module of the BRI process can operate at a capacity of 250–300 tons material per day. A module is composed of two gasifiers with a capacity of about 125 tons/day/each

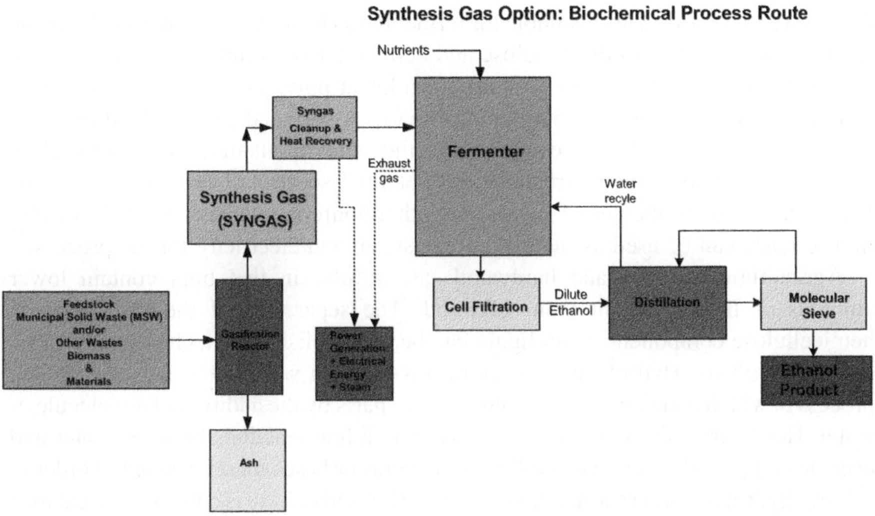

Synthesis Gas Option: Biochemical Process Route

FIGURE 10.12 BFD for Conversion of Syngas to Ethanol via Biochemical BRI Process.[29,34]

gasifier of waste and two fermenters. Each module will process some 85,000 tons of biomass per year and produce 7.0 million gallons of ethanol while generating 5 MW of power. This module can gasify a carbon-based material with a moisture content of less than 40 wt.%. Feedstocks need not be shredded or sorted to remove metal and glass. It is reported that any mixture of plastics, ties, manure, paper or yard wastes, construction debris, furniture, hazardous wastes, crop residues, timer slash, etc. are reported to be converted into syngas and then to ethanol.[34]

BRI reports that a mid-sized BRI, Inc. plant could process 1,000,000 tons of MSW, waste tires and/or biosolids per year and produce 80 million gallons/year of ethanol while generating 50 MW of power where 35 MW is excess for sale to the grid.[34]

PRODUCTION OF ETHANOL VIA A COMBINATION OF CHEMICAL AND BIO-CHEMICAL SYNTHESIS ROUTES USING BIOMASS (CELLULOSIC MATERIAL)

The combination of chemical and biochemical synthesis routes to ethanol does "not" involve the use of syngas but it is presented here to illustrate that this approach is different than the syngas to ethanol route. Much interest has been devoted to this approach so some understanding of the process is instructive for gathering proper perspective of various processes.

Several process alternatives exist for hydrolysis of cellulose materials for producing a feedstock (sugars) that can be further used in fermentation for the final production of ethanol. Lignocellulosic materials such as agricultural, hardwood, and softwood residues are potential sources of sugars. The cellulose and hemicellulose components of the lignocellulosic materials are essentially long molecular chains of sugars. These long chains of sugars are strongly held together by lignin. To make these long chains of sugars available for fermentation to ethanol, it is required that the lignin be separated from the cellulose and hemicellulose components and by doing so produce fermentable sugars. Cellulose is a linear polymer of glucose monomers. Hemicellulose is a branched polymer of primarily xylose and glucose. Hydrolysis of hemicellulose is readily achieved under mild acid or alkaline conditions. Both cellulose and hemicellulose fractions are potential sources of fermentable sugars. Lignin cannot be fermented to sugars, and other nonfermentables such as fermentation residues can be used as fuel to provide steam and electricity for the process.[35]

Agricultural residues and hardwoods are similar in that both contain lower amounts of lignin content than softwood. The separation of the cellulose and hemicellulose components from lignin can be accomplished by a chemical reaction called hydrolysis. Hydrolysis literally means reaction with water. It is a chemical process in which a molecule is cleaved into two parts by the addition of a molecule of water. However, under normal conditions, only a few reactions between water and organic compounds occur. Generally, strong acids or bases must be added in order to achieve hydrolysis where water has no effect. The acid or base is considered a catalyst to promote the hydrolysis reaction to occur under conditions in which the reaction products are fermentable sugars. Since softwood contains more lignin than hardwoods or agricultural residues, the hydrolysis step is more difficult. The hydrolysis step produces a number of sugars including pentose, which is particularly difficult to ferment. In addition, many of the hydrolysis reactions are also promoted by catalysts, called enzymes, which promote many organic reactions selectively. Thus, some processes are more suited to convert these materials to fermentable sugars for ethanol production via fermentation. The suitability of a process depends on the lignocellulosic feedstock material. Therefore, some processes will be discussed that have some advantages for particular feedstocks.[36]

As stated, hydrolysis with an acid can be used to chemically convert lignocellulosic biomass to fermentable sugars. A dilute acid in two steps followed by a third enzymatic step is suitable for softwoods. Straw and hardwood will most likely require just one step of dilute acid hydrolysis. The basic three-step process as mentioned above is shown in Fig. 10.13.[29]

The feedstock for this process of chemical and biochemical route to ethanol can be selected from softwood, hardwood, wood residues from sawmills, cultivated crop residues (corn stover, straw, grasses, etc.), and biomass wastes from agro-industry, forestry, or other origins.

Biomass is composed of cellulose, hemicellulose, and lignin. Cellulose and hemicellulose can be fermented to ethanol after pretreatment and hydrolysis. Ethanol is distilled to the final ethanol product, which can be very close to 100% ethanol when distillation and molecular sieve adsorption are used. The remaining nonfermentable residue can be converted to heat and electricity by thermal methods for use in the production process.[37,38]

Another process for conversion of lignocellulosic materials to ethanol is by Iogen Company. This process includes a steam explosion pretreatment and enzymes pioneered by Iogen as illustrated in Fig. 10.14. These enzymes are proprietary and manufactured by Iogen. The pretreatment step involves steam explosion with dilute

Option: Chemical & Biochemical Process Route to Ethanol

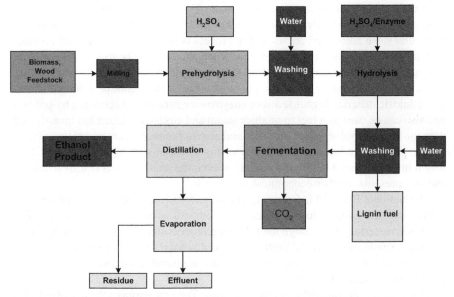

FIGURE 10.13 Simplified Process Diagram of Chemical and Biochemical Route to Ethanol.[29]

Option: Chemical & Biochemical Iogen Route to Ethanol

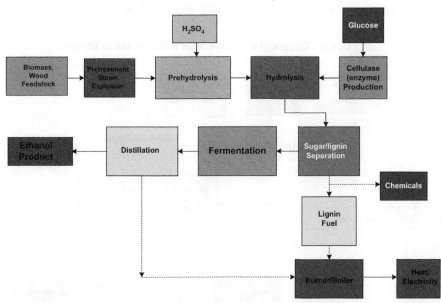

FIGURE 10.14 Process Block Flow Diagram for Conversion of Biomass to Ethanol via Chemical and Biochemical Iogen Route.[29]

acid at elevated temperature and pressure. The hydrolysis and fermentation steps are carried out at ambient temperature and pressure. Distillation is a conventional distillation step used in industry producing nearly 100% ethanol when distillation and molecular sieve adsorption are used. The process is suitable for agricultural residues such as wheat straw and corn stover. Hardwood residues can be used as a feedstock. A single pretreatment step is sufficient to make agricultural and hardwood residues efficiently hydrolyzed by enzymes. For separation of the lignin of softwoods from the cellulosic material, higher levels of enzymes are necessary. Lignin is a by-product and also can be used as a fuel to produce steam and electricity. Iogen has investigated uses for lignin as value-added products and also to produce steam and electricity.

The process produces large quantity of liquid waste with high BOD and COD, which must be treated before disposal. However, the technology is readily available and just presents another economical cost factor.[36]

The National Renewable Energy Laboratory (NREL) has developed a process for conversion of lignocellulosics materials to ethanol as shown in Fig. 10.15. The cellulosic material goes through the following process steps: pre-treatment, hydrolysis, and fermentation. Several pretreatment technologies can be used such as steam explosion and use of dilute acid. Simultaneous saccharification and co-fermentation (SSCF) are conducted in one vessel.[29]

Distillation is accomplished in two columns. The first removes the dissolved CO_2 and most of the water. The second column concentrates the ethanol to a near azeotropic composition. All the water from the near azeotropic mixture is removed by molecular sieve adsorption. Regeneration of the molecular adsorption column requires that an ethanol water mixture be recycled to distillation for recovery. The bottom of the first column contains all the unconverted insoluble and dissolved solids. The insoluble

Option: Chemical & Biochemical NREL Route to Ethanol

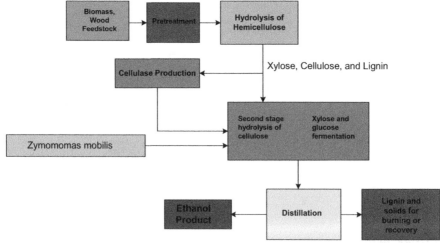

FIGURE 10.15 Process Block Flow Diagram for Conversion of Biomass to Ethanol via Chemical and Biochemical NREL Route.[29]

Option: Chemical & Biochemical BC International Corporation Route to Ethanol

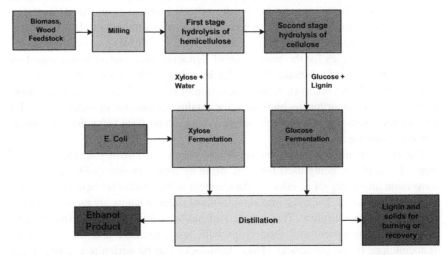

FIGURE 10.16 Process Block Flow Diagram for Conversion of Biomass to Ethanol via Chemical and Biochemical BCI Route.[29]

solids are separated by centrifugation and sent to the burner. The liquid from the centrifuge is concentrated using waste heat from distillation. The concentrated syrup is sent to the burner and the evaporated condensate is used as clean process water.[39]

BC International Corporation (BCI) developed a process for conversion of lignocellulosics materials to ethanol as illustrated in Fig. 10.16. The BCI technology was obtained from the University of Florida, which has an exclusive license for commercialization of this technology. BCI technology can release the sugar potential of cellulose and hemicellulose and produce ethanol by fermentation of both the glucose and nonglucose sugars in the cellulosic biomass. BCI's key technology comes from the genetically engineered strains of *Escherichia coli* bacteria, i.e., labeled KO11. This bacterium has the ethanol production genes of *Zymomonas* spliced into it. This recombinant organism can now ferment both hexose and pentose sugars with high efficiencies. In addition, this organism has been modified to increase its tolerance for ethanol, maintain these traits, increase pH and temperature tolerances, and include its ability to ferment short chains of sugars. Thus, the process can handle agricultural feedstocks and hardwoods that produce many different sugars. These bacteria can ferment nearly all the sugars released from the cellulosic biomass(s) into ethanol with subsequent higher efficiency and commercial feasibility. As shown in Fig. 10.16, xylose from first-stage hydrolysis is fermented separately to ethanol. Glucose from the second-stage hydrolysis stage is fermented to ethanol. These steps comprise two separate fermentation process; however, they use the same organism. The two-stage steps for dilute acid hydrolysis have been known for years and have been improved by BCI. The lignin and solids can be used in a thermal process to produce energy.[29] The BCI process produces lignin, the quantity depending on the feedstock. Current plans are to use lignin as a fuel to generate process heat for the plant. It is clear that BCI's advantage is the ability to ferment pentose and hexose sugars.

Significant amounts of waste are generated during the process. Besides lignin, there are extractables from the feedstock and cell biomass from the fermentation steps. Solids, besides lignin, need to be recovered from a treatment plant for energy values and the remainder need to be treated for BOD and COD prior to discharge.[36]

These technologies for the conversion of lignocellulosic materials to ethanol are presented here to inform/make the reader knowledgeable of such processes even though syngas is not involved in the production of ethanol. Professionals need to be aware of other important technologies for producing ethanol in order to develop a better understanding and possible future synergism of these technologies with the thermal technologies.

Current fermentation routes use sugar and starch from mainly agricultural food crops. These crops, being used for food, bring by nature an outcry for their use for conversion into energy, i.e., ethanol. An example is the current fermentation process from corn to ethanol. The most prevalent forms of sugar in nature are those existing as cellulose and hemicellulose. Thus, much work has been devoted to replacing the sugar and starch from agricultural food crops with less expensive and more abundant lignocellulose-based feedstocks. These feedstocks can be agriculture waste, forest residue, etc. This assessment has propelled the development of technologies for conversion of lignocellulosic materials to ethanol via a combination of chemical and biochemical synthesis routes using feedstocks of biomass (cellulosic materials).[37]

OXOSYNTHESIS (HYDROFORMYLATION): SYNGAS AND OLEFINIC HYDROCARBONS AND CHEMICAL SYNTHESIS

The oxosynthesis process is the hydroformylation of olefins with synthesis gas (CO and H_2). It serves as the route to C_3–C_{15} aldehydes that are then used for conversion to alcohols, acids, and other derivatives. World production of oxoalcohols was nearly 8.1 million tons/year for the year 2002. The oxosynthesis process is also known as hydroformylation.

Hydroformylation is an industrial synthesis route for many olefins. The synthesis route is used to produce solvents, synthetic detergents, flavorings, perfumes, high-value healthcare products, and value-added commodity chemicals. Oxoaldehyde products range from C_3 to C_{15}. The upper range of C_{11}–C_{14} is referred to as the detergent range, since these aldehydes are generally used as intermediates in the synthesis of alcohols with surfactancy appropriate for detergents. Reactants, catalyst, and products are all present in the same phase, typically as a liquid phase. Currently, hydroformylation processes comprise the fourth largest commercial use of syngas.[10,4] The major part of oxosynthesis process involves the hydroformylation of propylene to butyraldehyde, which is then converted to 2-ethylhexanol, which is a plasticizer alcohol used in the manufacturing of flexible PVC. Also, hydroformylation is a key step in the commercial production of vitamin A.

The oxosynthesis process is the reaction of CO and H_2 with an olefinic hydrocarbon to form an isomeric mixture of normal aldehydes and iso-aldehydes. The reaction is exothermic and is thermodynamically favored at ambient pressure and low

TABLE 10.14   Typical Carbon Monoxide (CO) and Oxosynthesis Gas

Component	CO	Oxosyngas
H_2/CO	—	~1
CO_2	—	<0.5 mol%
H_2	<0.1 mol%	~49 mol%
CO	>98.5 mol%	~49 mol%

Source: Ref. (10).

temperature. The reaction proceeds in the presence of homogeneous metal carbonyl catalysts. The "normal" or straight-chain isomer is the desired product according to Equation (10.19).[4]

$$RCH = CH_2 + CO + H_2 \rightleftharpoons RCH_2CH_2CHO \ (normal) + R(CH_3)CHCHO \ (branched)$$
(10.19)

Typically the H_2/CO ratio of 1 : 1 in the syngas mixture is desired for oxosynthesis. The feed gas impurities can significantly reduce the lifetimes of the catalyst. Catalyst poisoning can be from strong acids, HCN, organosulfur, H_2S, COS, O_2, and dienes. A typical feed composition and specifications of CO and oxosynthesis gas are shown in Table 10.14.[10]

The LP Oxo™ low-pressure oxo process was developed by Davy Processs Technology (DPT) and Union Carbide Corporation in 1971. This process is the world's leading process for the production of oxoalcohols from olefins. A simplified schematic of the oxoprocess is shown in Fig. 10.17.[4]

Synthesis Gas Option: LP Oxo Process Route

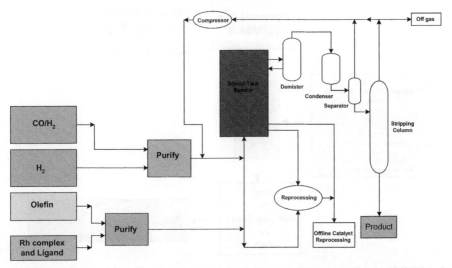

FIGURE 10.17   Process Block Flow Diagram, Syngas and Olefinic Hydrocarbon Oxosynthesis.[4]

The oxoaldehydes are intermediates for the production of alcohols, acids, and other chemical products. Some oxochemicals are *n*-butyraldehyde, *n*-butanol, 2-ethylhexanol, 2-ethylhexanol, 2-ethyl hexanoic acid, isobutyraldehyde, iosbutanol, *n*-propanol, propionic acid, and amyl alcohol.

Process improvements will continue to be developed for this important and valuable oxysynthesis process for the hydroformylation of olefins with synthesis gas (CO and H_2) in the manufacturing of many valuable industrial intermediates and/or high-value consumer and industrial products.

SLAG OR VITRIFIED SLAG OR ASH FROM GASIFICATION REACTOR AND SPECIALTY BY-PRODUCT OPTIONS

In the gasification of feedstocks, the organic materials in the feedstock are gasified to syngas and exit from the top of the gasifier. The inorganic/mineral materials of the feedstock are removed from the bottom of the gasifier as slag, vitrified slag, or ash and are by-products. The process conditions in the gasifier under which these inorganic/mineral materials (by-products) are formed influence the properties of these materials as to whether a slag, vitrified slag, or ash is produced. In addition, these bottom gasifier products can be processed to other by-products that may be of higher value. The objective of a process is to consider by-products as marketable products for reuse and sale in the marketplace. A common comment is to state the by-products are recyclable. Figure 10.18 illustrates simply where the slag, vitrified slag, or ash originates from the gasification process. The solid discharge from the bottom of the gasifier can be viewed as either a solid waste or as a by-product from the gasification

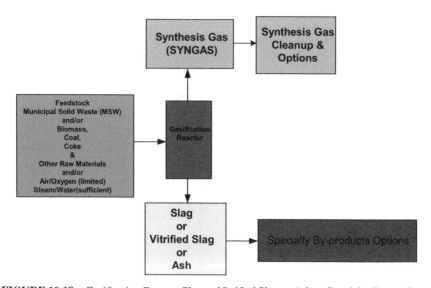

FIGURE 10.18   Gasification Reactor Slag or Vitrified Slag or Ash or Specialty By-products.

process. A positive approach is to use technology in the bottom solids discharge from the gasifier as by-products for productive use and sale in the marketplace. In this manner, the best environmental and most positive economic impact can be realized for companies, the country, and its citizens.

By-product slag or vitrified slag, i.e., rocklike by-product, will be discussed as to what use or value-added by-products can be made from this by-product (Specialty By-product Options), produced from the gasification of MSW. Some details will be discussed on the utilization of the rocklike by-product such as but not limited to metals, rock wool, floor tiles, roof tiles, insulation, landscaping blocks, road aggregate, or road material.[1]

Before looking for potential and/or applications for the by-products from gasification, properties of these by-products, slag, vitrified slag, and ash, and conditions in which such materials are produced, will be discussed. For example, if the gasifier is operated above the ash fusion temperature, it unequivocally shows that the solids discharged from the bottom of the gasifier are highly nonleachable. This fact has been shown by leachability data on the vitrified slag obtained from a Plasma Arc gasifier as shown in Table 10.15. The data is obtained by conducting the TCLP upon the vitrified slag. The TCLP is designed to simulate the leaching that a waste will undergo if disposed of in a sanitary landfill. The results are then compared with the standards set by constituents listed in 40 CFR §268.41 for land disposal restrictions. If the TCLP extract contains any one of the toxicity characteristic constituents in an amount equal to or exceeding the concentrations specified in 40 CFR §261.24, the waste possesses the characteristic of toxicity and is deemed a hazardous waste. Applicable TCLP standard limits are RCRA standards. For the data in Table 10.15, the Plasma Arc gasifier conditions under which the "vitrified slag" was produced by the gasification of MSW would be around a temperature range of 1,208–1,530 °C (2,206°F–2,876°F). Note, the toxicity concentrations as measured by the TCLP method on the vitrified slag produced by Plasma Arc gasification of MSW are all well below the RCRA permissible limits for the MSW sample analyzed.[40]

TABLE 10.15   TCLP Data on Vitrified Slag from Plasma Arc Gasification of Municipal Solid Waste (MSW)

Constituent	RCRA Limit (Permissible) (mg/l)	TCLP Measured Concentration (mg/l)
Arsenic	5.0	<0.1
Barium	100.0	0.47
Cadmium	1.0	<0.1
Chromium	5.0	<0.1
Lead	5.0	<0.1
Mercury	0.2	<0.1
Selenium	1.0	<0.1
Silver	5.0	<0.1

Source: Ref. (40).
Note: RCRA, Resource Conservation and Recovery Act; TCLP, Toxicity Characteristic Leaching Procedure.

Another interesting study adds further credibility for having the proper gasification conditions in the gasifier for the production of a slag that is nonleachable. Partially toxic ashes and fly ash from a typical coal-fired power plant were mixed together and processed in a plasma torch reactor. At the high temperature of the plasma torch reactor, the mixture of toxic ashes and fly ash was processed. The final product (vitrified material) obtained from the plasma torch processing is environmentally acceptable. The vitrified product properties included hardness and leaching and both test results were acceptable. When toxic ash, filter deposits, and residues are heated to sufficient temperatures, their elements (including minerals and toxic heavy metals) melt into a vitrified material. High temperatures above 1,700°K (1,427 °C or 2,600°F) are required for even partial vitrification and this temperature is not available in most incinerators. However, these high temperatures are easily reached by plasma arc furnaces (plasma torch reactor) or gasifiers. After vitrification, the mineral product is chemically stable and mechanically resistant. The product looks like a glassy basalt structured lava and is mechanically harder than basalt. The melting point of the mixture in the plasma arc reactor was between 1,550 and 1,600 °C. The bottom ash from a medical incinerator and the fly ash from a typical coal-fired power plant were mixed to form a 50/50 mixture, which was fed to a Plasma Arc reactor. Results from the experimental work are shown in Table 10.16. The Vicker's hardness varied between 480 and 520 depending on the mixture ratios of the bottom ash and fly ash. The conclusion from the tests was that the vitrified slag material is environmentally acceptable and can return to the environment as an aggregate in the construction industry.[41]

A study was conducted in Alaska on the use of Plasma Arc Gasification for the processing of MSW, construction and demolition debris, industrial waste, hazardous waste, and wood waste. Plasma torch systems were noted in the investigation to have significant advantages. In this study, Plasma Arc Gasifier of solid waste was presented as having environmental superiority with TCLP compositions as presented in Table 10.17 for a typical profile of the slag/vitrified glass from the gasifier. It is stated that the extremely tight physical and chemical bonds within the slag result in consistently low leachate characteristics.[42]

TABLE 10.16   Bottom Ash and Fly Ash Processed to Vitrified Slag by Plasma Arc Reactor

Metal Ion	[Concentration in Aqueous Extract (mg/l)]		
	Bottom Ash (mg/l)	Fly Ash (mg/l)	Vitrified Slag (mg/l)
Pb^{2+}	0.80	0.50	<0.005
Cu^{2+}	0.63	0.55	0.010
Ni^{2+}	0.84	0.28	<0.005
Zn^{2+}	0.52	0.66	<0.005
Cr^{3+}	0.05	2.26	<0.005
Mn^{2+}	0.42	0.78	<0.005

Source: Ref. (41).

TABLE 10.17   Typical Profile for Slag/Vitrified Glass, Plasma Arc Gasification of Solid Waste

Constituent	RCRA Limit (Permissible) (mg/l)	TCLP Composition (mg/l)
Lead	5.0	<0.001
Cadmium	1.0	<0.001
Mercury	0.2	<0.00005
Chromium	5.0	<0.002
Arsenic	5.0	<0.001

Source: Ref. (42).
Note: RCRA, Resource Conservation and Recovery Act; TCLP, Toxicity Characteristic Leaching Procedure.

Laboratory analysis of slag from the Wabash River gasifier on coal gasification (E-Gas™ gasification process) determined the slag to be nonleachable, nonhazardous material with regard to inorganic materials. Since this slag is in the vitrified state, it rarely fails the TCLP protocols for metals. The E-Gas™ coal gasifier is a slurry-feed, pressurized, upflow, entrained slagging gasifier with a two-stage operation. The gasifier operates at temperatures around 2,400–2,600°F (1,315–1,427 °C). Various feedstocks such as lignite, subbituminous coal, bituminous coal, and petroleum coke, have consistently produced a vitrified slag when processed by the E-Gas™ gasification process, which is deemed nonhazardous based upon TCLP test results. Consequently, this vitrified slag can be suitable for metal recovery. Also, the hardness of the vitrified slag makes it suitable as an abrasive. Other applications would be as a roadbed material and/or an aggregate in concrete formulations. Typical test results on the vitrified slag are shown in Table 10.18.[43]

TABLE 10.18   Vitrified Slag from E-Gas™ Gasifier, Test Results

Component	RCRA Limit (mg/l)	Leachate Concentration From TCLP Test (mg/l)
Arsenic	5.0	<0.06
Barium	100.0	0.32
Cadmium	1.0	<0.002
Chromium	5.0	<0.005
Lead	5.0	<0.08
Mercury	0.2	<0.0004
Selenium	1.0	<0.08
Silver	5.0	<0.002

Source: Ref. (43).
Note: RCRA, Resource Conservation and Recovery Act; TCLP, Toxicity Characteristic Leaching Procedure.

TABLE 10.19 Vitrified Slag from Lurgi Slagging Gasifier, Test Results

Component	RCRA Limit (mg/l)	Leachate Concentration from TCLP Test (mg/l)
Arsenic	5.0	0.00027
Barium	100.0	<0.2
Cadmium	1.0	0.000054
Chromium	5.0	0.0016
Lead	5.0	<0.0003
Mercury	0.2	0.00064
Selenium	1.0	<0.005
Silver	5.0	<0.0003

Source: Ref. (44).
Note: RCRA, Resource Conservation and Recovery Act; TCLP, Toxicity Characteristic Leaching Procedure.

Pittsburg #8 coal was processed on a Lurgi slagging gasifier and the slag from the gasifier had the leachate concentration from TCLP test shown in Table 10.19. The fused and solidified glassy particles are very inert to interaction with water.[44] One would expect the slagging gasifier operating temperatures of about 1,500–1,000 °C (2,732–3,632°F).

VITRIFIED SLAG, SLAG, AND ASHES: RESEARCH AND DEVELOPMENT (R&D), MARKETING, AND SALES

Process for Resolving Problems with Ashes

A process that has merit for resolving issues caused by other processes generating ashes is a process for plasma vitrification of incinerator ashes. A process was developed using thermal plasma as a solution for the treatment of fly ash and grate ash generated by the incineration of MSW. This technology has been proven in Japan where several commercial facilities have been installed. Incinerator ash is typically a hazardous material because incineration concentrates many of the heavy metal pollutants in the ash such as lead, zinc, and cadmium. In addition, the fly ash from the air pollution control devices can contain persistent organic pollutants (POPs) such as dioxins and furans. The plasma process destroys the POPs and reduces the ash to a vitreous slag that also captures a significant amount of heavy metal pollutants. The solid product is a vitreous slag that is highly resistant to leaching and can be used as a road building aggregate. The vitrified product is mechanically strong and environmentally stable with low leachability characteristics. It is reported that Japan uses about 75% of the vitrified product as a road construction material. In addition, a process has been developed to produce high-value glass ceramic tile products that have potential to be used as a quality building material such as granite and marble.[45,46]

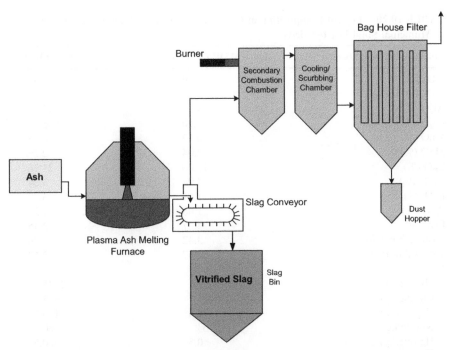

FIGURE 10.19 Simplified Process Flow Diagram for Plasma Vitrification of Ashes.[46]

A simplified block flow diagram for the plasma vitrification process is shown in Fig. 10.19. The Plasma Arc melting furnace is a cylindrical unit of welded construction and lined with high-quality refractory. Feed ash is fed to the plasma arc melting furnace and the temperature is maintained to melt the material at about 1,500 °C (2,732°F). The ash is melted and the molten slag material continuously overflows to slag conveyor where it is water granulated or cast.

The plasma arc reactor takes incinerator ashes and gasifies (destroys) the organic constituents and fuses the inorganic constituents to a vitrified slag. Information is available on the results obtained by processing MSW ash and sewage sludge waste (SSW) ash. Currently, over 100 million tonnes of MSW and 4 million tonnes of SSW are burnt (incinerated) annually worldwide. The ashes produced from the burning (incineration, mass burn) are low bulk density and contain leachable toxic heavy metals and persistent organics. Much of these ashes are landfilled.

An experimental study was undertaken to take the grate material and fly ash from an incinerator and process these materials in the plasma arc reactor. The reactor was configured to use either hollow graphite electrodes or plasma torches. The composition of feedstocks is shown in Table 10.20.[45]

All of the slags (vitrified slags) were produced in the plasma arc reactor at a temperature range of 1,350–1,600 °C (2,462–2,912°F). Composition of the vitrified slags produced from the feedstocks is shown in Table 10.21.[45]

TABLE 10.20 Typical Composition of Grate Ash and Fly Ash from Incineration of Municipal Solid Waste (MSW)

Constituent (Bulk Oxides)	MSW Grate Ash	MSW Fly Ash
SiO_2 (%)	47.34	25.81
TiO_2 (%)	0.93	1.83
Al_2O_3 (%)	9.38	13.43
Fe_2O_3 (%)	11.70	3.17
MnO (%)	0.12	0.14
MgO (%)	2.04	2.25
CaO (%)	13.00	21.76
Na_2O (%)	4.94	3.04
K_2O (%)	1.61	3.08
P_2O_5 (%)	1.08	1.50
LOI (%)	4.27	12.45
Total carbon (%)	3.42	2.30
Total sulfur (%)	0.57	2.62
Metals		
Pb (mg/kg)	2451	2945
Cu (mg/kg)	5613	675
Cd (mg/kg)	19.4	175
Ni (mg/kg)	163	60
Hg (mg/kg)	<0.5	<0.5
As (mg/kg)	41	80
Mn (mg/kg)	928	914
Zn (mg/kg)	4029	7129

Source: Ref. (45).

TABLE 10.21 Composition of Vitrified Slags Produced by Plasma Arc Reactor

Constituent	MSW Grate Ash (wt.%)	MSW Fly Ash (wt.%)	SSW Grate (wt.%)
SiO_2	48.72	38.17	48.77
TiO_2	1.01	2.69	0.94
Al_2O_3	13.85	18.65	16.36
Fe_2O_3	10.50	3.47	11.46
MnO	0.14	0.18	0.16
MgO	2.15	3.18	2.42
CaO	13.13	29.00	8.63
Na_2O	4.71	1.67	0.94
K_2O	1.47	0.77	1.71
P_2O_5	0.80	1.62	8.34
C	2.95	0.25	0.08
Cu	0.24	0.13	0.06
Zn	0.16	0.18	0.11
Ni	0.01	0.02	0.00
Pb	0.14	0.00	0.02

Source: Ref. (45).

TABLE 10.22   TCLP Leachate Tests from Untreated MSW fly ash, Vitrified Slag from MSW Grate Ash, MSW fly Ash, and SSE Grate Ash

Metal	TCLP Limit	Untreated MSW fly Ash (mg/l)	Vitrified Slag MSW Grate Ash (mg/l)	Vitrified Slag MSW fly Ash (mg/l)	Vitrified Slag SSW Grate Ash (mg/l)
Cd	1	0.1–1.3	<0.05	<0.05	<0.05
Cr	5	0.8	<0.05	<0.05–0.13	<0.05
As	5	0.15	<0.05	<0.05	<0.05
Pb	5	1.18–11.17	1.09–2.0	<0.05	<0.05
Cu	NLS	0.11	0.18–0.83	<0.05–0.26	0.37–0.56

Source: Ref. (45).
Note: NLS, no limit set.

Leachate, TCLP, tests were conducted on the vitrified slags produced from the feed materials, MSW grate ash, MSW fly ash, and SSW grate ash, created from an incineration operation. The vitrified slags were produced from the plasma arc reactor and the tests are shown in Table 10.22. Note, the TCLP tests shown are on untreated MSW fly ash, vitrified slag from MSW grate ash, vitrified MSW fly ash, and vitrified SSW grate ash.[45]

All leachate tests on the vitrified slags by TCLP method are well below the legislation limits/requirements (TCLP limit) as shown in Table 10.22. Note, the glaring difference in leachate test results comparing the untreated MSW fly ash and the vitrified slag of the MSW fly ash. The vitrified slag of MSW fly ash was processed in the plasma arc reactor.

From a commercial standpoint, the economics of the plasma arc vitrification process can be substantially improved if the vitrified slag can be used as decorative building tiles or an insulating foam. Simple cooling with air or water of the molten vitrified slag allows the product to be used as a low-cost road building aggregate. By contrast, if the slag product is modified in the process so as to produce a high-quality crystallized glass, then the properties of a crystallized glass can be compared with common high-grade construction materials as shown in Table 10.23.[45] Further process development needs to be done in using a vitrified slag from plasma arc processes for specialty high-value products for the marketplace.

TABLE 10.23   Physical Properties of a Potential High-Value Product: Recrystallized Slag

Property	Recrystallized Slag	Marble	Granite
Bend strength (kg/cm^2)	500	100	150
Bulk density (g/cm^3)	3	2.7	2.7
Mobs hardness	6–7	3	5–6
Acid resistance (%)	0.1	10.3	1
Thermal expansion coefficient (MK^{-1})	6.7	8	8.3

Source: Ref. (45).

Production of Road Material from Slag and Vitrified Slag

The following discussion will be developed for an understanding of the blast furnace slag process for producing granulated slag product currently used for a road material. Understanding of the granulated slag produced from blast furnace slag will possibly lead to insights on how one can potentially produce some form of granulated slag product from slag/vitrified slag originating from gasification plants.

In the United Kingdom (UK), the granulated slag-product is produced using the slag from blast furnace operation, i.e., production of molten iron in a blast furnace converting iron oxide feed materials at high temperatures. During the smelting operation, a slag is formed from the nonferrous components (gangue) of the iron oxide feed and the ash from the carbon-based fuels. One of the ingredients for feeding the blast furnace is metallurgical coke. In one zone of the blast furnace, the reduced iron-bearing materials start to melt. The iron starts to separate from the nonferrous components whereby these components drip through the metallurgical coke. Further, the nonferrous liquid absorbs metalloids from the percolating iron droplets. The iron settles through the slag layer and both liquids accumulate in the hearth. Both liquids are removed from the hearth by periodically running out a taphole and separated.

The slag is either run into large pits near the furnace for air cooling into aggregate product or forced quenching system such as a slag pelletizer or slag granulator.[47]

A slag granulation process is shown in Fig. 10.20. The molten slag from the blast furnace is at a temperature range 1,300–1,500 °C. The high water jet pressure and the thermal shock fragment the slag into granulated particles. The rapid cooling hinders the formation of crystals and a glassy material is produced. The granulated slag and water form a slurry and enter a dewatering drum. Dewatered granulated slag is discharged from the drum and onto a conveyor and into a stockpile. Final moisture content of the granulated slag-product will be below 8%. The stockpiled material (granulated slag-product) can be transferred to a cement mill for the production of

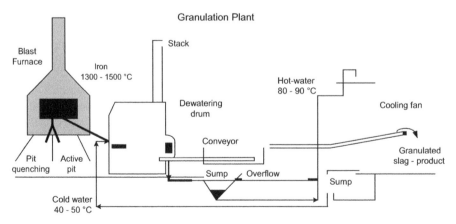

FIGURE 10.20 Process Flow Diagram for Production of Granulated Slag from Blast Furnace Slag.[47]

TABLE 10.24   Composition of Granulated
Slag

Constituent	% Mass (Dry Material)
SiO_2	27–39
Al_2O_3	8–20
CaO	38–50
MgO	<10

Source: Ref. (47).

slag cement or used in the production of slag bound mixtures (SBM) for road construction.[47]

The granulated slag (vitrified slag) has the appearance of sand but instead is glassy and fibrous. Any reactive properties of the granulated slag depend on such factors as temperature of slag before granulation, chemical composition, conditions during granulation such as flow rates and temperatures, and fines content. A typical composition of the granulated slag is shown in Table 10.24.

The authors state that granulated slag has been used as a hydraulic binder for road construction in European countries. Mixtures of granulated slag and limestone have been investigated. Also, SBM has been developed as a construction material. According to the authors, it is estimated that 65% of the French roads have a pavement layer of SBM.[47]

The process for producing granulated slag from blast furnace operations is interesting because there are similarities between the granulated slag produced and vitrified slag produced from gasification processes. In addition, vitrified slag composition from gasification processes, in several respects, is much like that of granulated slag composition from blast furnace plants. Further studies into the similarities and differences of these different slag process conditions and compositions would likely reveal and produce some scientific breakthroughs.

Production and Uses of Rock Wool, Stone Wool, and Mineral Wool

Rockwool is an inorganic mineral product that was first discovered in the islands of Hawaii. In 1840, American physicists exploring Hawaii discovered stonewool formed naturally during a volcanic eruption. Rockwool formations were created naturally as by-products of volcanic activity. Today, rockwool is typically made from rock or a combination of rock, limestone, and coke. The components of rockwool are melted at temperatures above 2,500°F. The molten solution is poured onto a spinning cylinder. As the molten solution flies off the cylinder due to centrifugal forces, it elongates and cools to form fibers. The resultant fibers (rockwool) are then pressed into sheets, cubes, blocks, or granulated. Granulated rockwool can be used in soil mixtures. Rockwool is inert, sterile, porous, and nondegradable.

FIGURE 10.21   Horticultural Rockwool Used For Hydroponic Growing.[48]

Horticultural rockwool has been used for hydroponic growing in Europe for years and is used to produce many greenhouse-grown vegetables. Rockwool's fibrous structure contains a high percentage of air space, which is about 20% even when wet. Since rockwool cannot bind nutrients and water, all nutrients and water in the rockwool are available to the plant. Since rockwool is inert, the irrigation water is used to control the pH. Pictures of horticultural rockwool used for hydroponic growing are shown in Fig. 10.21.[48]

In a hydroponic system, a substrate growing system is where the root zone is physically supported by media. Plants are fed by applying nutrients in an aqueous solution to the media. A substrate is considered as anything that supports the root structure except soil. The reason for using rockwool in a hydroponic growing system is to provide a nutrient solution in the root zone and maintaining proper volume of air (oxygen) in contact with the roots. The key is that rockwool as a substrate offers the safety of providing a reservoir of nutrients, water, and air that responds to changes in the feed made by the grower. With rockwool as a substrate, a plant can remove solution from a saturated rockwool as easily as it does from rockwool that has lost 50% or 70% of its moisture. Thus, plants grown in rockwool are not exposed to water stress until the rockwool is almost dry. Rockwool holds more water per unit volume than other inorganic substrates and thus has a greater buffering capacity. With the larger reserves of nutrient solution and excellent drainage capacity, rockwool tends to be the better substrate.[48]

In 1865, industrial rockwool was first created but only in crude form. Much development work came later in the 1970s in Europe. A simplified process flow diagram of stone wool manufacturing process is shown in Fig. 10.22. Stone wool is produced by melting igneous rocks and sometimes mixed with blast furnace slag in a cupola furnace at temperature near 1,500 °C. Coke is added to supply the source of heat energy. A continuous flow of molten lava is removed from the cupola furnace and enters some steel wheels rotating at a high speed. The spinning wheels at high speed spray the molten material into the air of the spinning chamber. In this manner, the material cools and solidifies into wool-like strands of fiber. Simultaneously, a small amount of oil and resin binder is sprayed on the wool-like strands so as to bind the

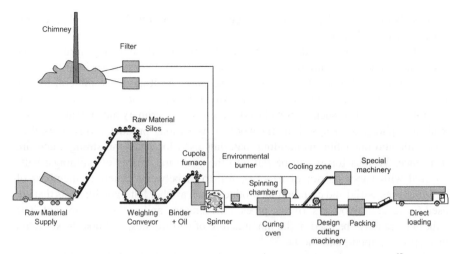

FIGURE 10.22 Process Flow Diagram for Production of Stone Wool.[49]

strands together. The wool-like material is collected on a conveyor belt and passes through a heated Curing oven where the texture of the wool-like material is compressed to density specifications and thickness for curing of the resin. At the end, the woolly material is cut and packaged. Some material is diverted to special machinery to produce specialty wool products such as pipe sections and when aluminum foil or wire netting is attached. A pipe section is made in special machinery by wrapping stone wool around a mandrel and is cured on the mandrel and then withdrawn.[49]

For raw materials, stone wool is made using basalt, diabase, other igneous rocks, and blast furnace slag. As mentioned earlier, coke is used to fuel the furnace and dolomite is used as a fluxing agent. The cupola furnace is fired by coke and oxygen to reach a temperature near 1,500 °C. In the curing oven, the stone wool is compressed for density control, which provides texture qualities such as rigidity and strength. Heat in the curing oven sets the resin binder by bonding the fibers together where strands touch one another. At the end, stone wool is cut to size and thickness for packaging. A slipstream from the curing oven is sent to special machinery for manufacture of custom stone wool products.[49]

Stone wool is made using raw materials of slag and basalt, melting, and fiberizing where droplets of vitreous furnace melt are spun into fibers by falling into rapidly rotating flywheels and injection of binders onto the fibers, producing a gray stone. The gray stone with binders is cured in a curing oven at about 200 °C where the material becomes gray-brown. Final stone wool product is cut into sizes and packaged.

Glass wool is made with raw materials of sand, limestone, and soda ash, melting (1,300–1,500 °C) and fiberizing where droplets of vitreous melt is drawn through tiny holes in rapidly rotating spinners forming fibers and injection of binders onto the fibers, producing a "white glass." The "white glass" with binders is cured in a curing oven at about 200 °C where the material becomes yellow. Final glass wool product is cut into sizes and packaged.

It should be also noted that mineral wool is often defined as any fibrous glassy substance made from minerals, typically natural rock materials such as basalt or diabase or mineral products such as slag and glass. Melting of the mineral feed is stated to be between 1,300 and 1650 °C. Mineral wool is made from feedstock of natural rock and slags such as iron blast furnace slag as a primary material and copper, lead, and phosphate slags. Mineral wool in its various product forms is used as insulation and other fibrous building materials used for structural strength and fire resistance. Various forms can be put into the structural spaces of buildings: batts covered with vapor barrier of paper or foil to fit between structural members of buildings, industrial and commercial products such as fiber felts, and blankets that are used for insulating boilers, ovens, pipes, refrigerators, and process equipment. The bulk fiber can also be used in the manufacture of ceiling tiles, wallboards, spray-on insulations, cement, and mortar.[50,51]

Mineral wool (rock and slag wool) insulation has physical and chemical properties that account for its utility. The fibers are noncombustible and have melting temperature in excess of 1,800–2,000°F, and are used to prevent the spread of fire. As a primary constituent of ceiling tile and sprayed fireproofing, rock and slag wool products supply fire protection as well as sound control and attenuation.[52]

Further studies into the similarities and differences of these different mineral wool-like products manufactured using feedstocks from current raw materials and slag sources, compared with slags from gasification plants, would likely lead to some scientific breakthroughs and creation of new technologies for the manufacture of new wool-like products for the marketplace.

Production of Aggregate

Road material/stones derived from a slag, using a high-temperature melting process (vitrification process, 1,450 °C) pass TCLP analysis and have been used as aggregates in asphalt and concrete. When the stones were used as a substitute for asphalt aggregate in constructing a public road in Kamagaya City, Japan, no difference was observed between the road construction area using these stones as substitute for coarse and fine aggregates compared with that using natural aggregates. When these stones were used as concrete aggregates, the results were better than using natural aggregates. Concrete aggregate strength tests are shown in Table 10.25 when substituting 100% of the stones for coarse aggregates.[53]

More Research and Development (R&D) efforts will undoubtedly expand this technology to improved product properties and extended use as road materials when using various mixtures of slags from various gasification processes with this technology.

Production of Flame-Resistant Foam

Graphite-impregnated foams (GIF) can be considered "inherently flame-resistant foam." An "inherently flame-resistant foam" is considered self-extinguishing and

TABLE 10.25   Concrete Aggregate Strength Test

	Result of Concrete Aggregate Strength Test		
	Compressive Strength (N/mm^2)	Flexural Strength (N/mm^2)	Tensile Strength (N/mm^2)
Base concrete	34.8	4.85	2.68
100% substitute for coarse aggregates	39.3	5.09	2.98

Source: Ref. (53).

highly resistant to combustion. This new technology is a niche market such as for aircraft seating. GIF technology produces foam that can meet airline safety standards for seats so as to reduce dependency on flame-retarded fabric. The question that arises then is "Can vitrified slag be processed in such manner as to produce a product similar to GIF?" Any R&D into this complex technology will be able to address the issue and determine an outcome.[54]

Destruction of Asbestos Wastes via Vitrification

Asbestos is a natural mineral that is fibrous and crystallized. Asbestos was valued as a construction material with good insulating properties and resistance to fire. However, upon inhalation, it may cause serious health problems such as asbestosis or pulmonary fibrosis, bronchio-pulmonary cancer, and pleural or peritoneal mesothelioma. The danger of asbestos lies with its "physical properties" and not with its chemical properties. The asbestos fibers settle in the lungs and remain for years after inhalation. Asbestos is a carcinogenic fiber and is classified as a hazardous waste. Much concern has been with the past use of asbestos as a construction material and then with the removal and disposal of asbestos material from buildings constructed years ago or being renovated or demolished. Procedures for removal of asbestos from buildings have been established. However once removed, the final disposal or resting place of such asbestos waste presents a long-term liability if one connects a future problem with the original source (origin) of the asbestos waste. Still, the landfill approach remains as a low-cost and effective means for final disposal. However, it would seem prudent and a better alternative for a final solution to asbestos disposal by eliminating the problem with the "liability" associated with the disposal of the asbestos waste. With this consideration and many other concerns, development of a process for the destruction of asbestos waste was undertaken. One such process for testing the ability to destroy asbestos waste was plasma torch technology. This process technology was considered because a high temperature is generated (greater than 1,600 °C) that can fuse the asbestos fibers together through vitrification, promoting the destruction of the asbestos fibers. In addition, not only the asbestos waste volume is destroyed via vitrification but the waste volume is reduced by 80% and the final vitrified product from the plasma torch vitrification process is inert and stable for recycling into road construction and maintenance.[55,56,57]

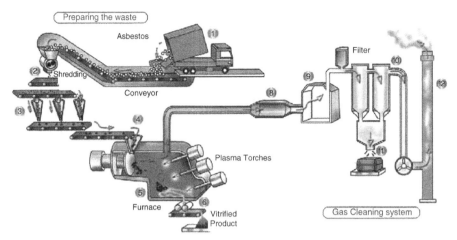

FIGURE 10.23 INERTAM—Asbestos Waste Treatment Process.[55,56,57]

The process for the destruction of asbestos waste via plasma torch vitrification was first developed by EADS, a French aerospace company. Later, Europlasma was founded to develop this plasma torch technology for industrial processes such as waste treatment. A treatment facility in Morcenx, France was built and run by Inertam, a subsidiary of Europlasma, SA. Current capacity of the INERTAM facility for treating asbestos waste is about 110 tonnes/day. A simplified flow diagram of the INTERTAM facility process is shown in Fig. 10.23.[55–57]

Asbestos waste arrives at the INERTAM treatment facility presorted. The waste is then crushed to proper size as feedstock to the plasma torch furnace. Waste is continuously fed to the plasma torch furnace, and output from the furnace is periodically removed every hour. After exiting the furnace, the vitrified product is stored for cooling. Europlasma labels the vitrified product as Cofalit. The Cofalit has been tested and shown to be nonhazardous and is sold for use as an aggregate for road construction or other construction products. The destructive transformation of the asbestos fibers is total and a mean mass balance for 1 tonne of asbestos wastes reduces to about 85% vitrified product, which is reduced in volume by approximately 80%.

Discussion of Potential Markets for the Vitrified Slag

Benefits of using vitrified slag as a by-product for sale in the marketplace would be a much better solution than the current practice of landfilling. Many potential applications for the use of slag/vitrified slag and by-products produced from slag/vitrified slag are listed in this discussion to facilitate the creation of new ideas by the reader.

Production of lightweight aggregates (LWA) from slag, used to make roof tile, lightweight block, and structural concrete, looks appealing as an excellent opportunity to develop a market. The economic incentive for developing this technology depends on the market prices of target applications for conventional LWAs made from expansible clays and sold for $40/ton and ultra-lightweight aggregates, ULWAs, made from expanded perlite and sold for $150/ton. Results have indicated that SLA is an excellent substitute for conventional LWA in roof tile, block, and structural concrete production. Slag-based near-ultra-lightweight material may also be used as a partial substitute for expanded perlite in agricultural and horticultural applications.[43,58]

Potential ideas and applications for slag and vitrified slag, as remembered from the literature, are listed below to generate interest in the creation of novel ideas and synergism of ideas for new by-products and processes for the marketplace:

- Construction structural backfill
- Asphalt paving aggregate—hot mix and seal coat aggregate
- Portland cement aggregate
- Asphalt shingle roofing granules
- Pipe bedding material
- Blasting grit
- Snow and ice control
- Mineral filler
- Road drainage media
- Water filtering medium
- Water-jet cutting—a new application for boiler slag
- Slag lightweight aggregate (SLA) and ULWA
- Decorative tiles
- Insulation material for steel beams
- Abrasive material
- Road sub-base material
- Insulation materials for construction industry
- Aggregate in cement – concrete and asphalt
- Roofing singles

The list of applications and ideas hopefully will help all to recognize that benefits of using vitrified slag as a by-product for sale in the marketplace would be a much better solution than the current practice of landfilling. Barriers to vitrified slag utilization can be institutional, regulatory, and legal. Institutional barriers typically are restrictions on use through requirements, standards, specifications, policies, procedures, or just attitudes of organizations and agencies involved in use or disposal. These can also include economic, marketing, environmental, public perception, and technical barriers. Some examples are local material transport requirements, opposition from established raw material marketers, unknown long-term effects on products made from vitrified slag, and product durability concerns.

Regulatory barriers to using slag/vitrified slag include federal, state and local legislation, and permitting requirements. Most states currently do not have specific regulations for dealing with new technologies and innovations in the marketplace, so matters are dealt with on a case-by-case basis.

Legal barriers include contract, patent, liability, and some regulatory issues. It is crucial to overcome barriers to vitrified slag use by demonstrating that the uses are technically safe, environmentally sound, socially beneficial, and commercially competitive. Improved specifications, fact sheets, and testing procedures need to be developed in collaboration with government and university researchers and standard-setting organizations of high professional standards and conduct.[43]

REFERENCES

1. Young, G.C., "Zapping MSW with Plasma Arc," Pollution Engineering, November 2006.
2. Young, G.C., "How Trash Can Power Ethanol Plants," Public Utilities Fortnightly, p. 72, February 2007.
3. Young, G.C., "From Waste Solids to Fuel," Pollution Engineering, February 2008.
4. Spath, P.L. and Dayton, D.C., "Preliminary screening—Technical and Economic Assessment of Synthesis Gas to Fuels and Chemicals with Emphasis on the Potential for Biomass-Derived Syngas," National Renewable Energy Laboratory (NREL), December 2003.
5. Khalid, W., "Update Catalyst Technology for Syngas Production," Hydrocarbon Processing, April 2009.
6. DOE/NETL-2007/1260, Baseline Technical and Economic Assessment of a Commercial Scale Fischer-Tropsch Liquids Facility, Final Report for Subtask 41817.401.01.08.001, National Energy Technology Laboratory (NETL), April 9, 2007.
7. Methanol Institute, 4100 N. Fairfax Drive, Suite 740, Arlington, VA 22203, www.methanol.org, April 2009.
8. Fitzpatrick, T., "LCM—The Low Cost Methanol Technology," Synetix, April 2009.
9. Methanex, www.methanex.com, Methanol, April 2009.
10. Higman, C. and van der Burgt, M., Gasification, 2nd edition, Elsevier, 2008.
11. Higman, C., "Methanol Production by Gasification of Heavy Residues," IChemE Conference, London, 22–23 November, 1995.
12. DOE/EI/10653—T1, "Coal Based Synthetic Fuel Technology Assessment Guides," Prepared for Energy Information Administration by Energy Resources Co., Inc., September 1981.
13. Indiana Coal to SNG, 2008, http://www.purdue.edu/dp/energy/pdf/CCTR-12-06-06-Jclark.pdf.
14. NETL, 2007. Industrial Size Gasification of Syngas, Substitute Natural Gas and Power Production. Report No. DOE/NETL-401/040607, http://www.netl.doe.gov/technologies/coalpower/gasification/pubs/systems_analyses.html.
15. Steinberg, M., 2005. Conversion of coal to substitute natural gas (SNG), http://www.hceco.com/HCEI105001.pdf.
16. Chandel, M. and Williams, E., "Synthetic Natural Gas (SNG): Technology, Environmental Implications and Economics," Duke University, January 2009.
17. Muller, W.D., "Gas Reduction," Ullmanns Encyclopedia of Industrial Chemistry, 5th edition, vol. A12, Weinheim, VCH Verlagsgesellschaft, 1989.

18. Gray, D. and Tomlinson, G., "Hydrogen From Coal," MTR 2002-31, Mitretek Technical Paper, July 2002.

19. Li, F. and Fan, L.-S., "Chemical Looping Gasification," Department of Chemical and Biomolecular Engineering, Ohio State University, October 2008.

20. ABS Energy Research, "The Hydrogen Economy," Hydrogen and Fuel Cells, Edition 1, 2006, www.absenergyresearch.com.

21. Schlinger, W.G., "Commercialization of the Texaco Coal Gasification Process," Conversion Engineering Conference, 1979.

22. Falsetti, J.S. and Skarbek, R.L., "Petroleum Coke Utilization With The Texaco Gasification Process," 1993 NPRA Annual Meeting, San Antonio, Texas, March 21—23, 1993.

23. NETL/DOE, "Hydrogen Production Facilities Plant Performance and Cost Comparisons," Final Report, Compilation of Letter Reports from June 1999 to July 2001, Parsons Infrastructure and Technology Group, Inc. March 2002.

24. Syntec Biofuel Research, Inc., Private Communication, Vancouver, Canada, August 2007.

25. Phillips, S., Aden, A., Jechura, J., Eggeman, T., and Dayton, D., "Thermochemical Ethanol via Indirect Gasification and Mixed Alcohol Synthesis of Lignocellulosic Biomass," National Renewable Energy Laboratory, Technical Report NREL/TP-510-41168, April, 2007.

26. Herman, R.G., "Chapter 7—Classical and Non-classical Routes for Alcohol Synthesis," New Trends in CO Activation, L. Guczi,ed., Elsevier, New York, pp. 265–349, 1991.

27. Forzatti, P., Tronconi, E., and Pasquon, I., "Higher Alcohol Synthesis," Catalysis Reviews—Science and Engineering, vol. 33 (1–2), pp. 109–168, 1991.

28. Quarderer, G.J., "Mixed Alcohols from Synthesis Gas," Proceedings from the 78th Spring National AIChE Meeting, April 6–10, New Orleans, LA, 1986.

29. Vessia, O., Finden, P., and Skreiberg, O., "Biofuels from lignocellulosic material, In the Norwegian context 2010, Technology, Potential and Costs," Norwegian University of Science and Technology (NTNU), December 20, 2005.

30. Skene, L., "Waste to Energy, The Creation of a BC Biorefinery & Sustainable Energy Cluster, Concept Overview & Project Development Strategy," Presented by New Energy Eco-Systems, Inc., August 8, 2006.

31. Younesi, H. et al., "Ethanol and acetate production from synthesis gas via fermentation process using anaerobic bacterium, *Clostridium ljungdahlii*," Biochemical Engineering Journal, vol. 27, pp. 110–119, 2005.

32. Eucar, E.R. et al., "Well to wheels analysis of future automotives fuels and power trains in the European context, well to tank report," EUCAR, CONCAWE and Joint Research Centre of the EU Commission, Appendix 1, 2003.

33. Schuetzle, D., Tamblyn, G., Tornatore, F., and MacDonald, T., "Assessment of Conversion Technologies for Bioalcohol Fuel Production," TSS Consultants and California Energy Commission, July 2006.

34. Bruce, W.F., "The Co-Production of Ethanol and Electricity from Carbon-based Wastes," A Report from Bioengineering Resources, Inc. (BRI, Inc.), Regarding a New Technology that Addresses Multiple Energy and Waste Disposal Solutions, March 2006.

35. Zessen, E. v., et al., "Ligno Cellulosic-Ethanol, A Second Opinion," Report 2GAVE-03.11, NOVEM, Netherlands Agency for Energy and Environment, May 2003.

36. Environment Canada, "Wood-Ethanol Report, Technology Review," 1999.

37. van Kasteren, J.M.N., Dizdarevic, D., van der Waall, W.R., Guo, J., and Verberne, R., "Bio-ethanol from bio-syngas," Eindhoven University of Technology (TU/e), Telos, Document 0456372-R02, December 5, 2005.

38. Etanolteknik AB (ETEK), webpage at 222.etek.se.

39. Wooley, R., Ruth, M., Sheehan, J., Ibsen, K., Majdeski, H., and Galvez, A., "Lignocellulosic Biomass to Ethanol Process Design and Economics Utilizing Co-Current Dilute Acid Prehydrolysis and Enzymatic Hydrolysis Current and Futuristic Scenarios," National Renewable Energy Laboratory (NREL), NREL/TP-580-26157, July 1999.

40. Circeo, L.J., Nemeth, J.C., Newsom, R.A., "Evaluation of Plasma Arc Technology for the Treatment of Municipal Solid Wastes in Georgia," Georgia Institute of Technology, Atlanta, Georgia, January 1997.

41. Cedzynska, K., Kolacinski, Z., Izydorczyk, M. and Sroczynski, W., "Plasma Vitrification of Waste Incinerator Ashes," Technical University of Lodz, Lodz, Poland, 1999.

42. "Technical Feasibility Study For Developing A Waste Reduction And Energy Recovery Facility In Southeast Alaska," Prepared by: DMC Technologies, Rexburg, Idaho, Prepared for: Silver Bay Logging, Inc., City of Wrangell, Alaska, June 15, 2003.

43. Ratafia-Brown, J., Manfredo, L., Hoffmann, J., and Ramezan, M., "Major Environmental Aspects of Gasification-Based Power Generation Technologies, Final Report," U. S. Department of Energy (DOE), National Energy Technology Laboratory (NETL), December 2002.

44. Heunisch, G.W. and Leaman, G.J., Jr., "Phase I: The Pipeline Gas Demonstration Plant. Analysis of Coal, B6-Products and Wastewaters from the Technical Support Program," DOE, FE-2542-23, August 1979.

45. Iddles, D.M., Chapman, C.D., Forde, A.J., and Heanley, C.P., "The Plasma Treatment of Incinerator Ashes," Tetronics, Ltd. , circa 2000.

46. Tetronics, Ltd., "Plasma Vitrification of Incinerator Ashes," 2005, www.tetronics.com.

47. Richardson, J.T.G, and Haynes, D.J., "Slag bound Materials in Composite Roads," SCI Lecture Papers Series, Society of Chemical Industry, UK, www.soci.org, 2000.

48. Dowgert, M.F., "Rockwool as a Substrate for Hydroponic Growing Systems," www.alternativegarden.com.

49. Rockwool International A/S, Environmental Report 2002.

50. Source Category Survey: Mineral Wool Manufacturing Industry, EPA-450/3-80-016, U.S. Environmental Protection Agency, Research Triangle Park, NC March 1980.

51. "The Facts On Rocks And Slag Wool," Pub. No. N 020, North American Insulation Manufacturers Association, Alexandria, VA.

52. North American Insulation Manufacturing Association (NAIMA), "FAQs About Mineral Wool (Rock and Slag Wool) Insulation," www.naima.org, 2006.

53. Nishida, K. et al., "Melting and Stone Production Using MSW Incinerated Ash," Tsukishima Kikai Co. , Ltd., Tokyo, Japan 1999.

54. U.S. Environmental Protection Agency, "Furniture Flame Retardancy Partnership: Environmental Profiles of Chemical Flame-Retardant Alternatives for Low-Density Polyurethane Foam," Vol. 1, EPA 742-R-05-002A, September 2005, www.wpa.gov/dfe.

55. Blary, F. and Rollin, M., "Vitrification of Asbestos Waste," INERTAM, circa 1995.

56. Waste Management World, "The Flexibility of Plasma," December 2007.

57. Europlasma, SA, "Asbestos melting unit," INERTAM, www.europlasma.com, 05/22/2009.
58. Douglas, F. and Chugh, Y.P., "Feasibility of Utilizing Integrated Gasification Combined Cycle (IGCC) Byproducts For Novel Materials Development," National Energy Technology Laboratory (NETL), DOE Cooperative Agreement: DE-AF26-04NT40752, August 2006.

CHAPTER 11

MSW Gasifiers and Process Equipment

As recognized by any in-depth discussion, gasification is an old technology used for the production of energy from solid materials. In general, gasification has been used for the conversion of carbonaceous materials into a gaseous product for the production of energy products and by-products in an oxygen-starved environment. The carbonaceous material, whether from municipal solid waste (MSW) or other sources, is processed in an oxygen-deficient environment with heat to produce a syngas comprised mostly of carbon monoxide (CO) and hydrogen (H_2). Chemical bonds of the carbonaceous materials (complex chemical compounds) break down with heat to produce the more simple and thermodynamically stable gaseous molecules of CO and H_2. The inorganic or "mineral" materials are converted to a solid rocklike material called slag or vitrified slag or ash.[1-3]

This chapter covers the various gasifiers and process equipment for the gasification of MSW to syngas and by-products. In addition, it includes a discussion of process equipment used in the gasification of coal because much of the gasification equipment used today was based upon its development over the years for use with coal gasification but has applications to MSW and other wastes as well. Thus, the discussion in this chapter will be on gasifiers and process equipment for the production of syngas (CO and H_2), and consequent downstream production of energy and/or chemicals, fuels, and materials through chemistry and associated by-products.

A typical process for the gasification of MSW to syngas, energy, products, and by-products is illustrated in Fig. 11.1. As mentioned, the organic components of MSW are converted into syngas, primary product, and the mineral components are converted into slag or vitrified slag or ash, a by-product. For some gasification processes, some (limited) amount of oxygen is supplied to the gasification reactor to provide heat in the form of combustion heat necessary for the oxygen-deficient reactions to produce syngas, which are endothermic reactions. Sufficient steam/water is supplied to the gasification reactor to promote some syngas reactions.

Municipal Solid Waste to Energy Conversion Processes: Economic, Technical, and Renewable Comparisons By Gary C. Young
Copyright © 2010 John Wiley & Sons, Inc.

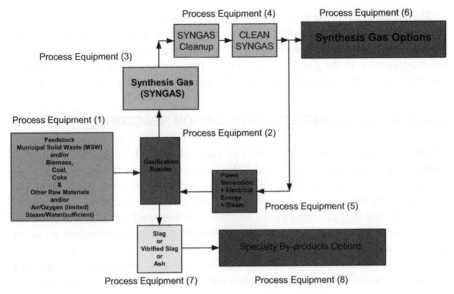

FIGURE 11.1   Process Equipment for the Gasification of MSW to Syngas, Energy, Products, and By-products.

Basic chemical reactions in the gasification process for producing syngas are:

$$C + O_2 \rightarrow CO_2 \quad \text{(exothermic)} \tag{11.1}$$

$$C + H_2O \rightleftharpoons CO + H_2 \quad \text{(endothermic)} \tag{11.2}$$

$$C + 2H_2 \rightleftharpoons CH_4 \quad \text{(exothermic)} \tag{11.3}$$

$$C + CO_2 \rightleftharpoons 2CO \quad \text{(endothermic)} \tag{11.4}$$

$$CO + H_2O \rightleftharpoons CO_2 + H_2 \quad \text{(exothermic)} \tag{11.5}$$

$$C_nH_m + nH_2O \rightleftharpoons nCO + (n + \tfrac{1}{2}m)H_2 \quad \text{(endothermic)} \tag{11.6}$$

where for the steam methane (CH_4) reforming reaction,, where $n = 1$ and $m = 4$, Equation (11.6) becomes,

$$CH_4 + H_2O \rightleftharpoons CO + 3H_2 \text{(endothermic)} \tag{11.7}$$

Note, an exothermic reaction produces heat (output) whereas an endothermic reaction requires heat (input). Also note, Equation (11.1) is a combustion reaction that is sometimes used in the gasification process to provide some internal (*in-situ*) process heat for the endothermic syngas reactions but only a (limited) amount of oxygen is supplied so the gasification reactor environment remains conducive for the production of syngas. The remaining process heat is supplied externally to the gasification reactor such as through the heat transfer surfaces of the reactor heating system.

This chapter covers the basic process equipment used for the gasification of MSW to syngas, energy, products, and by-products. Since the gasification of coal technology was one of the first technologies developed for syngas production, this technology has much application to the gasification of MSW. Thus, process equipment gasification used for coal gasification will be discussed as applicable to the gasification of MSW.

CONVENTIONAL GASIFIERS/GASIFICATION REACTORS

The predominant combustible products of gasification are CO and H_2 and a small amount of carbon dioxide (CO_2) and CH_4. Typically, the heat produced by a small amount of oxygen fed to the gasification reactor provides most of the heat via combustion to break down the chemical bonds in the hydrocarbon feedstock to be gasified. The heat provided by the limited oxidation reaction therefore raises the gasifier temperature along with other exothermic gasification reactions so as to drive the main endothermic gasification reactions.

Although there are many gasification reactors with various mechanical designs and operating conditions, gasification reactors can be classified as three basic types:

- Moving-bed gasifier/gasification reactors (also called fixed-bed gasification reactors)
- Fluidized-bed gasifier/gasification reactors
- Entrained-flow gasifier/gasification reactors.

These gasification reactors will be discussed followed by a description of commercial gasification reactor/systems.[4,5] The three basic types of gasifiers/gasification reactors are shown in Fig. 11.2.[6]

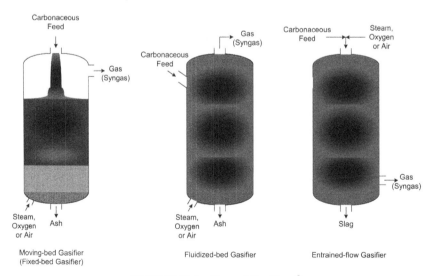

FIGURE 11.2 Type of Gasifiers.[6]

The moving-bed or fixed-bed gasifier is a typical Lurgi dry ash gasifier that is installed at Sasol coal to liquids (CTL) plants and the Great Plains Synfuels Plant. Providers of technology are BGL and Lurgi. Gasifier operates on any coal rank and outlet gas temperature of 800–1,200°F with the syngas exiting near the top of the gasifier. Ash handling is slagging and nonslagging. Fixed-bed or moving-bed gasification technology accounts for nearly 42% of the world's total installed gasification capacity. As discussed with coal as the carbonaceous feed, coal particles move slowly down the bed while reacting with the gases moving up through the bed. Feed enters the top of the reactor/gasifier; steam and oxygen or air enter the bottom while ash or slag is removed from the bottom. From top to bottom, zones can be envisioned as two types of chemical reactions taking place. At the top of the reactor/gasifier, the heated coal is heated and dried while cooling the syngas as it begins to exit the reactor. As it moves down the bed, the coal is further heated and devolatized as it proceeds down the bed called the carbonization zone. Next, a gasification zone is reached where the devolatized coal reacts with steam and CO_2. Near the bottom of the reactor, the zone is called the combustion zone, which is the hottest zone in the reactor where oxygen reacts with the remaining material. Where the temperature at the bottom of the reactor is moderate and below the ash-slagging temperature, the area is called a dry-ash gasifier. For example, it could be called a Lurgi dry ash gasifier. Where less steam is injected at the base of the reactor/gasifier, the combustion zone is hotter and above the ash-slagging temperature. Thus, this type of operation is referred to as a slagging gasification like that of British Gas/Lurgi or BGL gasifier. Moving-bed or fixed-bed gasifiers have these characteristics: there are low oxygen requirements, caking coals require design considerations, tars and oils are produced, and there is a limited ability to handle fines.[4]

Currently, fluidized-bed gasifiers account for about 2% of the installed capacity and appear to have potential for low-ranked coals. Technology is provided by HTW, KRW, KBR, and Winkler. The outlet gas temperature is 1,700–1,900°F and exits the top of the gasifier. Ash handling is nonslagging and exits the bottom of the gasifier. As discussed with coal as the carbonaceous feed, coal particles enter the side of the reactor while steam and oxygen or air enter the bottom and ash is removed from the bottom. The steam and oxygen or air entering the bottom suspend or fluidize the feed particles entering near the top of the reactor. Temperature is maintained below the ash-fusion temperature, thus avoiding clinker formation and possible defluidization of the bed. Fluidized-bed gasifiers have the following characteristics: they accept a wide range of solid feedstock; the temperature of the bed is uniform and moderate; there are moderate oxygen and steam requirements; and there is extensive char recycling.[4]

The entrained-flow gasifiers are licensed by ConocoPhillips, Future Energy, GE Energy, and Shell, which are the current market leaders. This gasification technology represents about 56% of the installed capacity. The entrained-flow gasifiers are popular because of the following reasons: reliable and proven design (widely used in the chemical industry), no internal moving parts, compact size, minimal by-products, and ability to supply syngas at higher pressures. The outlet gas temperature is 2,250–3,000°F and exits near the bottom of the gasifier. Ash is always slagging and is removed from the bottom of the reactor.[4–6] Feed, steam, and oxygen or air enter the top of the gasifier. Entrained-flow gasifiers operate at high temperatures well above the ash-slagging conditions and consequently have high carbon conversion. Entrained-flow gasifiers have the following characteristics: ability to gasify all coals regardless of coal rank, caking characteristics or quantity of coal fines, uniform temperature, short residence time in the reactor, the need for feed to be finely divided and homogenous, large oxygen requirement, high-temperature slagging operation, and possible entrainment of molten slag in the raw syngas.[4]

Commercial entrained flow gasifiers include the ChevronTexaco gasifier, Shell gasifier variant, Prenflo gasifier, the E-Gas™ (Destec) gasifier, and the Noell gasifier.[4]

ChevronTexaco Entrained-Flow Gasifier

The ChevronTexaco gasification technology utilizes a single-stage, downward-feed, entrained-flow gasifier as shown in Fig. 11.3.[4] A Coal water slurry (60–70% solids) enters the top of the gasifier with 95% oxygen as feed to the pressurized gasifier. The feed enters through a special injector at the top of the refractory lined gasifier. The fuel and oxygen react exothermally to produce a hot temperature from 2,200°F to 2,700°F and a pressure of >20 atmospheres. The gasification operation produces a raw syngas and molten ash (slag). The syngas passes downward where it is cooled by a radiant syngas cooler and high-pressure steam is produced. The syngas then passes over a

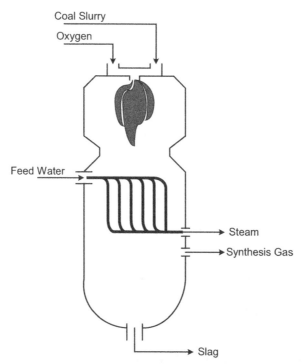

FIGURE 11.3 ChevronTexaco Gasifier.[4]

pool of water at the base of the cooler and exits the gasifier. The slag drops into the pool of water and is removed through a lock hopper. Metals and other ash constituents become bound into a glassy slag. Other modifications and configurations of this gasification process are made for specific process designs required for a particular feedstock. The ChevronTexaco technology has been around for more than 40 years and with various feedstocks.

E-Gas™ Entrained-Flow Gasifier

The design of the E-Gas™ gasifier was originally Destec's LGTI gasifier, Destec's Louisiana Gasification Technology, Inc., which was similar in size and operations. The E-Gas™ gasifier, as shown in Fig. 11.4, consists of two stages, a slagging first stage followed by an entrained-flow nonslagging second stage. The first stage comprises a horizontal refractory lined vessel whereby the carbonaceous fuel is partially combusted with oxygen to a temperature around 2,600°F and pressure of 400 psig. Preheated slurry and oxygen are fed into the first stage via two opposed mixing nozzles of proprietary design. The oxygen rate is controlled to maintain the temperature above the ash fusion point so as to form a slag. The highly exothermic reactions rapidly increase the temperature to about 2,400–2,600°F. In the second stage, nearly all the feed is converted to syngas consisting of CO, H_2, CO_2, and water

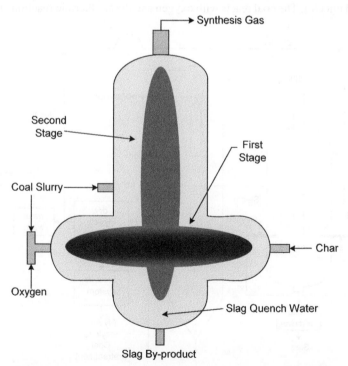

FIGURE 11.4   E-Gas™ Entrained-Flow Gasifier, ConocoPhillips.[7]

(H$_2$O). The ash from the coal and any added flux forms a molten slag that flows out from the bottom of the horizontal section into a quench water bath. The solidified slag exits from the bottom of the quench section. Ray syngas from the first stage flows upward into the vertical refractory lined second stage where additional slurry is injected. The feed material to the second stage is about 25% of the total coal slurry, which undergoes gasification/devolatilization and pyrolysis endothermic reactions. Any unreacted char is carried over with the syngas and leaves the gasifier at 1,900°F. Char formed is a function of the reactivity of the carbonaceous feed and any char generated is removed from the syngas in downstream processing and recycled back to the gasifier.[4,7] The E-Gas™ technology has been fully demonstrated at the Wabash River IGCC plant, which began operation in 1995. The Wabash River plant has gasified over two million tons of bituminous coal with 5.9% sulfur and petroleum coke with 7% sulfur during the several years of operation.[4,7]

Shell Entrained-Flow Gasifier

The Shell gasification process is a dry-feed pressurized entrained-flow gasifier as shown in Fig. 11.5, which can operate on a wide variety of feedstocks. Feed is conditioned by pulverizing and drying much the same as for a conventional pulverized coal boiler operation. The coal is pressurized in lock hoppers and fed into the gasifier by dense phase conveying. Oxygen (95% purity) is mixed with steam prior to feeding to the fuel injector. The coal reacts with oxygen and the exothermic reaction produces

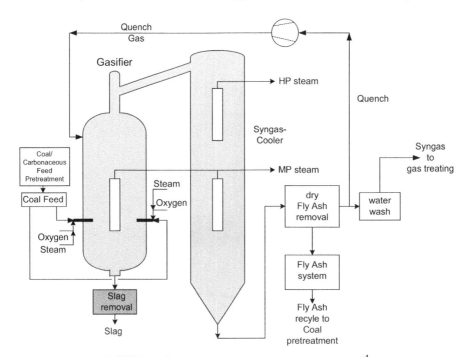

FIGURE 11.5   Shell Entrained-Flow Gasifier Process.[4]

a temperature between 2,700°F and 2,900°F and pressures between 350 and 650 psig. The typical syngas is principally H_2 and CO with a small amount of CO_2. The high-temperature gasification process converts the ash into molten slag, which moves downward in the refractory lined gasifier into a water bath where it solidifies and is removed by a lock hopper. The raw syngas exits the gasifier at around 2,500°F–3,000°F. Further cooling of the syngas occurs in the syngas cooler along with waste heat recovery where steam is created.[4,8]

Lurgi Dry-Ash Gasifier and British Gas/Lurgi Gasifier

The Lurgi dry-ash gasifier, as shown in Fig. 11.6, is a pressurized, dry-ash, moving-bed gasifier. A major characteristic is the use of a moving bed that uses steam and oxygen for gasification of lump coal rather than pulverized coal. It produces tars.

The Lurgi dry-ash Gasifier uses a high steam to oxygen ratio, which maintains the temperature in the dry-ash system so the ash does not melt. Lower temperatures means the gasifier is better on reactive coals such as lignites than on bituminous coals.[8] The sized coal moves downward through a coal lock and then through a rotating coal distributor, insuring uniform flow downward through the gasifier. Steam and oxygen are injected at the bottom of the gasifier, which move upwards through the bed and

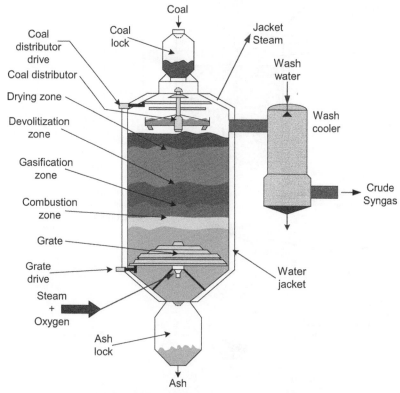

FIGURE 11.6 Lurgi Dry-Ash Gasifier.[4]

react with the coal producing a hot syngas. The hot syngas moves upwards and the coal moving downwards is dried, devolatized, and gasified. Syngas temperatures near the top are about 500–1,000°F. The temperature near the bottom is about 1,832°F. The raw syngas exits the gasifier at about 570–932°F and is quenched using recycled water to condense tar/oil. A water jacket cools the gasifier and produces steam for process use. Steam is injected at the base of the gasifier to keep the temperature from exceeding the melting temperature of the ash. Ash is removed by a rotating gate and removed by a lock hopper.[4,8]

The British Gas/Lurgi gasifier is shown in Fig. 11.7. The British Gas/Lurgi gasifier is a dry-feed, pressurized, moving bed, slagging gasifier. The British Gas/Lurgi gasifier operates at a lower steam to oxygen ratio than the Lurgi dry-ash gasifier and consequently at a higher temperature. The operation at higher temperature results in ash melting and formation of slag.

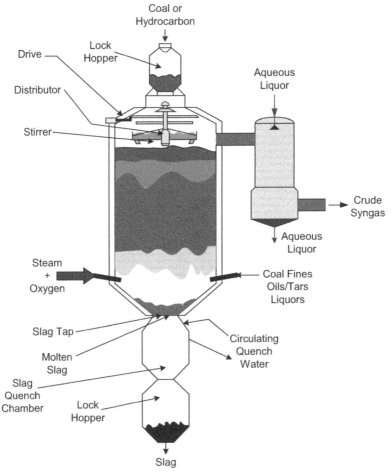

FIGURE 11.7 British Gas/Lurgi Gasifier.[4,8]

The reactor is water cooled and refractory lined. Oxygen and steam are introduced near the bottom sidewall through lances where oxidation and slag formation occur.

The feed enters the top of the gasifier and typically is a mixture of coarse coal, fines, briquettes, and flux. Coal moves downward and is dried and devolatilized. The coal descends and is converted into char and gasified into syngas. Slag is formed in the lower section and is quenched and removed by the lock hoppers. Syngas exits the gasifier at about 1,050°F near the top and passes into a quench vessel to about 300°F.

The British Gas/Lurgi gasifier was developed in the 1970s for the production of syngas with high CH_4 content for the manufacture of synthetic natural gas (SNG) from Coal. It was also developed and tested later for integrated gasification combined cycle (IGCC) power plants. The British Gas/Lurgi gasifiers have been used with feed from coal, sewage sludge, and refuse derived fuel (RDF).[4,8,9]

Prenflo Entrained Bed Gasifier

The Prenflo gasification process was developed by Uhde of Germany, formerly Krupp Uhde. The Prenflo gasifier is a pressurized, dry-feed, entrained-flow, slagging gasifier as shown in Fig. 11.8. Coal is ground to about 100 μm and pneumatically conveyed by

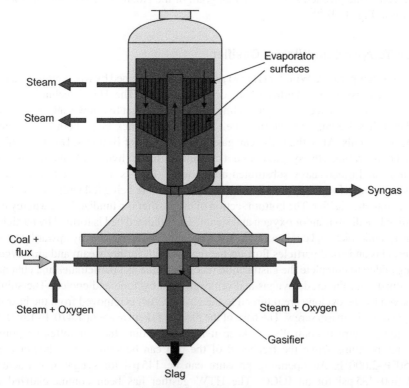

FIGURE 11.8   Prenflo Gasifier.[4]

nitrogen (N_2) to the gasifier. The internal surface of the gasifier structure is refractory lined. The coal is fed through injectors at the lower part of the gasifier with steam and oxygen. Syngas is produced at a temperature of up to 2,900°F. The syngas exiting is quenched with recycled clean syngas to lower the temperature to around 1,470°F. The syngas then flows upwards in a central distributor pipe and downward through the evaporator stages to reduce the syngas before exiting the gasifier to 716°F. Slag that is formed flows down the gasifier and is quenched with a water bath and granulated prior to being removed by a lock hopper system.[4]

Noell Entrained Flow Gasifier

The Noell gasification process was designed originally for the gasification of pulverized brown coal. The Noell gasifier is a pressurized, dry feed, entrained-flow, slagging gasifier. The oxygen to fuel ratio is adjusted to maintain the desired temperature in the gasifier so the inorganic matter melts and flows vertically downward with the gasification gas and leaves the gasifier through a special discharge unit. The gasification chamber is specially designed with a gas-tight membrane wall and is refractory lined. The liquid slag is handled by a specially designed system and removed from the bottom of the gasifier. Carbon conversion rates >99% have been achieved in the gasifier. A simplified diagram of the Noell entrained flow gasifier is shown in Fig. 11.9.[4,10]

High-Temperature Winkler Gasifier

The high-temperature Winkler (HTW) gasifier was developed by Rheinbraun and is a dry-feed, pressurized, fluidized-bed, dry-ash gasifier. The HTW technology can gasify a variety of feedstocks including all grades of reactive low-rank coals with higher ash softening temperature, i.e., brown coal, black coal, both caking and noncaking coals. Also, the HTW can gasify many forms of Biomass. Due to the high outlet temperatures, the syngas does not contain any higher hydrocarbons such as tars, phenols, and other heavy substituted aromatics.[8,4] Fuel is fine-grained coal and is pressurized in a lock hopper and then a charge bin before being fed with a screw-type feeder into the gasifier. The bottom section of the gasifier is a fluidized bed wherey the fluidized medium is air or oxygen and steam. The fluidized bed is formed by particles of ash, semi-coke, and coal and is maintained in the fluidized state by upward flow of gases. Gas and solid particles flow up the reactor with further steam and air/oxygen being added to complete the gasification reactions. Fine ash particulates and char are entrained with the crude syngas and removed by a cyclone and cooled. The solids removed by the cyclone are returned to the gasifier. Ash is removed from the base of the gasifier via an ash screw. The base of the gasifier is kept at about 1,470°F–1,650°F. The temperature is controlled to keep it from exceeding the ash softening point. The temperature above the freeboard of the bed can be significantly higher than 1,650°F–2,000°F. An operating pressure can be 145 psi for syngas manufacture and 360–435 psi for an IGCC. The HTW gasifier has been commercialized at 145 psi pressure for the gasification of lignite to syngas and conversion to methanol.

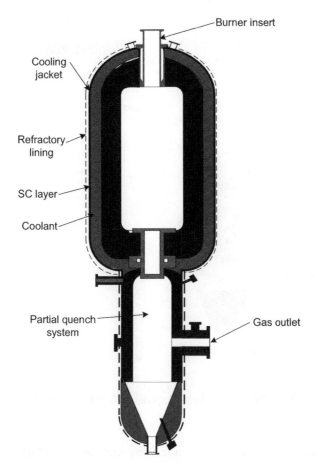

FIGURE 11.9   Noell Entrained Flow Gasifier.[4,10]

In Finland, the HTW gasifier is used to gasify peat for ammonia synthesis. A schematic of the HTW fluidized bed gasifier is shown in Fig. 11.10.[4,8]

KRW Fluidized Bed Gasifier

The KRW gasification process was developed by M.W. Kellogg Company, and is a pressurized, dry-feed, fluidized-bed, slagging process. The gasifier is illustrated in Fig. 11.11. The KRW IGCC technology is capable of gasifying many types of carbonaceous materials such as coals whether high-sulfur, high-ash, low-rank, or high-swelling. In addition, the KRW gasifier technology can gasify biowastes and refuse-derived wastes.

Coal and limestone are crushed to about $\frac{1}{4}$ in. material as a feedstock and transferred to the KRW fluidized-bed gasifier with a lock hopper system. Gasification

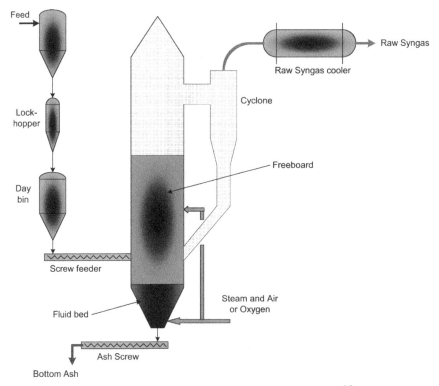

FIGURE 11.10   High-Temperature Winkler Gasifier.[4,8]

is the result of steam and air or oxygen reacting with the coal at high temperatures. The feedstock material and steam and air or oxygen are injected and mixed together at the bottom of the gasifier. The only solid waste mentioned was a mixture of ash and calcium sulfate.[4] When the coal enters the gasifier, it releases its volatile matter and oxidizes rapidly to supply the heat necessary for the endothermic gasification reactions. Rising gases upward through the bed with the internal recirculation of particulates in the fluid bed effectively transfers heat throughout the bed. With the rapid mixing of coal constituents and steam in a hot mixed fluidized bed, gasification of carbonaceous materials occurs and they are converted to syngas. The limestone material is calcined to CaO and then reacts with H_2S formed from the gasification of the coal leading to the formation of CaS.

As the char from the gasification of coal, particles become enriched in ash and begin to melt from the hot syngas. Melted particles recirculate into the fluid bed and agglomerate with other ash particles. Particles that escape with the syngas from the gasifier are separated from the syngas in a high-efficiency cyclone and recycled to the gasifier. After continued agglomeration, the particles begin to settle and gravitate to the bottom of the gasifier. In the end, the CaS forms $CaSO_4$ in the fluid-bed sulfator.[4]

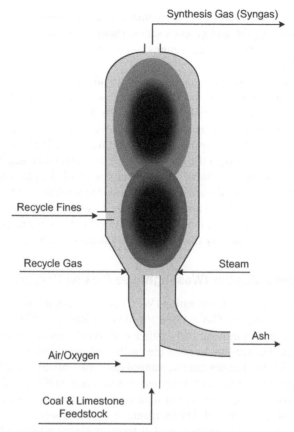

FIGURE 11.11   KRW Fluidized Bed Gasifier.[4]

PLASMA ARC GASIFICATION TECHNOLOGY

Plasma gasification has been used to treat waste products and incinerator ash into a nonhazardous glassy slag. Although application of plasma gasification to the management of MSW is relatively recent, the potential of plasma gasification to convert MSW to energy and commercial products is high. Earlier work has typically centered upon the use of Plasma Gasification technology for melting incinerator ash to a nonhazardous glassy vitrified slag and the destruction of hazardous or medical wastes at temperatures above 7,000°F. Only recently has the plasma technology been coupled with gasification technologies for the management of MSW.

Plasma technology was developed in the 1960s from the space program for development and testing/simulation of the intense kinetic heating that missiles and spacecrafts encountered during re-entry. A high-temperature gas was needed to test heat shield materials and classical gas combustion was not able to generate a sufficiently high temperature. Plasma arc technology was advanced significantly as a result of this need for space applications.

Plasma is called the fourth state of matter, which is distinctly different from the three others: solid, liquid, and gaseous states. Plasma can create an ionized gas by electrical forces whereby the temperatures can reach between 2,000°C and 5,000°C (3632°F–9,032°F).

Plasma is a hot ionized gas created by an electrical discharge. A hot ionized gas is created by plasma technology using an electrical discharge from an AC or DC source to heat the gas. The gas is typically air, oxygen, N_2, H_2, argon, or a combination of these gases. The hot ionized gas is created by a plasma torch. A plasma torch is one of two types, either a transferred torch or a nontransferred torch. The "transferred torch" creates an electric arc between the tip of the torch and a metal or slag at the bottom of the reactor or conductive lining of the reactor wall of the plasma arc gasifier. In a "nontransferred torch," the arc is located inside the torch itself where the plasma gas is created by passing the gas through the torch, is heated by the arc, and then exits the tip of the torch whereby the hot plasma gas is injected into the reactor. There are merits to both designs but only the Plasma Arc gasifiers will be discussed here.[11]

Alter Nrg Plasma Gasifier (Westinghouse Plasma Corporation) System

The Westinghouse Plasma Corporation (WPC) is a proprietary and proven plasma gasifier design, i.e., Plasma Gasification Vitrification Reactor (PGVR). The Plasma Gasification reactor is a moving-bed gasifier with WPC's proven industrial plasma torch technology. The feedstock enters the gasifier where it comes into contact with hot plasma gas. Several torches can be arranged circumferentially in the lower section of the gasifier to provide a more uniform heat flux as shown in Fig. 11.12. The amount of air or oxygen used in the torch is controlled to promote the endothermic gasification reactions of the organic material. The inorganic constituents are converted to a molten slag, which when typically quench-cooled forms a glassy nonhazardous slag. As shown, the hot plasma gas flows into the gasifier/reactor to gasify MSW and melt the inorganic materials.[11,12]

Alter Nrg Corporation uses a proprietary plasma-based gasification technology developed by a wholly owned subsidiary (WPC) for converting various feedstocks such as coal, petroleum coke, MSW, industrial waste, biomass, or biosolids into commercial syngas. The WPC technology is based upon a plasma cupola design where a cupola is a vertical shaft furnace that has been used in the foundry industry for remelting of scrap iron and steel. WPC cupola has been used for the production of cast iron at a General Motors plant in Defiance, Ohio since 1989.

WPC offers four models of plasma torch systems, each designed to operate over a wide range of power inputs. These torches are self-stabilized and nontransferred arc torches that can operate on many gases such as air, oxygen, N_2, etc. A wide variety of torches are available with power input from 80 kW to 2,400 kW. For example, four models of plasma torch systems are available for each torch: Marc 3a from 80 to 300 kW, Marc 3HC from 5 to 150 kW, Marc 11L from 300 to 800 kW, and Marc 11H from 700 to 2,400 kW. An illustration of a Westinghouse Plasma torch is shown in Fig. 11.13.[13] These torches have been in operation for many years. Consider that Hitachi Metals in Japan has used WPC to process MSW since 1999. WPC has

Alter Nrg Plasma Gasifier

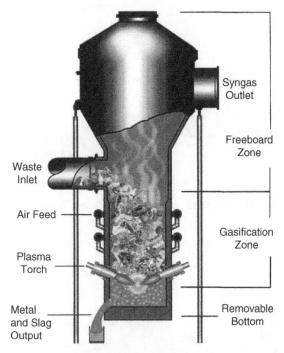

FIGURE 11.12   Alter Nrg Plasma Gasifier (Westinghouse Plasma Corporation).[12]

conducted over 100 pilot tests on a wide range of feedstocks at their Westinghouse Plasma Centre in Madison County, Pennsylvania.[12]

Another illustration of the WPC gasifier (Alter Nrg Plasma gasifier) is shown in Fig. 11.14. The gasification technology is a two-stage gasification/combustion process that permits efficient gas-to-gas heat transfer for a limited combustion, to supply some internal heat for the endothermic gasification reactions. The slag produced is a vitrified glassy material that is highly stable and resistant to leaching. Metals can be recovered and recycled to the metal industry. Difficult wastes such as tires, carpet, and sludge can be accepted as feedstock.[14] Hitachi Metals, Ltd. of Japan commissioned a plant in year 2002 for gasification of MSW and waste water sludge for district heating. A schematic of a plasma arc gasifier of Hitachi Metals, Ltd. is shown in Fig. 11.15.[15]

EUROPLASMA, Plasma Arc System

EUROPLASMA specializes in high-temperature industrial processes using plasma technology. EUROPLASMA uses a nontransferred arc plasma torch technology. The arc remains in the torch so only the plasma flow goes out of the torch. A plasma torch is composed of two metallic tubular electrodes, one upstream and one downstream. A gas stream passes between these electrodes, and an electrical arc flows between the negative and positive electrodes, which heats and ionizes the gas. The result is a high-

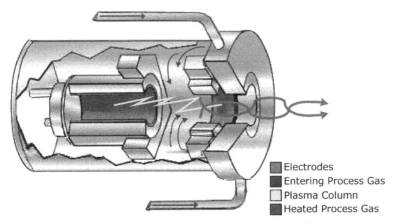

FIGURE 11.13 Westinghouse Plasma Torch.[13]

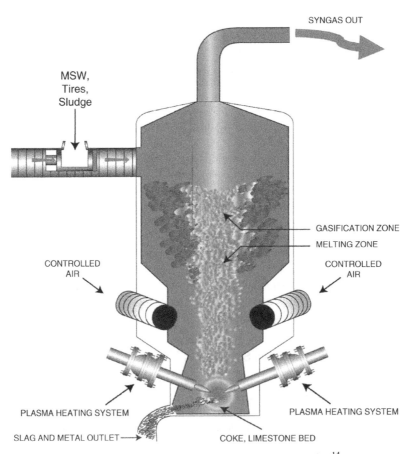

FIGURE 11.14 Westinghouse Plasma Torch Gasifier.[14]

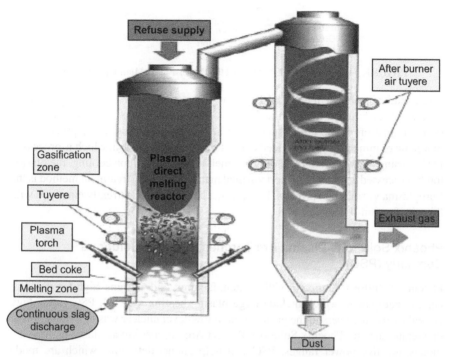

FIGURE 11.15   Westinghouse Plasma Assisted Gasifier, Hitachi Metals.[15]

temperature gas flow coming out from the downstream electrode as a plasma jet. One advantage of a plasma torch is the flexibility to operate, i.e., from 25% to 100% of power range. EUROPLASMA plasma torches are available in power ranges from 100 kW to 4 MW. A model A0025 has a power range from 5 to 25 kW, model C0500 from 100 to 500 kW, model C0700 from 200 to 700 kW, model D2000 from 500 to 2,000 kW, and model D4000 from 1,000 to 4,000 kW. Range of power and applications can be expressed as: 25–100 kW for gas treatment, biomass gasification; 100–300 kW for waste or biomass gasification, gas cleaning; 300–800 kW for ash melting, waste or biomass gasification and 800 kW–4 MW as ash melting, cupola air blast heating.[16]

The CHO-power process uses a high-temperature plasma torch to transform waste into a high-quality syngas by gasification. The syngas is then supplied to a gas turbine/ engine that produces electricity. The end-to-end electrical yield is possible to reach 35–40%. In preparation of the waste to form fuel (feedstock) for the CHO-process, the waste has metal and inert materials removed if necessary. Then the feedstock is coarsely crushed before conveying into the gasification plant. In the gasifier, the feed is first dried at about 200°C and then the organic components are converted to crude syngas at 600–800°C. The process is conducted in an atmosphere with absence of air. The gasification unit is designed based upon a robust grate furnace process design that has been proven commercially for about 50 years. The Europlasma, CHO-power plant

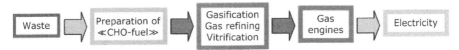

FIGURE 11.16 EUROPLASMA, CHO-Power Plant, Main Components.[17]

is schematically shown in Fig. 11.16. The crude syngas is refined using a plasma torch process to ensure complete dissociation of tar into syngas. Heat recovery and decontamination of the crude syngas are also done by a proprietary process. The inorganic or mineral part of the feedstock is transformed into ash, which is processed further using plasma torch technology to melt at 1,400°C. Upon cooling, the plasma torch–processed ash forms an inert vitrified material. The ash treatment method is the Europlasma vitrification process. Europlasma has six plants currently equipped with this vitrification process.[17]

Phoenix Solutions Plasma Arc Torches, Phoenix Solutions Company (PSC)

Phoenix Solutions Company (PSC) manufactures and designs high-temperature heating equipment, which include large plasma heating systems. PSC technology is used by its customers in the gasification of low-level nuclear waste or reduction of incinerator ash, or. Phoenix Solutions Plasma Arc Torches are available in three arc modes and five power ranges. PSC will help clients determine which arc mode (transferred, nontransferred, or convertible) and power range best fit the end-user's operating environment and processes. PSC's "transferred arc plasma torch" with one internal electrode transfers the arc of the plasma jet to the melt, resulting in localized and high heat. Transferred-arc torches can operate on a variety of gases such as argon, N_2, helium, H_2, air, CH_4, propane, or oxygen. Transferred-arc torches operate between 50 and 3,000 kW. PSC's "non-transferred arc torch" houses both front and rear internal electrodes, creating a jet of plasma from the end of the torch allowing it to be moved inside the furnace, vessel, reactor, or gasifier. Non-transferred plasma arc torches can operate on air, N_2, oxygen, H_2, CO, or CO_2 between 50 and 2,000 kW. PSC's convertible arc torch can transfer the arc on the fly from the working material to a front electrode installed on the torch, allowing both transferred and nontransferred operation.[18]

Torch power ranges depend on the size of the torch procured. A PT50 operates between 10 and 100 kW. This torch is suited for R&D and laboratory work. A PT150 operates between 100 and 300 kW. It is suited for pilot-scale projects and small industrial applications. PT200 operates between 200 and 900 kW and offers high-temperature operation well suited for applications processing wastes into fuel, producing silicon ingots and recovering high-value metals. The PT 250 operates between 800 and 3,000 kW and is used in applications such as waste pyrolysis and metal recovery. It is used in ash melting and low-level nuclear waste processing. The PT255 is the largest torch operating between 1,500 and 3,000 kW.[18]

PSC's plasma heating technology is used to generate electricity from MSW at the Plasco Energy Group facility in Ottawa, Canada installed on Plasco's conversion process. Also, Kobelco Eco-Solutions Company, Ltd. in Kobe, Japan uses PSC plasma heating technology to remediate ash from municipal incinerators reducing the volume and toxicity of the ash waste.

PyroGenesis Plasma-Based Waste to Energy

PyroGenesis builds plasma-based waste treatment systems, plasma torch systems, metal dross recovery systems, and custom high-temperature equipment. PyroGenesis develops and commercializes technologies based upon the use of intense energy found in plasma to convert waste into energy and useful by-products. PyroGenesis has developed a specialty system to treat various waste streams on board ships. PyroGenesis is a plasma waste technology provider and is capable of designing and manufacturing its own plasma torches.

PyroGenesis's waste treatment technologies include the "Plasma Arc Waste Destruction System (PAWDS)" for nonhazardous combustible ship-board applications. The PAWDS is capable of treating a variety of waste generated on board a ship such as paper, cardboard, food, food contaminated waste, plastics, textiles, wood, cabin waste, biosludge, and sludge oil. The effective destruction of these wastes eliminates the need for off-loading of waste in port. Also, PyroGenesis has the waste-to-energy "Plasma Resource Recovery System (PRRS)" for hazardous, industrial, clinical, and municipal solid wastes.

PyroGenesis designs and manufactures 50–500 kW plasma torch systems for its own use as well as for sale to third parties. These plasma torches supply high energy with temperatures near 5,000°C for gasification or vitrification of waste materials. As stated, the PyroGenesis' torches range in power from 50 to 500 kW.[19]

Integrated Environmental Technologies, LLC (InEnTec)

InEnTec (Integrated Environmental Technologies, LLC), has developed the proprietary Plasma Enhanced Melter, PEM™, system, which can convert many types of waste materials to Syngas which then can be converted to many clean fuels and/or generate electricity. A by-product is a solid nonleachable glasslike material that can be used to create items such as blasting grit or building materials and recoverable metals. The PEM™ system has proven effective, and commercial units are operational.

The diagram on the left of Fig. 11.17 illustrates the relative locations of the AC resistive heating and DC plasma-arc heating electrodes and the two areas where heating takes place. The DC plasma power is controlled by current and voltage. Adjusting the AC current flowing through the molten glass bath controls the joule heating power. The resistance of the molten glass determines the corresponding power added. The resistance of the glass is controlled by its composition. Thus, the operator will set the desired power as a set point and the AC power supply will achieve this power input by automatically adjusting the AC current to achieve that set point.

The syngas produced by the gasification of the feedstock/waste consists primarily of H_2 and CO, which is removed from the chamber as shown in Figs. 11.19 and 11.20.

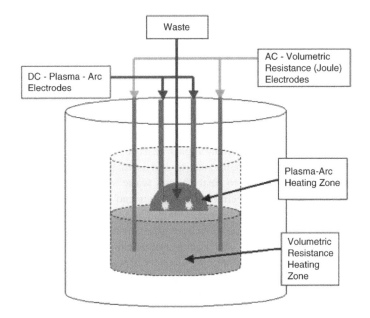

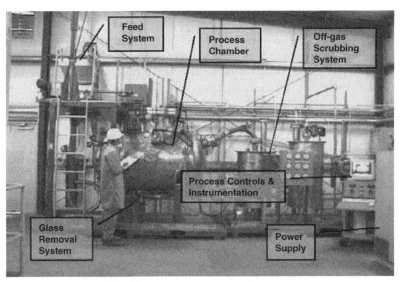

FIGURE 11.17 PEM™ Process System by Integrated Environmental Technologies, LLC (InEnTec).[20]

The syngas is further processed downstream via a gas scrubber system. The process chamber is maintained under a slight vacuum by the syngas vent system. The typical PEM™ system coverts waste solids feed to syngas, glass, and metals. Application of the syngas can be for electric power generation and production of SNG, methanol, ethanol, and chemical feedstock for petrochemical industry. The nonleachable solid

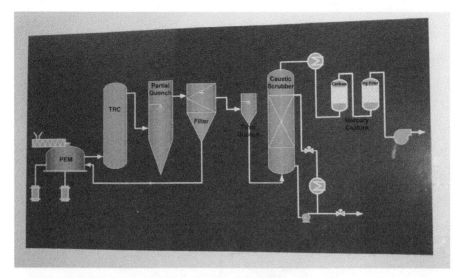

FIGURE 11.18   Plasma Enhanced Melter, PEM™ Simplified Process Flow Diagram.

glasslike material can be used for sand blasting media, construction materials such as aggregate and glass fiber, shingle materials, and road material. The metal can be recycled. A skid-mounted unit of the PEM™ process is shown to the right of Fig. 11.17.[20] A simplified process flow diagram of the PEM™ system is shown in Fig. 11.18. A process schematic for the PEM™ process chamber is shown in Fig. 11.19, which shows main locations of the chamber's key process elements. Also,

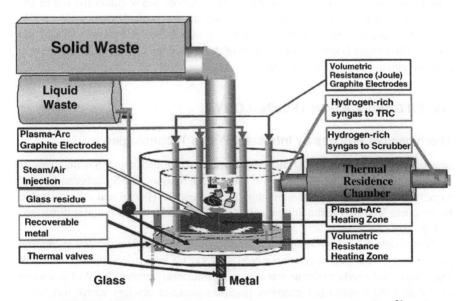

FIGURE 11.19   Process Schematic for the PEM™ Process Chamber.[21]

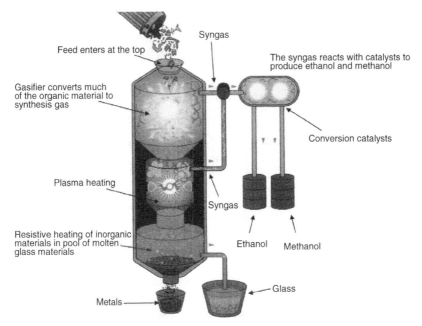

Syngas

Feed enters at the top

The syngas reacts with catalysts to produce ethanol and methanol

Gasifier converts much of the organic material to synthesis gas

Conversion catalysts

Plasma heating

Syngas

Resistive heating of inorganic materials in pool of molten glass materials

Ethanol   Methanol

Glass

Metals

FIGURE 11.20   Gasification, InEnTec Process, Liquid Fuels from Municipal Solid Waste.[22]

the process chamber accomplishes gasification of the organic constituents and vitrification of the mineral constituents simultaneously.[21]

Figure 11.20 illustrates the InEnTec gasification process for the production of liquid fuels such as ethanol and methanol from MSW. MSW enters the top of the plasma arc gasifier whereby the organic constituents are gasified to syngas (primarily H_2 and CO) and the high-temperature process produces vitrified glassy material and metals. The syngas is shown downstream of the gasifier as reacting with catalysts to produce ethanol and methanol fuels.[22]

OTHER GASIFICATION TECHNOLOGY

Thermoselect Process by Interstate Waste Technologies

Interstate Waste Technologies (IWT) has a process that transforms MSW into useable materials via gasification of waste into energy-rich Syngas. The Syngas can be used to generate electricity or to manufacture clean diesel or H_2 fuel and other valuable by-products. The IWT technology is called the thermoselect process and is illustrated in Fig. 11.21.

Thermoselect is a high-temperature gasification process. The technology has an initial pyrolysis (degassing) chamber at 570°F (300°C), which decomposes the MSW into syngas and a carbon char mixed with inorganic constituents of MSW. The carbon char enters the gasification chamber (high-temperature reactor) where oxygen is injected to complete the gasification of the carbon at 2,200°F (about 1,200°C) to syngas. The syngas leaves the top of the reactor at about 2,000°F and is then rapidly

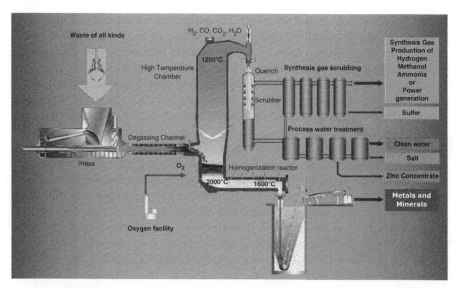

FIGURE 11.21   Thermoselect Process by Interstate Waste Technology.[23]

quench-cooled to below 200°F and cleaned. Cleaning is achieved through the use of acid scrubber followed by alkaline scrubber. Following these cleaning process units, the syngas is passed through a desulfurization process to convert hydrogen sulfide (H_2S) to elemental sulfur. By-products produced from the syngas cleaning process are marketable materials. The cleaned syngas can then be used as a fuel and combusted in a boiler, gas turbine, or reciprocating engine for power generation. The cleaned syngas is estimated to have a higher heating value of 256 Btu per cubic foot. Inorganic constituents in the MSW are mainly metals and silica and are heated in the bottom of the reactor where oxygen is added at >3,000°F whereby the inorganic material is converted to molten slag. The molten slag flows into a water bath and is recovered as a slag aggregate. Slag aggregate is a saleable product and so are the by-products produced from the cleaning of the syngas.[11,23]

IWT is the sole North American licensee of the thermoselect gasification technology. Thermoselect gasification can process several types of waste, which include MSW, construction waste, industrial waste, and sewage sludge. Thermoselect technology has been in commercial operation at several locations in Japan such as Chiba and Kurashiki processing MSW with other types of waste. Chiba has been operating for the longest time at a capacity of 290 tons/day and Kurashiki is one of the newest facilities at a capacity of 520 tons/day.[11,23]

Primenergy's Gasification System at Moderate Temperatures

Primenergy, L.L.C. is located in Tulsa, Oklahoma and with primary business of engineering, procurement, and construction of turnkey, biomass fueled energy conversion, and recovery facilities. Primary products are proprietary thermal oxidizers, gasification technology, and gas cleaning processes. Primenergy has extensive experience in the gasification and conversion of biomass feedstocks to recover energy in the form of steam, heat, and electricity. The gasification technology by Primenergy is air-blown, updraft, fixed-bed, and subatmospheric gasifier. The process of gasification is the thermal conversion of a solid biomass feed into a gas. The gasifier produces a gas that is typically syngas composed primarily of H_2 and CO. Air flow is controlled in the gasifier to maintain proper process operating temperature and the gasifier operates under oxygen-starved atmosphere. Negative pressure within the gasifier is controlled by an induced draft fan located downstream of the heat recovery unit. Commercial sizes vary from 30 to 550 tons/day of biomass. The energy from the gasification process is used in a variety of industrial needs such as hot air for material drying, process steam, and the generation of electricity. Also, Primenergy has access to steel fabrication, selection, and installation of refractory by a related company, Heater Specialists, Inc., which specializes in fabrication and refractory installation of fluid catalytic cracking systems. An installation for one of Primenergy's gasification system is shown in Fig. 11.22.[24] This facility has been in operation since 1996, and the biomass gasifier technology converts up to 600 ton per day of rice hulls for the customer with the capability of producing 6000 kW of electricity and 100,000 lb/hr steam.

Another application of particular interest is Primenergy's gasification technology used for the disposal of poultry litter. Past practice was to spread the poultry waste over agricultural fields. The intent was enrichment of the soil with this nutrient-laden material. However, poultry waste tends to decompose rapidly, releasing soil nutrients. Over time, these water-soluble leachable nutrients are dissolved in rainwater and get passed into watersheds, causing algae growth, lowering dissolved oxygen levels, and leading to deterioration of water quality.

Primenergy came up with a unique solution to the problem of poultry litter. Understanding that one of the primary soil nutrients in the droppings of the poultry litter was phosphorus, a potential problem existed. If the poultry litter is burned at

FIGURE 11.22 Primenergy's Gasification System.[24]

elevated temperatures such as in an incinerator, phosphate salts and phosphoric acid are formed, which can cause significant corrosion problems and fouling of heat recovery boilers. This problem would eliminate the litter as being a potential fuel source in a conventional steam generation plant. To resolve these potential problems, tests were conducted at the Tulsa, Oklahoma facility on poultry litter as an energy source at a rate of 1 ton/hr unit. The test showed that the recoverable energy from the bulk litter was approximately 5.000 Btus per pound. By gasification, the lower operating temperature vaporized less of the inorganic nutrients and evolved a syngas that could be "dry scrubbed" to remove any minor amounts of vaporized contaminants. The ash discharged from the gasifier was odor-free and biologically inert. An analytical analysis on the ash revealed that the inorganic soil nutrients and micronutrients remained in the ash. Thus, it was demonstrated that the poultry litter was an energy source and the ash produced from the gasifier had the necessary composition as a fertilizer. Thus, the gasification process had the potential for a complete solution to the poultry litter problem by providing energy and a fertilizer for the by-product ash. Working with a company in Maryland whose primary business expertise is the manufacture of process equipment for the fertilizer industry, tests were conducted on the by-product ash. The Maryland company had patents on a process of converting a feed material to fertilizer. One process would mix ash with other raw fertilizer ingredients such as a binder and water to produce a granulated fertilizer. The Maryland company's research with the litter ash proved that a refined product could be made and marketed as a granulated fertilizer with an estimated value of about $200.00 per ton. If used in conjunction with Primenergy's gasification system, the granulation system turns the ash from the poultry litter into a packageable, exportable, and granulated 5-30-5 fertilizer. The process is called the poultry litter gasification granulation process (PLGGP) and is illustrated in Fig. 11.23.[24] Currently, commercial economic considerations indicate that a plant capacity in the range of 160 to 250 tons/day (as received poultry litter) has potential for economic viability.[25]

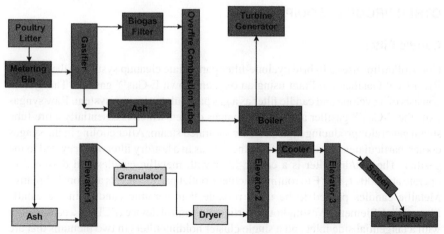

FIGURE 11.23   Poultry Litter Gasification Granulation Process (PLGGP).[24]

This process indicates that all solutions by application of gasification technology are "not" necessarily done at extremely high temperatures. Research into the constituents and properties of a particular biomass is essential for developing and determining a successful commercial process for use as a feedstock in the production of energy and marketable by-products.

Nexterra's Gasification System at Moderate Temperatures

Nexterra Energy Corporation develops gasification systems to generate energy from waste fuels for use by industrial and institutional customers. Nexterra's gasification technology uses biomass such as wood products from sawmills, panelboard plants, pulp, and paper mills. The Nexterra gasification technology uses fixed-bed updraft gasifier. Fuel sized to 3 inches or less is fed into the bottom of the refractory lined gasifier. Combustion air and/or oxygen and steam are also introduced near the bottom of the gasifier. The oxygen is used for the generation of heat following gasification of the feedstock by gasification reactions, principally endothermic reactions, in an oxygen-deficient atmosphere. The gasification reactions proceed at a moderate temperature range of 1,500–1,800°F as the feedstock is converted to syngas. Ash is discharged from the base of the gasification reactor/gasifier and is free-flowing because the gasification temperature in the fuel bed is kept below the ash melting point. The gasifier has flexibility for operation and can be stable from about 20% to 100% of design capacity. Nexterra gasifiers can accommodate a variety of wood fuels with a moisture content of up to 60%.

The cleaned syngas can be used as fuel for energy-producing equipment such as boilers, dryers, and kilns. The cleaned syngas can also be used in gas turbines or combustion engines for the production of electricity, steam, and/or hot water. The syngas is composed mainly of H_2, CO, and CH_4, with some amounts of pyrolysis vapor hydrocarbons.[26,27]

OTHER PROCESS EQUIPMENTS

Candle Filter

ConocoPhillips uses a hybrid cyclone-filter particulate cleanup system at the Wabash River Coal Gasification Plant using an oxygen-blown E-Gas™ gasifier. The system consists of a cyclone and candle filter as a gas particulate cleanup system. Raw syngas exits the E-Gas™ gasifier and enters a syngas cooler, which is essentially a fire-tube steam generator producing high-pressure saturated steam. After cooling in the syngas cooler, particulates are removed from the syngas in a hot/dry filter and recycled to the gasifier. The hot/dry filter is a candle filter with metallic candles and designed to operate at 1,000–1,800°F to routinely achieve outlet particle levels at about 0.1 ppmw. Metallic candles proved to be more reliable than ceramic candles in the candle filter.[28–31] A Siemens-Westinghouse particulate control device (PCD) is a candle filter with a tangential side inlet and a single cluster holding filters in two plenums that are back-pulsed separately. It is designed to operate at 1,000–1,800°F.[31]

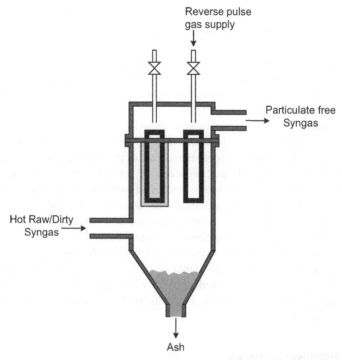

FIGURE 11.24 Schematic of a Candle Filter Unit Based Dry Cleaning System.[28,32]

This new emerging technology of a candle filter is in simplicity the accepted basic concept of a dust collector, i.e., a baghouse with filter bags and periodic reverse pulse gas jet cleaning of the bags. In the case of high-temperature gasifier using the candle filter for particulate removal, the unit comprises multiple particulate filter elements in the shape of a candle arranged similar to the simplistic diagram shown in Fig. 11.24.[32] The hot raw "dirty" syngas enters on the outer surface of the candle filter elements where the particles from the sygnas collect and form a cake of ash as the clean syngas flows into the center of the candle filter element and passes out from the top of the unit. The buildup of the ash cake/layer on the outer surface of the filter element causes the pressure drop across the filter element to increase due to the resistance created by the gas passing through the dust cake layer. Periodically, a gas or inert gas is back-pulsed through the candle filter element to dislodge the ash cake/dust layer from the outer surface of the candle filter element. The dislodged ash layer is withdrawn from the bottom of the unit.[28,32]

Pressure Swing Adsorption (PSA) Units

PSA is a technology to separate a constituent from a gas under pressure depending on the constituent's molecular characteristics and adsorption potential for an adsorbent material. PSA is an energy-efficient noncryogenic way to purify a gas.

Special adsorbents in a packed bed (column) are used to typically adsorb the undesirable components in the gas stream at high pressure and pass through the desirable component out from the exit of the packed bed. The process then swings to a lower pressure to desorb the undesirable components from the bed. Adsorptive materials such as zeolites preferably adsorb the undesired gaseous components onto their surface at high pressure. Activated carbon is also a well-known adsorbent material.

As an example, H_2 gas is to be recovered from a gaseous mixture of H_2, CO_2, CO, and H_2O. The gaseous mixture is passed through an adsorbent bed. The impurities (CO_2, CO, and H_2O) are adsorbed onto the surface of the adsorbent bed. The H_2 passes through the bed and out from the exit of the bed, i.e., out from the top of the packed bed column. Before the adsorbent bed is saturated with the impurities, the bed is depressurized allowing the impurities to be desorbed from the adsorbent and flow out from the bed. This pressurizing/depressurizing cycle or "pressure swing" of the adsorbent bed gives the process its name, PSA. Two vessels are used so while one vessel is pressurizing (adsorbing), the other bed is depressurizing and regenerating its bed (desorbing). H_2 purity of $>99 + \%$ is achieved. For a PSA unit producing oxygen from the air, a product purity of oxygen (95% purity) is typical for use in gasification plants. N_2 gas can be produced from air with a PSA unit with a purity of 95–99.5%.[33–37]

Mercury Removal Systems

Eastman Gasification Services Company (a subsidiary of Eastman Chemical Company) have a process for vapor-phase removal of mercury from syngas with removal of >90–95%. Eastman Chemical approach to the removal of mercury from the vapor phase has been with activated carbon beds for 20 years. These beds are located upstream of the acid gas removal (AGR) process. Beds are changed every 18–24 months with mercury removal of 90–95%.[38–40]

Main Sulfur Removal Technologies

Selexol™ process is licensed by UOP, LLC. The Selexol™ solvent absorbs the acid gases from the Syngas at pressure range from 300 to 2,000 psia. The rich solvent is then reduced in pressure and/or sent to a steam stripper to recover the acid gases. The Selexol™ process can selectively recover H_2S and CO_2 as separate streams. Thus, the H_2S can be sent to a Claus plant for conversion to elemental sulfur. The CO_2 can be recovered and used as a raw material locally and/or sold as a marketable product. The Selexol™ process uses dimethyl ethers of polyethylene glycol (DEPE) (Union Carbide's Selexol™ solvent) for AGR via physical absorption with removal efficiency of $99 + \%$ sulfur removal. The Selexol™ solvent is chemically inert and is not subject to degradation. The Selexol™ process also removes COS, mercaptans, ammonia, HCN, and metal carbonyls.[38,41–43]

Rectisol™ process is licensed by Linde, AG. The Rectisol™ process uses cold methanol at about $-40°F$, which absorbs the acid gases from syngas at relatively high pressures from 400 to 1,000 psia. The rich solvent is then reduced in pressure and/or steam stripped to release and recover the acid gases. The Rectisol™ process can selectively recover H_2S and CO_2 as separate streams. Thus, the H_2S can be sent to a Claus plant for conversion to elemental sulfur. The CO_2 can be recovered and used as a raw material locally and/or sold as a marketable product. The Rectisol™ process for AGR uses cold methanol and is a physical absorption process recovering between 99.5% and 99.9 + % sulfur removal and complete CO_2 removal is possible. The Rectisol™ process is used by Eastman Chemical.[38,44,43]

Methyldiethanolamine (MDEA) is a chemical absorption process that absorbs the acid gases from syngas at moderate operating temperature and has low AGR cost. MDEA process has 98% to 99 + % sulfur removal and can remove CO_2. The process consists basically of two columns, one column (absorber) for absorbing the H_2S and CO_2 with MDEA and the other column (regenerator) for stripping off the H_2S and CO_2 from the MDEA. In the absorber, the downflowing MDEA liquid absorbs the H_2S and CO_2 from the upflowing acid gas. In the regenerator, the downflowing MDEA liquid rich in H_2S and CO_2 is stripped by upflowing lean MDEA vapors. The H_2S stream is usually sent to a Claus process for recovery of elemental sulfur.[38,43]

COMBUSTION TURBINE FOR SYNGAS AND GAS ENGINE FOR SYNGAS

Siemens-Westinghouse Syngas Combustion Turbine for Syngas

In a combined cycle power plant, the hot exhaust gas from the combustion gas turbine is used by a downstream heat recovery steam generator (HRSG) for use in a steam turbine. Thus, both the combustion gas turbine and the steam turbine are used to provide power for a generator in the production of electricity with the highest efficiency. When the combined cycle power plant is associated with gasification technology for the production of the fuel gas to be used by the combustion gas turbine, the system is called IGCC.[45–50]

For example, gasification technology could be used to convert coal and other fuels into a synthetic gas (syngas), which is used by the IGCC system to produce power. Current state-of-the-art gas turbine and steam turbine system is well suited for IGCC power plants. A Siemens' gas turbine and steam turbine are illustrated in Figs. 11.25 and 11.26, respectively.[45]

Siemens and former Westinghouse gas turbine division have many years of experience with syngas. Siemens large E-class gas turbine and F-class gas turbine are for use with syngas in IGCC projects. Syngas can be produced from a wide variety of fuels such as coal, biomass, petroleum coke, and refinery residues. The purified syngas provides a useable source of fuel for the gas turbine. Gas turbines operating on syngas have low NO_x emissions and therefore have a positive influence upon a power plant. An illustration of a pictorial internal view of Siemens' gas turbine is shown in Fig. 11.27.[46]

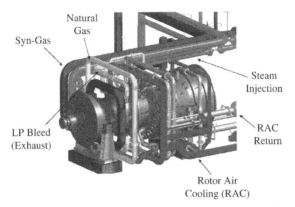

FIGURE 11.25 Siemens SGT6-5000F Gas Turbine.[45]

General Electric (GE) Combustion Turbine for Syngas

For a combustion turbine, a couple of typical gaseous fuel sources are available depending on the fuel source. One common source is a medium-heating-value fuel from 200 Btu/SCF to typically 300 Btu/SCF produced from oxygen-blown gasification of coal. This manufactured fuel is commonly referred to as syngas or synthesis gas and contains predominantly H_2 and CO.

These gas mixtures typically produce high flame temperatures. For an air-blown gasification of coal or biomass, the typical gaseous fuel produced has a low-heating-value fuel of about 100–200 Btu/SCF. A gas turbine obviously must be of a design to operate reliably and efficiently on the gaseous fuel and/or fuels available for a particular customer's application.[51]

The IGCC power plants are highly efficient commercially in the production of electricity at cost below that of a conventional solid fuel plant. An IGCC power plant has a gaseous fuel supplying a gas turbine producing electricity and the exhaust gases

FIGURE 11.26 Siemens SST6-5000 Steam Turbine.[45]

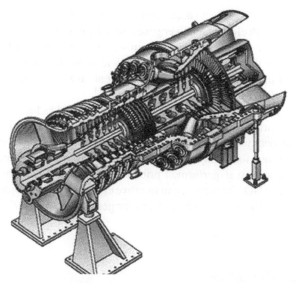

FIGURE 11.27   Siemens Gas Turbine.[46]

from the gas turbine are fed to an HRSG that produces steam for a steam turbine producing electricity. The combined electricity from the gas turbine followed by the steam turbine makes the IGCC an efficient process for the production of electricity. With a coal gasification plant producing syngas for use in the IGCC power plant, the synthetic gas (syngas) has a low heating value; therefore, the mass flow and thus power output of the gas turbine is much higher for an IGCC application. For a GE model 7FA gas turbine on syngas would be rated at about 197 MW but for an IGCC operation it would be 280 MW. IGCC SO_x, NO_x, and particle emissions are fractions of those of a conventional pulverized coal (PC) boiler power plant.[52,53] An illustration of a GE model MS 7001FA gas turbine is shown in Fig. 11.28.[54]

The GE model MS 7001FA gas turbine was used in the clean coal technology (CCT) demonstration program in cooperation between government and industry to demonstrate a new generation of innovative coal gasification process and power plant.

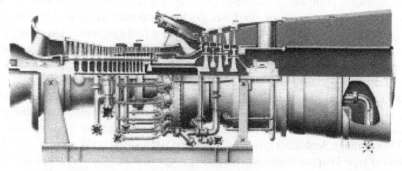

FIGURE 11.28   General Electric Model MS 7001FA Gas Turbine.[54]

The Wabash River CCT project successfully demonstrated the E-Gas™ coal gasification technology with electrical power generation. Emissions of SO_2 and NO_x were far below regulatory requirements. SO_2 emissions averaged about 0.1 lb/million Btu compared with the allowable limit of 1.2 lb/million Btu. NO_x emissions were 0.15 lb/million Btu, which met current requirements for a coal-fired power plant. Particulate emissions were below detectable limit.[54]

The cleaned syngas at the Wabash River project was used as fuel in the GE model MS 7001FA gas turbine where the gas was combusted to generate electricity. The Wabash River project was the first to use an "F" machine on syngas. Advanced gas turbine design allows for the syngas combustion at firing temperature of 2,350°F. Syngas humidification was performed prior to combustion to control combustion temperature, reduce NO_x formation, and improve efficiency. Gas turbine exhaust heat was supplied to the HRSG to produce steam for generation of electricity in a steam generator.[54,55]

GE Gas Engine for Syngas

GE Jenbacher is a manufacturer of gas-fueled reciprocating engines, packaged generator sets, and cogeneration systems for power generation.[56] These gas engines are primarily designed to operate on fuel gas such as natural gas, biogas, landfill gas, and syngas.

Various gases of extremely low calorific value (BTU value) with high water content and varying composition can be used as fuel to run the GE Jenbacher engines.

GE Jenbacher combustion engines can use biogas as fuels created by anaerobic fermentation of biological materials. Biogas includes mainly CH_4 and CO_2 as the constituents. A biogas can originate from the anaerobic fermentation of organic materials. Landfill gas is created by the decomposition/fermentation of the organic substances such as from MSW, consists mainly of CH_4, CO_2, and N_2 (N_2), and has a calorific value of about 480 Btu/SCF. Typical composition can be 45–60% volume CH_4, 40–60% volume CO_2 and 2–5% volume N_2. Digester gas (sewage or biogas) has a calorific value of about 690 Btu/SCF. Note that natural gas is typically 950–1150 Btu/SCF. A simplified schematic of a process generating biogas and using a GE Jenbacher combustion engine to generate electricity is shown in Fig. 11.29.

Syngas or synthetic gas or synthesis gas has a calorific value of 125–350 Btu/SCF. GE Jenbacher engines can run on syngas as a fuel gas, which has a very low calorific value. A simplified schematic of a gasification process generating syngas and using a GE Jenbacher combustion engine to generate electricity is shown in Fig. 11.30.

GE's Jenbacher gas engines range in power from 0.25 to 3 MW and operate with natural gas or a variety of other gases such as biogas, landfill gas, coal mine gas, sewage gas, and industrial combustible waste gases. The Jenbacher type 2 operates in the 350 kW range. Jenbacher type 3 runs between 500 and 1,100 kW power range. The Jenbacher type 4 engine operates in the power range 800–1,500 kW. A Jenbacher type 6 engine runs in the power range of 1.8–3 MW. The new Jenbacher J624 GS can operate with electrical output of 4 MW.[56,57]

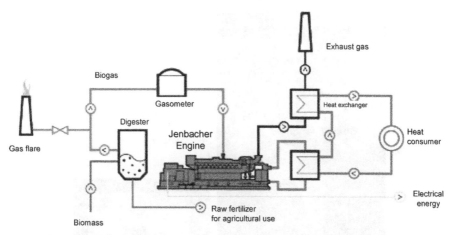

FIGURE 11.29 Schematic of Process Producing Biogas and Using a GE Jenbacher Combustion Engine to Generate Electricity.[56,57]

NONCONTACT SOLIDS FLOW METER FOR WASTE SOLIDS (RayMas® Meter)

The feedstocks for gasifiers are diverse solid particulate materials with properties including particulate size, composition, density, and flow characteristics. The feedstock could be sticky due to moisture content, therefore posing a problem for instrumentation such as meters for measuring mass flow rate. Some new metering technology has emerged over the past several years to resolve the problems associated with measuring mass flow of solids (particulate mass flow) through process lines and ducts.

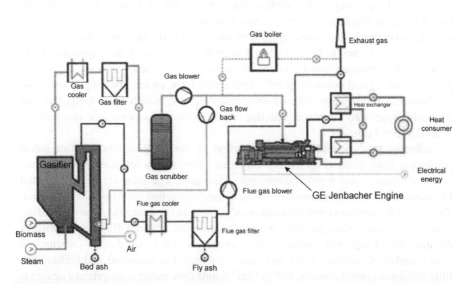

FIGURE 11.30 Schematic of Gasification Process Producing Syngas and Using a GE Jenbacher Combustion Engine to Generate Electricity.[56,57]

Sensor

Flow Tube

Control Panel/Instrumentation

FIGURE 11.31 RayMas® Noncontact Solids Flow Meter, 6-inch diameter Gravity Flow Type Metering Material, 0.5 to 1.5 micron range Mineral Powder.

This new development in mass flow metering technology is known as the RayMas® meter, i.e., proprietary and patented noncontact solid flow metering technology. Simply put, the RayMas® meter consists of the following components: a flow tube, a sensor, and controls/instrumentation. A picture of an early version of the meter with components is shown in Fig. 11.31, a 6-inch diameter gravity flow type meter. As shown, a fine mineral powder is conveyed by a screw feeder into the top of the Raymas® meter and drops by gravity flow through the meter, discharging out from the bottom of the meter into another screw feeder, which then feeds the fine mineral powder into a furnace. The fine mineral powder has a particle size ranging from 0.5 to 1.5 μm.

Referring to the simplified representation of the RayMas® noncontact solids particulate flow meter (Fig. 11.32), the sensor transmits a low-energy microwave to the particulate matter passing through a flow tube. The intensity of the reflected Doppler-shifted energy is measured by the sensor converting it into a 4–20 mA signal. This signal is then converted by a unique algorithm in the controller/instrumentation to mass flow rate, such as pounds per hour of particulate material. RayMas® measures the intensity of reflected Doppler-shifted energy, which is based upon velocity and flow quantity of material. Thus only moving material is measured and buildup has little influence upon the meter. In RayMas® solids flow meter, a proprietary design is used coupled with a unique algorithm to provide a meter with mass flow rate and

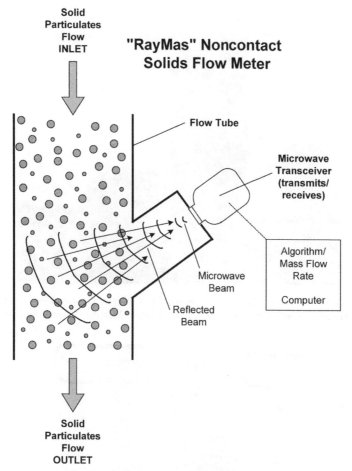

FIGURE 11.32   RayMas® Solids Flow Meter, Simplified Schematic Representation.

totalizer (quantity) measurement capabilities for each particular material to be metered. Gravity flow type and pneumatic flow type meters are available, but experience has shown that the gravity flow type is the most easily adapted to applications.

The 6-inch diameter and a 8-inch diameter gravity flow meter are shown in Fig. 11.33. The 8-inch diameter meter is being used to meter a corn by-product material at flow rates from 8000 to 20,000 lbs/hr.

For another gravity flow application, a 20-inch diameter meter has been installed for metering a wet corn fibrous material at rates between 15,000 and 40,000 lbs/hour. The 20-inch diameter meter is shown in Fig. 11.34.

A 4-inch diameter meter is shown in Fig. 11.35 for a pneumatic conveying (P.C.) type of RayMas® noncontact solids flow meter.

The RayMas® meter is a new technology and of unique and proprietary design. Final accuracy of the RayMas® meter depends on the material being measured and the

Flow Tube ⟶

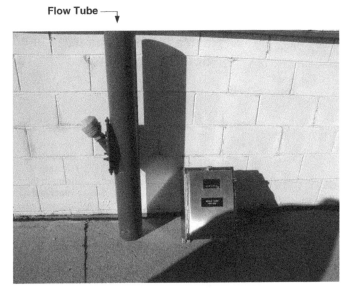

FIGURE 11.33 RayMas® Noncontact Solids Flow Meter, Gravity Flow Type; 6-inch Diameter Metering Whole Corn, Similar 8-inch Diameter Metering Ground Processed Corn.

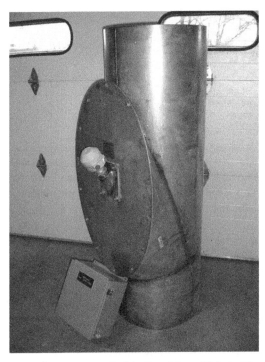

FIGURE 11.34 RayMas® Noncontact Solids Flow Meter, 20-inch Diameter, Gravity Flow Type, Product Material Metered—Wet Corn Material.

FIGURE 11.35   RayMas® Noncontact Solids Flow Meter, Pneumatic Conveying (P.C.) Type.

particular installation. Gravity flow applications produce a more reliable mass flow reading compared with pneumatic flow applications. A pneumatic flow application has more variables for a particular installation, such as gas flow conditions and suspension of particulates in the conveying line.

The industrial market has many applications for segments such as chemical, plastics, petroleum, utility (power plants), food, pharmaceutical, cement, mining, precision farming, and others for a gravity type mass meter or a PC type mass flow meter applications. As the new Raymas® noncontact solids mass flow meter improves from refinements and upgrades via technology, the market potential and need for such process innovation in metering instrumentation will continue to be of much interest.[58-64]

In the research and development (R&D) of the RayMas® noncontact solids flow meter, many experimental runs were made comparing weighing scale readings with the particular RayMas® meter calibrated on a laboratory test stand. Some tests were conducted to develop the technology and other runs were made for calibration of a meter for a particular customer's industrial application. The following cases were obtained experimentally, and illustrate the future potential for the RayMas® noncontact solids flow meter. Typically, a particulate solids material was passed through the

FIGURE 11.36 Calibration Equipment for a RayMas® Solids Flow Meter, Gravity Flow Types.

meter and recordings of the meter were compared with those from a weighing scale. The experimental apparatus used for some of the experimental tests and calibrations is shown in Fig. 11.36 for use with a gravity flow type meter.

Case #1 A small portable RayMas® noncontact solids flow meter (gravity type) was designed, developed, and tested on a test stand. Some high-density polyethylene plastic beads were metered into the top of the 2-1/2 inch diameter vertical gravity flow meter and passed through the flow tube into a bin on a weighing scale. From the mass flow rate of beads and the corresponding microwave sensor output, a unique algorithm was computed for the high-density polyethylene plastic beads and for that particular meter. The meter was calibrated for a mass rate range of 0 lbs/hr (no flow) to 1,817 lbs/hr. The beads were 5/32 inch in diameter with a bulk density from 35.78 to 36.14 lbs/ft³. Results from this initial development work were successful as shown in Table 11.1.

Case #2 A RayMas® gravity flow meter was designed and calibrated for metering green oats. Green oats have the following properties: bulk density ranging from 34.93 lb/ft³ (loose) to 37.67 lb/ft³ (tapped) and particle shape of about $^1/_2$ in. length and 1/8 in. width in the center and tapering toward the ends to a point. The RayMas®

TABLE 11.1   Portable RayMas® Meter Gravity Flow Type, 2-1/2 inch diameter Meter, Calibration Runs

Run No.	RayMas® Meter Reading (lbs/hour)	Weighing Scale Reading (lbs/hour)	Error (%)
1	91.0	91.1	0.11
2	148.9	149.7	0.53
3	259.2	258.7	−0.19
4	419.0	420.0	0.24
5	681.9	680.6	−0.19
6	1053.4	1053.9	0.05

Note: Meter: RayMas® Solids Flow Meter, 2-1/2 inch diameter, gravity flow type, aluminum product metered: high-density polyethylene plastic beads, 5/32 in. diameter, bulk density (loose) −35.78 lb/ft³, bulk density (tamped)—36.14 lb/ft³. Average absolute weighing error $= 0.22 +/-\%$. The maximum range of mass flow through the meter was designed from 0 to 1817 lbs/hr. This mass rate does not necessarily represent the maximum permissible rate through this meter.

meter is shown as fabricated in Fig. 11.37. The meter mass flow rate data comparing the RayMas® meter readings with the weighing scale readings are shown in Fig. 11.38. The cross-sectional diameter of the flow tube of the meter were 12-1/2 in. × 22-1/4 in. I.D.

FIGURE 11.37   RayMas® Noncontact Solids Flow Meter, Calibrated for Metering Green Oats, 12-1/2 inch × 22-1/4 inch I.D., Gravity Flow Type.

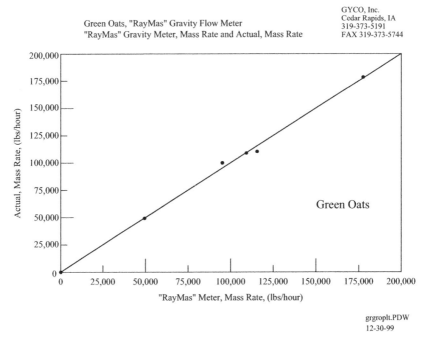

FIGURE 11.38 RayMas® Noncontact Solid Flow Meter, Gravity Type, Metering Green Oats.

After installation, test runs were made with the RayMas® in a production plant. The results of these runs are shown in Table 11.2.

The work in designing and calibrating the RayMas® meter for the product, green oats, demonstrated the flexibility in the design of the flow tube and microwave sensor when proper design criteria were followed.[58–63]

Case #3 A fine powder, an ore oxide, needed to be metered into a high-temperature process operation at a production plant. The ore oxide had a fine crystalline powder having a particle size ranging from 0.5 to 1.5 μm. Flowing properties of this powder are poor to caking. The bulk density ranged from 103.8 lb/ft^3 (loose) to 129.6 lbs/ft^3 (tapped). The meter was designed and fabricated for this application. The installed meter is shown in Fig. 11.31. The meter mass flow rate data comparing the RayMas® meter readings with the weighing scale readings are shown in Fig. 11.39.

Case #4 Whole corn was to be metered using a gravity flow type RayMas®. The whole corn had the following properties: bulk density ranging from 56.3 lbs/ft^3 (loose) to 58.1 lbs/ft^3 (tapped). This whole corn was rather free flowing and had typical particle dimensions as 3/8 in. width × 9/16 in. length and 3/16 in. thickness. Whole corn was clean with minimal dust.

TABLE 11.2   RayMas® Noncontact Solids Flow Meter, Metering Green Oats

Run Number	Run Time (hours)	Mass Rate (lbs/hour)	RayMas® Meter Totalizer (Pounds)	Customer Scale Totalizer (Pounds)	Error (%)
1	168	0–114,010	19,153,740	19,449,500	1.52
2	28	0–110,161	3,084,500	3,047,400	−1.22
3	—	—	13,850,160	14,035,300	1.32
4	—	—	33,216,080	34,136,500	2.70
5	—	—	40,799,370	41,084,000	0.69

Note: Average absolute deviation from customer scale is 1.49%±.

	Run Time	RayMas® Meter Totalizer	Customer Scale Totalizer	Truck Scale (Weights and Measures)
6	28 hours	3,083,900 lbs	3,084,500 lbs	3,047,400 lbs

Note: For Runs 1 and 2, the mass rate was estimated from the totalizer reading and the hours of the run. Operations at the customer's plant could process green oats from no mass flow rate to maximum rate of the meter calibrated. At a test facility, the RayMas® meter was calibrated for mass rate ranging from 0 to 175,000 lbs/hour.

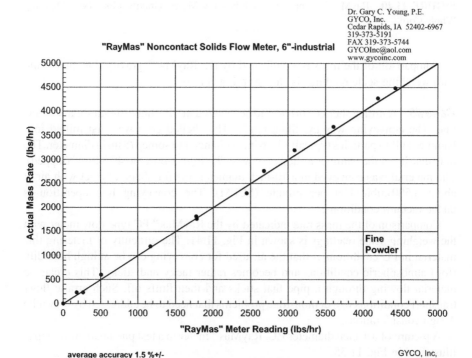

Dr. Gary C. Young, P.E.
GYCO, Inc.
Cedar Rapids, IA 52402-6967
319-373-5191
FAX 319-373-5744
GYCOInc@aol.com
www.gycoinc.com

"RayMas" Noncontact Solids Flow Meter, 6"-industrial

average accuracy 1.5 %+/-

GYCO, Inc,
rmpow1.PDW

FIGURE 11.39   RayMas® Non-Contact Solids Flow Meter, 6-inch Diameter Gravity Type, Fine Powder Product Metered: Fine Ore Powder, Particle Size Range Between 0.5 to 1.5 micron.

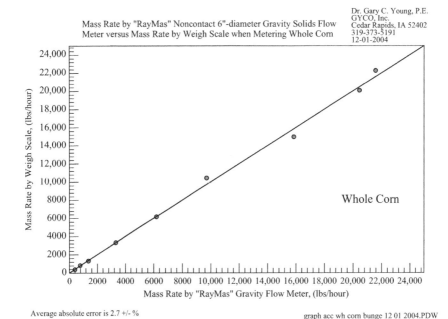

Dr. Gary C. Young, P.E.
GYCO, Inc.
Cedar Rapids, IA 52402
319-373-5191
12-01-2004

Mass Rate by "RayMas" Noncontact 6"-diameter Gravity Solids Flow Meter versus Mass Rate by Weigh Scale when Metering Whole Corn

Average absolute error is 2.7 +/- %   graph acc wh corn bunge 12 01 2004.PDW

FIGURE 11.40   RayMas® Noncontact Solids Flow Meter, Gravity Flow Type Metering Whole Corn.

Comparison of the mass rate indicated by the RayMas® gravity flow meter and with weighing scale readings is shown in Fig. 11.40.

Case #5   A milled shelled corn was to be metered in a PC line, 10—in.-diameter PC line. The conveying was by dilute phase. The shelled corn was first milled in a hammer mill to particles typically 1/16 in. in diameter to some 1/8 in. in diameter. The bulk density of the hammer milled corn was 37.4 lb/ft^3 (loose) to 49.8 lb/ft^3 (tapped). The material was conveyed in a 10-inch diameter line for about 400 feet with air at about 3,500 cubic feet per minute (SCFM). The conveying line operation is under vacuum conditions.

Comparison of the mass rate indicated by the RayMas® PC type flow meter with the weighing scale readings is shown in Fig. 11.41. The difficulty of metering this material in PC conditions is that the air used for conveying can be at high humidity (fog) atmospheric conditions and becomes rather tacky and sticky. This creates a material flowing through a pipe that sticks and then fluffs off. Such erratic flow/ transport conditions will demonstrate problems that need to be addressed by flow tube design considerations.

A picture of a 4-inch diameter P.C. RayMas® meter in a test pneumatic test loop is illustrated in Fig. 11.35.

The RayMas® noncontact solids flow meter technology is in its infancy and of unique and proprietary design.[58–63] The potential for this meter in the industrial

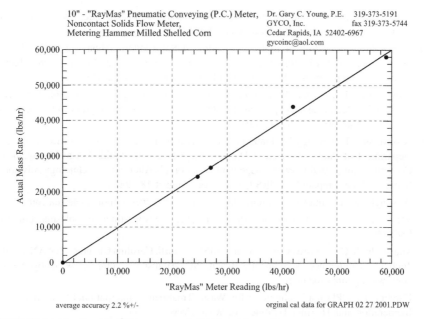

10" - "RayMas" Pneumatic Conveying (P.C.) Meter,   Dr. Gary C. Young, P.E.   319-373-5191
Noncontact Solids Flow Meter,   GYCO, Inc.   fax 319-373-5744
Metering Hammer Milled Shelled Corn   Cedar Rapids, IA 52402-6967
gycoinc@aol.com

average accuracy 2.2 %+/-   orginal cal data for GRAPH 02 27 2001.PDW

FIGURE 11.41 RayMas® Noncontact Solid Flow Meter, Pneumatic Conveying (P.C.) Type, 10-inch Diameter, Metering a Hammer Milled Shelled Corn.

market is substantial in the chemical, plastics, petroleum utility (power plants), food, pharmaceutical, cement, mining, and farming for both gravity and pneumatic type flow meters. As this new technology develops, so will the accuracy, ease of installation, and calibration of the RayMas® meter.

REFERENCES

1. Young, G.C., "Zapping MSW with Plasma Arc," Pollution Engineering, November 2006.
2. Young, G.C., "How Trash Can Power Ethanol Plants," Public Utilities Fortnightly, p. 72, February 2007.
3. Young, G.C."From Waste Solids to Fuel, " Pollution Engineering, February 2008.
4. Ratafia-Brown, J., Manfredo, L., Hoffmann, J., and Ramezan, M., "Major Environmental Aspects Of Gasification-Based Power Generation Technologies, Final Report," Gasification Technologies Program, National Energy Technology Laboratory (NETL) and U.S. Department of Energy (DOE), December 2002.
5. U.S. Department of Energy (DOE), "Combined Heat and Power Market Potential for Opportunity Fuels," Distributed Energy Program Report, August 2004.
6. Process Energy Solutions (PES), Patterson, New York, May 2009.
7. U.S. Department of Energy (DOE), National Energy Technology Laboratory (NETL), "Wabash River Coal Gasification Repowering Project; A DOE Assessment, DOE/NETL-2—2/1164, January 2002.

8. U.K. Department of Trade and Industry, "Gasification of Solid And Liquid Fuels For Power Generation—Status Report," December 1998.

9. Oliver, R. "Application of BGL Gasification of Solid Hydrocarbons for IGCC Power Generation," Global Energy, Inc. Technical paper presented at 2000 Gasification Technologies Conference, San Francisco, California, October 8–11, 2000.

10. Gaudig, U., "Experience During Erection and Commissioning the Gasification Plant Seal Sands, UK," Technical paper presented at 2001 Gasification Technologies Conference, San Francisco, California, October 7–10, 2001.

11. "Conversion Technology Evaluation Report," Prepared for: The County of Los Angeles, Department of Public Works and The Los Angeles County Solids Waste Management Committee/Integrated Waste Management Task Force's Alternative Technology Advisory Subcommittee; Prepared by URS Corporation, August 18, 2005.

12. Westinghouse Plasma Corporation, Madison, PA, www.westinghouse-plasma.com.

13. Westinghouse Plasma Corporation, Plasma Torches, www.westinghouse-plasma.com, 05/08/2009.

14. Martin, Jr., R.C., "Plasma Gasification of MSW," North Carolina Chapter—SWANA, Fall Technical Conference, Wrightsville Beach, NC, August 27, 2008, www.jacobyenergy.com, www.geoplasma.com.

15. Carabin, P., "Plasma Technology for Waste Treatment," International Conference on Incineration and Thermal Treatment of Waste, May 14, 2007.

16. EUROPLASMA, 21 RUE Daugere, 33 520 Bruges, France, www.europlasma.com.

17. Europlasma, CHO-Power plant, www.cho-power.com, 05/08/2009.

18. Phoenix Solutions Company, Minneapolis, Minnesota, Plasma torches, www.phoenixsolutionsco.com, 05/08/2009.

19. PyroGenesis, Inc., Montreal, Quebec, Canada, www.pyrogenesis,com, 05/08/2009.

20. Integrated Environmental Technologies, LLC, InEnTech, www.inentec.com, 05/08/2009.

21. "Environmental Technology Verification Report for the Plasma Enhanced Melter™," Prepared by Environmental Technology Evaluation Center (EvTEC), CERF/IIEC Report: #40633, May 2002.

22. Surma, J., "Reducing Greenhouse Gas Emissions through Conversion of Waste into Clean Fuel," InEnTec, November 21, 2008.

23. Interstate Waste Technologies, Inc., http://iwtonline.com/technology/index.html, 05/08/2009.

24. Primenergy, L.L.C., Tulsa, Oklahoma, www.primenergy.com, 05/09/2009.

25. McQuigg, K., Primenergy, LLC, Tulsa, Oklahoma, Private communication, May 11, 2009.

26. Nexterra Energy Corporation, www.nexterra.ca, May 12, 2009.

27. Editorial Comment, Sawn & Quartered, "Following Oil Money, When the oil patch starts investing in wood residues, it's time to take our own biomass potential seriously," Canadian Wood Products, January/February 2007.

28. National Energy Technology Laboratory (NETL), www.netl.doe.gov, 2007.

29. Longanbach, J.R., Smith, P.V., and Wheeldon, J., "Operation of the Transport Gasifier and Particulate Control Device at the Power Systems Development Facility," 1999.

30. Vimalchand, P., Leonard, R.F., Pinkston, T.E., Smith, P.V., and Wheeldon, J.M., "Power Systems Development Facility: Status and Operation of a Transport Reactor System with a Westinghouse Candle Filter," 1998.

31. Leonard, R., Rogers, L., Vimalchand, P., Liu Guohai, Smith, P.V., and Longanabach, J., "Development Status of the Transport Gasifier at the PSDF," KBR Paper #1707, October 7-10, 2001.

32. Sharma, S. and Thambinulthu, K., "Research Aimed at Lower Emissions," Centre for Low Emission Technology, Annual Report, cLET, Level 3, Technology Transfer Centre, Technology Court, Pullenvale, OLD Australia 4069, March 2006, www.clet.net.

33. Praxair, Inc., www.praxair.com, 2007.

34. Universal Industrial Gases, Inc., www.uigi.com, 2007.

35. The Linde Group and BOC Gases, www.linde.com, www.boc-gases.com, 2009.

36. NATCO, Hydrogen Purity by Pressure Swing Adsorption (PSA), www.natcogroup.com, 2009.

37. UOP, A Honeywell Company, www.uop.com, 2009.

38. Lilly, R.D., "Eastman Gasification Overview," Eastman Gasification Services Company (subsidiary of Eastman Chemical Company), April 2005.

39. Rutkowski, M.D., Klett, M.G., and Maxwell, R.C., "The Cost of Mercury Removal in an IGCC Plant," Gasification Technologies Public Policy Workshop, Washington, D.C., October 1, 2002.

40. Stiegel, G.J., Longanbach, J.R., Klett, M.G., Maxwell, R.C., and Rutkowski, M.D., "The Cost of Mercury Removal in an IGCC Plant," Final Report, Prepared for U.S. D.O.E. and NETL by Parsons Infrastructure and Technology Group, Inc., September 2002.

41. UOP, LLC, Gas Processing, SELEXOL™, 2002.

42. Sweny, J.W., "Synthetic Fuel Gas Purification by the SELEXOL™ Process," 1973.

43. Gallaspy, D., "Eastman Gasification Overview, GTC Regulators Workshop, April 12, 2005.

44. Rectisol™ Process, The Linde Group, www.linde-engineering.com, 5/10/2009.

45. Feller, G.J. and Gadde, S., "Introducing The SGCC6-5000F 2 × 1 Reference Power Block For IGCC Applications," Siemens Power Generation, Inc., 2006.

46. Siemens Power Technology, "Siemens Combined Cycle Power Plant (SCC™)," Siemens Power Generation, Inc., www.power-technology.com, 05/12/2009.

47. Siemens Power Technology, "Integrated Gasification Combined Cycle (IGCC)," www.power-technology.com, 05/12/2009.

48. Poloczek, V. and Hermsmeyer, H., "Modern Gas Turbines with High Fuel Flexibility," Siemens AG, Energy Sector, Germany 2006.

49. Gadde, S., Xia J. and McQuiggan, G., "Advanced F Class Gas Turbines Can Be A Reliable Choice For IGCC Applications," Siemens Power Generation, Inc., 2006.

50. Gadde, S., Wu J., Gulati A., McQuiggan, G., Doestlin, B. and Prade, B., "Syngas Capable Combustion Systems Development For Advanced Gas Turbines," Siemens Power Generation, Inc. 2005.

51. Phillips, J., "Different Types of Gasifiers and Their integration with Gas Turbines," EPRI/Advanced Coal Generation, 2006.

52. GE Energy, "IGCC Benefits," www.gepower.com, May 11, 2009.

53. GE Energy, "Gas Turbine and Combined Cycle Products," www.gepower.com, May 2007.

54. U.S. Department of Energy (DOE) and Wabash River Coal Gasification Project Joint Venture, "Clean Coal Technology, The Wabash River Coal Gasification Repowering Project, A 262 MWe Commercial Scale Integrated Gasification Combined Cycle Power Plant, An Update," Topical Report Number 20, September 2000.

55. U.S. Department of Energy (DOE) and National Energy Technology Laboratory (NETL), "Wabash River Coal Gasification Repowering Project: A DOE Assessment," DOE/NETL-2002/1164, January 2002.

56. GE Energy, (GE Jenbacher, division of GE Energy), News Release, www.gepower.com, May 3, 2004.

57. GE Energy (GE Jenbacher, division of GE Energy), www.gepower.com, May 11, 2009.

58. Young, G.C., U.S. Patent No. 5,986,553, RayMas® meter, GYCO, Inc., gycoinc@aol.com.

59. Young, G.C., U.S. Patent No. 6,404,344, RayMas® meter, GYCO, Inc., gycoinc@aol.com.

60. Young, G.C., U.S. Patent No. 7,310,046, RayMas® meter, GYCO, Inc., gycoinc@aol.com.

61. Young, G.C., Canadian Patent No. 2,207,132, RayMas® meter, GYCO, Inc., gycoinc@aol.com.

62. Young, G.C., Canadian Patent No. 2,322,375, RayMas® meter, GYCO, Inc., gycoinc@aol.com.

63. Young, G.C., Canadian Patent No. 2,555,546, RayMas® meter, GYCO, Inc., gycoinc@aol.com.

64. "RayMas" is a registered trademark of GYCO, Inc., gycoinc@aol.com.

CHAPTER 12

Other Renewable Energy Sources

WIND ENERGY: INTRODUCTION

As a historical account of wind energy and its origin, the Chinese first invented windmills in about 200 B.C. according to most experts. About 500–900 A.D., windmills sprang up across Persia for grinding grain and pumping water. Between 1941 and 1945, a Vermont wind generator, called the Smith–Putnam machine, is credited as being the first to supply power to the local electrical grid. In 1981–1984, California witnessed the addition of 6,870 turbines by wind energy speculators. In 1992, the Federal Energy Policy Act offered power companies a 1.7-cent tax credit for each kilowatt-hour of wind-generated electricity they produce and offer customers. Currently, the Federal tax credit is 1.8 cents per kilowatt-hour for wind-generated energy. In addition, the state of Iowa has recently enacted an additional state tax credit equal to 1 cent per kilowatt-hour of wind energy generated by utilities and smaller independent power producers.[1,2]

Currently there is interest in using wind energy to generate electricity via windmills (wind turbines), which on a large scale is called a wind farm. A wind turbine consists of blades, which are turned by air currents passing over them much like a propeller whereby the air current lifts the blades in a similar manner as the wind lifts an airplane's wing. On these turbines, the blades are mounted on a pole or tower, which is 100 feet or more in height. The rotating blades turn a gearbox inside a housing, called a nacelle, which in turn generates the electricity. In summary for a wind turbine, the wind turns the blades, which turns the shaft and generates electricity as illustrated in Fig. 12.1.[3]

Figure 12.1 is a pictorial representation of a GE Power 1.5 MW Series Wind Turbine. Over 5,000 units are in operation worldwide and this 1.5 MW unit is one of the world's widely used wind turbines in its class. The 1.5 MW machine is active yaw and pitch regulated with power/torque control and an asynchronous generator. All the nacelle components are joined on a common structure. The generator and gearbox are supported by elastomeric elements to minimize noise emissions. In a wind turbine,

Municipal Solid Waste to Energy Conversion Processes: Economic, Technical, and Renewable Comparisons By Gary C. Young
Copyright © 2010 John Wiley & Sons, Inc.

1. Nacelle
2. Heat Exchanger
3. Generator
4. Control Panel
5. Main Frame
6. Impact Noise Insulation
7. Hydraulic Parking Brake
8. Gearbox
9. Impact Noise Insulation
10. Yaw Drive
11. Yaw Drive
12. Rotor Shaft
13. Oil Cooler
14. Pitch Drive
15. Rotor Hub
16. Nose Cone

FIGURE 12.1 GE, 1.5 MW Series Wind Turbine Components.[3]

the nacelle refers to the structure that houses the generating components, gearbox, drive train, rotor shaft, yaw drive, pitch drive, etc.[3]

Wind is a form of solar energy because winds are caused by the uneven heating of the atmosphere by the sun, the nonuniform surface of the earth, and the rotation of the earth. Wind flow patterns vary due to the earth's terrain, bodies of water, and vegetation. Wind energy is a process that describes how wind is used to generate mechanical power or electricity. Wind turbines convert the kinetic energy of the wind into mechanical power, which then is converted to electricity by a generator. The amount of wind capacity that is converted to electricity by the turbines in a farm is called the "capacity factor." By definition, the "capacity factor" is the amount of system capacity converted to energy for the time period under consideration. Typically, large wind turbines are classified as being 100 kilowatts (kW) or megawatts (MW) in capacity. Small wind turbines usually are listed as below 100 kW in size. Some of the basic components are shown in Fig. 12.2 and listed below.[4]

a. *Anemometer*: Measures the wind speed and transmits wind speed data to the controller.

b. *Blades*: Typically, a turbine will have two or three blades whereby the wind blowing over the blades causes the blades to lift and rotate.

c. *Brake*: A disc brake that can be applied to stop the rotor in emergencies.

d. *Controller*: The controller typically starts the turbine at wind speeds from 8 to 16 miles per hour and shuts down the unit at speeds above 55 mph. The low wind speed shutdown is due to the generator design limits, and the high wind speed can result in damage to the unit.

e. *Gear box*: Gears connect the low-speed shaft to the high-speed shaft to increase rotational speeds from about 30 rpm to 60 rpm to near 1,000 to 1,800 rpm as required by most generators to produce electricity.

f. *Generator*: The generator is typically an induction generator that produces 60-cycle AC electricity.

g. *High-speed shaft*: The high-speed shaft drives the generator.

h. *Low-speed shaft*: The low-speed shaft is turned by the rotor at about 30–60 rpm.

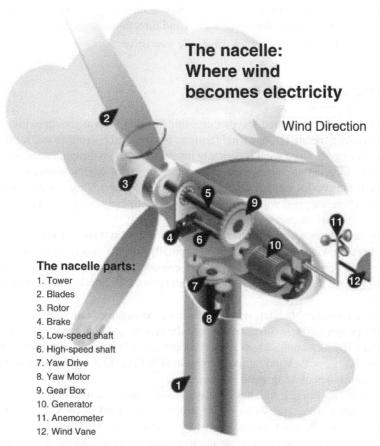

**The nacelle:
Where wind
becomes electricity**

Wind Direction

The nacelle parts:
1. Tower
2. Blades
3. Rotor
4. Brake
5. Low-speed shaft
6. High-speed shaft
7. Yaw Drive
8. Yaw Motor
9. Gear Box
10. Generator
11. Anemometer
12. Wind Vane

FIGURE 12.2   Nacelle, Wind Becomes Electricity.[4]

i. *Nacelle*: The nacelle is located on top of the tower, and covers and contains the gear box, shafts, generator, controller, and brake.

j. *Pitch*: Blades are turned or pitched out of the wind to control the rotor speed and keep the rotor from turning in wind speeds too low or too high for producing electricity.

k. *Rotor*: The blades and hub are called the rotor.

l. *Tower*: Towers are used to elevate the nacelle into winds with higher velocity because wind speed increases with height. Taller towers enable the turbine to capture more energy and generate more electricity.

m. *Wind direction*: An upwind turbine operates facing the wind. A downward turbine faces away from the wind.

n. *Wind vane*: The wind vane measures wind direction and communicates this data to the yaw drive to orient the turbine with respect to the wind.

o. *Yaw drive*: Upwind turbines face into the wind and the yaw drive is used to keep the rotor facing into the wind as the wind direction changes.

p. *Yaw motor*: The yaw motor drives the yaw drive.

BIG WIND SYSTEMS TO ENERGY

A number of turbines located together constitute a wind farm as shown in Fig. 12.3. The electricity generated from a wind farm can be collected by a utility and sent to the regular utility grid. Since wind is typically intermittent, when the wind does not blow sufficiently to generate energy/electricity, other forms of energy sources must be available to the grid for a steady supply of energy to customers. The regular utility grid is also therefore supplied with electricity from power plants using natural gas, coal, hydro, and/or nuclear energy.[5]

Many are interested by a profit motive or environmental motive for using wind energy because wind energy is an environmentally friendly source of energy and is considered as "green energy." Consequently, considerable interest in wind farms has attracted the attention of citizens, environmental groups, politicians, and commercial companies.

With the considerable interest in wind farms, an economic reality check is needed to focus a proper business direction for both the private and public sectors. Thus, the following examples and/or cases for the economic of large-scale wind farms were computed. Large-scale wind farms are presented in Figs. 12.3 and 12.4. Figure 12.4 shows the turbines of a big wind farm up close.[6,5]

MidAmerican Energy, Inc. has built the world's largest wind energy generation project in Iowa. The capital cost of the project is reported to be $323 million; it consists of 180–200 wind turbines and has a capacity of 310 MW. An additional capital cost of $15 million will be for interconnecting and development, which adds

FIGURE 12.3   Big Wind Farm.[5]

FIGURE 12.4   Big Wind Farm/Turbines.[6]

up to a total capital investment of $338 million. Landowners will be paid $4,000 per turbine annually for easements. These economic data/parameters were considered for the following preliminary economic assessment of the MidAmerican project.[7,6]

Many with a profit motive and/or an environmental motive brandish about the economy of wind farms to generate electricity because wind energy is an environmentally friendly source of energy/electricity or "green power." Thus, wind farms attracts the attention of many, including regular citizens, environmental groups, politicians, and commercial companies. With this diverse interest, a sense of direction is needed to bring a "reality check" on the commercial viability as to the "economics" of wind farms. A reality check can be formulated in our opinion by considering the following example for a large-scale wind farm.

In a published article, MidAmerican Energy, Inc. plans to build the world's largest wind energy generation project in Iowa. As reported, the project will consist of 180–200 wind turbines and have a capacity of 310 MW of electricity. The project is reported to cost $323 million. An additional capital cost of $15 million would be used for interconnecting and development costs giving a total estimated capital cost of $338 million for the economic evaluation. Landowners will be paid $4,000 per turbine annually for turbine easements. With this data as a basis, consider the economics of this project as discussed but first with a detailed method of financial analysis presented:[7]

Economic Example and Cases

Example: Wind Farm Financial Analysis, Before Taxes
   (detailed methodology):
INPUT: Financial Parameters

Installed Capacity	310.00 MW
Capital Investment	$338,000,000.00
Financing period	20 years

Bond interest rate	5.50%
Payments per year	2
Capacity factor	35.0%
Federal tax credit	1.80 cents/kWh
State tax credit	1.00 cents/kWh
Green tags	2.00 cents/kWh
**Selling price of energy	0.1380 cents/kWh
O&M	1.9000 cents/kWh
Number of turbines	200
Easement cost	$4,000.00/year/turbine

OUTPUT:

Production   950,460,000 kWh/year

(310 MW × 1000 kW/MW × (35.0/100) × 24 hours/day × 365 days/year)

Revenues: Expenditures:

Energy Sales $1,311,634.80/year   Bond payment $28,075,303.71/year

(950,460,000 kWh/year × 0.1380 ¢/kWh/100)

Federal tax credit $17,108,280.00/year   O&M $18,058,740.00/year

(950,460,000 kWh/year × 1.80 ¢/kWh/100)   (950,460,000 kWh/year × 1.900 ¢/kwh/100)

State tax credit $9,504,600.00/year   Easement cost $800,000.00/year

(950,460,000 kWh/year × 1.0 ¢/kWh/100)   ($4,000.00/year/turbine × 200 turbines)

Green tags $19,009,200.00/year

(950,460,000 kWh/year × 2.00 ¢/kWh/100)

Total Revenue(s) $46,933,714.80/year = ?   Total Expenditure(s) $46,934,043.71/year

(Note, Total Revenue(s) = Total Expenditure(s) or the difference is zero at breakeven.)

Breakeven is determined by varying the selling price of energy until "total revenue(s) minus total expenditure(s)" is zero. In the example case above, the break-even selling price of energy is 0.1380 cents/kWh (before taxes).

Economic Cases 1A, 1B, 1C, 2A, 2B, 3, 4, 5, 6, and 7 will be presented with input financial parameters and output computations. After listing these economic cases, a discussion of each case will follow.

Case-1A: Simple Cash Flow Analysis

Wind Farm Financial Analysis Before Taxes:

INPUT: Financial Parameters

Installed Capacity	310.00 MW
Capital Investment	$338,000,000.00
Financing period	20 years
Bond interest rate	5.50%
Payments per year	2
Capacity factor	35.0%
Federal tax credit	0.00 cents/kWh
State tax credit	0.00 cents/kWh
Green tags	0.00 cents/kWh
**Selling price of energy	3.37780 cents/kWh
O&M	0.33980 cents/kWh

| Number of turbines | 200 |
| Easement cost | $4,000.00/year/turbine |

OUTPUT:

Production: 950,460,000 kWh/year

Revenues:		Expenditures:	
Energy Sales	$32,104,637.88/year	Bond payment	$28,075,303.71/year
			2.9539 cents/kWh
Federal tax credit	$0.00/year	O&M	$3,229,663.08/year
			0.3398 cents/kWh
State tax credit	$0.00/year	Easement cost	$800,000.00/year
			0.0842 cents/kWh
Green tags	$0.00/year		
Total Revenue(s)	$32,104,637.88/year = ?	Total Expenditure(s)	
			$32,104,966.79/year

Case-1B: Simple Cash Flow Analysis

Wind Farm Financial Analysis Before Taxes

INPUT: Financial Parameters

Installed Capacity	310.00 MW
Capital Investment	$338,000,000.00
Financing period	20 years
Bond interest rate	5.50%
Payments per year	12
Capacity factor	35.0%
Federal tax credit	0.00 cents/kWh
State tax credit	0.00 cents/kWh
Green tags	0.00 cents/kWh
**Selling price of energy	3.35900 cents/kWh
O&M	0.33980 cents/kWh
Number of turbines	200
Easement cost	$4,000.00/year/turbine

OUTPUT

Production   950,460,000 kWh/year

Revenues:		Expenditures:	
Energy Sales	$31,925,951.40/year	Bond payment	$27,900,709.21/year
			2.9355 cents/kWh
Federal tax credit	$0.00/year	O&M	$3,229,663.08/year
			0.3398 cents/kWh
State tax credit	$0.00/year	Easement cost	$800,000.00/year
			0.0842 cents/kWh
Green tags	$0.00/year		
Total Revenue(s)	$31,925,951.40/year = ?	Total Expenditure(s)	
			$31,930,372.29/year

Case-1C: Simple Cash Flow Analysis

Wind Farm Financial Analysis Before Taxes:
INPUT: Financial Parameters

Installed Capacity	310.00 MW
Capital Investment	$338,000,000.00
Financing period	20 years
Bond interest rate	5.50%
Payments per year	2
Capacity factor	20.0%
Federal tax credit	0.00 cents/kWh
State tax credit	0.00 cents/kWh
Green tags	0.00 cents/kWh
**Selling price of energy	5.65640 cents/kWh
O&M	0.33980 cents/kWh
Number of turbines	200
Easement cost	$4,000.00/year/turbine

OUTPUT

Production   543,120,000 kWh/year

Revenues:		Expenditures:	
Energy Sales	$30,721,039.68/year	Bond payment	$28,075,303.71/year
			5.1693 cents/kWh
Federal tax credit	$0.00/year	O&M	$1,845,521.46/year
			0.3398 cents/kWh
State tax credit	$0.00/year	Easement cost	$800,000.00/year
			0.1473 cents/kWh
Green tags	$0.00/year		
Total Revenue(s)	$30,721,039.68/year	= ?	Total Expenditure(s) $30,720,825.47/year

Case-2A: Simple Cash Flow Analysis

Wind Farm Financial Analysis Before Taxes:
INPUT: Financial Parameters

Installed Capacity:	310.00 MW
Capital Investment:	$338,000,000.00
Financing period:	20 years
Bond interest rate:	5.50%
Payments per year:	2
Capacity factor:	35.0%
Federal tax credit:	0.00 cents/kWh
State tax credit:	0.00 cents/kWh
Green tags:	0.00 cents/kWh
**Selling price of energy:	3.86800 cents/kWh
O&M:	0.83000 cents/kWh
Number of turbines:	200
Easement cost:	$4,000.00/year/turbine

OUTPUT
Production	950,460,000 kWh/year		
Revenues:		Expenditures:	
Energy Sales	$36,763,792.80/year	Bond payment	$28,075,303.71/year
			2.9539 cents/kWh
Federal tax credit	$0.00/year	O&M	$7,888,818.00/year
			0.8300 cents/kWh
State tax credit	$0.00/year	Easement cost	$800,000.00/year
			0.0842 cents/kWh
Green tags	$0.00/year		
Total Revenue(s)	$36,763,792.80/year = ?		Total Expenditure(s) $36,764,121.71/year

Case-2B: Simple Cash Flow Analysis

Wind Farm Financial Analysis Before Taxes:
INPUT: Financial Parameters

Installed Capacity	310.00 MW
Capital Investment	$338,000,000.00
Financing period	20 years
Bond interest rate	5.50%
Payments per year	2
Capacity factor	35.0%
Federal tax credit	1.80 cents/kWh
State tax credit	1.00 cents/kWh
Green tags	0.00 cents/kWh
**Selling price of energy	1.06800 cents/kWh
O&M	0.83000 cents/kWh
Number of turbines	200
Easement cost	$4,000.00/year/turbine

OUTPUT:
Production	950,460,000 kWh/year		
Revenues:		Expenditures:	
Energy Sales	$10,150,912.80/year	Bond payment	$28,075,303.71/year
			2.9539 cents/kWh
Federal tax credit	$17,108,280.00/year	O&M	$7,888,818.00/year
			0.8300 cents/kWh
State tax credit	$9,504,600.00/year	Easement cost	$800,000.00/year
			0.0842 cents/kWh
Green tags	$0.00/year		
Total Revenue(s)	$36,763,792.80/year = ?		Total Expenditure(s) $36,764,121.71/year

Case-3: Simple Cash Flow Analysis

Wind Farm Financial Analysis Before Taxes:
INPUT: Financial Parameters

Installed Capacity	310.00 MW
Capital Investment	$338,000,000.00
Financing period	20 years
Bond interest rate	5.50%
Payments per year	2
Capacity factor	35.0%
Federal tax credit	1.80 cents/kWh
State tax credit	1.00 cents/kWh
Green tags	0.00 cents/kWh
**Selling price of energy	1.06800 cents/kWh
O&M	0.83000 cents/kWh
Number of turbines	200
Easement cost	$4,000.00/year/turbine

OUTPUT

Production 950,460,000 kWh/year

Revenues:		Expenditures:	
Energy Sales	$10,150,912.80/year	Bond payment	$28,075,303.71/year 2.9539 cents/kWh
Federal tax credit	$17,108,280.00/year	O&M	$7,888,818.00/year 0.8300 cents/kWh
State tax credit	$9,504,600.00/year	Easement cost	$800,000.00/year 0.0842 cents/kWh
Green tags	$0.00/year		
Total Revenue(s)	$36,763,792.80/year	= ?	Total Expenditure(s) $36,764,121.71/year

Case-4: Simple Cash Flow Analysis

Wind Farm Financial Analysis Before Taxes:
INPUT: Financial Parameters

Installed Capacity	310.00 MW
Capital Investment	$338,000,000.00
Financing period	20 years
Bond interest rate	5.50%
Payments per year:	2
Capacity factor	35.0%
Federal tax credit	1.80 cents/kWh
State tax credit	1.00 cents/kWh
Green tags	2.00 cents/kWh
**Selling price of energy	-0.93200 cents/kWh
O&M	0.83000 cents/kWh
Number of turbines	200
Easement cost	$4,000.00/year/turbine

Production 950,460,000 kWh/year

Revenues:		Expenditures:	
Energy Sales	−$8,858,287.20/year	Bond payment	$28,075,303.71/year
			2.9539 cents/kWh
Federal tax credit	$17,108,280.00/year	O&M	$7,888,818.00/year
			0.8300 cents/kWh
State tax credit	$9,504,600.00/year	Easement cost	$800,000.00/year
			0.0842 cents/kWh
Green tags	$19,009,200.00/year		
Total Revenue(s)	$36,763,792.80/year	= ?	Total Expenditure(s)
			$36,764,121.71/year

Case-5: Simple Cash Flow Analysis

Wind Farm Financial Analysis Before Taxes:
INPUT: Financial Parameters

Installed Capacity	310.00 MW
Capital Investment	$338,000,000.00
Financing period	20 years
Bond interest rate	5.50%
Payments per year	2
Capacity factor	35.0%
Federal tax credit	1.80 cents/kWh
State tax credit	1.00 cents/kWh
Green tags	1.00 cents/kWh
**Selling price of energy	0.23800 cents/kWh
O&M	1.00000 cents/kWh
Number of turbines	200
Easement cost	$4,000.00/year/turbine

OUTPUT:

Production 950,460,000 kWh/year

Revenues:		Expenditures:	
Energy Sales	$2,262,094.80/year	Bond payment	$28,075,303.71/year
			2.9539 cents/kWh
Federal tax credit	$17,108,280.00/year	O&M	$9,504,600.00/year
			1.0000 cents/kWh
State tax credit	$9,504,600.00/year	Easement cost	$800,000.00/year
			0.0842 cents/kWh
Green tags	$9,504,600.00/year		
Total Revenue(s)	$38,379,574.80/year	= ?	Total Expenditure(s)
			$38,379,903.71/year

Case-6: Internal Rate of Return (IRR)

Capital Cost of Project	$338,000,000
Finance Structure	75% debt, 5.5% interest, 20 year, 25% equity

Wind turbines (quantity)	200
Annual easement cost	$4,000 per turbine per year
Operation/Maintenance cost	$3,230,000 per year
Project life	20 years
Depreciation	MACRS tax depreciation
Power Capacity	310,000 KW
Capacity factor	35% (typical value for analysis but site specific)
Selling price of energy	3.4 cents/KW-hour.
Federal Tax Credit	0.0 cents/KWH sold to utility customers

Computations:

Electrical power generated per year:
310,000 KW × 0.35 × 24 hrs/day × 365 days/year = 950,463,000 KW-hours/year
IRR (after taxes) = 4.26%, (with no Federal Tax Credit)

Case-7: IRR with Tax Credit

Capital Cost of Project	$338,000,000
Finance Structure	75% debt, 5.5% interest, 20 year, 25% equity
Wind turbines (quantity)	200
Annual easement cost	$4,000 per turbine per year
Operation/Maintenance cost	$3,230,000 per year
Project life	20 years
Depreciation	MACRS tax depreciation
Power Capacity	310,000 KW
Capacity factor	35% (typical value for analysis but site specific)
Selling price of energy	3.4 cents/KW-hour.
Federal Tax Credit	1.8 cents/KWH sold to utility customers

Computations:

Electrical power generated per year:
310,000 KW × 0.35 × 24 hrs/day × 365 days/year = 950,463,000 KW-hours/year
IRR (after taxes) = 27.1% (with the Federal Tax Credit of 1.8 cents/kWh)

Note: The Federal tax credit reduces the taxes paid by the utility company by $17,108,300 per year, i.e., 950,463,000 KW-hour/year × $0.018/kWh.

Discussion of Economics For the Large Wind Farm Cases

Case-1A presented, in our opinion, represents a simple approximate method to generate a cash flow analysis to obtain a break-even cost for wind-generated electric power for a large wind farm of 310 MW as about 3.4 cents/kWh. In our opinion, this represents an approximation of the best case. A capacity factor of 35% was used. Case-1B is similar to Case-1A except the bond payment was made in terms of 12 payments per year rather than the usual two payments per year. If a more detailed economic analysis is done with MACRS, the break-even price for wind energy becomes 3.4 cents/kWh as shown in Case-6. This just illustrates that the approximation method can be useful. For another site where the capacity factor is lower and if a site-specific case has a capacity factor of 20%, the break-even cost increases and the value becomes about 5.7 cents/kWh by the approximation method of analysis as shown in Case-1C.[8]

The economic analysis of Case-6 yields a IRR (after taxes) of 4.26% when the selling price of wind farm generated energy is 3.4 cents/kWh with *no* federal tax credit. Case-7 was taken but with a federal tax credit of 1.8 cents/kWh of wind farm energy generated for customers with rest being the same as Case-6. This federal tax credit becomes annually, $17.1 million per year, i.e., 950,463,000 kWh/year × $0.018/kWh. This Federal tax subsidy in effect increases the IRR (after taxes) to 27.1%. If for this same case, the debt was increased to 60% at 7.0% interest and with a 40% equity, the IRR (after taxes) would be reduced to 13.8% considering the Federal tax subsidy but with the Operation/Maintenance cost increased from $3.23 million to $5.00 million per year.[9]

Case-3 is the same as Case-2A except a federal tax credit of 1.8 cents/kWh and a state tax credit of 1.0 cents/kWh were applied. Case-3 has a break-even energy selling price of 1.068 cents/kWh with the federal tax credit and the state tax credit. Compare Case-3 with a break-even energy selling price of 1.068 cents/kWh with tax credits with the break-even energy selling price of 3.868 cents/kWh without tax credits.

Case-4 is the same as Case-3 except an additional revenue source of green tags at 2.0 cents/kWh was considered. Note, the break-even selling price of energy is −0.932 cents/kWh or approximately zero cost of producing electricity from the wind farm.

Case-5 was computed like Case-4 except the green tags were considered at 1.0 cents/kWh and the O&M costs at 1.00 cents/kWh. The result is a break-even selling price of energy of 0.2380 cents/kWh.

These analyses illustrate the economics of wind farming and the importance of the wind farm capacity factor and also show that a tax subsidy can generate much cash to the bottom line of a utility in some specific and properly chosen wind farm site.

Another factor to consider is a matter of "green credits." Customers, if they wish, can be charged by the utility an additional cents/kWh to their existing rate for purchasing from this energy source (wind energy). This additional revenue can make a project more profitable. Another factor of importance in the state of Iowa is the 1.0 cent/kWh state tax credit for wind energy. Also, there is a federal tax credit of 1.8 cents/kWh. Another credit for wind energy is "green power" or "green tags." A credit for green tags has typically been about 2.0 cents/kWh and is considered a revenue source for a wind farm. Green power, whether it is a delivered product or a green tag, is that electricity generated by a source that has lower emissions and/or fewer environmental impacts reduces the need for electricity generated by a source that causes higher emissions and/or has greater environmental impacts.

The reader should be aware that any economic analysis should be site/case specific and the economics presented in this article are presented only as information to the public.

It should be remembered that the downside to wind farms for producing electricity is the unreliability of the wind. Thus, wind generation of electricity is just another

FIGURE 12.5   Eagles Flying Among Wind Turbines.[11]

form of energy source to generate power for use in the grid system along with the other sources of energy power plants listed previously such as natural gas, coal, hydro, and/or nuclear.

Another concern by some environmentalists is the perceived damage associated with avian mortality, i.e., birds and fowl. Figure 12.5 shows eagles flying among wind turbines.[10]

Figure 12.6 shows an eagle flying among the wind turbine in the Altamont Pass Wind Resource Area (APWRA) as reported in a California Energy Commission study.[11,12]

As wind energy expands across the country/United States, the effects of commercial wind farms upon the bird population have been investigated. These inquiries into the effects upon the bird population have been driven by: (a) potential litigation from the death of a bird protected by the migratory Bird Treaty Act and/or the Endangered

FIGURE 12.6   Golden Eagle at APWRA by Daniel Driscoll.[12]

Species act and (b) the effect of bird mortality upon the bird populations. Logically the facts should determine a proper course of action. Thus, the following discussion is devoted to the presentation of facts as published.[13]

Two major studies were conducted in the Tehachapi Pass and the San Gorgonio Pass located in the wind resource area (WRA). Results from these studies have deferred from results learned in the Altamont Pass Wind Resource Area (APWRA). Results indicated that some birds collide with wind turbines and raptors appear to have a higher degree of risk for collision. However, the number of collisions is small in the Tehachapi Pass and San Gorgonio Pass WRA than in the Altamont Pass WRA.

A study of the Ponnequin Wind Farm in Colorado was conducted by National Renewable Energy Laboratory (NREL) to complement the study done by the locally investor-owned utility, Xcel Energy, preconstruction site study. About 23 avian species were identified on the site and seven raptor species were discovered: Northern harrier, Swainson's hawk, Rough-legged hawk, Ferruginous hawk, Golden eagle, American kestrel, and Prairie falcon. No raptor carcasses were found in the one-year study period.[13]

For Green Mountain Power's wind turbine development in Searsburg, Vermont, a study was conducted on avian behavior such as night migration of songbirds, daytime migration of hawks, and breeding. This study was conducted both during preconstruction and postconstruction periods. The study focused on the impacts of wind turbines on migrating birds. During the study period, no avian fatalities were found.[13]

A study was conducted prior to the construction of the Conservation and Renewable Energy System's (CARES) wind facility for Goldendale, Washington. The purpose of the study was to record the patterns of WRA use by birds before and after the wind facility construction. Few bird fatalities were found.[13]

A long-term field study was conducted on the ecology of Golden eagles (*Aquila chrysaetos*) in the vicinity of the APWRA. The facility lies east of San Francisco Bay in California and consists of about 6,500 wind turbines of 190 km^2 of rolling grassland as pictured in Figs. 12.6 and 12.7.[11]

In the WRA, to estimate survival rates, 179 eagles were tagged with radio transmitters equipped with mortality sensors to function for 4 years. The tagged sample included 79 juveniles, 45 subadults, 17 floaters (nonterritorial adults), and 38 breeders. Juveniles are eagles less than 1 year old. Subadults are one, two, and three years of age. Floaters are adults without breeding territories. Floaters are an indication that all of the habitat suitable for breeding is occupied by territorial pairs. For a floater to mate, it must wait for a vacancy or evict a territorial owner. As stated, the population grows until all serviceable breeding locations are occupied by pairs, at which point floaters begin to accumulate." "Because of environmental restrictions to the size of the breeding segment, there is a resulting stabilization in floater numbers and overall populations size, a phenomenon know as Moffat's equilibrium."[11]

Of 61 recorded deaths of radio-tagged eagles during the 4-year investigation, 54% resulted from electrical generation or transmission. Of these deaths, 38% were caused by wind turbine blade strikes and 16% were by electrocutions on distribution lines. Of the turbine strike fatalities, 19 were subadults, 3 were floaters, and only 1 was

The Altamont Pass Wind Resource Area (photo by Daniel Driscoll)

FIGURE 12.7   Wind Farm in The Altamont Pass Wind Resource Area (APWRA).[11]

a breeder. Among the species of raptors killed at Altamont Pass, the most likely to be impacted was the Golden eagle. The breeding and recruitment rates of Golden eagles are naturally slow, which makes for greater susceptibility to population decline. However, floaters (nonterritorial adults) have a tendency to replenish those lost. As stated, "floaters safeguard the breeding segment by quickly replacing breeders that have died."[11,14]

From the studies conducted and summarized in the NREL report, data suggest that the most significant U.S. avian and wind turbine interaction problem occurs in the APWRA. With the studies conducted to date, it has been learned over the past several years that avian issues should not be a concern for wind farm development because the potential problems can be addressed. The avian and wind turbine problems can be identified and dealt with for the particular wind farm site. Also, it is important to consider the number of bird fatalities in proportion to the population size. For example, the absolute number of bird fatalities in a wind farm site may not be important compared to the local population in the area. As stated, "for example, if 10 birds of a particular species are killed, it is important to know if this is out of a local population of 30 or 3000. The overall impact of 10 birds being killed will be different depending on the size of the local population."[13]

Economy of Scale Associated With Wind Farms

Wind turbines convert the kinetic energy of the wind into mechanical power, which then is converted to electricity by a generator. Typically, large wind turbines are classified as having 100 kilowatts (kW) to megawatts (MW) of power. Small wind turbines usually are listed as below 100 kW in size.[10] Whether considering large or small turbines and/or combination of many turbines as a small or large wind farm, economy of scale is associated with the size of the facility. The economy of scale becomes apparent by review of Table 12.1 developed from commercial wind farms.

TABLE 12.1   Commercial Wind Farms, Enron Wind, California, Year 2001

Item	Cabazon Site	Sky River Site	Victory Garden Site	251
Installed capacity (MW)	39.75	77	22	18.4
Production (MWh)	107,000	200,000	50,000	38,000
Capacity factor, listed (%)	33.3	32.5	27.9	22.8
Capacity factor, computed (%)	30.7	29.7	25.9	23.6
Operating expenses (cents/kWh)	3.3327	1.8060	1.8780	3.3763
Total expense (cents/kWh) (Operating expense + D&A)	5.2206	5.6375	4.5940	6.1026

Economy of scale is noticeable in Table 12.1 but discrepancies are apparent. Site location and many other capital and financial factors could influence the data present in the table. However, from the data in Table 12.1 and other known data, a figure can be constructed of estimated operation and maintenance (O&M) costs associated with various wind farm capacities. Figure 12.8 was constructed accordingly as an estimate for the variation of O&M with wind Farm capacity.

Figure 12.9 shows installed capacity (MW) and wind farm production (MWh/year).

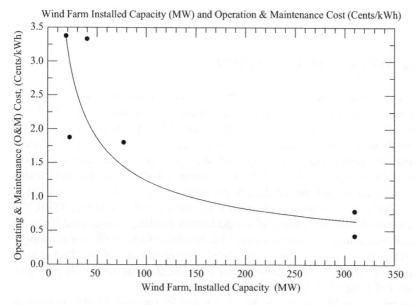

Wind Farm Installed Capacity (MW) and Operation & Maintenance Cost (Cents/kWh)

Dr. Gary C. Young, Ph.D., P.E.
B-T-E, Inc.
WindFarmData,prostat,07 01 2009.pdw

FIGURE 12.8   Wind Farm Installed Capacity (MW) and O&M Cost (cents/kWh).

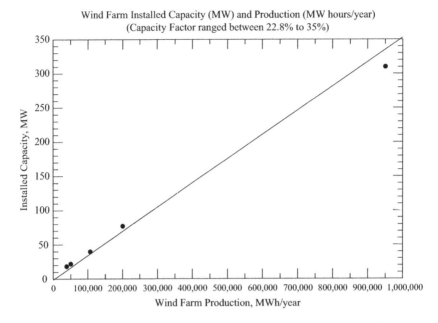

Wind Farm Installed Capacity (MW) and Production (MW hours/year)
(Capacity Factor ranged between 22.8% to 35%)

Dr. Gary C. Young, Ph.D., P.E.
B-T-E, Inc.
WindCFvsCentspkWh,07 01 2009.pdw

FIGURE 12.9 Wind Farm Production and Installed Capacity.

SMALL WIND SYSTEMS TO ENERGY

Small wind turbines usually are listed as below 100 kW in size. A typical small wind turbine is a turbine of 10 kW capacity. The amount of this turbine capacity that is converted into energy is called the "capacity factor." The capacity factor is defined as the amount of system capacity converted to energy for the time period under consideration. Depending upon a specific site location, the capacity factor can typically vary between 20% and 35%. For example, a 10 kW wind turbine with a 35% capacity factor will convert 3.5 kW to energy or 30,660 kWh/year, i.e., 10 kW 0.35 × 24 h/day × 365 days/year. A small wind turbine is shown in Fig. 12.10.[5,15]

When considering the viability of a wind farm or turbine(s), large wind farms are more economical than small wind farms. This economic fact is typical of economics and is referred to as the "economy of scale." The economics of small wind farm systems will be shown as less economically attractive than the previously presented economics of large wind farm systems.[5]

Economic Cases 1S, 2S, 3S, 4S, 5S, 6S, 7S, and 8S for small wind farm systems will be presented with input financial parameters and output computations. After listing these economic cases, a discussion of each case will follow.

FIGURE 12.10   Small Wind Turbine.[15]

Case-1S: Simple Cash Flow Analysis

Small Wind Farm Financial Analysis Before Taxes:
INPUT: Financial Parameters

Installed Capacity	10.00 kW
Capital Investment	$45,039.54
Grant, % of Investment	0.0%
Investment—Grant	$45,039.54
Financing period	20 years
Bond interest rate	5.50%
Payments per year	2
Capacity factor	35.0%
Federal tax credit	0.00 cents/kWh
State tax credit	0.00 cents/kWh
Green tags	0.00 cents/kWh
**Selling price of energy	15.7000 cents/kWh
O&M	3.500 cents/kWh
Number of turbines	1
Easement cost	$0.00/year/turbine

OUTPUT

Production	30,660 kWh/year		
Revenues:		Expenditures:	
Energy Sales	$4,813.62/year	Bond payment	$3,741.12/year
			12.2029 cents/kWh
Federal tax credit	$0.00/year	O&M	$1,073.10/year
			3.500 cents/kWh
State tax credit	$0.00/year	Easement cost	$0.00/year
			0.000 cents/kWh
Green tags	$0.00/year		
Total Revenue(s)	$4,813.62/year	= ?	Total Expenditure(s)
			$4,814.22/year

Case-2S: Simple Cash Flow Analysis

Small Wind Farm Financial Analysis Before Taxes:
INPUT: Financial Parameters

Installed Capacity	10.00 kW
Capital Investment	$45,039.54
Grant, % of Investment	0.0%
Investment—Grant	$45,039.54
Financing period	20 years
Bond interest rate	5.50%
Payments per year	2
Capacity factor	23.2%
Federal tax credit	0.00 cents/kWh
State tax credit	0.00 cents/kWh
Green tags	0.00 cents/kWh
**Selling price of energy:	21.9000 cents/kWh
O&M	3.500 cents/kWh
Number of turbines	1
Easement cost	$0.00/year/turbine

OUTPUT:

Production	20,323 kWh/year		
Revenues:		Expenditures:	
Energy Sales	$4,450.78/year	Bond payment	$3,741.12/year
			18.4081 cents/kWh
Federal tax credit	$0.00/year	O&M	$711.31/year
			3.500 cents/kWh
State tax credit	$0.00/year	Easement cost	$0.00/year
			0.000 cents/kWh
Green tags	$0.00/year		
Total Revenue(s)	$4,450.78/year	= ?	Total Expenditure(s)
			$4,452.43/year

Case-3S: Simple Cash Flow Analysis

Small Wind Farm Financial Analysis Before Taxes:
INPUT: Financial Parameters

Installed Capacity	10.00 kW
Capital Investment	$45,039.54
Grant, % of Investment	50.0%
Investment—Grant:	$22,519.77
Financing period	20 years
Bond interest rate	5.50%
Payments per year	2
Capacity factor	35.0%
Federal tax credit	0.00 cents/kWh
State tax credit	0.00 cents/kWh
Green tags	0.00 cents/kWh
**Selling price of energy:	9.6000 cents/kWh
O&M	3.500 cents/kWh
Number of turbines	1
Easement cost	$0.00/year/turbine

OUTPUT:

Production   30,660 kWh/year

Revenues:		Expenditures:	
Energy Sales	$2,943.36/year	Bond payment	$1,870.56/year
			6.1010 cents/kWh
Federal tax credit	$0.00/year	O&M	$1,073.10/year
			3.500 cents/kWh
State tax credit	$0.00/year	Easement cost	$0.00/year
			0.000 cents/kWh
Green tags	$0.00/year		
Total Revenue(s)	$2,943.36/year = ?		Total Expenditure(s) $2,943.66/year

Case-4S: Simple Cash Flow Analysis

Small Wind Farm Financial Analysis Before Taxes:
INPUT: Financial Parameters

Installed Capacity	10.00 kW
Capital Investment	$45,039.54
Grant, % of Investment	50.0%
Investment—Grant:	$22,519.77
Financing period	20 years
Bond interest rate	5.50%
Payments per year	2
Capacity factor	35.0%
Federal tax credit	1.80 cents/kWh
State tax credit	0.00 cents/kWh
Green tags	0.00 cents/kWh

**Selling price of energy	7.8000 cents/kWh		
O&M	3.500 cents/kWh		
Number of turbines	1		
Easement cost	$0.00/year/turbine		

OUTPUT:

Production	30,660 kWh/year		
Revenues:		Expenditures:	
Energy Sales	$2,391.48/year	Bond payment	$1,870.56/year
			6.1010 cents/kWh
Federal tax credit	$551.88/year	O&M	$1,073.10/year
			3.500 cents/kWh
State tax credit	$0.00/year	Easement cost	$0.00/year
			0.000 cents/kWh
Green tags	$0.00/year		
Total Revenue(s)	$2,943.36/year	= ?	Total Expenditure(s) $2,943.66/year

Case-5S: Simple Cash Flow Analysis

Small Wind Farm Financial Analysis Before Taxes:
INPUT: Financial Parameters

Installed Capacity	10.00 kW
Capital Investment	$45,039.54
Grant, % of Investment	50.0%
Investment—Grant	$22,519.77
Financing period	20 years
Bond interest rate	5.50%
Payments per year	2
Capacity factor	35.0%
Federal tax credit	1.80 cents/kWh
State tax credit	1.00 cents/kWh
Green tags	2.00 cents/kWh
**Selling price of energy	4.8000 cents/kWh
O&M	3.500 cents/kWh
Number of turbines	1
Easement cost	$0.00/year/turbine

OUTPUT:

Production	30,660 kWh/year		
Revenues:		Expenditures:	
Energy Sales	$1,471.68/year	Bond payment	$1,870.56/year
			6.1010 cents/kWh
Federal tax credit	$551.88/year	O&M	$1,073.10/year
			3.500 cents/kWh
State tax credit	$306.60/year	Easement cost	$0.00/year
			0.000 cents/kWh

Green tags $613.20/year

Total Revenue(s) $2,943.36/year   = ?   Total Expenditure(s)
                                                                                 $2,943.66/year

Case-6S: Simple Cash Flow Analysis

Small Wind Farm Financial Analysis Before Taxes:
INPUT: Financial Parameters

Installed Capacity	10.00 kW
Capital Investment	$45,039.54
Grant, % of Investment	50.0%
Investment—Grant	$22,519.77
Financing period	20 years
Bond interest rate	5.50%
Payments per year	2
Capacity factor	23.2%
Federal tax credit	1.80 cents/kWh
State tax credit	1.00 cents/kWh
Green tags	2.00 cents/kWh
**Selling price of energy	7.9050 cents/kWh
O&M	3.500 cents/kWh
Number of turbines	1
Easement cost	$0.00/year/turbine

OUTPUT:

Production   20,323 kWh/year

Revenues:		Expenditures:	
Energy Sales	$1,606.55/year	Bond payment	$1,870.56/year
			9.2041 cents/kWh
Federal tax credit	$365.82/year	O&M	$711.31/year
			3.500 cents/kWh
State tax credit	$203.23/year	Easement cost	$0.00/year
			0.000 cents/kWh
Green tags	$406.46/year		

Total Revenue(s) $2,582.06/year   = ?   Total Expenditure(s)
                                                                      $2,581.87/year

Case-7S: Simple Cash Flow Analysis

Small Wind Farm Financial Analysis Before Taxes:
INPUT: Financial Parameters

Installed Capacity	5.00 kW
Capital Investment	$29,715.01
Grant, % of Investment	50.0%
Investment—Grant	$14,857.51
Financing period	20 years
Bond interest rate	5.50%
Payments per year	2

Capacity factor	35.0%
Federal tax credit	0.00 cents/kWh
State tax credit	0.00 cents/kWh
Green tags	0.00 cents/kWh
**Selling price of energy	11.5500 cents/kWh
O&M	3.500 cents/kWh
Number of turbines	1
Easement cost	$0.00/year/turbine

OUTPUT:

Production 15,330 kWh/year

Revenues:		Expenditures:	
Energy Sales	$1,770.62/year	Bond payment	$1,234.11/year 8.0503 cents/kWh
Federal tax credit	$0.00/year	O&M	$536.55/year 3.500 cents/kWh
State tax credit	$0.00/year	Easement cost	$0.00/year 0.000 cents/kWh
Green tags	$0.00/year		
Total Revenue(s) $1,770.62/year	= ?		Total Expenditure(s) $1,770.66/year

Case-8S: Simple Cash Flow Analysis

Small Wind Farm Financial Analysis Before Taxes:

INPUT: Financial Parameters

Installed Capacity	5.00 kW
Capital Investment	$29,715.01
Grant, % of Investment	50.0%
Investment—Grant:	$14,857.51
Financing period	20 years
Bond interest rate	5.50%
Payments per year	2
Capacity factor	35.0%
Federal tax credit	1.80 cents/kWh
State tax credit	1.00 cents/kWh
Green tags	2.00 cents/kWh
**Selling price of energy	6.7500 cents/kWh
O&M	3.500 cents/kWh
Number of turbines	1
Easement cost	$0.00/year/turbine

OUTPUT:

Production 15,330 kWh/year

Revenues:		Expenditures:	
Energy Sales	$1,034.78/year	Bond payment	$1,234.11/year 8.0503 cents/kWh
Federal tax credit	$275.94/year	O&M	$536.55/year 3.500 cents/kWh

State tax credit $153.30/year   Easement cost $0.00/year
                                                                         0.000 cents/kWh

Green tags $306.60/year

Total Revenue(s) $1,770.62/year   = ?   Total Expenditure(s)
                                                       $1,770.66/year

Discussion of Economics for the Small Wind Farm Cases

In Case-1S, a small wind turbine with a capacity of 10 kW is installed at a capital investment of $45,039. A bond is used to finance the project at 5.50% interest for a period of 20 years by making two payments per year. The O&M costs are 3.50 cents/kWh of energy production. With a capacity factor of 35%, the energy production becomes 30,660 kWh/year. The expenditures become: bond payment— $3,741.12/year and O&M—$1,073.10/year. By trial and error on the selling price of energy, 15.70 cents/kWh is break-even selling price whereby revenues equal expenditures. Note, "no" federal tax credit, state tax credit, and green tags were used as revenue.

For Case-2S, a small 10 kW wind turbine was installed with all the same parameters as Case-1S, except the capacity factor is 23.2%. A break-even selling price of energy was computed to be 21.90 cents/kWh. Note, the break-even selling price was 15.70 cents/kWh at a capacity factor of 35% but at a capacity factor of 23.2%, the break-even selling price became 21.90%. The capacity factor is a very important economic parameter.

In Case-3S, a small 10 kW wind turbine was installed with all the same parameters as Case-1S, except a grant was obtained for 50% of the capital investment. Thus, only $22,519.77 needed to be financed. The break-even selling price was then computed to be 9.60 cents/kWh. With a small wind turbine, one can conclude the importance of grants to help finance a project.

Case-4S, a wind turbine with 10 kW capacity was installed with the same parameters as in Case-3S, except a Federal tax credit of 1.80 cents/kWh for energy production was obtained and considered as revenue. The break-even selling price of energy was then 7.80 cents/kWh. This case illustrates the importance of tax credits.

In Case-5S, a 10 kW wind turbine was installed like in Case-4S, except an additional state tax credit of 1.00 cents/kWh and a 2.00 cents/kWh credit for green tags were obtained. The break-even price was determined to be 4.80 cents/kWh. This case shows the importance of having federal tax credit, state tax credit, and green tags.

For Case-6S, a 10 kW wind turbine project was the same as Case-5S, except the capacity factor was 23.2%. The result was a break-even selling price of energy of 7.905 cents/kWh. Note how the capacity factor at 35% equates to a break-even price of 4.80 cents/kWh in the previous case but when the capacity factor is reduced to 23.2%, the break-even price increases to 7.905 cents/kWh. Once again, capacity factor is an importance economic parameter.

Case-7S installed a 5 kW wind turbine but with all other economic parameters being the same as Case-3S. The break-even energy price is 11.550 cents/kWh. Note,

TABLE 12.2 Break-Even Selling Prices of Energy for Large and Small Wind Systems

Case Number	System Size (kW)	Capacity Factor (%)	Federal Tax Credit (¢/kWh)	State Tax Credit (¢/kWh)	Green Tags (¢/kWh)	Grant (%)	Break-Even Energy Price (¢/kWh)
1-S	10	35	0.00	0.00	0.00	0.0	15.70
2-S	10	23.2	0.00	0.00	0.00	0.0	21.90
3-S	10	35	0.00	0.00	0.00	50.0	9.600
4-S	10	35	1.80	0.00	0.00	50.0	7.800
5-S	10	35	1.80	1.00	2.00	50.0	4.800
6-S	10	23.2	1.80	1.00	2.00	50.0	7.905
7-S	5	35	0.00	0.00	0.00	50.0	11.55
8-S	5	35	1.08	1.00	2.00	50.0	6.750
3	310 MW	35	1.80	1.00	0.00	0.0	1.068

Case-3S had a break-even selling price of 9.600 cents/kWh when a 10 kW wind turbine was installed. This case just illustrates the importance of economy of scale even for small wind turbines and should be taken into consideration.

In Case-8S, a 5 kW wind turbine was installed with all parameters the same as Case-7S, except these additional revenue sources were considered: Federal tax credit of 1.80 cents/kWh, state tax credit of 1.00 cents/kWh, and green tags of 2.00 cents/kWh. The break-even selling price of energy is therefore 6.75 cents/kWh.

A summary of small wind turbine cases is shown in Table 12.2 for comparing the small wind systems previously discussed with one case from a large wind system.

Table 12.2 clearly illustrates the importance of economy of scale when considering a business plan for a wind energy system.

The reader should be aware that any economic analysis should be site/case specific and the economics presented in this article are presented only as information to the public. Also, it should be remembered that the downside to wind farms for producing electricity is the unreliability of the wind. Thus, wind generation of electricity is just one other form of energy source to generate power for use in the grid system along with the other sources of energy power plants listed previously such as natural gas, coal, hydro, and/or nuclear.

HYDROELECTRIC ENERGY: INTRODUCTION

Most of the energy in the United States is produced by fossil fuel and nuclear power plants. However, hydroelectricity is important to the United States because it represents 7.1% of the total power produced. As shown in Fig. 12.11, various sources contribute to the electricity production in the United States. For example, as a percentage contribution of the total electricity produced in the United States for year 2006, coal is 48.9%, nuclear is 19.3%, natural Gas is 20.0%, and hydroelectric is 7.1%.[16]

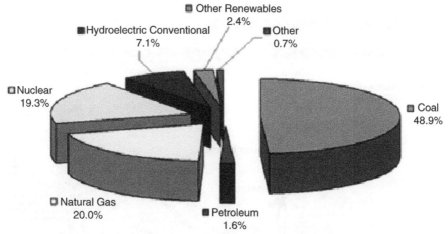

Sources of electricity in the USA 2006. Data from
http://www.eia.doe.gov/cneaf/electricity/epa/epat1p1.html

FIGURE 12.11   Sources of Electricity in the USA, 2006.[16]

Hydroelectric power is produced in theory by having a dam on a river with a large drop in elevation. Water is stored behind the dam and then flows at the base of the dam through a conduit called the penstock into the turbine. Blades in the turbine are turned by the water and rotate a shaft called the generator shaft. The generator shaft turns a generator that produces the electricity. The electricity is then sent to the power transmission cables into the electrical grid system that supplies the customers. The waster exits the turbine and returns through a conduit called the tailrace into the river past the dam. The rather simple process is illustrated in Fig. 12.12.

A diagram of a hydroelectric generator is shown in Fig. 12.13 as supplied by the U.S. Army Corps of Engineers. The hydraulic turbine converts the energy of the flowing water into mechanical energy. The hydroelectric generator converts the mechanical energy into electricity. In summary, a hydroelectric plant uses falling water to turn a turbine and resultant generator to produce electricity, i.e., hydroelectric energy.

Hydroelectric power and system have several advantages:

- Fuel is not burned, thereby minimizing pollution.
- Water used in the hydroelectric plant is provided free by nature.
- Hydroelectric power plays a role in reducing greenhouse gas emissions.
- Hydropower has low operation and maintenance costs.
- Hydroelectric power is a proven technology over time and is reliable.
- Hydroelectric energy is renewable because rainfall renews the water in the dam reservoir.
- A related positive feature of the power plant and dam is the scenic and recreation area provided by the reservoir for business and tourism activity in the local area.

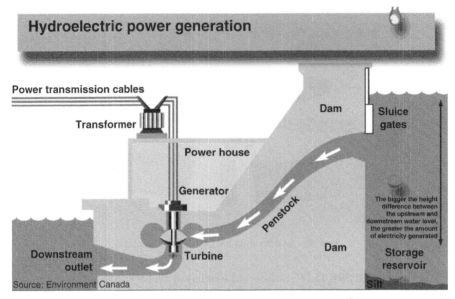

FIGURE 12.12 Hydroelectric Power Generation.[17]

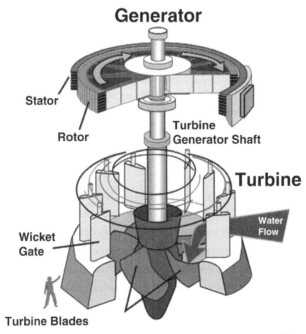

FIGURE 12.13 Hydroelectric Turbine and Generator.[18]

- It contributes to the storage of drinking water and water for irrigation.
- The vulnerability to floods and droughts is reduced by the storage of water in the reservoir.
- Electric rates can be stabilized and be more reliable.
- Hydroelectric power plants do not release pollutants into the air. Hydro energy can be a substitute for generation of energy by fossil fuel power plants.

Hydroelectric power and system have some disadvantages:

- It has high investment costs.
- It depends upon rain, precipitation.
- It causes some inundation of the land and wildlife habitat.
- A modification of the fish habitat occurs in some cases.
- Some movement of fish is restricted.
- In some cases, changes in reservoir and stream water quality may be affected.
- Local populations may be displaced in some cases.

Hydroelectric power is site specific because it requires large amounts of water and land where a dam can be built. Many of the best locations for hydroelectric power plants have been taken already in the United States. Thus, the trend in the future is likely to be with small-scale hydroelectric plants that can generate electricity for a local community. Small-scale hydroelectric plants with or without reservoirs are typically run-of-the-river plants. A run-of-the-river operation uses daily river flows to produce the hydroelectric energy, thereby reducing large fluctuations in the reservoir from accumulation and depletions. Consequently, a hydroelectric plant with no reservoir capacity is called a run-of-the-river plant.

Small-scale or low-head hydroelectric power plants are defined as plants having an electrical capacity of less than 15 MW (small scale) or dam height of less than 20 meters (low head). 20 meters is 65.6 feet. Small hydro plants have also been reported as those plants producing up to 10 MW or projects up to 30 MW. The major advantage of a hydroelectric plant is elimination of the cost of fuel. Also, hydroelectric plants tend to have a very long economic life such as up to 50 years.[19]

The following represents a preliminary economic assessment of several small-scale and low-head hydroelectric power plant projects, i.e., with potential in the state of Iowa.

HYDROELECTRIC MILL DAM: NASHUA, IOWA

The economics of a low-head dam and hydroenergy at Nashua Mill Dam, Nashua, Iowa is discussed as a source of green energy and how a low-head dam can improve a city utility's bottom line. Many communities have become more interested in a source of additional revenue during economic downturns where tax revenues have been reduced substantially. Since many rural areas have existing low-head dams constructed in the early 1900s, a logical approach to increasing revenues is to refurbish these old assets by utilization of existing but unused powerhouse for the generation of hydroelectric energy for use by the municipal government. Any excess energy from the hydroelectric power plant would be sold to the existing grid system currently supplying power to the surrounding area. With today's high cost of electricity, a small community could find this hydroelectric generation of power attractive. Thus, some interest has been generated recently from citizens, environmental groups, politicians, and commercial companies.Low-head dams, which have a potential of generating additional revenues to be used for revitalizing a declining rural community, have been of particular interest to civic authorities of cities. Finally, a Federal incentive payment of 1.8 ¢/kWh is available for the generation of renewable energy and interest-free Clean Renewable Energy Bonds (CREBs) as a result of the recently enacted Energy Policy Act of 2005. In addition, the Federal government enacted the American Recovery and Reinvestment Act of 2009, which provided a grant for 30% of the project cost of a qualified hydropower facility. These energy bills have made many past potential hydro sites currently economically attractive.[20,21]

With this interest, a sense of direction is needed to bring a reality check as to the economic potential of existing low-head dams for generation of hydroelectric energy.

Much useful information can be obtained in publications on low-head dams. Small-scale or low-head power plants have been defined as having an electrical capacity of less than 15 MW (small-scale) or a dam height of less than 65.6 feet (low-head). With fuel costs escalating over the years, economic conditions are beginning to favor a return to small-scale and low-head hydroelectric power generation. This is particularly true for municipalities with existing low-head dams, which typically had abandoned hydropower generation around the 1950s or so.[22,19]

Consider the following as typical of a low-head hydroelectric power generation facility. A picture of such a typical existing low-head dam facility with potential is shown in Fig. 12.14 and is the Nashua Mill Dam and Powerhouse, Nashua, Iowa.

Such an existing low-head dam facility was investigated as to its economic feasibility for hydroelectric generation of power for the local city government use. The $2.675 million project will consist of one turbine and have a capacity of 600 kW. This capital cost includes generating equipment, controls, switchgear, transmission line, transformers, and for an upgrade to the existing dam. Design flow is 416 cfs, and design head is 17 feet. This run-of-the-river facility would exist on the Cedar River in Iowa and would generate an average of about 3,966,700 kWh/year of electricity based upon a 20-year flow curve. Maintenance and operating (M&O) costs were considered to be $9,917 per year plus a capital budget reserve of $12,500 per year, since it was considered to be integrated with the existing city municipal operations. The combined M&O would total 0.565 cents/kWh.[23]

Current purchase price of electricity for the municipal government use has an average value of 5.940 cents per kWh. Capital is considered at a cost of 5.75% interest, including the cost of financing and insurance for 20 years. This amounts to $226,800 per year for the capital cost of the project. Revenues and expenditures were escalated

FIGURE 12.14   Nashua Mill Dam and Powerhouse, Nashua, Iowa.

at 2% per year. The capital reserve of $12,500 per year was included at a fixed value, since it is considered to be in an interest-bearing account.

Discussion of the Nashua Hydroelectric Economic Analyses

Case #1: Nashua Hydroelectric Facility   A revenue to the hydrofacility results from the city purchasing the electrical energy at 5.940 cents/kWh, which is the current purchase price of energy from the utility company. Additional cash flow results when excess energy is generated beyond the needs of the local city government and sold to the existing utility company at 4.500 cents/kWh. A federal incentive payment provides for a cash flow of 1.8 ¢/kWh for renewable energy generated by the hydrofacility. Green tags for the generation of green energy are considered to have a market value of 1.0 ¢/kWh. As a result of these revenues, a positive cash flow begins at year 1 for $47,142 and increases to $81,356 at year 10. Federal incentive payments from the 2005 Energy Bill end at year 10. Even with no Federal incentive payments for generation of renewable energy, the project continues to generate positive cash flows. After year 20, the bond has been paid so as to generate an additional positive cash flow of $226,800.00 per year, from paying off the bond, for the life of the physical plant, which is typically for another 30 years. Details are presented in the Table 12.3.

When financing over 30 years, the positive cash flow amounts to about twice the amount for 20 years but much more interest is paid out for the life of the project. For example, the financing expenditures (bond + insurance + interest) for 30 years would be $187,460 per year compared with the previous 20 years financing amount of $226,800 per year. Using the 30-year amortization, positive cash flow to the city for these years would be: year 1, $86,482; year 5, $100,938; year 10, $120,696. In comparison, the 20-year amortization for these years is: year 1, $47,142; year 5,

TABLE 12.3 Nashua Mill Dam and Powerhouse Economics (Case #1)

Nashua Hydroelectric Project, CASE # 1
Cedar Rapids, Iowa
Dr. Gary C. Young, P.E.

Nashua Mill Dam Hydroelectric Facility
Nashua, Iowa

PRELIMINARY

Capital Investment	$2,675,000.00			
Interest	5.75	%, average over life of project including financing & insurance		
Bond	20	years and payments semi-annually		
O & M	0.250	cents per kWh of production	O&M including Capital Budget Reserve	0.565 cents per kWh of production
Hydrofacility Production:	3,966,655	kWh/year; average for 20-years run-of-the-river		

Energy Usage from Hydrofacility:

Wells + Water plant + Sewer Plant	471,703	kWh/year	or	$28,019.16 $, in YR-1 (Includes City Hall, Civic facilities, and Street Lights)
City Hall + Civic Facilities	0	kWh/year	or	$0.00 $, in YR-1
Street Lights	0	kWh/year	or	0.0000 cents/KWh, YR-1
Purchased Energy	3,494,952	kWh/year	at	4.5000 cents/KWh, YR-1
Green Tags	3,966,655	kWh/year	at	1.0000 cents/kWh (NOTE: The green tags could be worth more with time but were considered constant in this economic evaluation.)
Fed Incentive revenue	3,966,655	kWh/year	at	1.8000 cents/kWh (NOTE: For a ten year period at beginning of project; also there is an inflation factor that can be used here but was neglected in this ecnomic analysis.)
Increase per year of energy sales price "and" O&M Expenses	2.00	% increase per year		(Note: Capital Budget Reserve is not increased annually because it is expected to be in an interest bearing accou[nt])

Year--->	1	2	3	4	5	6	7	8	9	10
Bond Interest, %	5.750	5.750	5.750	5.750	5.750	5.750	5.750	5.750	5.750	5.750
Energy Prices (cents/kWh):										
Wells + Water plant + Sewer Plant	5.940	6.059	6.180	6.304	6.430	6.558	6.689	6.823	6.960	7.099
City Hall + Civic Facilities	0.000	0.000	0.000	0.000	0.000	0.000	0.000	0.000	0.000	0.000
Street Lights	0.000	0.000	0.000	0.000	0.000	0.000	0.000	0.000	0.000	0.000
Purchased Energy	4.500	4.590	4.682	4.775	4.871	4.968	5.068	5.169	5.272	5.378
Green Tags	1.000	1.000	1.000	1.000	1.000	1.000	1.000	1.000	1.000	1.000
Fed Incentive revenue	1.800	1.800	1.800	1.800	1.800	1.800	1.800	1.800	1.800	1.800
Revenues:										
Wells + Water plant + Sewer Plant	$28,019.16	$28,579.54	$29,151.13	$29,734.16	$30,328.84	$30,935.42	$31,554.13	$32,185.21	$32,828.91	$33,485.49
City Hall + Civic Facilities	$0.00	$0.00	$0.00	$0.00	$0.00	$0.00	$0.00	$0.00	$0.00	$0.00
Street Lights	$0.00	$0.00	$0.00	$0.00	$0.00	$0.00	$0.00	$0.00	$0.00	$0.00
Purchased Energy	$157,272.84	$160,418.30	$163,626.66	$166,899.20	$170,237.18	$173,641.92	$177,114.76	$180,657.06	$184,270.20	$187,955.60
Green Tags	$39,666.55	$39,666.55	$39,666.55	$39,666.55	$39,666.55	$39,666.55	$39,666.55	$39,666.55	$39,666.55	$39,666.55
Fed Incentive revenue	$71,399.79	$71,399.79	$71,399.79	$71,399.79	$71,399.79	$71,399.79	$71,399.79	$71,399.79	$71,399.79	$71,399.79
Total Revenues	$296,358.34	$300,064.18	$303,844.14	$307,699.69	$311,632.36	$315,643.68	$319,735.23	$323,908.60	$328,165.45	$332,507.43

Expenditures:

	11	12	13	14	15	16	17	18	19	20
Bond + Insurance + Interest:	$226,800.16	$226,800.16	$226,800.16	$226,800.16	$226,800.16	$226,800.16	$226,800.16	$226,800.16	$226,800.16	$226,800.16
O & M	$9,916.64	$10,114.97	$10,317.27	$10,523.62	$10,734.09	$10,948.77	$11,167.74	$11,391.10	$11,618.92	$11,851.30
Capital Budget Reserve	$12,500.00	$12,500.00	$12,500.00	$12,500.00	$12,500.00	$12,500.00	$12,500.00	$12,500.00	$12,500.00	$12,500.00
Total Expenditures	$249,216.80	$249,415.13	$249,617.43	$249,823.78	$250,034.25	$250,248.93	$250,467.90	$250,691.26	$250,919.08	$251,151.46
(Revenues - Expenditures)	$47,141.54	$50,649.05	$54,226.71	$57,875.92	$61,598.11	$65,394.75	$69,267.32	$73,217.35	$77,246.37	$81,355.97
Year---->	11	12	13	14	15	16	17	18	19	20
Bond Interest, %	5.750	5.750	5.750	5.750	5.750	5.750	5.750	5.750	5.750	5.750
Energy Prices (cents/kWh):										
Wells + Water plant + Sewer Plant	7.241	7.386	7.533	7.684	7.838	7.994	8.154	8.317	8.484	8.653
City Hall + Civic Facilities	0.000	0.000	0.000	0.000	0.000	0.000	0.000	0.000	0.000	0.000
Street Lights	0.000	0.000	0.000	0.000	0.000	0.000	0.000	0.000	0.000	0.000
Purchased Energy	5.485	5.595	5.707	5.821	5.938	6.056	6.178	6.301	6.427	6.556
Green Tags	1.000	1.000	1.000	1.000	1.000	1.000	1.000	1.000	1.000	1.000
Fed Incentive revenue	0.000	0.000	0.000	0.000	0.000	0.000	0.000	0.000	0.000	0.000
Revenues:										
Wells + Water plant + Sewer Plant	$34,155.20	$34,838.30	$35,535.07	$36,245.77	$36,970.69	$37,710.10	$38,464.30	$39,233.59	$40,018.26	$40,818.63
City Hall + Civic Facilities	$0.00	$0.00	$0.00	$0.00	$0.00	$0.00	$0.00	$0.00	$0.00	$0.00
Street Lights	$0.00	$0.00	$0.00	$0.00	$0.00	$0.00	$0.00	$0.00	$0.00	$0.00
Purchased Energy	$191,714.71	$195,549.01	$199,459.99	$203,449.19	$207,518.17	$211,668.54	$215,901.91	$220,219.94	$224,624.34	$229,116.83
Green Tags	$39,666.55	$39,666.55	$39,666.55	$39,666.55	$39,666.55	$39,666.55	$39,666.55	$39,666.55	$39,666.55	$39,666.55
Fed Incentive revenue	$0.00	$0.00	$0.00	$0.00	$0.00	$0.00	$0.00	$0.00	$0.00	$0.00
Total Revenues	$265,536.46	$270,053.86	$274,661.61	$279,361.51	$284,155.41	$289,045.19	$294,032.76	$299,120.08	$304,309.15	$309,602.01
Expenditures:										
Bond + Insurance + Interest:	$226,800.16	$226,800.16	$226,800.16	$226,800.16	$226,800.16	$226,800.16	$226,800.16	$226,800.16	$226,800.16	$226,800.16
O & M	$12,088.33	$12,330.09	$12,576.69	$12,828.23	$13,084.79	$13,346.49	$13,613.42	$13,885.69	$14,163.40	$14,446.67
Capital Budget Reserve	$12,500.00	$12,500.00	$12,500.00	$12,500.00	$12,500.00	$12,500.00	$12,500.00	$12,500.00	$12,500.00	$12,500.00
Total Expenditures	$251,388.49	$251,630.25	$251,876.85	$252,128.39	$252,384.95	$252,646.65	$252,913.58	$253,185.85	$253,463.56	$253,746.83
(Revenues - Expenditures)	$14,147.98	$18,423.61	$22,784.75	$27,233.12	$31,770.46	$36,398.54	$41,119.18	$45,934.24	$50,845.59	$55,855.18
Total (Revenues - Expenditures)	$982,485.74	20 year, analysis								
Average Annual (Revenue - Expenditures)	$49,124.29 /year									

$61,598; year 10, $81,356. The average annual net revenue (revenue–expenditures) in 20 years would be $49,124/year, before taxes.

Financial considerations discussed previously have been done using capital costs for "new" equipment. If "used" equipment is considered, an estimated capital savings is initially realized and the corresponding cash flow with 20-year amortization for year 1 is $86,482. The cash flow for year 1 with "new" equipment was $47,142. Also, the capital investment was reduced by 17.4% by using used equipment. As a word of caution, the "used" equipment case may not realize the full economics due to possible lower energy generation efficiencies, reduced on-stream time, and higher O&M costs.

For a 20-year financing, Case #1, economic details are presented in Table 12.3 with annual cash flows: a Federal incentive payment of 1.8 cents/kWh for production in the first 10 years and green tags at 1.00 cents/kWh for 20 years. The average annual net revenue (revenue–expenditures) over 20 years would be $49,124/year, before taxes.

Case #2: Nashua Hydroelectric Facility   For a 20-year financing, Case #2, economic details are presented in Table 12.4 with annual cash flows. Case #2 is similar to Case #1, but "no" Federal incentive payment of 1.8 cents/kWh for production was considered. As before, green tags were taken at 1.00 cents/kWh for 20 years. However, the economic analysis used a Federal government grant of 30% of the project cost as provided by the enacted American Recovery and Reinvestment Act of 2009 for a qualified hydropower facility. Thus, the project facility cost was reduced from $2,675,000.00 to $1,872,500.00 as presented in Table 12.4.[20]

For Case #2, positive cash flows for the 20-year financing in these years are: year 1, $43,781; year 5, $58,238; year 10, $77,996. The average annual net revenue (revenue–expenditures) over 20 years would be $81,464/year, before taxes. One key point in this economic analysis is that the 30% grant is eligible for the private sector.

Case #3: Nashua Hydroelectric Facility   For a 20 year financing, Case #3, economic details are presented in Table 12.5 with annual cash flows. Case #3 is similar to Case #1 with Federal incentive payment of 1.8 cents/kWh for production in the first 10 years. As before, green tags were taken at 1.00 cents/kWh for 20 years. However, the economic analysis used a Federal government program of CREBs provision for financing the project cost as provided Energy Policy Act of 2005.[20] CREBs provide for interest-free "zero percent interest" financing of the project for a qualified hydropower facility. Thus, the project facility cash flows are presented in Table 12.5.

For Case #3, positive cash flows for the 20-year financing in these years are: year 1, $140,192; year 5, $154,648; year 10, $174,406. The average annual net revenue (revenue–expenditures) over 20 years would be $142,174/year, with no taxes for a local community facility.

The Nashua economic cases illustrate the importance of selling "excess" power from the hydrofacility to a grid system. The excess power is the amount of electrical energy produced from the hydrofacility above the city's usage per year. If this excess power is sold at 4.500 cents/kWh to the grid system, cash flow to the city and hydrofacility is attractive. For some areas of the country, utilities may be hard-nosed and resist fair and equitable negotiations. When this occurs, a rural community could

TABLE 12.4 Nashua Mill Dam and Powerhouse Economics (Case #2)

Nashua Hydroelectric Project, CASE # 2
Cedar Rapids, Iowa
Dr. Gary C. Young, P.E.

Nashua Mill Dam Hydroelectric Facility
Nashua, Iowa

Capital Investment	$1,872,500.00		
Interest	5.75	%, average over life of project including financing & insurance	
Bond	20	years and payments semi-annually	
O & M	0.250	cents per kWh of production	
Hydrofacility Production:	3,966,655	kWh/year; average for 20-years run-of-the-river	
		O&M including Capital Budget Reserve	0.565 cents per kWh of production

Energy Usage from Hydrofacility:

Wells + Water plant + Sewer Plant	471,703	kWh/year	or	$28,019.16 $, in YR-1 (includes City Hall, Civic facilities, and Street Lights)
City Hall + Civic Facilities	0	kWh/year	or	$0.00 $, in YR-1
Street Lights	0	kWh/year	or	0.0000 cents/kWh, YR-1
Purchased Energy	3,494,952	kWh/year	at	4.5000 cents/kWh, YR-1
Green Tags	3,966,655	kWh/year	at	1.0000 cents/kWh

(NOTE: The green tags could be worth more with time but were considered constant in this economic evaluation.)

Fed Incentive revenue	3,966,655	kWh/year	at	1.8000 cents/kWh

(NOTE: For a ten year period at beginning of project; also there is an inflation factor that can be used here but was neglected in this economic analysis.)

Increase per year of energy sales price "and" O&M Expenses	2.00	% increase per year

(Note: Capital Budget Reserve is not increased annually because it is expected to be in an interest bearing accou...)

Year---->	1	2	3	4	5	6	7	8	9	10
Bond Interest, %	5.750	5.750	5.750	5.750	5.750	5.750	5.750	5.750	5.750	5.750
Energy Prices (cents/kWh):										
Wells + Water plant + Sewer Plant	5.940	6.059	6.180	6.304	6.430	6.558	6.689	6.823	6.960	7.099
City Hall + Civic Facilities	0.000	0.000	0.000	0.000	0.000	0.000	0.000	0.000	0.000	0.000
Street Lights	0.000	0.000	0.000	0.000	0.000	0.000	0.000	0.000	0.000	0.000
Purchased Energy	4.500	4.590	4.682	4.775	4.871	4.968	5.068	5.169	5.272	5.378
Green Tags	1.000	1.000	1.000	1.000	1.000	1.000	1.000	1.000	1.000	1.000
Fed Incentive revenue	0.000	0.000	0.000	0.000	0.000	0.000	0.000	0.000	0.000	0.000
Revenues:										
Wells + Water plant + Sewer Plant	$28,019.16	$28,579.54	$29,151.13	$29,734.16	$30,328.84	$30,935.42	$31,554.13	$32,185.21	$32,828.91	$33,485.49
City Hall + Civic Facilities	$0.00	$0.00	$0.00	$0.00	$0.00	$0.00	$0.00	$0.00	$0.00	$0.00
Street Lights	$0.00	$0.00	$0.00	$0.00	$0.00	$0.00	$0.00	$0.00	$0.00	$0.00
Purchased Energy	$157,272.84	$160,418.30	$163,626.66	$166,899.20	$170,237.18	$173,641.92	$177,114.76	$180,657.06	$184,270.20	$187,955.60
Green Tags	$39,666.55	$39,666.55	$39,666.55	$39,666.55	$39,666.55	$39,666.55	$39,666.55	$39,666.55	$39,666.55	$39,666.55
Fed Incentive revenue	$0.00	$0.00	$0.00	$0.00	$0.00	$0.00	$0.00	$0.00	$0.00	$0.00
Total Revenues	$224,958.55	$228,664.39	$232,444.35	$236,299.90	$240,232.57	$244,243.89	$248,335.44	$252,508.81	$256,765.66	$261,107.64

(continued)

TABLE 12.4 (*Continued*)

Expenditures:										
Bond + Insurance + Interest:	$158,760.11	$158,760.11	$158,760.11	$158,760.11	$158,760.11	$158,760.11	$158,760.11	$158,760.11	$158,760.11	$158,760.11
O & M	$9,916.64	$10,114.97	$10,317.27	$10,523.62	$10,734.09	$10,948.77	$11,167.74	$11,391.10	$11,618.92	$11,851.30
Capital Budget Reserve	$12,500.00	$12,500.00	$12,500.00	$12,500.00	$12,500.00	$12,500.00	$12,500.00	$12,500.00	$12,500.00	$12,500.00
Total Expenditures	$181,176.75	$181,375.08	$181,577.38	$181,783.73	$181,994.20	$182,208.88	$182,427.85	$182,651.21	$182,879.03	$183,111.41
(Revenues - Expenditures)	$43,781.80	$47,289.31	$50,866.97	$54,516.18	$58,238.37	$62,035.01	$65,907.58	$69,857.61	$73,886.63	$77,996.23
Year---->	11	12	13	14	15	16	17	18	19	20
Bond Interest, %	5.750	5.750	5.750	5.750	5.750	5.750	5.750	5.750	5.750	5.750
Energy Prices (cents/kWh):										
Wells + Water plant + Sewer Plant	7.241	7.386	7.533	7.684	7.838	7.994	8.154	8.317	8.484	8.653
City Hall + Civic Facilities	0.000	0.000	0.000	0.000	0.000	0.000	0.000	0.000	0.000	0.000
Street Lights	0.000	0.000	0.000	0.000	0.000	0.000	0.000	0.000	0.000	0.000
Purchased Energy	5.485	5.595	5.707	5.821	5.938	6.056	6.178	6.301	6.427	6.556
Green Tags	1.000	1.000	1.000	1.000	1.000	1.000	1.000	1.000	1.000	1.000
Fed Incentive revenue	0.000	0.000	0.000	0.000	0.000	0.000	0.000	0.000	0.000	0.000
Revenues:										
Wells + Water plant + Sewer Plant	$34,155.20	$34,838.30	$35,535.07	$36,245.77	$36,970.69	$37,710.10	$38,464.30	$39,233.59	$40,018.26	$40,818.63
City Hall + Civic Facilities	$0.00	$0.00	$0.00	$0.00	$0.00	$0.00	$0.00	$0.00	$0.00	$0.00
Street Lights	$0.00	$0.00	$0.00	$0.00	$0.00	$0.00	$0.00	$0.00	$0.00	$0.00
Purchased Energy	$191,714.71	$195,549.01	$199,459.99	$203,449.19	$207,518.17	$211,668.54	$215,901.51	$220,219.94	$224,624.34	$229,116.83
Green Tags	$39,666.55	$39,666.55	$39,666.55	$39,666.55	$39,666.55	$39,666.55	$39,666.55	$39,666.55	$39,666.55	$39,666.55
Fed Incentive revenue	$0.00	$0.00	$0.00	$0.00	$0.00	$0.00	$0.00	$0.00	$0.00	$0.00
Total Revenues	$265,536.46	$270,053.86	$274,661.61	$279,361.51	$284,155.41	$289,045.19	$294,032.76	$299,120.08	$304,309.15	$309,602.01
Expenditures:										
Bond + Insurance + Interest:	$158,760.11	$158,760.11	$158,760.11	$158,760.11	$158,760.11	$158,760.11	$158,760.11	$158,760.11	$158,760.11	$158,760.11
O & M	$12,088.33	$12,330.09	$12,576.69	$12,828.23	$13,084.79	$13,346.49	$13,613.42	$13,885.69	$14,163.40	$14,446.67
Capital Budget Reserve	$12,500.00	$12,500.00	$12,500.00	$12,500.00	$12,500.00	$12,500.00	$12,500.00	$12,500.00	$12,500.00	$12,500.00
Total Expenditures	$183,348.44	$183,590.20	$183,836.80	$184,088.34	$184,344.90	$184,606.60	$184,873.53	$185,145.80	$185,423.51	$185,706.78
(Revenues - Expenditures)	$82,188.03	$86,463.66	$90,824.80	$95,273.17	$99,810.51	$104,438.59	$109,159.23	$113,974.29	$118,885.64	$123,895.23
Total (Revenues - Expenditures)	$1,629,288.84									
Average Annual (Revenue - Expenditures)	$81,464.44 /year									

20 year, analysis

290

TABLE 12.5 Nashua Mill Dam and Powerhouse Economics (Case #3)

Nashua Hydroelectric Project, CASE # 3
Cedar Rapids, Iowa
Dr. Gary C. Young, P.E.

Nashua Mill Dam Hydroelectric Facility
Nashua, Iowa

Capital Investment	$2,675,000.00			
Interest	0.00	%, average over life of project including financing & insurance		
Bond	20	years and payments semi-annually		
O & M	0.250	cents per kWh of production	O&M including Capital Budget Reserve	0.565 cents per kWh of production
Hydrofacility Production:	3,966,655	kWh/year; average for 20-years run-of-the-river		

Energy Usage from Hydrofacility:

Wells + Water plant + Sewer Plant	471,703	kWh/year	or	$28,019.16 $, in YR-1 (Includes City Hall, Civic facilities, and Street Lights)
City Hall + Civic Facilities	0	kWh/year	or	$0.00 $, in YR-1
Street Lights	0	kWh/year	or	0.0000 cents/kWh, YR-1
Purchased Energy	3,494,952	kWh/year	at	4.5000 cents/kWh, YR-1
Green Tags	3,966,655	kWh/year	at	1.0000 cents/kWh (NOTE: The green tags could be worth more with time but were considered constant in this economic evaluation.)
Fed Incentive revenue	3,966,655	kWh/year	at	1.8000 cents/kWh (NOTE: For a ten year period at beginning of project; also there is an inflation factor that can be used here but was neglected in this economic analysis.)
Increase per year of energy sales price "and" O&M Expenses	2.00	% increase per year		(Note: Capital Budget Reserve is not increased annually because it is expected to be in an interest bearing accou...

Year--->	1	2	3	4	5	6	7	8	9	10
Bond Interest, %	5.750	5.750	5.750	5.750	5.750	5.750	5.750	5.750	5.750	5.750
Energy Prices (cents/kWh):										
Wells + Water plant + Sewer Plant	5.940	6.059	6.180	6.304	6.430	6.558	6.689	6.823	6.960	7.099
City Hall + Civic Facilities	0.000	0.000	0.000	0.000	0.000	0.000	0.000	0.000	0.000	0.000
Street Lights	0.000	0.000	0.000	0.000	0.000	0.000	0.000	0.000	0.000	0.000
Purchased Energy	4.500	4.590	4.682	4.775	4.871	4.968	5.068	5.169	5.272	5.378
Green Tags	1.000	1.000	1.000	1.000	1.000	1.000	1.000	1.000	1.000	1.000
Fed Incentive revenue	1.800	1.800	1.800	1.800	1.800	1.800	1.800	1.800	1.800	1.800
Revenues:										
Wells + Water plant + Sewer Plant	$28,019.16	$28,579.54	$29,151.13	$29,734.16	$30,328.84	$30,935.42	$31,554.13	$32,185.21	$32,828.91	$33,485.49
City Hall + Civic Facilities	$0.00	$0.00	$0.00	$0.00	$0.00	$0.00	$0.00	$0.00	$0.00	$0.00
Street Lights	$0.00	$0.00	$0.00	$0.00	$0.00	$0.00	$0.00	$0.00	$0.00	$0.00
Purchased Energy	$157,272.84	$160,418.30	$163,626.66	$166,899.20	$170,237.18	$173,641.92	$177,114.76	$180,657.06	$184,270.20	$187,955.60
Green Tags	$39,666.55	$39,666.55	$39,666.55	$39,666.55	$39,666.55	$39,666.55	$39,666.55	$39,666.55	$39,666.55	$39,666.55
Fed Incentive revenue	$71,399.79	$71,399.79	$71,399.79	$71,399.79	$71,399.79	$71,399.79	$71,399.79	$71,399.79	$71,399.79	$71,399.79
Total Revenues	$296,358.34	$300,064.18	$303,844.14	$307,699.69	$311,632.36	$315,643.68	$319,735.23	$323,908.60	$328,165.45	$332,507.43

(continued)

TABLE 12.5 (*Continued*)

Expenditures:

Bond + Insurance + Interest	$133,750.14	$133,750.14	$133,750.14	$133,750.14	$133,750.14	$133,750.14	$133,750.14	$133,750.14	$133,750.14	$133,750.14
O & M	$9,916.64	$10,114.97	$10,317.27	$10,523.62	$10,734.09	$10,948.77	$11,167.74	$11,391.10	$11,618.92	$11,851.30
Capital Budget Reserve	$12,500.00	$12,500.00	$12,500.00	$12,500.00	$12,500.00	$12,500.00	$12,500.00	$12,500.00	$12,500.00	$12,500.00
Total Expenditures	$156,166.78	$156,365.11	$156,567.41	$156,773.76	$156,984.23	$157,198.91	$157,417.88	$157,641.24	$157,869.06	$158,101.44
(Revenues - Expenditures)	$140,191.56	$143,699.07	$147,276.73	$150,925.94	$154,648.13	$158,444.77	$162,317.34	$166,267.37	$170,296.39	$174,405.99
Year---->	11	12	13	14	15	16	17	18	19	20
Bond Interest, %	5.750	5.750	5.750	5.750	5.750	5.750	5.750	5.750	5.750	5.750
Energy Prices (cents/kWh):										
Wells + Water plant + Sewer Plant	7.241	7.386	7.533	7.684	7.838	7.994	8.154	8.317	8.484	8.653
City Hall + Civic Facilities	0.000	0.000	0.000	0.000	0.000	0.000	0.000	0.000	0.000	0.000
Street Lights	0.000	0.000	0.000	0.000	0.000	0.000	0.000	0.000	0.000	0.000
Purchased Energy	5.485	5.595	5.707	5.821	5.938	6.056	6.178	6.301	6.427	6.556
Green Tags	1.000	1.000	1.000	1.000	1.000	1.000	1.000	1.000	1.000	1.000
Fed Incentive revenue	0.000	0.000	0.000	0.000	0.000	0.000	0.000	0.000	0.000	0.000

Revenues:

Wells + Water plant + Sewer Plant	$34,155.20	$34,838.30	$35,535.07	$36,245.77	$36,970.69	$37,710.10	$38,464.30	$39,233.59	$40,018.26	$40,818.63
City Hall + Civic Facilities	$0.00	$0.00	$0.00	$0.00	$0.00	$0.00	$0.00	$0.00	$0.00	$0.00
Street Lights	$0.00	$0.00	$0.00	$0.00	$0.00	$0.00	$0.00	$0.00	$0.00	$0.00
Purchased Energy	$191,714.71	$195,549.01	$199,459.99	$203,449.19	$207,518.17	$211,668.54	$215,901.91	$220,219.94	$224,624.34	$229,116.83
Green Tags	$39,666.55	$39,666.55	$39,666.55	$39,666.55	$39,666.55	$39,666.55	$39,666.55	$39,666.55	$39,666.55	$39,666.55
Fed Incentive revenue	$0.00	$0.00	$0.00	$0.00	$0.00	$0.00	$0.00	$0.00	$0.00	$0.00
Total Revenues	$265,536.46	$270,053.86	$274,661.61	$279,361.51	$284,155.41	$289,045.19	$294,032.76	$299,120.08	$304,309.15	$309,602.01

Expenditures:

Bond + Insurance + Interest	$133,750.14	$133,750.14	$133,750.14	$133,750.14	$133,750.14	$133,750.14	$133,750.14	$133,750.14	$133,750.14	$133,750.14
O & M	$12,088.33	$12,330.09	$12,576.69	$12,828.23	$13,084.79	$13,346.49	$13,613.42	$13,885.69	$14,163.40	$14,446.67
Capital Budget Reserve	$12,500.00	$12,500.00	$12,500.00	$12,500.00	$12,500.00	$12,500.00	$12,500.00	$12,500.00	$12,500.00	$12,500.00
Total Expenditures	$158,338.47	$158,580.23	$158,826.83	$159,078.37	$159,334.93	$159,596.63	$159,863.56	$160,135.83	$160,413.54	$160,696.81
(Revenues - Expenditures)	$107,198.00	$111,473.63	$115,834.77	$120,283.14	$124,820.48	$129,448.56	$134,169 20	$138,984.26	$143,895.61	$148,905.20
Total (Revenues - Expenditures)	$2,843,486.14	20 year, analysis								
Average Annual (Revenue - Expenditures)	$142,174.31 /year									

look for a technology company wishing to have a relatively fixed and low-cost source of energy to locate in their community using electrical power from the hydroelectric facility. In this way, the new company and the community will prosper by utilization of the hydro facilities and keeping the local energy assets and generated wealth at home. This approach is innovative as it brings in new residents (company's employees), increased tax base, growth in housing, schools, new jobs, and new business opportunities for the community.[22]

HYDROELECTRIC MILL DAM: DELHI, IOWA

The Delhi Mill dam hydroelectric facility is located on the Maquoketa River (South Fork) about 1.4 miles south of the town of Delhi, Iowa. Looking downstream from left to right, the facility comprises a 60-foot wide concrete reinforced earth fill embankment abutting a dolomitic limestone bluff on the north side of the river, a 61-foot-wide reinforced concrete powerhouse, an 86-foot-long concrete gated spillway with three 25-foot-wide by 17-foot-high vertical sluice gates and a 495-foot-long zoned embankment initiating on the right abutment (south side) of the Maquoketa River. County road X31 crosses the dam, spillway, and the powerhouse. The total height of the structure from the streambed to the top of the roadway is 58.5 feet. A commercial substation is located just 100 feet of the powerhouse on the north side of the river. Two turbines were installed into the powerhouse in 1928 and used for generation of electricity until 1973. Consider this facility to be a typical low-head hydroelectric power generation application. A picture of such a typical existing low-head dam facility with potential is shown in Figs. 12.15 and 12.16 (the Delhi Mill Dam/Spillway, Powerhouse, and Roadway, Delhi, Iowa (Fig. 12.17).

FIGURE 12.15   Delhi Mill Dam and Spillway, Delhi, Iowa.

FIGURE 12.16   Delhi Mill Dam and Facility, Delhi, Iowa.

The existing low-head dam facility was investigated as to its potential economic feasibility for hydroelectric generation of power. The local owners are the Lake Delhi Recreation Association. The $2.792 million project will consist of two turbines and have a capacity of about 750 kW each for a total of about 1500 kW. This capital cost includes generating equipment, controls, switchgear, transmission line, and transformers. Each will have a hydraulic capacity of 276 cfs. The estimated average head is 37.5 feet. Operation of the facility is run-of-the-river. The powerhouse on the Delhi Mill Dam facility is located on the Maquoketa River in Iowa and would generate an average of about 3,472,500 kWh/year of electricity. O&M costs were considered to be $17,362 per year plus a capital budget reserve of $12,500 per year. The combined O&M costs would total 0.860 cents/kWh.

Capital is considered at a cost of 5.75% interest, including the cost of financing and insurance for 20 years. This amounts to $236,709 per year for the capital cost of the project. Revenues and expenditures were escalated at 2% per year. The capital reserve of $12,500 per year was included at a fixed value, since it is considered to be an interest-bearing account.

Discussion of the Delhi Hydroelectric Economic Analyses

Case #1: Delhi Hydroelectric Facility   A revenue to the hydrofacility results by the city purchasing the electrical energy at 5.000 cents/kWh, which is a reasonable purchase price of energy from a utility company. A Federal incentive payment provides for a cash flow of 1.8 ¢/kWh for renewable energy generated by the hydrofacility. Green tags for the generation of green energy are considered to have a market value of 1.0 ¢/kWh. As a result of these revenues, a positive cash

FIGURE 12.17   Delhi Roadway, Powerhouse, Spillway, and Dam.

flow begins at year 1 for $4,284 and increases to $34,790 at year 10. Federal incentive payments from the Energy Bill 2005 end at year 10. Even with no Federal incentive payments for generation of renewable energy, the project continues to generate positive cash flows. After year 20, the bond has been paid so as to generate an additional positive cash flow of $236,700.00 per year, from paying off the bond, for the life of the physical plant, which is typically for another 30 years.

For a 20-year financing, Case #1, economic details are presented in Table 12.6 with annual cash flows. Federal incentive payment of 1.8 cents/kWh was used for production in the first 10 years and green tags were set at 1.00 cents/kWh for 20 years. The average annual net revenue (revenue–expenditures) in 20 years would be $6,607/year, before taxes, which is basically breakeven until the bond is paid off.

Case #2: Delhi Hydroelectric Facility   For a 20-year financing, Case #2, economic details are presented in Table 12.7 with annual cash flows. Case #2 is similar to Case #1, but "no" Federal incentive payment of 1.8 cents/kWh for

TABLE 12.6 Delhi Mill Dam and Powerhouse Economics (Case #1)

Lake Delhi Hydroelectric Project, CASE # 1
Cedar Rapids, Iowa
Dr. Gary C. Young, P.E.

Lake Delhi Mill Dam Hydroelectric Facility
Delhi, Iowa

PRELIMINARY

Capital Investment	$2,791,865.92				
Interest	5.75	%, average over life of project including financing & insurance			
Bond	20	years and payments semi-annually			
O & M	0.500	cents per kWh of production	O&M including Capital Budget Reserve	0.860 cents per kWh of production	
Hydrofacility Production:	3,472,500	kWh/year; average for 20-years run-of-the-river			

Energy Usage from Hydrofacility:

Wells + Water plant + Sewer Plant	0	kWh/year	or	$0.00 $, in YR-1	(Includes City Hall, Civic facilities, and Street Lights)
City Hall + Civic Facilities	0	kWh/year	or	$0.00 $, in YR-1	
Street Lights	0	kWh/year	or	0.0000 cents/kWh, YR-1	
Purchased Energy	3,472,500	kWh/year	at	5.0000 cents/kWh, YR-1	
Green Tags	3,472,500	kWh/year	at	1.0000 cents/kWh	(NOTE: The green tags could be worth more with time but were considered constant in this economic evaluation.)
Fed Incentive revenue	3,472,500	kWh/year	at	1.8000 cents/kWh	(NOTE: For a ten year period at beginning of project; also there is an inflation factor that can be used here but was neglected in this economic analysis.)

Increase per year of energy sales price "and" O&M Expenses	2.00	% increase per year	(Note: Capital Budget Reserve is not increased annually because it is expected to be in an interest bearing accou[nt])

Year---->	1	2	3	4	5	6	7	8	9	10
Bond Interest, %	5.750	5.750	5.750	5.750	5.750	5.750	5.750	5.750	5.750	5.750
Energy Prices (cents/kWh):										
Wells + Water plant + Sewer Plant	0.000	0.000	0.000	0.000	0.000	0.000	0.000	0.000	0.000	0.000
City Hall + Civic Facilities	0.000	0.000	0.000	0.000	0.000	0.000	0.000	0.000	0.000	0.000
Street Lights	0.000	0.000	0.000	0.000	0.000	0.000	0.000	0.000	0.000	0.000
Energy Sold	5.000	5.100	5.202	5.306	5.412	5.520	5.631	5.743	5.858	5.975
Green Tags	1.000	1.000	1.000	1.000	1.000	1.000	1.000	1.000	1.000	1.000
Fed Incentive revenue	1.800	1.800	1.800	1.800	1.800	1.800	1.800	1.800	1.800	1.800
Revenues:										
Wells + Water plant + Sewer Plant	$0.00	$0.00	$0.00	$0.00	$0.00	$0.00	$0.00	$0.00	$0.00	$0.00
City Hall + Civic Facilities	$0.00	$0.00	$0.00	$0.00	$0.00	$0.00	$0.00	$0.00	$0.00	$0.00
Street Lights	$0.00	$0.00	$0.00	$0.00	$0.00	$0.00	$0.00	$0.00	$0.00	$0.00
Purchased Energy	$173,625.00	$177,097.50	$180,639.45	$184,252.24	$187,937.28	$191,696.03	$195,529.95	$199,440.55	$203,429.36	$207,497.95
Green Tags	$34,725.00	$34,725.00	$34,725.00	$34,725.00	$34,725.00	$34,725.00	$34,725.00	$34,725.00	$34,725.00	$34,725.00
Fed Incentive revenue	$62,505.00	$62,505.00	$62,505.00	$62,505.00	$62,505.00	$62,505.00	$62,505.00	$62,505.00	$62,505.00	$62,505.00
Total Revenues	$270,855.00	$274,327.50	$277,869.45	$281,482.24	$285,167.28	$288,926.03	$292,759.95	$296,670.55	$300,659.36	$304,727.95

Expenditures:	Year 11	Year 12	Year 13	Year 14	Year 15	Year 16	Year 17	Year 18	Year 19	Year 20
Bond + Insurance + Interest:	$236,708.65	$236,708.65	$236,708.65	$236,708.65	$236,708.65	$236,708.65	$236,708.65	$236,708.65	$236,708.65	$236,708.65
O & M	$17,362.50	$17,709.75	$18,063.95	$18,425.22	$18,793.73	$19,169.60	$19,553.00	$19,944.05	$20,342.94	$20,749.79
Capital Budget Reserve	$12,500.00	$12,500.00	$12,500.00	$12,500.00	$12,500.00	$12,500.00	$12,500.00	$12,500.00	$12,500.00	$12,500.00
Total Expenditures	$266,571.15	$266,918.40	$267,272.60	$267,633.87	$268,002.38	$268,378.25	$268,761.65	$269,152.70	$269,551.59	$269,958.44
(Revenues - Expenditures)	$4,283.85	$7,409.10	$10,596.86	$13,848.37	$17,164.91	$20,547.78	$23,998.31	$27,517.84	$31,107.77	$34,769.50
Year---->	11	12	13	14	15	16	17	18	19	20
Bond Interest, %	5.750	5.750	5.750	5.750	5.750	5.750	5.750	5.750	5.750	5.750
Energy Prices (cents/kWh):										
Wells + Water plant + Sewer Plant	0.000	0.000	0.000	0.000	0.000	0.000	0.000	0.000	0.000	0.000
City Hall + Civic Facilities	0.000	0.000	0.000	0.000	0.000	0.000	0.000	0.000	0.000	0.000
Street Lights	0.000	0.000	0.000	0.000	0.000	0.000	0.000	0.000	0.000	0.000
Energy Sold	6.095	6.217	6.341	6.468	6.597	6.729	6.864	7.001	7.141	7.284
Green Tags	1.000	1.000	1.000	1.000	1.000	1.000	1.000	1.000	1.000	1.000
Fed Incentive revenue	0.000	0.000	0.000	0.000	0.000	0.000	0.000	0.000	0.000	0.000
Revenues:										
Wells + Water plant + Sewer Plant	$0.00	$0.00	$0.00	$0.00	$0.00	$0.00	$0.00	$0.00	$0.00	$0.00
City Hall + Civic Facilities	$0.00	$0.00	$0.00	$0.00	$0.00	$0.00	$0.00	$0.00	$0.00	$0.00
Street Lights	$0.00	$0.00	$0.00	$0.00	$0.00	$0.00	$0.00	$0.00	$0.00	$0.00
Purchased Energy	$211,647.91	$215,880.86	$220,198.48	$224,602.45	$229,094.50	$233,676.39	$238,349.92	$243,116.92	$247,979.25	$252,936.84
Green Tags	$34,725.00	$34,725.00	$34,725.00	$34,725.00	$34,725.00	$34,725.00	$34,725.00	$34,725.00	$34,725.00	$34,725.00
Fed Incentive revenue	$0.00	$0.00	$0.00	$0.00	$0.00	$0.00	$0.00	$0.00	$0.00	$0.00
Total Revenues	$246,372.91	$250,605.86	$254,923.48	$259,327.45	$263,819.50	$268,401.39	$273,074.92	$277,841.92	$282,704.25	$287,663.84
Expenditures:										
Bond + Insurance + Interest:	$236,708.65	$236,708.65	$236,708.65	$236,708.65	$236,708.65	$236,708.65	$236,708.65	$236,708.65	$236,708.65	$236,708.65
O & M	$21,164.79	$21,588.09	$22,019.85	$22,460.25	$22,909.45	$23,367.64	$23,834.99	$24,311.69	$24,797.93	$25,293.88
Capital Budget Reserve	$12,500.00	$12,500.00	$12,500.00	$12,500.00	$12,500.00	$12,500.00	$12,500.00	$12,500.00	$12,500.00	$12,500.00
Total Expenditures	$270,373.44	$270,796.74	$271,228.50	$271,668.90	$272,118.10	$272,576.29	$273,043.64	$273,520.34	$274,006.58	$274,502.53
(Revenues - Expenditures)	-$24,000.53	-$20,190.87	-$16,305.02	-$12,341.44	-$8,298.60	-$4,174.90	$31.28	$4,321.57	$8,697.68	$13,161.31
Total (Revenues - Expenditures)	$132,144.75	20 year, analysis								
Average Annual (Revenue - Expenditures)	$6,607.24 /year									

production was considered. As before, green tags were taken at 1.00 cents/kWh for 20 years. However, the economic analysis used a Federal government grant of 30% of the project cost as provided by the enacted American Recovery and Reinvestment Act of 2009 for a qualified hydropower facility. Thus, the project facility cost was reduced from $2,791,900.00 to $1,954,300.00 as presented in Table 12.7.[20]

For Case #2, positive cash flows for the 20-year financing in these years are: year 1, $12,791; year 5, $25,672; year 10, $43,277. The average annual net revenue (revenue–expenditures) over 20 years would be $46,367/year, before taxes. One key point in this economic analysis is that the 30% grant is eligible for the private sector.

Case #3: Delhi Hydroelectric Facility   For a 20-year financing, Case #3, economic details are presented in Table 12.8 with annual cash flows. Case #3 is similar to Case #1 with Federal incentive payment of 1.8 cents/kWh for production in the first 10 years. As before, green tags were taken at 1.00 cents/kWh for 20 years. However, the economic analysis used a Federal government program of CREBs provision for financing the project cost as provided in the Energy Policy Act of 2005.[20] CREBs provide for interest-free "zero percent interest" financing of the project for a qualified hydropower facility. Thus, the project facility cash flows are presented in Table 12.8.

For Case #3, positive cash flows for the 20-year financing in these years are: year 1, $101,399; year 5, $114,280; year 10, $131,885. The average annual net revenue (revenue–expenditures) over 20 years would be $103,722/year, with no taxes for a local community facility.

The Delhi economic cases illustrate the importance of selling power from the hydrofacility to a grid system. If this power is sold at 5.000 cents/kWh to the grid system, cash flow to the Recreation Association and hydrofacility is attractive. For some areas of the country, utilities may be hard-nosed and resist fair and equitable negotiations. When this occurs, a rural community could look for a technology company wishing to have a relatively fixed and low-cost source of energy to locate in their community using electrical power from the hydroelectric facility. In this way, the new company and the community will prosper by utilization of the hydrofacilities and keeping the local energy assets and generated wealth at home. This approach is innovative by bringing in new residents (company's employees), increased tax base, growth in housing, schools, new jobs, and new business opportunities for the community.[22]

HYDROELECTRIC MILL DAM: FORT DODGE, IOWA

The economic viability of low-head dam and hydroenergy at Fort Dodge Mill Dam, Fort Dodge, Iowa is discussed as a source of green energy and how a low-head dam can improve a city utility's bottom line. Many communities have become more interested in a source of additional revenue during economic downturns where tax revenues have been reduced substantially. Since many rural areas have existing low-head dams constructed in the early 1900s, a logical approach to increasing revenues is

TABLE 12.7 Delhi Mill Dam and Powerhouse Economics (Case #2)

Lake Delhi Hydroelectric Project, CASE # 2
Cedar Rapids, Iowa
Dr. Gary C. Young, P.E.

Lake Delhi Mill Dam Hydroelectric Facility
Delhi, Iowa

Capital Investment	$1,954,306.14			
Interest	5.75	%, average over life of project including financing & insurance		
Bond	20	years and payments semi-annually		
O & M	0.500	cents per kWh of production	O&M including Capital Budget Reserve	
Hydrofacility Production:	3,472,500	kWh/year; average for 20-years run-of-the-river		0.860 cents per kWh of production

Energy Usage from Hydrofacility:

Wells + Water plant + Sewer Plant	0	kWh/year	or	$0.00 $, in YR-1	(includes City Hall, Civic facilities, and Street Lights)
City Hall + Civic Facilities	0	kWh/year	or	$0.00 $, in YR-1	
Street Lights	0	kWh/year	or	0.0000 cents/kWh, YR-1	
Purchased Energy	3,472,500	kWh/year	at	5.0000 cents/kWh, YR-1	
Green Tags	3,472,500	kWh/year	at	1.0000 cents/kWh	
Fed Incentive revenue	3,472,500	kWh/year	at	1.8000 cents/kWh	

(NOTE: The green tags could be worth more with time but were considered constant in this economic evaluation.)

(NOTE: For a ten year period at beginning of project; also there is an inflation factor that can be used here but was neglected in this ecnomic analysis.)

Increase per year of energy sales price "and" O&M Expenses 2.00 % increase per year (Note: Capital Budget Reserve is not increased annually because it is expected to be in an interest bearing accou

Year---->	1	2	3	4	5	6	7	8	9	10
Bond Interest, %	5.750	5.750	5.750	5.750	5.750	5.750	5.750	5.750	5.750	5.750
Energy Prices (cents/kWh):										
Wells + Water plant + Sewer Plant	0.000	0.000	0.000	0.000	0.000	0.000	0.000	0.000	0.000	0.000
City Hall + Civic Facilities	0.000	0.000	0.000	0.000	0.000	0.000	0.000	0.000	0.000	0.000
Street Lights	0.000	0.000	0.000	0.000	0.000	0.000	0.000	0.000	0.000	0.000
Energy Sold	5.000	5.100	5.202	5.306	5.412	5.520	5.631	5.743	5.858	5.975
Green Tags	1.000	1.000	1.000	1.000	1.000	1.000	1.000	1.000	1.000	1.000
Fed Incentive revenue	0.000	0.000	0.000	0.000	0.000	0.000	0.000	0.000	0.000	0.000
Revenues:										
Wells + Water plant + Sewer Plant	$0.00	$0.00	$0.00	$0.00	$0.00	$0.00	$0.00	$0.00	$0.00	$0.00
City Hall + Civic Facilities	$0.00	$0.00	$0.00	$0.00	$0.00	$0.00	$0.00	$0.00	$0.00	$0.00
Street Lights	$0.00	$0.00	$0.00	$0.00	$0.00	$0.00	$0.00	$0.00	$0.00	$0.00
Purchased Energy	$173,625.00	$177,097.50	$180,639.45	$184,252.24	$187,937.28	$191,696.03	$195,529.95	$199,440.55	$203,429.36	$207,497.95
Green Tags	$34,725.00	$34,725.00	$34,725.00	$34,725.00	$34,725.00	$34,725.00	$34,725.00	$34,725.00	$34,725.00	$34,725.00
Fed Incentive revenue	$0.00	$0.00	$0.00	$0.00	$0.00	$0.00	$0.00	$0.00	$0.00	$0.00
Total Revenues	$208,350.00	$211,822.50	$215,364.45	$218,977.24	$222,662.28	$226,421.03	$230,254.95	$234,165.55	$238,154.36	$242,222.95

(*continued*)

TABLE 12.7 (Continued)

Expenditures:

	11	12	13	14	15	16	17	18	19	20
Bond + Insurance + Interest:	$165,696.05	$165,696.05	$165,696.05	$165,696.05	$165,696.05	$165,696.05	$165,696.05	$165,696.05	$165,696.05	$165,696.05
O & M	$17,362.50	$17,709.75	$18,063.95	$18,425.22	$18,793.73	$19,169.60	$19,553.00	$19,944.05	$20,342.94	$20,749.79
Capital Budget Reserve	$12,500.00	$12,500.00	$12,500.00	$12,500.00	$12,500.00	$12,500.00	$12,500.00	$12,500.00	$12,500.00	$12,500.00
Total Expenditures	$195,558.55	$195,905.80	$196,260.00	$196,621.27	$196,989.78	$197,365.65	$197,749.05	$198,140.10	$198,538.99	$198,945.84
(Revenues - Expenditures)	$12,791.45	$15,916.70	$19,104.46	$22,355.97	$25,672.51	$29,055.38	$32,505.91	$36,025.44	$39,615.37	$43,277.10
Year---->	11	12	13	14	15	16	17	18	19	20
Bond Interest, %	5.750	5.750	5.750	5.750	5.750	5.750	5.750	5.750	5.750	5.750

Energy Prices (cents/kWh):

	11	12	13	14	15	16	17	18	19	20
Wells + Water plant + Sewer Plant	0.000	0.000	0.000	0.000	0.000	0.000	0.000	0.000	0.000	0.000
City Hall + Civic Facilities	0.000	0.000	0.000	0.000	0.000	0.000	0.000	0.000	0.000	0.000
Street Lights	0.000	0.000	0.000	0.000	0.000	0.000	0.000	0.000	0.000	0.000
Energy Sold	6.095	6.217	6.341	6.468	6.597	6.729	6.864	7.001	7.141	7.284
Green Tags	1.000	1.000	1.000	1.000	1.000	1.000	1.000	1.000	1.000	1.000
Fed Incentive revenue	0.000	0.000	0.000	0.000	0.000	0.000	0.000	0.000	0.000	0.000

Revenues:

	11	12	13	14	15	16	17	18	19	20
Wells + Water plant + Sewer Plant	$0.00	$0.00	$0.00	$0.00	$0.00	$0.00	$0.00	$0.00	$0.00	$0.00
City Hall + Civic Facilities	$0.00	$0.00	$0.00	$0.00	$0.00	$0.00	$0.00	$0.00	$0.00	$0.00
Street Lights	$0.00	$0.00	$0.00	$0.00	$0.00	$0.00	$0.00	$0.00	$0.00	$0.00
Purchased Energy	$211,647.91	$215,880.86	$220,198.48	$224,602.45	$229,094.50	$233,676.39	$238,349.92	$243,116.92	$247,979.25	$252,938.84
Green Tags	$34,725.00	$34,725.00	$34,725.00	$34,725.00	$34,725.00	$34,725.00	$34,725.00	$34,725.00	$34,725.00	$34,725.00
Fed Incentive revenue	$0.00	$0.00	$0.00	$0.00	$0.00	$0.00	$0.00	$0.00	$0.00	$0.00
Total Revenues	$246,372.91	$250,605.86	$254,923.48	$259,327.45	$263,819.50	$268,401.39	$273,074.92	$277,841.92	$282,704.25	$287,663.84

Expenditures:

	11	12	13	14	15	16	17	18	19	20
Bond + Insurance + Interest:	$165,696.05	$165,696.05	$165,696.05	$165,696.05	$165,696.05	$165,696.05	$165,696.05	$165,696.05	$165,696.05	$165,696.05
O & M	$21,164.79	$21,588.09	$22,019.85	$22,460.25	$22,909.45	$23,367.64	$23,834.99	$24,311.69	$24,797.93	$25,293.88
Capital Budget Reserve	$12,500.00	$12,500.00	$12,500.00	$12,500.00	$12,500.00	$12,500.00	$12,500.00	$12,500.00	$12,500.00	$12,500.00
Total Expenditures	$199,360.84	$199,784.14	$200,215.90	$200,656.30	$201,105.50	$201,563.69	$202,031.04	$202,507.74	$202,993.98	$203,489.93
(Revenues - Expenditures)	$47,012.07	$50,821.73	$54,707.58	$58,671.16	$62,714.00	$66,837.70	$71,043.88	$75,334.17	$79,710.28	$84,173.91

Total (Revenues - Expenditures) $927,346.75   20 year, analysis

Average Annual (Revenue - Expenditures) $46,367.34 /year

TABLE 12.8 Delhi Mill Dam and Powerhouse Economics (Case #3)

Lake Delhi Hydroelectric Project, CASE # 3   Lake Delhi Mill Dam Hydroelectric Facility
Cedar Rapids, Iowa   Delhi, Iowa
Dr. Gary C. Young, P.E.

Capital Investment	$2,791,865.92		
Interest	0.00	%, average over life of project including financing & insurance	
Bond	20	years and payments semi-annually	
O & M	0.500	cents per kWh of production	
Hydrofacility Production:	3,472,500	kWh/year; average for 20-years run-of-the-river	

O&M including Capital Budget Reserve   (Includes City Hall, Civic facilities, and Street Lights)
0.860 cents per kWh of production

Energy Usage from Hydrofacility:

Wells + Water plant + Sewer Plant	0	kWh/year	or	$0.00 $, in YR-1
City Hall + Civic Facilities	0	kWh/year	or	$0.00 $, in YR-1
Street Lights	0	kWh/year	or	0.0000 cents/kWh, YR-1
Purchased Energy	3,472,500	kWh/year	at	5.0000 cents/kWh, YR-1
Green Tags	3,472,500	kWh/year	at	1.0000 cents/kWh

(NOTE: The green tags could be worth more with time but were considered constant in this economic evaluation.)

Fed Incentive revenue	3,472,500	kWh/year	at	1.8000 cents/kWh

(NOTE: For a ten year period at beginning of project; also there is an inflation factor that can be used here but was neglected in this economic analysis.)

Increase per year of energy sales price "and" O&M Expenses   2.00   % increase per year

(Note: Capital Budget Reserve is not increased annually because it is expected to be in an interest bearing accou[nt])

Year--->	1	2	3	4	5	6	7	8	9	10
Bond Interest, %	5.750	5.750	5.750	5.750	5.750	5.750	5.750	5.750	5.750	5.750
Energy Prices (cents/kWh):										
Wells + Water plant + Sewer Plant	0.000	0.000	0.000	0.000	0.000	0.000	0.000	0.000	0.000	0.000
City Hall + Civic Facilities	0.000	0.000	0.000	0.000	0.000	0.000	0.000	0.000	0.000	0.000
Street Lights	0.000	0.000	0.000	0.000	0.000	0.000	0.000	0.000	0.000	0.000
Energy Sold	5.000	5.100	5.202	5.306	5.412	5.520	5.631	5.743	5.858	5.975
Green Tags	1.000	1.000	1.000	1.000	1.000	1.000	1.000	1.000	1.000	1.000
Fed Incentive revenue	1.800	1.800	1.800	1.800	1.800	1.800	1.800	1.800	1.800	1.800
Revenues:										
Wells + Water plant + Sewer Plant	$0.00	$0.00	$0.00	$0.00	$0.00	$0.00	$0.00	$0.00	$0.00	$0.00
City Hall + Civic Facilities	$0.00	$0.00	$0.00	$0.00	$0.00	$0.00	$0.00	$0.00	$0.00	$0.00
Street Lights	$0.00	$0.00	$0.00	$0.00	$0.00	$0.00	$0.00	$0.00	$0.00	$0.00
Purchased Energy	$173,625.00	$177,097.50	$180,639.45	$184,252.24	$187,937.28	$191,696.03	$195,529.95	$199,440.55	$203,429.36	$207,497.95
Green Tags	$34,725.00	$34,725.00	$34,725.00	$34,725.00	$34,725.00	$34,725.00	$34,725.00	$34,725.00	$34,725.00	$34,725.00
Fed Incentive revenue	$62,505.00	$62,505.00	$62,505.00	$62,505.00	$62,505.00	$62,505.00	$62,505.00	$62,505.00	$62,505.00	$62,505.00
Total Revenues	$270,855.00	$274,327.50	$277,869.45	$281,482.24	$285,167.28	$288,926.03	$292,759.95	$296,670.55	$300,659.36	$304,727.95

(continued)

TABLE 12.8 (*Continued*)

302

	11	12	13	14	15	16	17	18	19	20
Expenditures:										
Bond + Insurance + Interest:	$139,593.44	$139,593.44	$139,593.44	$139,593.44	$139,593.44	$139,593.44	$139,593.44	$139,593.44	$139,593.44	$139,593.44
O & M	$17,362.50	$17,709.75	$18,063.95	$18,425.22	$18,793.73	$19,169.60	$19,553.00	$19,944.05	$20,342.94	$20,749.79
Capital Budget Reserve	$12,500.00	$12,500.00	$12,500.00	$12,500.00	$12,500.00	$12,500.00	$12,500.00	$12,500.00	$12,500.00	$12,500.00
Total Expenditures	$169,455.94	$169,803.19	$170,157.39	$170,518.66	$170,887.17	$171,263.04	$171,646.44	$172,037.49	$172,436.38	$172,843.23
(Revenues - Expenditures)	$101,399.06	$104,524.31	$107,712.07	$110,963.58	$114,280.12	$117,662.99	$121,113.52	$124,633.05	$128,222.98	$131,884.71
Year--->	11	12	13	14	15	16	17	18	19	20
Bond Interest, %	5.750	5.750	5.750	5.750	5.750	5.750	5.750	5.750	5.750	5.750
Energy Prices (cents/kWh):										
Wells + Water plant + Sewer Plant	0.000	0.000	0.000	0.000	0.000	0.000	0.000	0.000	0.000	0.000
City Hall + Civic Facilities	0.000	0.000	0.000	0.000	0.000	0.000	0.000	0.000	0.000	0.000
Street Lights	0.000	0.000	0.000	0.000	0.000	0.000	0.000	0.000	0.000	0.000
Energy Sold	6.095	6.217	6.341	6.468	6.597	6.729	6.864	7.001	7.141	7.284
Green Tags	1.000	1.000	1.000	1.000	1.000	1.000	1.000	1.000	1.000	1.000
Fed Incentive revenue	0.000	0.000	0.000	0.000	0.000	0.000	0.000	0.000	0.000	0.000
Revenues:										
Wells + Water plant + Sewer Plant	$0.00	$0.00	$0.00	$0.00	$0.00	$0.00	$0.00	$0.00	$0.00	$0.00
City Hall + Civic Facilities	$0.00	$0.00	$0.00	$0.00	$0.00	$0.00	$0.00	$0.00	$0.00	$0.00
Street Lights	$0.00	$0.00	$0.00	$0.00	$0.00	$0.00	$0.00	$0.00	$0.00	$0.00
Purchased Energy	$211,647.91	$215,880.86	$220,198.48	$224,602.45	$229,094.50	$233,676.39	$238,349.92	$243,116.92	$247,979.25	$252,938.84
Green Tags	$34,725.00	$34,725.00	$34,725.00	$34,725.00	$34,725.00	$34,725.00	$34,725.00	$34,725.00	$34,725.00	$34,725.00
Fed Incentive revenue	$0.00	$0.00	$0.00	$0.00	$0.00	$0.00	$0.00	$0.00	$0.00	$0.00
Total Revenues	$246,372.91	$250,605.86	$254,923.48	$259,327.45	$263,819.50	$268,401.39	$273,074.92	$277,841.92	$282,704.25	$287,663.84
Expenditures:										
Bond + Insurance + Interest:	$139,593.44	$139,593.44	$139,593.44	$139,593.44	$139,593.44	$139,593.44	$139,593.44	$139,593.44	$139,593.44	$139,593.44
O & M	$21,164.79	$21,588.09	$22,019.85	$22,460.25	$22,909.45	$23,367.64	$23,834.99	$24,311.69	$24,797.93	$25,293.88
Capital Budget Reserve	$12,500.00	$12,500.00	$12,500.00	$12,500.00	$12,500.00	$12,500.00	$12,500.00	$12,500.00	$12,500.00	$12,500.00
Total Expenditures	$173,258.23	$173,681.53	$174,113.29	$174,553.69	$175,002.89	$175,461.08	$175,928.43	$176,405.13	$176,891.37	$177,387.32
(Revenues - Expenditures)	$73,114.68	$76,924.34	$80,810.19	$84,773.77	$88,816.61	$92,940.31	$97,146.49	$101,436.78	$105,812.89	$110,276.52
Total (Revenues - Expenditures)	$2,074,448.95									
Average Annual (Revenue - Expenditures)	$103,722.45 /year		20 year, analysis							

to refurbish these old assets by utilization of existing but unused powerhouse for the generation of hydroelectric energy for use by the municipal government. Any excess energy from the hydroelectric power plant would be sold to the existing grid system currently supplying power to the surrounding area. With today's high cost of electricity, a small community could find this hydroelectric generation of power attractive. Thus, some interest has been generated recently from citizens, environmental groups, politicians, and commercial companies. Particular interest lies with civic authorities of cities with low-head dams, which have a potential of generating additional revenues to be used for revitalizing a declining rural community. Finally, a Federal incentive payment of 1.8 ¢/kWh is available for the generation of renewable energy and interest-free (CREBs as a result of the recently enacted Energy Policy Act of 2005. In addition, the Federal government enacted the American Recovery and Reinvestment Act of 2009, which provided a grant for 30% of the project cost of a qualified hydropower facility. These energy bills have made many past potential hydrosites currently economically attractive.[20,21]

With this interest, a sense of direction is needed to bring a reality check as to the economic potential of existing low-head dams for generation of hydroelectric energy. Much useful information can be obtained in publications on low-head dams. Small-scale or low-head power plants have been defined as having an electrical capacity of less than 15 MW (small-scale) or a dam height of less than 65.6 feet (low-head). With fuel costs escalating over the years, economic conditions are beginning to favor a return to small-scale and low-head hydroelectric power generation. This is particularly true for municipalities with existing low-head dams, which typically had abandoned hydropower generation around the 1950s or so.[22,19]

Pictures of such a typical existing low-head dam facility with potential are shown in Figs. 12.18, 12.19, and 12.20 for the Fort Dodge Mill Dam, Gates and Powerhouse, Fort Dodge, Iowa.

The Fort Dodge hydroelectric facility was constructed in 1916 and used to generate electrical power until 1971 when the Iowa Illinois Gas & Electric Company decommissioned the facility. The facility was decommissioned because lower energy cost was available from other sources. Since that time, fundamentally, the economic picture in the energy business has changed dramatically. Thus, new thinking

FIGURE 12.18   Fort Dodge Mill Dam, Gates and Powerhouse, Fort Dodge, Iowa.

FIGURE 12.19   Fort Dodge Powerhouse and Mill Dam Gates, Fort Dodge, Iowa.

FIGURE 12.20   Fort Dodge Mill Dam, Gates and Powerhouse, Fort Dodge, Iowa

outside-the-box and a fresh look at the economic viability of this hydroelectric facility provided the logical atmosphere to work hard at developing a business plan to make the Fort Dodge Mill Dam hydroelectric facility economically viable without raising taxes on the public.

As a brief description of the site, the Fort Dodge Mill Dam hydropower facility is a powerhouse and a concrete dam, 342 feet long and 18 feet high, consisting of a 230-foot-long overflow spillway and five (15 feet × 20 feet) tainter gates. As mentioned earlier, the facility generated hydropower from about 1916 to 1971. The turbines, generators, and electrical equipment were removed, but the trash racks and stop log guides are in place. The reservoir is about 90 acres with negligible storage. From USGS data, the estimated maximum flow of the Des Moines River at the plant will be based upon stream flow records from 1981 to 2000:

Maximum flow	30,200 cfs (7/14/1993)
Minimum flow	20 cfs (12/24/1989)
Average flow of the stream per day	2,644 cfs (from 1981 to 2000)

The economic assessment presented here includes the installation of two new 700 kW Kaplan turbine generator units and associated equipment in the existing powerhouse. The installed capacity will be 1.4 MW. The average annual generation is estimated to be 7,506,117 kWh per year. The type of operation will be run-of-the-river. Normal operating gross head will be 14.5 feet. Upstream and downstream silt from the powerhouse area will need to be removed. Also, a transmission line will need to be connected with the powerhouse grid owned by MidAmerican Energy Company. The facility is located on the Des Moines River, which is a navigable waterway of the United States.

The estimated capital requirements of the project cost for placing the Fort Dodge hydroelectric facility into operation are:

- Equipment, Site, and Construction $6,000,000.00
- Dam Refurbishment $1,000,000.00

           Total estimated project cost $7,000,000.00

As mentioned earlier, the estimated annual generation would be 7,506,117 kWh/year. The methodology used to estimate the energy production from the Fort Dodge hydroelectric facility will be presented at the end. Lastly, the production of hydroelectric energy from the facility would be for the City of Fort Dodge, Iowa.

Discussion of the Fort Dodge Hydroelectric Economic Analyses

Case #1: Fort Dodge Hydroelectric Facility A revenue source to the hydrofacility results from the city replacing the electrical energy from these sources of the city currently purchasing power from the local utility:

a. Six wells + Water plant + Sewer plant: 6,443,964 kWh/year at 6.094 cents/kWh
b. City Hall + Library + Blanden: 730,960 kWh/year at 7.416 cents/kWh
c. Street lights: 641,784 kWh/year at 2.669 cents/kWh

Other sources of revenue are from enacted Federal energy regulations. Federal incentive payment provides for a cash flow of 1.8 ¢/kWh for renewable energy generated by the hydrofacility for the first 10 years. Green tags for the generation of green energy are considered to have a market value of 1.0 ¢/kWh.

For expenditures, the project cost of $7,000,000.00 was financed with bonds for 20 years at 5.75% and payment semiannually. O&M cost was selected at 0.25 kWh/year plus a capital budget reserve of $12,500.00/year, which results in a total O&M cost of 0.583 cents/kWh. O&M cost was based upon an understanding that the current hydrofacility was operated and maintained by city services and that a modern renovated hydrofacility would lend itself to economy of scale by integration into the city's currently available services. When a shortage of energy results from the hydrofacility, any purchased energy for the city's users listed would come from the local utility at 6.091 cents/kWh. Any excess energy generated from the hydrofacility would be sold to the local utility at 4.500 cents/kWh.

Energy production from the Fort Dodge hydrofacility was estimated on a daily basis from 1981 to 2000. Consequently, daily, monthly, and yearly production of energy was available; a summary of monthly and yearly energy production from the hydrofacility is shown in Table 12.12. The methodology for computation of these daily energy production estimates is presented at the end and in Table 12.13.

As a result of these economic parameters, a positive cash is generated of $15,811/year over the 20-year period of the project. Essentially, this case is a break-even so the project generates sufficient cash to operate and maintain the dam and hydrofacility without funds being required from the taxpayers. After year 20, the bond has been paid so as to generate an additional positive cash flow of $593,495.74 per year, after paying off the bond. Also, the life of a hydrophysical plant is typically for another 30 years or more, which consequently generates much cash for future generations.

For a 20-year financing, Case #1, economic details are presented in Table 12.9 with annual cash flows. Federal incentive payment of 1.8 cents/kWh was used for production in the first 10 years and green tags were set at 1.00 cents/kWh for 20 years. The average annual net revenue (revenue–expenditures) in 20 years would be $15,811.27/year, before taxes (city pays no taxes), which is basically breakeven until the bond is paid off.

Case #2: Fort Dodge Hydroelectric Facility For a 20-year financing, Case #2, economic details are presented in Table 12.10 with annual cash flows. Case #2 is similar to Case #1, but "no" Federal incentive payment of 1.8 cents/kWh for production was considered. As before, green tags were taken at 1.00 cents/kWh for 20 years. O&M cost was taken at 1.000 cents/kWh plus a capital budget reserve of $25,000.00 per year, giving a total of 1.333 cents/kWh. However, the economic analysis used a Federal government grant of 30% of the project cost as provided by the enacted American Recovery and Reinvestment Act of 2009 for a qualified hydropower facility. Thus, the project facility cost was reduced from $7,000,000.00 to $4,900,000.00 as presented in Table 12.10.[20]

For Case #2, overall, positive cash flows exist for the 20-year financing. The average annual net revenue (revenue–expenditures) over 20 years would be $65,423/year, before taxes. One key point in this economic analysis is that the 30% grant is eligible for the private sector.

Case #3: Fort Dodge Hydroelectric Facility For a 20-year financing, Case #3, economic details are presented in Table 12.11 with annual cash flows. Case #3 is similar to Case #1 with Federal incentive payment of 1.8 cents/kWh for production in the first 10 years. As before, green tags were taken at 1.00 cents/kWh for 20 years. O&M cost was taken at 0.500 cents/kWh plus a capital budget reserve of $25,000.00 per year, giving a total of 0.833 cents/kWh. However, the economic analysis used a Federal government program of CREBs provision for financing the project cost as

TABLE 12.9 Fort Dodge Mill Dam and Powerhouse Economics (Case #1)

Fort Dodge Hydroelectric Project, CASE #1
Cedar Rapids, Iowa
Dr. Gary C. Young, P.E.

Fort Dodge Mill Dam Hydroelectric Facility
Fort Dodge, Iowa

Capital Investment	$7,000,000.00		
[Capital Investment - GRANT] =	$7,000,000.00		
Grant =	0.00	% =	$0.00 GRANT
Interest	5.75	%, average over life of project including financing & insurance	
Bond	20	years and payments semi-annually	
	2	payment per year	
O & M	0.250	cents per kWh of production	O&M Including Capital Budget Reserve 0.583 cents/kWh of production
Hydrofacility Production:	7,506,117	kWh/year; average for 20-years run-of-the-river; (O&M based upon this production)	

Energy Usage from Hydrofacility:

Six Wells + Water plant + Sewer Plant	6,443,964	kWh/year	
City Hall + Library + Blanden	730,960	kWh/year	
Street Lights	641,784	kWh/year	
Excess Energy SOLD, YR-1	0	kWh/year	or $392,700.13 $, in YR-1
Excess Energy PURCHASED, YR-2	1,336,428	kWh/year	or $54,204.39 $, in YR-1
Green Tags, YR-1	5,560,159	kWh/year	or 2.6689 cents/kWh, YR-1
Fed Incentive revenue, YR-1	5,560,159	kWh/year	at **4.5000** cents/kWh, Yr-1; Negotiated with utility company, selling price when excess energy is positive
			at **6.0910** cents/kWh, Yr-1; Negotiated with utility company, selling price when excess energy is negative
			at 1.0000 cents/kWh (Note: The green tags have a typical market value between 1.0 to 2.0 cents/kWh)
			at 1.8000 cents/kWh (Note: First 10-years of project, no inflation factor was used)

Increase per year of energy sales price "and" O&M Expenses   2.00   % increase per year   (NOTE: Capital Budget Reserve is not increased annually because it is expected to be in an interest bearing account.)

Year--->	1	2	3	4	5	6	7	8	9	10
Bond Interest, %	5.750	5.750	5.750	5.750	5.750	5.750	5.750	5.750	5.750	5.750

Energy Prices (cents/kWh):

	1	2	3	4	5	6	7	8	9	10
Six Wells + Water plant + Sewer Plant	6.094	6.216	6.340	6.467	6.596	6.728	6.863	7.000	7.140	7.283
City Hall + Library + Blanden	7.416	7.564	7.715	7.869	8.027	8.187	8.351	8.518	8.688	8.862
Street Lights	2.669	2.722	2.777	2.832	2.889	2.947	3.006	3.066	3.127	3.190
Excess Energy (+) Selling Price	4.500	4.590	4.682	4.775	4.871	4.968	5.068	5.169	5.272	5.378
Excess Energy (-) Purchase Price	6.091	6.213	6.337	6.464	6.593	6.725	6.859	6.997	7.137	7.279
Green Tags	1.000	1.000	1.000	1.000	1.000	1.000	1.000	1.000	1.000	1.000
Fed Incentive revenue	1.800	1.800	1.800	1.800	1.800	1.800	1.800	1.800	1.800	1.800
Yearly Production, kWh/YR	5,560,159	9,153,136	9,917,632	8,436,527	8,598,157	9,866,657	6,626,189	3,533,065	1,412,662	3,612,490

(continued)

TABLE 12.9 (*Continued*)

Revenues:										
Six Wells + Water plant + Sewer Plant	$392,700.13	$400,554.13	$408,565.22	$416,736.52	$425,071.25	$433,572.67	$442,244.13	$451,089.01	$460,110.79	$469,313.01
City Hall + Library + Blanden	$54,204.39	$55,288.48	$56,394.25	$57,522.13	$58,672.57	$59,846.03	$61,042.95	$62,263.81	$63,509.08	$64,779.26
Street Lights	$17,128.57	$17,471.14	$17,820.57	$18,176.98	$18,540.52	$18,911.33	$19,289.56	$19,675.35	$20,068.85	$20,470.23
Excess Energy SOLD	$0.00	$61,342.05	$98,361.06	$29,599.06	$38,063.95	$101,848.92	$0.00	$0.00	$0.00	$0.00
Green Tags	$55,601.59	$91,531.36	$99,176.32	$84,365.27	$85,981.57	$98,666.57	$66,261.89	$35,330.65	$14,126.62	$36,124.90
Fed Incentive revenue	$100,082.86	$164,756.45	$178,517.38	$151,857.49	$154,766.83	$177,599.83	$119,271.40	$63,595.17	$25,427.92	$65,024.82
Total Revenues	$619,717.55	$790,943.61	$858,834.79	$758,257.45	$781,096.69	$890,445.35	$708,109.92	$631,953.98	$583,243.26	$655,712.22
Expenditures:										
Bond + Insurance + Interest	$593,495.74	$593,495.74	$593,495.74	$593,495.74	$593,495.74	$593,495.74	$593,495.74	$593,495.74	$593,495.74	$593,495.74
Excess Energy PURCHASED	$137,446.40	$0.00	$0.00	$0.00	$0.00	$0.00	$81,663.12	$299,711.27	$457,029.69	$306,038.01
O & M	$18,765.29	$19,140.60	$19,523.41	$19,913.88	$20,312.16	$20,718.40	$21,132.77	$21,555.42	$21,986.53	$22,426.26
Capital Budget Reserve	$25,000.00	$25,000.00	$25,000.00	$25,000.00	$25,000.00	$25,000.00	$25,000.00	$25,000.00	$25,000.00	$25,000.00
Total Expenditures	$774,707.43	$637,636.34	$638,019.15	$638,409.62	$638,807.90	$639,214.14	$721,291.63	$939,762.43	$1,097,511.96	$946,960.01
(Revenues - Expenditures)	-$154,989.89	$153,307.27	$220,815.64	$119,847.83	$142,288.79	$251,231.21	-$13,181.70	-$307,808.45	-$514,268.70	-$291,247.79
Year--->	11	12	13	14	15	16	17	18	19	20
Bond Interest, %	5.750	5.750	5.750	5.750	5.750	5.750	5.750	5.750	5.750	5.750
Energy Prices (cents/kWh):										
Six Wells + Water plant + Sewer Plant	7.429	7.577	7.729	7.883	8.041	8.202	8.366	8.533	8.704	8.878
City Hall + Library + Blanden	9.039	9.220	9.405	9.593	9.785	9.980	10.180	10.384	10.591	10.803
Street Lights	3.253	3.318	3.385	3.453	3.522	3.592	3.664	3.737	3.812	3.888
Excess Energy (+) Selling Price	5.485	5.595	5.707	5.821	5.938	6.056	6.176	6.301	6.427	6.556
Excess Energy (-) Purchase Price	7.425	7.573	7.725	7.879	8.037	8.198	8.362	8.529	8.699	8.873
Green Tags	1.000	1.000	1.000	1.000	1.000	1.000	1.000	1.000	1.000	1.000
Fed Incentive revenue	0.000	0.000	0.000	0.000	0.000	0.000	0.000	0.000	0.000	0.000
Yearly Production, kWh/YR	7,810,812	11,115,698	10,061,600	10,071,809	9,764,614	9,105,206	7,359,394	8,216,684	6,425,422	3,473,929
Revenues:										
Six Wells + Water plant + Sewer Plant	$478,699.27	$488,273.25	$498,038.72	$507,999.49	$518,159.48	$528,522.67	$539,093.12	$549,874.99	$560,872.49	$572,089.94
City Hall + Library + Blanden	$66,074.85	$67,396.35	$68,744.27	$70,119.16	$71,521.54	$72,951.97	$74,411.01	$75,899.23	$77,417.22	$78,965.56
Street Lights	$20,879.64	$21,297.23	$21,723.17	$22,157.64	$22,600.79	$23,052.80	$23,513.86	$23,984.14	$24,463.82	$24,953.10
Excess Energy SOLD	$0.00	$0.00	$0.00	$0.00	$0.00	$0.00	$0.00	$0.00	$0.00	$0.00
Green Tags	$78,108.12	$111,156.98	$100,616.00	$100,718.09	$97,646.14	$91,052.06	$73,598.94	$82,166.84	$64,254.22	$34,739.29
Fed Incentive revenue	$0.00	$184,584.57	$128,117.96	$131,274.61	$115,659.93	$78,036.69	$0.00	$25,202.83	$0.00	$0.00
Total Revenues	$643,761.87	$872,708.38	$817,240.13	$832,268.99	$825,587.88	$793,616.20	$710,616.94	$757,128.03	$727,007.74	$710,747.88

Expenditures:

Bond + Insurance + Interest:	$593,495.74	$593,495.74	$593,495.74	$593,495.74	$593,495.74	$593,495.74	$593,495.74	$593,495.74	$593,495.74
Excess Energy PURCHASED	$437.77	$0.00	$0.00	$0.00	$0.00	$0.00	$0.00	$0.00	$0.00
O & M	$22,874.79	$23,332.28	$23,798.93	$24,274.91	$24,760.41	$25,255.61	$25,760.73	$26,275.94	$26,801.46	$27,337.49
Capital Budget Reserve	$25,000.00	$25,000.00	$25,000.00	$25,000.00	$25,000.00	$25,000.00	$25,000.00	$25,000.00	$25,000.00
Total Expenditures	$641,808.30	$641,828.02	$642,294.67	$642,770.65	$643,256.15	$643,751.35	$682,453.60	$644,771.68	$766,331.40	$1,031,186.98
(Revenues - Expenditures)	$1,953.57	$230,880.36	$174,945.46	$189,498.34	$182,331.73	$149,864.84	$28,163.34	$112,356.35	-$39,323.66	-$320,439.10

Total (Revenues - Expenditures) $316,225.44   20 year, analysis
Avg. Annual (Revenues - Expenditures) $15,811.27 /year
Annual Average Production = 7,506,117 kWh/Year

TABLE 12.10 Fort Dodge Mill Dam and Powerhouse Economics (Case #2)

Fort Dodge Hydroelectric Project, CASE #2
Cedar Rapids, Iowa
Dr. Gary C. Young, P.E.

Fort Dodge Mill Dam Hydroelectric Facility
Fort Dodge, Iowa

PRELIMINARY

Capital Investment	$7,000,000.00		
[Capital Investment - GRANT] =	$4,900,000.00		
Grant =	30.00	%	= $2,100,000.00 GRANT
Interest	5.75	%	average over life of project including financing & insurance
Bond	20		years and payments semi-annually
	2		payment per year
O & M	1.000		cents per kWh of production

Hydrofacility Production: 7,506,117 kWh/year, average for 20-years run-of-the-river (**O&M based upon this production**)

O&M Including Capital Budget Reserve 1.333 cents/kWh of production

Energy Usage from Hydrofacility:

Six Wells + Water plant + Sewer Plant	6,443,964	kWh/year	or	$392,700.13 $, in YR-1
City Hall + Library + Blanden	730,960	kWh/year	or	$54,204.39 $, in YR-1
Street Lights	641,784	kWh/year	or	2.6689 cents/kWh, YR-1
Excess Energy SOLD, YR-1	0	kWh/year	at	4.5000 cents/kWh, Yr-1: Negotiated with utility company, selling price when excess energy is positive
Excess Energy PURCHASED, YR-2	1,336,428	kWh/year	at	6.0910 cents/kWh, Yr-1: Negotiated with utility company, selling price when excess energy is negative
Green Tags, YR-1	5,560,159	kWh/year	at	1.0000 cents/kWh (Note: The green tags have a typical market value between 1.0 to 2.0 cents/kW
Fed Incentive revenue, YR-1	5,560,159	kWh/year	at	1.8000 cents/kWh (Note: First 10-years of project, no inflation factor was used)

Increase per year of energy sales price "and" O&M Expenses 2.00 % increase per year (NOTE: Capital Budget Reserve is not increased annually because it is expected to be in an interest bearing account.)

Year---->	1	2	3	4	5	6	7	8	9	10
Bond Interest, %	5.750	5.750	5.750	5.750	5.750	5.750	5.750	5.750	5.750	5.750
Energy Prices (cents/kWh):										
Six Wells + Water plant + Sewer Plant	6.094	6.216	6.340	6.467	6.596	6.728	6.863	7.000	7.140	7.283
City Hall + Library + Blanden	7.416	7.564	7.715	7.869	8.027	8.187	8.351	8.518	8.688	8.862
Street Lights	2.669	2.722	2.777	2.832	2.889	2.947	3.006	3.066	3.127	3.190
Excess Energy (+) Selling Price	4.500	4.590	4.682	4.775	4.871	4.968	5.068	5.169	5.272	5.378
Excess Energy (-) Purchase Price	6.091	6.213	6.337	6.464	6.593	6.725	6.859	6.997	7.137	7.279
Green Tags	1.000	1.000	1.000	1.000	1.000	1.000	1.000	1.000	1.000	1.000
Fed Incentive revenue	0.000	0.000	0.000	0.000	0.000	0.000	0.000	0.000	0.000	0.000
Yearly Production, kWh/YR	5,560,159	9,153,136	9,917,632	8,436,527	8,598,157	9,866,657	6,626,189	3,533,065	1,412,662	3,612,490

	11	12	13	14	15	16	17	18	19	20
Revenues:										
Six Wells + Water plant + Sewer Plant	$392,700.13	$400,554.13	$408,565.22	$416,736.52	$425,071.25	$433,572.67	$442,244.13	$451,089.01	$460,110.79	$469,313.01
City Hall + Library + Blanden	$54,204.39	$55,288.48	$56,394.25	$57,522.13	$58,672.57	$59,846.03	$61,042.95	$62,263.81	$63,509.08	$64,779.26
Street Lights	$17,128.57	$17,471.14	$17,820.57	$18,176.98	$18,540.52	$18,911.33	$19,289.56	$19,675.35	$20,068.85	$20,470.23
Excess Energy SOLD	$0.00	$61,342.05	$98,361.06	$29,599.06	$38,063.95	$101,848.92	$0.00	$0.00	$0.00	$0.00
Green Tags	$55,601.59	$91,531.36	$99,176.32	$84,365.27	$85,981.57	$98,666.57	$66,261.89	$35,330.65	$14,126.62	$36,124.90
Fed Incentive revenue	$0.00	$0.00	$0.00	$0.00	$0.00	$0.00	$0.00	$0.00	$0.00	$0.00
Total Revenues	$519,634.68	$626,187.16	$680,317.41	$606,399.96	$626,329.86	$712,845.52	$588,838.52	$568,358.81	$557,815.35	$590,687.40
Expenditures:										
Bond + Insurance + Interest	$415,447.02	$415,447.02	$415,447.02	$415,447.02	$415,447.02	$415,447.02	$415,447.02	$415,447.02	$415,447.02	$415,447.02
Excess Energy PURCHASED	$137,446.40	$0.00	$0.00	$0.00	$0.00	$0.00	$81,663.12	$299,711.27	$457,029.69	$306,038.01
O & M	$75,061.17	$76,562.39	$78,093.64	$79,655.52	$81,248.63	$82,873.60	$84,531.07	$86,221.69	$87,946.13	$89,705.05
Capital Budget Reserve	$25,000.00	$25,000.00	$25,000.00	$25,000.00	$25,000.00	$25,000.00	$25,000.00	$25,000.00	$25,000.00	$25,000.00
Total Expenditures	$652,954.59	$517,009.41	$518,540.66	$520,102.54	$521,695.65	$523,320.62	$606,641.21	$826,379.98	$985,422.84	$836,190.08
(Revenues - Expenditures)	-$133,319.91	$109,177.75	$161,776.75	$86,297.43	$104,634.22	$189,524.90	-$17,802.69	-$258,021.17	-$427,607.49	-$245,502.68
Year--->	11	12	13	14	15	16	17	18	19	20
Bond Interest, %	5.750	5.750	5.750	5.750	5.750	5.750	5.750	5.750	5.750	5.750
Energy Prices (cents/kWh):										
Six Wells + Water plant + Sewer Plant	7.429	7.577	7.729	7.883	8.041	8.202	8.366	8.533	8.704	8.878
City Hall + Library + Blanden	9.039	9.220	9.405	9.593	9.785	9.980	10.180	10.384	10.591	10.803
Street Lights	3.253	3.318	3.385	3.453	3.522	3.592	3.664	3.737	3.812	3.888
Excess Energy (+) Selling Price	5.485	5.595	5.707	5.821	5.938	6.056	6.178	6.301	6.427	6.556
Excess Energy (-) Purchase Price	7.425	7.573	7.725	7.879	8.037	8.198	8.362	8.529	8.699	8.873
Green Tags	1.000	1.000	1.000	1.000	1.000	1.000	1.000	1.000	1.000	1.000
Fed Incentive revenue	0.000	0.000	0.000	0.000	0.000	0.000	0.000	0.000	0.000	0.000
Yearly Production, kWh/YR	7,810,812	11,115,698	10,061,600	10,071,809	9,764,614	9,105,206	7,359,894	8,216,684	6,425,422	3,473,929
Revenues:										
Six Wells + Water plant + Sewer Plant	$478,699.27	$488,273.25	$498,038.72	$507,999.49	$518,159.48	$528,522.67	$539,093.12	$549,874.99	$560,872.49	$572,089.94
City Hall + Library + Blanden	$66,074.85	$67,396.35	$68,744.27	$70,119.16	$71,521.54	$72,951.97	$74,411.01	$75,899.23	$77,417.22	$78,965.56
Street Lights	$20,879.64	$21,297.23	$21,723.17	$22,157.64	$22,600.79	$23,052.80	$23,513.86	$23,984.14	$24,463.82	$24,953.10
Excess Energy SOLD	$0.00	$184,584.57	$128,117.96	$131,274.61	$115,659.93	$78,036.69	$0.00	$25,202.83	$0.00	$0.00
Green Tags	$78,108.12	$111,156.98	$100,616.00	$100,718.09	$97,646.14	$91,052.06	$73,598.94	$82,166.84	$64,254.22	$34,739.29
Fed Incentive revenue	$0.00	$0.00	$0.00	$0.00	$0.00	$0.00	$0.00	$0.00	$0.00	$0.00
Total Revenues	$643,761.87	$872,708.38	$817,240.13	$832,268.99	$825,587.88	$793,616.20	$710,616.94	$757,128.03	$727,007.74	$710,747.88

(continued)

311

TABLE 12.10 (*Continued*)

Expenditures:									
Bond + Insurance + Interest:	$415,447.02	$415,447.02	$415,447.02	$415,447.02	$415,447.02	$415,447.02	$415,447.02	$415,447.02	$415,447.02
Excess Energy PURCHASED	$437.77	$0.00	$0.00	$0.00	$0.00	$0.00	$38,197.13	$0.00	$121,034.20	$385,353.75
O & M	$91,499.15	$93,329.13	$95,195.71	$97,099.63	$99,041.62	$101,022.45	$103,042.90	$105,103.76	$107,205.84	$109,349.95
Capital Budget Reserve	$25,000.00	$25,000.00	$25,000.00	$25,000.00	$25,000.00	$25,000.00	$25,000.00	$25,000.00	$25,000.00	$25,000.00
Total Expenditures	$532,383.94	$533,776.15	$535,642.73	$537,546.65	$539,488.64	$541,469.47	$581,687.05	$545,550.78	$668,687.06	$935,150.72
(Revenues - Expenditures)	$111,377.93	$338,932.23	$281,597.39	$294,722.34	$286,099.24	$252,146.72	$128,929.88	$211,577.25	$58,320.69	-$224,402.84

Total (Revenues - Expenditures)	$1,308,457.94	20 year, analysis
Avg. Annual (Revenues - Expenditures)	$65,422.90 /year	
Annual Average Production =	7,506,117 kWh/Year	

TABLE 12.11 Fort Dodge Mill Dam and Powerhouse Economics (Case #3)

Fort Dodge Hydroelectric Project, CASE #3
Cedar Rapids, Iowa
Dr. Gary C. Young, P.E.

Fort Dodge Mill Dam Hydroelectric Facility
Fort Dodge, Iowa

Capital Investment	$7,000,000.00		
[Capital Investment - GRANT] =	$7,000,000.00		
Grant =	0.00	% =	$0.00 GRANT
Interest	0.00	%, average over life of project including financing & insurance	
Bond	20	years and payments semi-annually	
	2	payment per year	
O & M	0.500	cents per kWh of production	
Hydrofacility Production:	7,506,117	kWh/year; average for 20-years run-of-the-river; (O&M based upon this production)	

O&M Including Capital Budget Reserve 0.833 cents/kWh of production

Energy Usage from Hydrofacility:

Six Wells + Water plant + Sewer Plant	6,443,964	kWh/year	or	$392,700.13 $, in YR-1
City Hall + Library + Blanden	730,960	kWh/year	or	$54,204.39 $, in YR-1
Street Lights	641,784	kWh/year	or	2.6689 cents/kWh, YR-1
Excess Energy SOLD, YR-1	0	kWh/year	at	4.5000 cents/kWh, Yr-1: Negotiated with utility company, selling price when excess energy is positive
Excess Energy PURCHASED, YR-2	1,336,428	kWh/year	at	6.0910 cents/kWh, Yr-1: Negotiated with utility company, selling price when excess energy is negative
Green Tags, YR-1	5,560,159	kWh/year	at	1.0000 cents/kWh (Note: The green tags have a typical market value between 1.0 to 2.0 cents/kW
Fed Incentive revenue, YR-1	5,560,159	kWh/year	at	1.8000 cents/kWh (Note: First 10-years of project, no inflation factor was used)

Increase per year of energy sales price "and" O&M Expenses 2.00 % increase per year (NOTE: Capital Budget Reserve is not increased annually because it is expected to be in an interest bearing account.)

Year---->	1	2	3	4	5	6	7	8	9	10
Bond Interest, %	0.000	0.000	0.000	0.000	0.000	0.000	0.000	0.000	0.000	0.000
Energy Prices (cents/kWh):										
Six Wells + Water plant + Sewer Plant	6.094	6.216	6.340	6.467	6.596	6.728	6.863	7.000	7.140	7.283
City Hall + Library + Blanden	7.416	7.564	7.715	7.869	8.027	8.187	8.351	8.518	8.688	8.862
Street Lights	2.669	2.722	2.777	2.832	2.889	2.947	3.006	3.066	3.127	3.190
Excess Energy (+) Selling Price	4.500	4.590	4.682	4.775	4.871	4.968	5.068	5.169	5.272	5.378
Excess Energy (-) Purchase Price	6.091	6.213	6.337	6.464	6.593	6.725	6.859	6.997	7.137	7.279
Green Tags	1.000	1.000	1.000	1.000	1.000	1.000	1.000	1.000	1.000	1.000
Fed Incentive revenue	1.800	1.800	1.800	1.800	1.800	1.800	1.800	1.800	1.800	1.800
Yearly Production, kWh/YR	5,560,159	9,153,136	9,917,632	8,436,527	8,598,157	9,866,657	6,626,189	3,533,065	1,412,662	3,612,490

(continued)

TABLE 12.11 (*Continued*)

	11	12	13	14	15	16	17	18	19	20
Revenues:										
Six Wells + Water plant + Sewer Plant	$392,700.13	$400,554.13	$408,565.22	$416,736.52	$425,071.25	$433,572.67	$442,244.13	$451,089.01	$460,110.79	$469,313.01
City Hall + Library + Blanden	$54,204.39	$55,288.48	$56,394.25	$57,522.13	$58,672.57	$59,846.03	$61,042.95	$62,263.81	$63,509.08	$64,779.26
Street Lights	$17,128.57	$17,471.14	$17,820.57	$18,176.98	$18,540.52	$18,911.33	$19,289.56	$19,675.35	$20,068.85	$20,470.23
Excess Energy SOLD	$0.00	$61,342.05	$98,361.06	$29,599.06	$38,063.95	$101,848.92	$0.00	$0.00	$0.00	$0.00
Green Tags	$55,601.59	$91,531.36	$99,176.32	$84,365.27	$85,981.57	$98,666.57	$66,261.89	$35,330.65	$14,126.62	$36,124.90
Fed Incentive revenue	$100,082.86	$164,756.45	$178,517.38	$151,857.49	$154,766.83	$177,599.83	$119,271.40	$63,595.17	$25,427.92	$65,024.82
Total Revenues	$619,717.55	$790,943.61	$858,834.79	$758,257.45	$781,096.69	$890,445.35	$708,109.92	$631,953.98	$583,243.26	$655,712.22
Expenditures:										
Bond + Insurance + Interest:	$350,000.36	$350,000.36	$350,000.36	$350,000.36	$350,000.36	$350,000.36	$350,000.36	$350,000.36	$350,000.36	$350,000.36
Excess Energy PURCHASED	$137,446.40	$0.00	$0.00	$0.00	$0.00	$0.00	$81,663.12	$299,711.27	$457,029.69	$306,038.01
O & M	$37,530.59	$38,281.20	$39,046.82	$39,827.76	$40,624.31	$41,436.80	$42,265.53	$43,110.85	$43,973.06	$44,852.52
Capital Budget Reserve	$25,000.00	$25,000.00	$25,000.00	$25,000.00	$25,000.00	$25,000.00	$25,000.00	$25,000.00	$25,000.00	$25,000.00
Total Expenditures	$549,977.35	$413,281.56	$414,047.18	$414,828.12	$415,624.67	$416,437.16	$498,929.01	$717,822.47	$876,003.12	$725,890.90
(Revenues - Expenditures)	$69,740.20	$377,662.05	$444,787.60	$343,429.33	$365,472.02	$474,008.19	$209,180.91	-$85,868.49	-$292,759.85	-$70,178.67
Year--->	11	12	13	14	15	16	17	18	19	20
Bond Interest, %	0.000	0.000	0.000	0.000	0.000	0.000	0.000	0.000	0.000	0.000
Energy Prices (cents/kWh):										
Six Wells + Water plant + Sewer Plant	7.429	7.577	7.729	7.883	8.041	8.202	8.366	8.533	8.704	8.878
City Hall + Library + Blanden	9.039	9.220	9.405	9.593	9.785	9.980	10.180	10.384	10.591	10.803
Street Lights	3.253	3.318	3.385	3.453	3.522	3.592	3.664	3.737	3.812	3.888
Excess Energy (+) Selling Price	5.485	5.595	5.707	5.821	5.938	6.056	6.178	6.301	6.427	6.556
Excess Energy (-) Purchase Price	7.425	7.573	7.725	7.879	8.037	8.198	8.362	8.529	8.699	8.873
Green Tags	1.000	1.000	1.000	1.000	1.000	1.000	1.000	1.000	1.000	1.000
Fed Incentive revenue	0.000	0.000	0.000	0.000	0.000	0.000	0.000	0.000	0.000	0.000
Yearly Production, kWh/YR	7,810,812	11,115,698	10,061,600	10,071,809	9,764,614	9,105,206	7,359,894	8,216,684	6,425,422	3,473,929
Revenues:										
Six Wells + Water plant + Sewer Plant	$478,699.27	$488,273.25	$498,038.72	$507,999.49	$518,159.48	$528,522.67	$539,093.12	$549,874.99	$560,872.49	$572,089.94
City Hall + Library + Blanden	$66,074.85	$67,396.35	$68,744.27	$70,119.16	$71,521.54	$72,951.97	$74,411.01	$75,899.23	$77,417.22	$78,965.56
Street Lights	$20,879.64	$21,297.23	$21,723.17	$22,157.64	$22,600.79	$23,052.80	$23,513.86	$23,984.14	$24,463.82	$24,953.10
Excess Energy SOLD	$0.00	$184,584.57	$128,117.96	$131,274.61	$115,659.93	$78,036.69	$0.00	$25,202.83	$0.00	$0.00
Green Tags	$78,108.12	$111,156.98	$100,616.00	$100,718.09	$97,646.14	$91,052.06	$73,598.94	$82,166.84	$64,254.22	$34,739.29
Fed Incentive revenue	$0.00	$0.00	$0.00	$0.00	$0.00	$0.00	$0.00	$0.00	$0.00	$0.00
Total Revenues	$643,761.87	$872,708.38	$817,240.13	$832,268.99	$825,587.88	$793,616.20	$710,616.94	$757,128.03	$727,007.74	$710,747.88

Expenditures:

Bond + Insurance + Interest:	$350,000.36	$350,000.36	$350,000.36	$350,000.36	$350,000.36	$350,000.36	$350,000.36	$350,000.36	$350,000.36
Excess Energy PURCHASED	$437.77	$0.00	$0.00	$0.00	$0.00	$0.00	$38,197.13	$0.00	$121,034.20	$385,353.75
O & M	$45,749.57	$46,664.57	$47,597.86	$48,549.81	$49,520.81	$50,511.23	$51,521.45	$52,551.88	$53,602.92	$54,674.98
Capital Budget Reserve	$25,000.00	$25,000.00	$25,000.00	$25,000.00	$25,000.00	$25,000.00	$25,000.00	$25,000.00	$25,000.00	$25,000.00
Total Expenditures	$421,187.71	$421,664.93	$422,598.22	$423,550.17	$424,521.17	$425,511.59	$464,718.94	$427,552.24	$549,637.48	$815,029.09
(Revenues - Expenditures)	$222,574.17	$451,043.45	$394,641.91	$408,718.81	$401,066.71	$368,104.61	$245,897.99	$329,575.79	$177,370.27	-$104,281.20

Total (Revenues - Expenditures) $4,730,185.79   20 year, analysis
Avg. Annual (Revenues - Expenditures) $236,509.29 /year
Annual Average Production = 7,506,117 kWh/Year

315

provided by Energy Policy Act of 2005.[20] CREBs provide for interest-free "zero percent interest" financing of the project for a qualified hydropower facility. Thus, the project facility cash flows are presented in Table 12.11.

For Case #3, positive cash flows for the 20-year financing in these years are: year 1, $69,740 and year 5, $365,472. The average annual net revenue (revenue–expenditures) over 20 years would be $236,509/year, with no taxes for a local community facility.

For Fort Dodge Mill Dam hydroelectric facility in Fort Dodge, Iowa, the economic cases illustrate the importance of using power from the hydrofacility for the benefit of the local community. With the proper economically viable business plan selected, this hydrofacility can pay for itself without a tax burden upon the local taxpayers. For some areas of the country, utilities may be hard nosed and resist fair and equitable negotiations. When this occurs, an alternative business plan may be to look for a technology company wishing to have a relatively fixed and low-cost source of energy to locate in their community using electrical power from the hydroelectric facility. In this way, the new company and the community will prosper by utilization of the hydrofacilities and keeping the local energy assets and generated wealth at home. This approach is innovative in bringing in new residents (company's employees), increased tax base, growth in housing, schools, new jobs, and new business opportunities for the community.[22]

Daily Flow and Production Methodology, Fort Dodge Mill Dam Hydroelectric Facility

FORT DODGE MILL HYDROELECTRIC PROJECT
Daily Flow and Production Tables
Formulas Used in Calculations
The text below shows how the daily flow data from the USGS was analyzed from 01/01/1981 to 12/31/2000 to obtain a daily power generation and kWh energy production. The daily kWh energy production was summed and presented as monthly figures followed by addition of monthly production figures to annual production.

The recognized equation, Equation (12.1), for computation of power generation from water was used:[19]

$$KW = QHE_t E_{gtot}/11.8 \qquad (12.1)$$

where,

KW = electrical power generated by a turbine, kilowatts (kW)
Q = water flow through the turbine, cubic feet per second (cfs)
H = elevation through which water drops minus losses (feet) (Note, H is the effective head, which is the static head minus the losses of the water passage). (H may not be known for preliminary computations, so "conservative" values for turbine operation with flow ranges were used in these computations when the elevation through which water drops was used for H.)

E_t = turbine efficiency (fraction)

E_{gtot} = generator (E_{gen}) × gear ratio (E_{gear}) × transformer (E_{trans}), electromechanical efficiency(s), (fraction)

The two turbines and related electromechanical efficiencies used are:

	Capacity (kW)	Flow Capacity Minimum (q_{min}) (cfs)	Flow Capacity Maximum (q_{max}) (cfs)	E_{gen} (%)	E_{gear} (%)	E_{trans} (%)
Turbine #1	700	268.0	670.0	94.5	97.0	98.0
Turbine #2	700	268.0	670.0	94.5	97.0	98.0

For a given turbine, it can operate from 100% of design flow to a percentage of design flow.

The Kaplan type turbine can operate to 40% of design flow or 268 cfs (670 × 40/100). Thus, each of these turbines can operate from 268 cfs to 670 cfs. (Note, these limitations are conservative and were specifically used so as to be conservative in estimating energy production for a preliminary economic analysis when all facts of a project case are unknown until final engineering is completed in the future if the project is considered economically viable by preliminary economic analysis.) With this limitation on flow requirements of each turbine, the daily run-of-the-river flow data in the fourth column of Table 12.13 must be considered as to how the two turbines can operate at this daily flow rate. For example, five possible cases arise to analyze the daily data. Consider these definitions for flow rates: Q = run-of-the-river flow (cfs); Q_1 = flow through turbine #1 (cfs); and Q_2 = flow through turbine #2 (cfs). Conditions of flow for these two turbines to operate are:

Condition 1   IF($Q < q_{min}$, $Q_1 = 0$, $Q_2 = 0$)
Condition 2   IF(AND($Q \geq q_{min}$,$Q \leq q_{max}$), $Q_1 = Q$, $Q_2 = 0$
Condition 3   IF(AND($Q > q_{max}$, $Q \leq q_{max} + q_{min}$), $Q_1 = Q - q_{min}$, $Q_2 = q_{min}$
Condition 4   IF(AND($Q > q_{max} + q_{min}$, $Q \leq q_{max} + q_{max}$), $Q_1 = Q - q_{max}$, $Q_2 = q_{max}$
Condition 5   IF($Q > q_{max} + q_{max}$, $Q_1 = q_{max}$, $Q_2 = q_{max}$)

These "five conditions" allow the turbines to operate properly, and the daily flow data is analyzed and computed accordingly.

Before proceeding, the turbine efficiency is a function of flow rate through the turbine.

Thus, Turbine #1 efficiency depends upon flow rate, Q_1, and Turbine #2 efficiency depends upon flow rate, Q_2. Turbine efficiency for the Kaplan type is taken as:

Q_1 or Q_2 (cfs)	67.0	100.5	134.0	167.5	201.0	234.5	268.0	301.5	335.0
E_t in %	49.2	55.5	61.3	66.6	71.5	75.8	79.6	83.0	85.8
Q_1 or Q_2 (cfs)	368.5	402.0	596.3	469.0	502.5	536.0	569.5	603.0	636.5
E_t in %	88.2	90.0	91.0	92.2	92.6	92.5	91.9	90.7	89.1
Q_1 or Q_2, (cfs)	670.0								
E_t in %	87.0								

TABLE 12.12 Summary of Potential Energy Generation of Fort Dodge Mill Dam, Fort Dodge, Iowa

Fort Dodge Mill Dam
Fort Dodge, Iowa

10/30/2003
Dr. Gary C. Young, Ph.D, P.E.
GYCO, Inc.
Cedar Rapids, Iowa

Subject: Opinion of Potential Energy Generation based upon the past years of USGS data.
Years from 1981 through 2000
1400 KW Electrical Power Generation Capacity
14.5 FT., Elevation through which water drops

SUMMARY TABLE

Year	January kWh	February kWh	March kWh	April kWh	May kWh	June kWh	July kWh	August kWh	September kWh	October kWh	November kWh	December kWh	Total Annual kWh
1981	0	143,036	137,392	328,336	584,086	768,470	944,047	772,190	507,893	571,416	467,850	335,443	5,560,159
1982	0	216,093	957,910	927,010	957,910	927,010	957,910	526,226	840,237	957,910	927,010	957,910	9,153,136
1983	950,217	735,946	957,910	927,010	957,910	927,010	957,910	637,809	509,127	767,321	894,578	694,884	9,917,632
1984	502,440	620,866	957,910	927,010	957,910	927,010	957,910	692,337	228,047	333,592	634,674	696,821	8,436,527
1985	463,383	115,091	880,863	927,010	957,910	917,074	609,477	220,466	805,218	957,910	904,354	839,401	8,598,157
1986	414,476	337,775	957,910	927,010	957,910	927,010	957,910	867,552	726,945	957,910	927,010	907,239	9,866,657
1987	584,065	435,516	740,123	927,010	951,146	735,718	771,976	416,413	380,634	235,534	149,297	298,757	6,626,189
1988	44,982	178,228	931,907	892,599	957,675	467,580	18,085	27,842	7,467	6,700	0	0	3,533,065
1989	0	0	443,232	364,873	471,004	121,263	12,290	0	0	0	0	0	1,412,662
1990	0	0	123,373	75,765	586,787	917,449	899,475	725,705	278,024	0	5,912	0	3,612,490
1991	0	68,409	930,153	927,010	957,910	927,010	957,910	888,034	359,564	233,873	603,029	957,910	7,810,812
1992	953,946	896,109	957,910	927,010	957,910	925,892	957,910	957,910	874,265	857,138	927,010	922,688	11,115,698
1993	565,575	454,432	724,759	927,010	957,910	927,010	957,910	957,910	807,568	867,974	836,190	836,190	10,061,600
1994	556,048	513,181	724,759	927,010	957,910	927,010	957,910	953,693	807,568	950,403	876,568	686,598	10,071,809
1995	421,045	395,110	838,761	927,010	957,910	927,010	957,910	897,813	777,848	957,910	920,052	786,235	9,764,614
1996	488,179	733,070	827,326	927,010	957,910	927,010	941,586	925,598	465,120	380,416	711,503	820,478	9,105,206
1997	496,075	565,006	957,910	927,010	957,910	927,010	957,910	706,691	254,522	232,616	208,080	169,154	7,359,894
1998	70,969	394,446	913,894	927,010	957,910	927,010	873,057	641,348	265,802	502,615	927,010	815,613	8,216,684
1999	376,227	749,084	954,974	927,010	957,910	927,010	957,910	525,408	49,889	0	0	0	6,425,422
2000	0	92,270	87,443	23,594	365,561	896,890	919,412	512,557	53,970	0	476,743	45,489	3,473,929
Averages:	344,381 kWh / month	382,183 kWh / month	761,979 kWh / month	779,516 kWh / month	866,350 kWh / month	843,723 kWh / month	826,321 kWh / month	642,675 kWh / month	455,958 kWh / month	493,059 kWh / month	571,433 kWh / month	538,541 kWh / month	7,506,117 kWh / year

Reference: Spreadsheet; FDMD, Flow data & energy analysis, USGS data, 10-28-2003
Results are based upon U.S. Geological Survey data.
USGS; Des Moines River at Fort Dodge, IA

TABLE 12.13 Fort Dodge Mill Dam Flow and Production Computations, Fort Dodge, Iowa

U.S. Geological Survey
National Water Information System
Retrieved: 2003-03-31 21:19:23 EST
#
This file contains published daily mean streamflow data.
#
Further Descriptions of the dv_cd column can be found at:
http://waterdata.usgs.gov/nwis/help?codes_help#dv_cd
#
This information includes the following fields:
#
agency_cd   Agency Code
site_no   USGS station number
dv_dt   date of daily mean streamflow
dv_va   daily mean streamflow value, in cubic-feet per second
dv_cd   daily mean streamflow value qualification code
#
Sites in this file include:
USGS 05480500 Des Moines River at Fort Dodge, IA
#

Gary C. Young, Ph.D., P.E.
GYCO, Inc.
Cedar Rapids, Iowa

$kW = Q * H * Et * Egtot / 11.8$

where:
- kW = electrical power generated by turbine
- Q = water flow through the turbine, cfs
- H = elevation through which the water drops, feet
- Et = turbine efficiency, %
- $Egtot$ = generator * gear ratios * transformer electro-mechanical efficiencies, %

Turbines:

	(Input) Capacity KW	(computed) Capacity cfs (min)	(Input) Capacity cfs (max)	(Input) Egen %	(Input) Egear %	(Input) Etrans %
Turbine #1	700	268.0	670.0	94.5	97.0	98.0
Turbine #2	700	268.0	670.0	94.5	97.0	98.0

H	14.5 feet	(input)

agency_cd	site_no	dv_dt	dv_va	dv_cd	Run of River (cfs)	computed condition-1 IF(Q<=qmin, Q1=0, Q2=0)	computed condition-2 IF(AND(Q >=qmin, Q <=qmax), Q1=Q, Q2=0	(computed) condition-3 IF(AND(Q >qmax, Q2=qmin	(computed) condition-4 IF(AND(Q> qmax+qmin, Q<=qmax +qmin), Q1=Q-qmin, Q2=qmin	(computed) condition-5 IF(Q>qmax, Q1=qmax, Q2=qmax)	(INPUT) (CONDIT- IONAL) Flow thru Turbine 1 Q1 (cfs)	(input) Et, Turbine Efficiency Turbine 1 (%)	(comp- uted) Egtot, Efficiency Turbine 1 (%)	(comp- uted) Power from Turbine 1 (kW)	(comp- uted) Energy from Turbine 1 * 24 hrs (kWh)	(INPUT) (CONDIT- IONAL) Flow thru Turbine 2 Q2 (cfs)	(input) Et, Turbine Efficiency Turbine 2 (%)	(comp- uted) Egtot, Efficiency Turbine 2 (%)	Turbine 2 Power from Turbine 2 (kW)	(comp- uted) Energy from Turbine 2 * 24 hrs (kWh)	(comp- uted) Total Power from Turbines 1&2 (kW)	(computed) Total Energy from Turbines 1&2 * 24 hrs (kWh)
5s	15s	10d	12n	3s																		
USGS	5480500	1/1/1981	198			TRUE	FALSE	FALSE	FALSE	FALSE	0	0.0	89.83	0.0	0.0	0	0.0	89.83	0.0	0.0	0.0	0.0
USGS	5480500	1/2/1981	196			TRUE	FALSE	FALSE	FALSE	FALSE	0	0.0	89.83	0.0	0.0	0	0.0	89.83	0.0	0.0	0.0	0.0
USGS	5480500	1/3/1981	192			TRUE	FALSE	FALSE	FALSE	FALSE	0	0.0	89.83	0.0	0.0	0	0.0	89.83	0.0	0.0	0.0	0.0
USGS	5480500	1/4/1981	190			TRUE	FALSE	FALSE	FALSE	FALSE	0	0.0	89.83	0.0	0.0	0	0.0	89.83	0.0	0.0	0.0	0.0
USGS	5480500	1/5/1981	210			TRUE	FALSE	FALSE	FALSE	FALSE	0	0.0	89.83	0.0	0.0	0	0.0	89.83	0.0	0.0	0.0	0.0

(continued)

319

TABLE 12.13 (Continued)

USGS	5480500	1/6/1981	170	TRUE	FALSE	FALSE	FALSE	FALSE	0	0.0	89.83	0.0	0	0.0	89.83	0.0	0.0	0.0
USGS	5480500	1/7/1981	196	TRUE	FALSE	FALSE	FALSE	FALSE	0	0.0	89.83	0.0	0	0.0	89.83	0.0	0.0	0.0
USGS	5480500	1/8/1981	168	TRUE	FALSE	FALSE	FALSE	FALSE	0	0.0	89.83	0.0	0	0.0	89.83	0.0	0.0	0.0
USGS	5480500	1/9/1981	154	TRUE	FALSE	FALSE	FALSE	FALSE	0	0.0	89.83	0.0	0	0.0	89.83	0.0	0.0	0.0
USGS	5480500	1/10/1981	156	TRUE	FALSE	FALSE	FALSE	FALSE	0	0.0	89.83	0.0	0	0.0	89.83	0.0	0.0	0.0
USGS	5480500	1/11/1981	160	TRUE	FALSE	FALSE	FALSE	FALSE	0	0.0	89.83	0.0	0	0.0	89.83	0.0	0.0	0.0
USGS	5480500	1/12/1981	142	TRUE	FALSE	FALSE	FALSE	FALSE	0	0.0	89.83	0.0	0	0.0	89.83	0.0	0.0	0.0
USGS	5480500	1/13/1981	140	TRUE	FALSE	FALSE	FALSE	FALSE	0	0.0	89.83	0.0	0	0.0	89.83	0.0	0.0	0.0
USGS	5480500	1/14/1981	140	TRUE	FALSE	FALSE	FALSE	FALSE	0	0.0	89.83	0.0	0	0.0	89.83	0.0	0.0	0.0
USGS	5480500	1/15/1981	134	TRUE	FALSE	FALSE	FALSE	FALSE	0	0.0	89.83	0.0	0	0.0	89.83	0.0	0.0	0.0
USGS	5480500	1/16/1981	124	TRUE	FALSE	FALSE	FALSE	FALSE	0	0.0	89.83	0.0	0	0.0	89.83	0.0	0.0	0.0
USGS	5480500	1/17/1981	126	TRUE	FALSE	FALSE	FALSE	FALSE	0	0.0	89.83	0.0	0	0.0	89.83	0.0	0.0	0.0
USGS	5480500	1/18/1981	130	TRUE	FALSE	FALSE	FALSE	FALSE	0	0.0	89.83	0.0	0	0.0	89.83	0.0	0.0	0.0
USGS	5480500	1/19/1981	134	TRUE	FALSE	FALSE	FALSE	FALSE	0	0.0	89.83	0.0	0	0.0	89.83	0.0	0.0	0.0
USGS	5480500	1/20/1981	134	TRUE	FALSE	FALSE	FALSE	FALSE	0	0.0	89.83	0.0	0	0.0	89.83	0.0	0.0	0.0
USGS	5480500	1/21/1981	134	TRUE	FALSE	FALSE	FALSE	FALSE	0	0.0	89.83	0.0	0	0.0	89.83	0.0	0.0	0.0
USGS	5480500	1/22/1981	136	TRUE	FALSE	FALSE	FALSE	FALSE	0	0.0	89.83	0.0	0	0.0	89.83	0.0	0.0	0.0
USGS	5480500	1/23/1981	138	TRUE	FALSE	FALSE	FALSE	FALSE	0	0.0	89.83	0.0	0	0.0	89.83	0.0	0.0	0.0
USGS	5480500	1/24/1981	144	TRUE	FALSE	FALSE	FALSE	FALSE	0	0.0	89.83	0.0	0	0.0	89.83	0.0	0.0	0.0
USGS	5480500	1/25/1981	154	TRUE	FALSE	FALSE	FALSE	FALSE	0	0.0	89.83	0.0	0	0.0	89.83	0.0	0.0	0.0
USGS	5480500	1/26/1981	168	TRUE	FALSE	FALSE	FALSE	FALSE	0	0.0	89.83	0.0	0	0.0	89.83	0.0	0.0	0.0
USGS	5480500	1/27/1981	168	TRUE	FALSE	FALSE	FALSE	FALSE	0	0.0	89.83	0.0	0	0.0	89.83	0.0	0.0	0.0
USGS	5480500	1/28/1981	160	TRUE	FALSE	FALSE	FALSE	FALSE	0	0.0	89.83	0.0	0	0.0	89.83	0.0	0.0	0.0
USGS	5480500	1/29/1981	150	TRUE	FALSE	FALSE	FALSE	FALSE	0	0.0	89.83	0.0	0	0.0	89.83	0.0	0.0	0.0
USGS	5480500	1/30/1981	148	TRUE	FALSE	FALSE	FALSE	FALSE	0	0.0	89.83	0.0	0	0.0	89.83	0.0	0.0	0.0
USGS	5480500	1/31/1981	138	TRUE	FALSE	FALSE	FALSE	FALSE	0	0.0	89.83	0.0	0	0.0	89.83	0.0	0.0	0.0
								Total- January 1981	0	0.0		0.0						0.0

Agency	Station	Date	Value	Flag1	Flag2	Flag3	Flag4	Flag5	n	%	Mean		Vol		Cum
USGS	5480500	2/1/1981	140	TRUE	FALSE	FALSE	FALSE	FALSE	0	0.0	89.83	0.0	0.0	0.0	0.0
USGS	5480500	2/2/1981	136	TRUE	FALSE	FALSE	FALSE	FALSE	0	0.0	89.83	0.0	0.0	0.0	0.0
USGS	5480500	2/3/1981	130	TRUE	FALSE	FALSE	FALSE	FALSE	0	0.0	89.83	0.0	0.0	0.0	0.0
USGS	5480500	2/4/1981	130	TRUE	FALSE	FALSE	FALSE	FALSE	0	0.0	89.83	0.0	0.0	0.0	0.0
USGS	5480500	2/5/1981	132	TRUE	FALSE	FALSE	FALSE	FALSE	0	0.0	89.83	0.0	0.0	0.0	0.0
USGS	5480500	2/6/1981	130	TRUE	FALSE	FALSE	FALSE	FALSE	0	0.0	89.83	0.0	0.0	0.0	0.0
USGS	5480500	2/7/1981	132	TRUE	FALSE	FALSE	FALSE	FALSE	0	0.0	89.83	0.0	0.0	0.0	0.0
USGS	5480500	2/8/1981	128	TRUE	FALSE	FALSE	FALSE	FALSE	0	0.0	89.83	0.0	0.0	0.0	0.0
USGS	5480500	2/9/1981	122	TRUE	FALSE	FALSE	FALSE	FALSE	0	0.0	89.83	0.0	0.0	0.0	0.0
USGS	5480500	2/10/1981	120	TRUE	FALSE	FALSE	FALSE	FALSE	0	0.0	89.83	0.0	0.0	0.0	0.0
USGS	5480500	2/11/1981	115	TRUE	FALSE	FALSE	FALSE	FALSE	0	0.0	89.83	0.0	0.0	0.0	0.0
USGS	5480500	2/12/1981	115	TRUE	FALSE	FALSE	FALSE	FALSE	0	0.0	89.83	0.0	0.0	0.0	0.0
USGS	5480500	2/13/1981	116	TRUE	FALSE	FALSE	FALSE	FALSE	0	0.0	89.83	0.0	0.0	0.0	0.0
USGS	5480500	2/14/1981	118	TRUE	FALSE	FALSE	FALSE	FALSE	0	0.0	89.83	0.0	0.0	0.0	0.0
USGS	5480500	2/15/1981	168	TRUE	FALSE	FALSE	FALSE	FALSE	0	0.0	89.83	0.0	0.0	0.0	0.0
USGS	5480500	2/16/1981	238	TRUE	FALSE	FALSE	FALSE	FALSE	0	0.0	89.83	0.0	0.0	0.0	0.0
USGS	5480500	2/17/1981	300	FALSE	TRUE	FALSE	FALSE	FALSE	300	82.8	89.83	0.0	274.3	0.0	6,582.3
USGS	5480500	2/18/1981	527	FALSE	TRUE	FALSE	FALSE	FALSE	527	92.6	89.83	0.0	538.5	0.0	12,923.0
USGS	5480500	2/19/1981	602	FALSE	TRUE	FALSE	FALSE	FALSE	602	90.8	89.83	0.0	603.3	0.0	14,479.7
USGS	5480500	2/20/1981	656	FALSE	TRUE	FALSE	FALSE	FALSE	656	88.0	89.83	0.0	637.1	0.0	15,290.3
USGS	5480500	2/21/1981	642	FALSE	TRUE	FALSE	FALSE	FALSE	642	88.8	89.83	0.0	629.5	0.0	15,108.9
USGS	5480500	2/22/1981	636	FALSE	TRUE	FALSE	FALSE	FALSE	636	89.2	89.83	0.0	626.0	0.0	15,024.7
USGS	5480500	2/23/1981	575	FALSE	TRUE	FALSE	FALSE	FALSE	575	91.7	89.83	0.0	582.1	0.0	13,970.9
USGS	5480500	2/24/1981	439	FALSE	TRUE	FALSE	FALSE	FALSE	439	91.5	89.83	0.0	443.3	0.0	10,639.2
USGS	5480500	2/25/1981	431	FALSE	TRUE	FALSE	FALSE	FALSE	431	91.2	89.83	0.0	433.9	0.0	10,414.8
USGS	5480500	2/26/1981	423	FALSE	TRUE	FALSE	FALSE	FALSE	423	90.9	89.83	0.0	424.5	0.0	10,188.3
USGS	5480500	2/27/1981	397	FALSE	TRUE	FALSE	FALSE	FALSE	397	89.8	89.83	0.0	393.4	0.0	9,440.5
USGS	5480500	2/28/1981	381	FALSE	TRUE	FALSE	FALSE	FALSE	381	88.9	89.83	0.0	373.9	0.0	8,973.3
												0.0	**Total- February 1981**		**143,036.1**

(continued)

TABLE 12.13 (*Continued*)

USGS	5480500	3/1/1981	364	FALSE	FALSE	TRUE	FALSE	FALSE	364	87.9	89.83	353.0	89.83	8,473.0	0	89.83	0.0	353.0	0.0	8,473.0
USGS	5480500	3/2/1981	353	FALSE	FALSE	TRUE	FALSE	FALSE	353	87.1	89.83	339.5	89.83	8,147.9	0	89.83	0.0	339.5	0.0	8,147.9
USGS	5480500	3/3/1981	338	FALSE	FALSE	TRUE	FALSE	FALSE	338	86.0	89.83	321.0	89.83	7,703.8	0	89.83	0.0	321.0	0.0	7,703.8
USGS	5480500	3/4/1981	332	FALSE	FALSE	TRUE	FALSE	FALSE	332	85.6	89.83	313.6	89.83	7,526.1	0	89.83	0.0	313.6	0.0	7,526.1
USGS	5480500	3/5/1981	317	FALSE	FALSE	TRUE	FALSE	FALSE	317	84.3	89.83	295.1	89.83	7,082.6	0	89.83	0.0	295.1	0.0	7,082.6
USGS	5480500	3/6/1981	313	FALSE	FALSE	TRUE	FALSE	FALSE	313	84.0	89.83	290.2	89.83	6,964.6	0	89.83	0.0	290.2	0.0	6,964.6
USGS	5480500	3/7/1981	318	FALSE	FALSE	TRUE	FALSE	FALSE	318	84.4	89.83	296.3	89.83	7,112.2	0	89.83	0.0	296.3	0.0	7,112.2
USGS	5480500	3/8/1981	300	FALSE	FALSE	TRUE	FALSE	FALSE	300	82.8	89.83	274.3	89.83	6,582.3	0	89.83	0.0	274.3	0.0	6,582.3
USGS	5480500	3/9/1981	306	FALSE	FALSE	TRUE	FALSE	FALSE	306	83.4	89.83	281.6	89.83	6,758.5	0	89.83	0.0	281.6	0.0	6,758.5
USGS	5480500	3/10/1981	308	FALSE	FALSE	TRUE	FALSE	FALSE	308	83.5	89.83	284.1	89.83	6,817.4	0	89.83	0.0	284.1	0.0	6,817.4
USGS	5480500	3/11/1981	301	FALSE	FALSE	TRUE	FALSE	FALSE	301	82.9	89.83	275.5	89.83	6,611.7	0	89.83	0.0	275.5	0.0	6,611.7
USGS	5480500	3/12/1981	299	FALSE	FALSE	TRUE	FALSE	FALSE	299	82.7	89.83	273.0	89.83	6,553.0	0	89.83	0.0	273.0	0.0	6,553.0
USGS	5480500	3/13/1981	294	FALSE	FALSE	TRUE	FALSE	FALSE	294	82.3	89.83	266.9	89.83	6,406.6	0	89.83	0.0	266.9	0.0	6,406.6
USGS	5480500	3/14/1981	281	FALSE	FALSE	TRUE	FALSE	FALSE	281	81.0	89.83	251.2	89.83	6,028.1	0	89.83	0.0	251.2	0.0	6,028.1
USGS	5480500	3/15/1981	272	FALSE	FALSE	TRUE	FALSE	FALSE	272	80.0	89.83	240.3	89.83	5,768.1	0	89.83	0.0	240.3	0.0	5,768.1
USGS	5480500	3/16/1981	268	FALSE	FALSE	TRUE	FALSE	FALSE	268	79.6	89.83	235.5	89.83	5,653.1	0	89.83	0.0	235.5	0.0	5,653.1
USGS	5480500	3/17/1981	270	FALSE	FALSE	TRUE	FALSE	FALSE	270	79.8	89.83	237.9	89.83	5,710.5	0	89.83	0.0	237.9	0.0	5,710.5
USGS	5480500	3/18/1981	263	FALSE	FALSE	FALSE	FALSE	FALSE	0	0.0	89.83	0.0	89.83	0.0	0	89.83	0.0	0.0	0.0	0.0
USGS	5480500	3/19/1981	255	FALSE	FALSE	FALSE	FALSE	FALSE	0	0.0	89.83	0.0	89.83	0.0	0	89.83	0.0	0.0	0.0	0.0
USGS	5480500	3/20/1981	245	FALSE	FALSE	FALSE	FALSE	FALSE	0	0.0	89.83	0.0	89.83	0.0	0	89.83	0.0	0.0	0.0	0.0
USGS	5480500	3/21/1981	246	FALSE	FALSE	FALSE	FALSE	FALSE	0	0.0	89.83	0.0	89.83	0.0	0	89.83	0.0	0.0	0.0	0.0
USGS	5480500	3/22/1981	251	FALSE	FALSE	FALSE	FALSE	FALSE	0	0.0	89.83	0.0	89.83	0.0	0	89.83	0.0	0.0	0.0	0.0
USGS	5480500	3/23/1981	253	FALSE	FALSE	FALSE	FALSE	FALSE	0	0.0	89.83	0.0	89.83	0.0	0	89.83	0.0	0.0	0.0	0.0
USGS	5480500	3/24/1981	253	FALSE	FALSE	FALSE	FALSE	FALSE	0	0.0	89.83	0.0	89.83	0.0	0	89.83	0.0	0.0	0.0	0.0
USGS	5480500	3/25/1981	251	FALSE	FALSE	FALSE	FALSE	FALSE	0	0.0	89.83	0.0	89.83	0.0	0	89.83	0.0	0.0	0.0	0.0
USGS	5480500	3/26/1981	256	FALSE	FALSE	FALSE	FALSE	FALSE	0	0.0	89.83	0.0	89.83	0.0	0	89.83	0.0	0.0	0.0	0.0
USGS	5480500	3/27/1981	259	FALSE	FALSE	FALSE	FALSE	FALSE	0	0.0	89.83	0.0	89.83	0.0	0	89.83	0.0	0.0	0.0	0.0
USGS	5480500	3/28/1981	253	FALSE	FALSE	FALSE	FALSE	FALSE	0	0.0	89.83	0.0	89.83	0.0	0	89.83	0.0	0.0	0.0	0.0
USGS	5480500	3/29/1981	285	FALSE	FALSE	FALSE	FALSE	FALSE	285	81.4	89.83	256.0	89.83	6,144.2	0	89.83	0.0	256.0	0.0	6,144.2
USGS	5480500	3/30/1981	327	FALSE	FALSE	TRUE	FALSE	FALSE	327	85.2	89.83	307.4	89.83	7,378.2	0	89.83	0.0	307.4	0.0	7,378.2
USGS	5480500	3/31/1981	347	FALSE	FALSE	FALSE	FALSE	FALSE	347	86.7	89.83	332.1	89.83	7,970.3	0	89.83	0.0	332.1	0.0	7,970.3
								Total- March 1981						**137,392.1**			**0.0**			**137,392.1**

Agency	Site	Date	Value						Value	%	Factor	Value	Value		Factor			Value	Value
USGS	5480500	4/1/1981	371	FALSE	FALSE	TRUE	FALSE	FALSE	371	88.3	89.83	8,679.4	361.6	0	89.83	0.0	0.0	361.6	8,679.4
USGS	5480500	4/2/1981	335	FALSE	FALSE	TRUE	FALSE	FALSE	335	85.8	89.83	7,614.9	317.3	0	89.83	0.0	0.0	317.3	7,614.9
USGS	5480500	4/3/1981	441	FALSE	FALSE	TRUE	FALSE	FALSE	441	91.5	89.83	10,695.0	445.6	0	89.83	0.0	0.0	445.6	10,695.0
USGS	5480500	4/4/1981	531	FALSE	FALSE	TRUE	FALSE	FALSE	531	92.5	89.83	13,016.6	542.4	0	89.83	0.0	0.0	542.4	13,016.6
USGS	5480500	4/5/1981	518	FALSE	FALSE	TRUE	FALSE	FALSE	518	92.6	89.83	12,708.6	529.5	0	89.83	0.0	0.0	529.5	12,708.6
USGS	5480500	4/6/1981	528	FALSE	FALSE	TRUE	FALSE	FALSE	528	92.6	89.83	12,946.5	539.4	0	89.83	0.0	0.0	539.4	12,946.5
USGS	5480500	4/7/1981	523	FALSE	FALSE	TRUE	FALSE	FALSE	523	92.6	89.83	12,828.3	534.5	0	89.83	0.0	0.0	534.5	12,828.3
USGS	5480500	4/8/1981	519	FALSE	FALSE	TRUE	FALSE	FALSE	519	92.6	89.83	12,732.7	530.5	0	89.83	0.0	0.0	530.5	12,732.7
USGS	5480500	4/9/1981	502	FALSE	FALSE	TRUE	FALSE	FALSE	502	92.6	89.83	12,315.2	513.1	0	89.83	0.0	0.0	513.1	12,315.2
USGS	5480500	4/10/1981	487	FALSE	FALSE	TRUE	FALSE	FALSE	487	92.5	89.83	11,933.2	497.2	0	89.83	0.0	0.0	497.2	11,933.2
USGS	5480500	4/11/1981	471	FALSE	FALSE	TRUE	FALSE	FALSE	471	92.3	89.83	11,513.1	479.7	0	89.83	0.0	0.0	479.7	11,513.1
USGS	5480500	4/12/1981	489	FALSE	FALSE	TRUE	FALSE	FALSE	489	92.5	89.83	11,984.8	499.4	0	89.83	0.0	0.0	499.4	11,984.8
USGS	5480500	4/13/1981	479	FALSE	FALSE	TRUE	FALSE	FALSE	479	92.4	89.83	11,724.7	488.5	0	89.83	0.0	0.0	488.5	11,724.7
USGS	5480500	4/14/1981	453	FALSE	FALSE	TRUE	FALSE	FALSE	453	91.9	89.83	11,026.6	459.4	0	89.83	0.0	0.0	459.4	11,026.6
USGS	5480500	4/15/1981	434	FALSE	FALSE	TRUE	FALSE	FALSE	434	91.3	89.83	10,499.2	437.5	0	89.83	0.0	0.0	437.5	10,499.2
USGS	5480500	4/16/1981	440	FALSE	FALSE	TRUE	FALSE	FALSE	440	91.5	89.83	10,667.1	444.5	0	89.83	0.0	0.0	444.5	10,667.1
USGS	5480500	4/17/1981	442	FALSE	FALSE	TRUE	FALSE	FALSE	442	91.6	89.83	10,722.8	446.8	0	89.83	0.0	0.0	446.8	10,722.8
USGS	5480500	4/18/1981	427	FALSE	FALSE	TRUE	FALSE	FALSE	427	91.1	89.83	10,301.8	429.2	0	89.83	0.0	0.0	429.2	10,301.8
USGS	5480500	4/19/1981	423	FALSE	FALSE	TRUE	FALSE	FALSE	423	90.9	89.83	10,188.3	424.5	0	89.83	0.0	0.0	424.5	10,188.3
USGS	5480500	4/20/1981	422	FALSE	FALSE	TRUE	FALSE	FALSE	422	90.9	89.83	10,159.9	423.3	0	89.83	0.0	0.0	423.3	10,159.9
USGS	5480500	4/21/1981	394	FALSE	FALSE	TRUE	FALSE	FALSE	394	89.6	89.83	9,353.3	389.7	0	89.83	0.0	0.0	389.7	9,353.3
USGS	5480500	4/22/1981	419	FALSE	FALSE	TRUE	FALSE	FALSE	419	90.8	89.83	10,074.4	419.8	0	89.83	0.0	0.0	419.8	10,074.4
USGS	5480500	4/23/1981	433	FALSE	FALSE	TRUE	FALSE	FALSE	433	91.3	89.83	10,471.1	436.3	0	89.83	0.0	0.0	436.3	10,471.1
USGS	5480500	4/24/1981	415	FALSE	FALSE	TRUE	FALSE	FALSE	415	90.6	89.83	9,960.0	415.0	0	89.83	0.0	0.0	415.0	9,960.0
USGS	5480500	4/25/1981	408	FALSE	FALSE	TRUE	FALSE	FALSE	408	90.3	89.83	9,758.9	406.6	0	89.83	0.0	0.0	406.6	9,758.9
USGS	5480500	4/26/1981	391	FALSE	FALSE	TRUE	FALSE	FALSE	391	89.5	89.83	9,265.9	386.1	0	89.83	0.0	0.0	386.1	9,265.9
USGS	5480500	4/27/1981	408	FALSE	FALSE	TRUE	FALSE	FALSE	408	90.3	89.83	9,758.9	406.6	0	89.83	0.0	0.0	406.6	9,758.9
USGS	5480500	4/28/1981	558	FALSE	FALSE	TRUE	FALSE	FALSE	558	92.1	89.83	13,619.4	567.5	0	89.83	0.0	0.0	567.5	13,619.4
USGS	5480500	4/29/1981	471	FALSE	FALSE	TRUE	FALSE	FALSE	471	92.3	89.83	11,513.1	479.7	0	89.83	0.0	0.0	479.7	11,513.1
USGS	5480500	4/30/1981	427	FALSE	FALSE	TRUE	FALSE	FALSE	427	91.1	89.83	10,301.8	429.2	0	89.83	0.0	0.0	429.2	10,301.8
		Total- April 1981										**328,335.5**					**0.0**		**328,335.5**

(continued)

TABLE 12.13 (*Continued*)

Agency	Station	Date	Q1	f1	f2	f3	f4	Q2	Pct	89.83	A1	B1	V1	V2	V3	V4	89.83	A2	V5	V6
USGS	5480500	5/1/1981	406	TRUE	FALSE	FALSE	FALSE	406	90.2	89.83	0.0	0	9,701.2	404.2	0.0	0.0	89.83	0.0	404.2	9,701.2
USGS	5480500	5/2/1981	384	TRUE	FALSE	FALSE	FALSE	384	89.1	89.83	0.0	0	9,061.3	377.6	0.0	0.0	89.83	0.0	377.6	9,061.3
USGS	5480500	5/3/1981	376	TRUE	FALSE	FALSE	FALSE	376	88.6	89.83	0.0	0	8,826.5	367.8	0.0	0.0	89.83	0.0	367.8	8,826.5
USGS	5480500	5/4/1981	489	TRUE	FALSE	FALSE	FALSE	489	92.5	89.83	0.0	0	11,984.8	499.4	0.0	0.0	89.83	0.0	499.4	11,984.8
USGS	5480500	5/5/1981	551	FALSE	TRUE	FALSE	FALSE	551	92.3	89.83	0.0	0	13,468.1	561.2	0.0	0.0	89.83	0.0	561.2	13,468.1
USGS	5480500	5/6/1981	683	FALSE	TRUE	FALSE	FALSE	415	90.6	89.83	79.6	268	9,960.0	415.0	5,653.1	235.5	89.83	79.6	650.5	15,613.1
USGS	5480500	5/7/1981	852	FALSE	TRUE	FALSE	FALSE	584	91.4	89.83	79.6	268	14,147.5	589.5	5,653.1	235.5	89.83	79.6	825.0	19,800.6
USGS	5480500	5/8/1981	911	FALSE	TRUE	FALSE	FALSE	643	88.8	89.83	79.6	268	15,122.5	630.1	5,653.1	235.5	89.83	79.6	865.7	20,775.6
USGS	5480500	5/9/1981	880	FALSE	TRUE	FALSE	FALSE	612	90.4	89.83	79.6	268	14,651.6	610.5	5,653.1	235.5	89.83	79.6	846.0	20,304.7
USGS	5480500	5/10/1981	815	FALSE	TRUE	FALSE	FALSE	547	92.3	89.83	79.6	268	13,380.1	557.5	5,653.1	235.5	89.83	79.6	793.0	19,033.2
USGS	5480500	5/11/1981	746	FALSE	TRUE	FALSE	FALSE	478	92.4	89.83	79.6	268	11,698.4	487.4	5,653.1	235.5	89.83	79.6	723.0	17,351.5
USGS	5480500	5/12/1981	695	FALSE	TRUE	FALSE	FALSE	427	91.1	89.83	0.0	0	10,301.8	429.2	0.0	0.0	89.83	0.0	664.8	15,954.9
USGS	5480500	5/13/1981	668	TRUE	FALSE	FALSE	FALSE	668	87.2	89.83	0.0	0	15,428.7	642.9	0.0	0.0	89.83	0.0	642.9	15,428.7
USGS	5480500	5/14/1981	643	TRUE	FALSE	FALSE	FALSE	643	88.8	89.83	0.0	0	15,122.5	630.1	0.0	0.0	89.83	0.0	630.1	15,122.5
USGS	5480500	5/15/1981	641	TRUE	FALSE	FALSE	FALSE	641	88.9	89.83	0.0	0	15,095.1	629.0	0.0	0.0	89.83	0.0	629.0	15,095.1
USGS	5480500	5/16/1981	631	FALSE	FALSE	FALSE	FALSE	631	89.4	89.83	0.0	0	14,951.7	623.0	0.0	0.0	89.83	0.0	623.0	14,951.7
USGS	5480500	5/17/1981	612	FALSE	FALSE	FALSE	FALSE	612	90.4	89.83	0.0	0	14,651.6	610.5	0.0	0.0	89.83	0.0	610.5	14,651.6
USGS	5480500	5/18/1981	590	FALSE	FALSE	FALSE	FALSE	590	91.2	89.83	0.0	0	14,261.4	594.2	0.0	0.0	89.83	0.0	594.2	14,261.4
USGS	5480500	5/19/1981	548	FALSE	FALSE	FALSE	FALSE	548	92.3	89.83	0.0	0	13,402.2	558.4	0.0	0.0	89.83	0.0	558.4	13,402.2
USGS	5480500	5/20/1981	516	FALSE	FALSE	FALSE	FALSE	516	92.6	89.83	0.0	0	12,660.3	527.5	0.0	0.0	89.83	0.0	527.5	12,660.3
USGS	5480500	5/21/1981	474	TRUE	FALSE	FALSE	FALSE	474	92.3	89.83	0.0	0	11,592.8	483.0	0.0	0.0	89.83	0.0	483.0	11,592.8
USGS	5480500	5/22/1981	461	FALSE	FALSE	FALSE	FALSE	461	92.1	89.83	0.0	0	11,244.5	468.5	0.0	0.0	89.83	0.0	468.5	11,244.5
USGS	5480500	5/23/1981	718	FALSE	TRUE	FALSE	FALSE	450	91.8	89.83	79.6	268	10,944.2	456.0	5,653.1	235.5	89.83	79.6	691.6	16,597.3
USGS	5480500	5/24/1981	2680	FALSE	FALSE	TRUE	FALSE	670	87.0	89.83	87.0	670	15,450.2	643.8	15,450.2	643.8	89.83	87.0	1,287.5	30,900.3
USGS	5480500	5/25/1981	3110	FALSE	FALSE	TRUE	FALSE	670	87.0	89.83	87.0	670	15,450.2	643.8	15,450.2	643.8	89.83	87.0	1,287.5	30,900.3
USGS	5480500	5/26/1981	2580	FALSE	FALSE	TRUE	FALSE	670	87.0	89.83	87.0	670	15,450.2	643.8	15,450.2	643.8	89.83	87.0	1,287.5	30,900.3
USGS	5480500	5/27/1981	2260	FALSE	FALSE	TRUE	FALSE	670	87.0	89.83	87.0	670	15,450.2	643.8	15,450.2	643.8	89.83	87.0	1,287.5	30,900.3
USGS	5480500	5/28/1981	2060	FALSE	FALSE	TRUE	FALSE	670	87.0	89.83	87.0	670	15,450.2	643.8	15,450.2	643.8	89.83	87.0	1,287.5	30,900.3
USGS	5480500	5/29/1981	1920	FALSE	FALSE	TRUE	FALSE	670	87.0	89.83	87.0	670	15,450.2	643.8	15,450.2	643.8	89.83	87.0	1,287.5	30,900.3
USGS	5480500	5/30/1981	1690	FALSE	FALSE	TRUE	FALSE	670	87.0	89.83	87.0	670	15,450.2	643.8	15,450.2	643.8	89.83	87.0	1,287.5	30,900.3
USGS	5480500	5/31/1981	1390	FALSE	FALSE	TRUE	FALSE	670	87.0	89.83	87.0	670	15,450.2	643.8	15,450.2	643.8	89.83	87.0	1,287.5	30,900.3
												Total- May 1981	**415,260.3**		**168,826.0**					**564,086.3**

Agency	Site	Date	V1	B1	B2	B3	B4	B5	V2	V3	V4	V5	V6	V7	V8	V9	V10	V11	V12	V13
USGS	5480500	6/1/1981	1170	FALSE	TRUE	FALSE	FALSE	FALSE	500	92.6	89.83	511.0	12,265.0	670	87.0	89.83	643.8	15,450.2	1,154.8	27,715.1
USGS	5480500	6/2/1981	1040	FALSE	TRUE	FALSE	FALSE	FALSE	370	88.2	89.83	360.4	8,649.9	670	87.0	89.83	643.8	15,450.2	1,004.2	24,100.1
USGS	5480500	6/3/1981	935	FALSE	FALSE	TRUE	FALSE	FALSE	667	87.3	89.83	642.4	15,417.8	268	79.6	89.83	235.5	5,653.1	878.0	21,070.9
USGS	5480500	6/4/1981	850	FALSE	FALSE	TRUE	FALSE	FALSE	582	91.5	89.83	587.9	14,108.8	268	79.6	89.83	235.5	5,653.1	823.4	19,761.9
USGS	5480500	6/5/1981	779	FALSE	FALSE	TRUE	FALSE	FALSE	511	92.6	89.83	522.4	12,538.3	268	79.6	89.83	235.5	5,653.1	756.0	18,191.4
USGS	5480500	6/6/1981	710	FALSE	FALSE	TRUE	FALSE	FALSE	442	91.6	89.83	446.8	10,722.8	268	79.6	89.83	235.5	5,653.1	682.3	16,375.9
USGS	5480500	6/7/1981	651	FALSE	FALSE	FALSE	TRUE	FALSE	651	88.3	89.83	634.5	15,228.0	0	0.0	89.83	0.0	0.0	634.5	15,228.0
USGS	5480500	6/8/1981	612	FALSE	FALSE	FALSE	TRUE	FALSE	612	90.4	89.83	610.5	14,651.6	0	0.0	89.83	0.0	0.0	610.5	14,651.6
USGS	5480500	6/9/1981	635	FALSE	FALSE	FALSE	TRUE	FALSE	635	89.2	89.83	625.4	15,010.3	0	0.0	89.83	0.0	0.0	625.4	15,010.3
USGS	5480500	6/10/1981	666	FALSE	FALSE	FALSE	TRUE	FALSE	666	87.3	89.83	641.9	15,406.8	0	0.0	89.83	0.0	0.0	641.9	15,406.8
USGS	5480500	6/11/1981	650	FALSE	FALSE	FALSE	TRUE	FALSE	650	88.4	89.83	634.0	15,215.2	0	0.0	89.83	0.0	0.0	634.0	15,215.2
USGS	5480500	6/12/1981	609	FALSE	TRUE	FALSE	FALSE	FALSE	609	90.5	89.83	608.4	14,601.0	0	0.0	89.83	0.0	0.0	608.4	14,601.0
USGS	5480500	6/13/1981	1100	FALSE	TRUE	FALSE	FALSE	FALSE	430	91.2	89.83	432.8	10,386.6	670	87.0	89.83	643.8	15,450.2	1,076.5	25,836.8
USGS	5480500	6/14/1981	2640	FALSE	FALSE	FALSE	FALSE	TRUE	670	87.0	89.83	643.8	15,450.2	670	87.0	89.83	643.8	15,450.2	1,287.5	30,900.3
USGS	5480500	6/15/1981	3580	FALSE	FALSE	FALSE	FALSE	TRUE	670	87.0	89.83	643.8	15,450.2	670	87.0	89.83	643.8	15,450.2	1,287.5	30,900.3
USGS	5480500	6/16/1981	4220	FALSE	FALSE	FALSE	FALSE	TRUE	670	87.0	89.83	643.8	15,450.2	670	87.0	89.83	643.8	15,450.2	1,287.5	30,900.3
USGS	5480500	6/17/1981	4280	FALSE	FALSE	FALSE	FALSE	TRUE	670	87.0	89.83	643.8	15,450.2	670	87.0	89.83	643.8	15,450.2	1,287.5	30,900.3
USGS	5480500	6/18/1981	3850	FALSE	FALSE	FALSE	FALSE	TRUE	670	87.0	89.83	643.8	15,450.2	670	87.0	89.83	643.8	15,450.2	1,287.5	30,900.3
USGS	5480500	6/19/1981	3240	FALSE	FALSE	FALSE	FALSE	TRUE	670	87.0	89.83	643.8	15,450.2	670	87.0	89.83	643.8	15,450.2	1,287.5	30,900.3
USGS	5480500	6/20/1981	2940	FALSE	FALSE	FALSE	FALSE	TRUE	670	87.0	89.83	643.8	15,450.2	670	87.0	89.83	643.8	15,450.2	1,287.5	30,900.3
USGS	5480500	6/21/1981	2700	FALSE	FALSE	FALSE	FALSE	TRUE	670	87.0	89.83	643.8	15,450.2	670	87.0	89.83	643.8	15,450.2	1,287.5	30,900.3
USGS	5480500	6/22/1981	2450	FALSE	FALSE	FALSE	FALSE	TRUE	670	87.0	89.83	643.8	15,450.2	670	87.0	89.83	643.8	15,450.2	1,287.5	30,900.3
USGS	5480500	6/23/1981	2120	FALSE	FALSE	FALSE	FALSE	TRUE	670	87.0	89.83	643.8	15,450.2	670	87.0	89.83	643.8	15,450.2	1,287.5	30,900.3
USGS	5480500	6/24/1981	8250	FALSE	FALSE	FALSE	FALSE	TRUE	670	87.0	89.83	643.8	15,450.2	670	87.0	89.83	643.8	15,450.2	1,287.5	30,900.3
USGS	5480500	6/25/1981	11800	FALSE	FALSE	FALSE	FALSE	TRUE	670	87.0	89.83	643.8	15,450.2	670	87.0	89.83	643.8	15,450.2	1,287.5	30,900.3
USGS	5480500	6/26/1981	9920	FALSE	FALSE	FALSE	FALSE	TRUE	670	87.0	89.83	643.8	15,450.2	670	87.0	89.83	643.8	15,450.2	1,287.5	30,900.3
USGS	5480500	6/27/1981	9470	FALSE	FALSE	FALSE	FALSE	TRUE	670	87.0	89.83	643.8	15,450.2	670	87.0	89.83	643.8	15,450.2	1,287.5	30,900.3
USGS	5480500	6/28/1981	8730	FALSE	FALSE	FALSE	FALSE	TRUE	670	87.0	89.83	643.8	15,450.2	670	87.0	89.83	643.8	15,450.2	1,287.5	30,900.3
USGS	5480500	6/29/1981	7840	FALSE	FALSE	FALSE	FALSE	TRUE	670	87.0	89.83	643.8	15,450.2	670	87.0	89.83	643.8	15,450.2	1,287.5	30,900.3
USGS	5480500	6/30/1981	7590	FALSE	FALSE	FALSE	FALSE	TRUE	670	87.0	89.83	643.8	15,450.2	670	87.0	89.83	643.8	15,450.2	1,287.5	30,900.3
		Total- June 1981											435,854.8					331,615.6		768,470.4

(continued)

325

TABLE 12.13 (*Continued*)

USGS	5480500	7/1/1981	7280	FALSE	FALSE	FALSE	TRUE	670	87.0	89.83	643.8	15,450.2	670	87.0	89.83	643.8	15,450.2	1,287.5	30,900.3
USGS	5480500	7/2/1981	6200	FALSE	FALSE	FALSE	TRUE	670	87.0	89.83	643.8	15,450.2	670	87.0	89.83	643.8	15,450.2	1,287.5	30,900.3
USGS	5480500	7/3/1981	5330	FALSE	FALSE	FALSE	TRUE	670	87.0	89.83	643.8	15,450.2	670	87.0	89.83	643.8	15,450.2	1,287.5	30,900.3
USGS	5480500	7/4/1981	4640	FALSE	FALSE	FALSE	TRUE	670	87.0	89.83	643.8	15,450.2	670	87.0	89.83	643.8	15,450.2	1,287.5	30,900.3
USGS	5480500	7/5/1981	4130	FALSE	FALSE	FALSE	TRUE	670	87.0	89.83	643.8	15,450.2	670	87.0	89.83	643.8	15,450.2	1,287.5	30,900.3
USGS	5480500	7/6/1981	3630	FALSE	FALSE	FALSE	TRUE	670	87.0	89.83	643.8	15,450.2	670	87.0	89.83	643.8	15,450.2	1,287.5	30,900.3
USGS	5480500	7/7/1981	3370	FALSE	FALSE	FALSE	TRUE	670	87.0	89.83	643.8	15,450.2	670	87.0	89.83	643.8	15,450.2	1,287.5	30,900.3
USGS	5480500	7/8/1981	3060	FALSE	FALSE	FALSE	TRUE	670	87.0	89.83	643.8	15,450.2	670	87.0	89.83	643.8	15,450.2	1,287.5	30,900.3
USGS	5480500	7/9/1981	2730	FALSE	FALSE	FALSE	TRUE	670	87.0	89.83	643.8	15,450.2	670	87.0	89.83	643.8	15,450.2	1,287.5	30,900.3
USGS	5480500	7/10/1981	2410	FALSE	FALSE	FALSE	TRUE	670	87.0	89.83	643.8	15,450.2	670	87.0	89.83	643.8	15,450.2	1,287.5	30,900.3
USGS	5480500	7/11/1981	2120	FALSE	FALSE	FALSE	TRUE	670	87.0	89.83	643.8	15,450.2	670	87.0	89.83	643.8	15,450.2	1,287.5	30,900.3
USGS	5480500	7/12/1981	1840	FALSE	FALSE	FALSE	TRUE	670	87.0	89.83	643.8	15,450.2	670	87.0	89.83	643.8	15,450.2	1,287.5	30,900.3
USGS	5480500	7/13/1981	1590	FALSE	FALSE	FALSE	TRUE	670	87.0	89.83	643.8	15,450.2	670	87.0	89.83	643.8	15,450.2	1,287.5	30,900.3
USGS	5480500	7/14/1981	1360	FALSE	FALSE	FALSE	TRUE	670	87.0	89.83	643.8	15,450.2	670	87.0	89.83	643.8	15,450.2	1,287.5	30,900.3
USGS	5480500	7/15/1981	1180	FALSE	FALSE	FALSE	FALSE	510	92.6	89.83	521.4	12,513.8	670	87.0	89.83	643.8	15,450.2	1,165.2	27,963.9
USGS	5480500	7/16/1981	1120	FALSE	FALSE	TRUE	FALSE	450	91.8	89.83	456.0	10,944.2	670	87.0	89.83	643.8	15,450.2	1,099.8	26,394.3
USGS	5480500	7/17/1981	1140	FALSE	FALSE	TRUE	FALSE	470	92.2	89.83	478.6	11,486.4	670	87.0	89.83	643.8	15,450.2	1,185.1	26,936.6
USGS	5480500	7/18/1981	1200	FALSE	FALSE	TRUE	FALSE	530	92.5	89.83	541.4	12,993.3	670	87.0	89.83	643.8	15,450.2	1,122.4	28,443.4
USGS	5480500	7/19/1981	1430	FALSE	FALSE	FALSE	TRUE	670	87.0	89.83	643.8	15,450.2	670	87.0	89.83	643.8	15,450.2	1,287.5	30,900.3
USGS	5480500	7/20/1981	1790	FALSE	FALSE	FALSE	TRUE	670	87.0	89.83	643.8	15,450.2	670	87.0	89.83	643.8	15,450.2	1,287.5	30,900.3
USGS	5480500	7/21/1981	1820	FALSE	FALSE	FALSE	TRUE	670	87.0	89.83	643.8	15,450.2	670	87.0	89.83	643.8	15,450.2	1,287.5	30,900.3
USGS	5480500	7/22/1981	1920	FALSE	FALSE	FALSE	TRUE	670	87.0	89.83	643.8	15,450.2	670	87.0	89.83	643.8	15,450.2	1,287.5	30,900.3
USGS	5480500	7/23/1981	2400	FALSE	FALSE	FALSE	TRUE	670	87.0	89.83	643.8	15,450.2	670	87.0	89.83	643.8	15,450.2	1,287.5	30,900.3
USGS	5480500	7/24/1981	2580	FALSE	FALSE	FALSE	TRUE	670	87.0	89.83	643.8	15,450.2	670	87.0	89.83	643.8	15,450.2	1,287.5	30,900.3
USGS	5480500	7/25/1981	2970	FALSE	FALSE	FALSE	TRUE	670	87.0	89.83	643.8	15,450.2	670	87.0	89.83	643.8	15,450.2	1,287.5	30,900.3
USGS	5480500	7/26/1981	3120	FALSE	FALSE	FALSE	TRUE	670	87.0	89.83	643.8	15,450.2	670	87.0	89.83	643.8	15,450.2	1,287.5	30,900.3
USGS	5480500	7/27/1981	3000	FALSE	FALSE	FALSE	TRUE	670	87.0	89.83	643.8	15,450.2	670	87.0	89.83	643.8	15,450.2	1,287.5	30,900.3
USGS	5480500	7/28/1981	3080	FALSE	FALSE	FALSE	TRUE	670	87.0	89.83	643.8	15,450.2	670	87.0	89.83	643.8	15,450.2	1,287.5	30,900.3
USGS	5480500	7/29/1981	3250	FALSE	FALSE	FALSE	TRUE	670	87.0	89.83	643.8	15,450.2	670	87.0	89.83	643.8	15,450.2	1,287.5	30,900.3
USGS	5480500	7/30/1981	3350	FALSE	FALSE	FALSE	TRUE	670	87.0	89.83	643.8	15,450.2	670	87.0	89.83	643.8	15,450.2	1,287.5	30,900.3
USGS	5480500	7/31/1981	3190	FALSE	FALSE	FALSE	TRUE	670	87.0	89.83	643.8	15,450.2	670	87.0	89.83	643.8	15,450.2	1,287.5	30,900.3
							Total- July 1981					465,092.0					478,954.9		944,046.9

Agency	Station	Date																		
USGS	5480500	8/1/1981	2880	FALSE	FALSE	FALSE	FALSE	TRUE	670	87.0	89.83	643.8	15,450.2	670	87.0	89.83	643.8	15,450.2	1,287.5	30,900.3
USGS	5480500	8/2/1981	3150	FALSE	FALSE	FALSE	FALSE	TRUE	670	87.0	89.83	643.8	15,450.2	670	87.0	89.83	643.8	15,450.2	1,287.5	30,900.3
USGS	5480500	8/3/1981	4760	FALSE	FALSE	FALSE	FALSE	TRUE	670	87.0	89.83	643.8	15,450.2	670	87.0	89.83	643.8	15,450.2	1,287.5	30,900.3
USGS	5480500	8/4/1981	3740	FALSE	FALSE	FALSE	FALSE	TRUE	670	87.0	89.83	643.8	15,450.2	670	87.0	89.83	643.8	15,450.2	1,287.5	30,900.3
USGS	5480500	8/5/1981	3060	FALSE	FALSE	FALSE	FALSE	TRUE	670	87.0	89.83	643.8	15,450.2	670	87.0	89.83	643.8	15,450.2	1,287.5	30,900.3
USGS	5480500	8/6/1981	2740	FALSE	FALSE	FALSE	FALSE	TRUE	670	87.0	89.83	643.8	15,450.2	670	87.0	89.83	643.8	15,450.2	1,287.5	30,900.3
USGS	5480500	8/7/1981	2430	FALSE	FALSE	FALSE	FALSE	TRUE	670	87.0	89.83	643.8	15,450.2	670	87.0	89.83	643.8	15,450.2	1,287.5	30,900.3
USGS	5480500	8/8/1981	2110	FALSE	FALSE	FALSE	FALSE	TRUE	670	87.0	89.83	643.8	15,450.2	670	87.0	89.83	643.8	15,450.2	1,287.5	30,900.3
USGS	5480500	8/9/1981	1840	FALSE	FALSE	FALSE	FALSE	TRUE	670	87.0	89.83	643.8	15,450.2	670	87.0	89.83	643.8	15,450.2	1,287.5	30,900.3
USGS	5480500	8/10/1981	1580	FALSE	FALSE	FALSE	FALSE	TRUE	670	87.0	89.83	643.8	15,450.2	670	87.0	89.83	643.8	15,450.2	1,287.5	30,900.3
USGS	5480500	8/11/1981	1350	FALSE	FALSE	FALSE	FALSE	TRUE	670	87.0	89.83	643.8	15,450.2	670	87.0	89.83	643.8	15,450.2	1,287.5	30,900.3
USGS	5480500	8/12/1981	1150	FALSE	FALSE	FALSE	TRUE	FALSE	480	92.4	89.83	489.6	8,059.1	670	87.0	89.83	643.8	15,450.2	1,133.4	27,201.1
USGS	5480500	8/13/1981	1020	FALSE	FALSE	FALSE	TRUE	FALSE	350	86.9	89.83	335.8	5,883.4	670	87.0	89.83	643.8	15,450.2	979.6	23,509.2
USGS	5480500	8/14/1981	946	FALSE	TRUE	FALSE	FALSE	FALSE	276	80.5	89.83	245.1	14,390.3	268	79.6	89.83	235.5	5,653.1	888.9	21,333.6
USGS	5480500	8/15/1981	865	FALSE	TRUE	FALSE	FALSE	FALSE	597	91.0	89.83	599.6	13,828.9	268	79.6	89.83	235.5	5,653.1	835.1	20,043.4
USGS	5480500	8/16/1981	836	FALSE	TRUE	FALSE	FALSE	FALSE	568	91.9	89.83	576.2	15,215.2	268	79.6	89.83	235.5	5,653.1	811.8	19,482.0
USGS	5480500	8/17/1981	918	FALSE	FALSE	FALSE	TRUE	FALSE	650	88.4	89.83	634.0	6,641.0	670	87.0	89.83	643.8	15,450.2	869.5	20,868.3
USGS	5480500	8/18/1981	972	FALSE	FALSE	FALSE	TRUE	FALSE	302	83.0	89.83	276.7	7,289.4	670	87.0	89.83	643.8	15,450.2	920.5	22,091.2
USGS	5480500	8/19/1981	994	FALSE	TRUE	FALSE	FALSE	FALSE	324	84.9	89.83	303.7	15,350.0	268	79.6	89.83	235.5	5,653.1	947.5	22,739.6
USGS	5480500	8/20/1981	929	FALSE	TRUE	FALSE	FALSE	FALSE	661	92.6	89.83	639.6	12,899.4	268	79.6	89.83	235.5	5,653.1	875.1	21,003.0
USGS	5480500	8/21/1981	794	FALSE	FALSE	TRUE	FALSE	FALSE	526	91.7	89.83	537.5	13,970.9	0	0.0	89.83	0.0	0.0	773.0	18,552.5
USGS	5480500	8/22/1981	575	FALSE	FALSE	TRUE	FALSE	FALSE	575	92.1	89.83	582.1	13,619.4	0	0.0	89.83	0.0	0.0	582.1	13,970.9
USGS	5480500	8/23/1981	558	FALSE	FALSE	TRUE	FALSE	FALSE	558	92.3	89.83	567.5	13,468.1	0	0.0	89.83	0.0	0.0	567.5	13,619.4
USGS	5480500	8/24/1981	551	FALSE	FALSE	TRUE	FALSE	FALSE	551	92.4	89.83	561.2	13,268.4	0	0.0	89.83	0.0	0.0	561.2	13,468.1
USGS	5480500	8/25/1981	542	FALSE	TRUE	FALSE	FALSE	FALSE	542	91.7	89.83	552.9	10,861.4	268	79.6	89.83	235.5	5,653.1	552.9	13,268.4
USGS	5480500	8/26/1981	715	FALSE	FALSE	FALSE	TRUE	FALSE	447	79.7	89.83	452.6	5,681.8	670	87.0	89.83	643.8	15,450.2	688.1	16,514.5
USGS	5480500	8/27/1981	939	FALSE	FALSE	FALSE	TRUE	FALSE	269	79.7	89.83	236.7	15,338.3	670	87.0	89.83	643.8	15,450.2	880.5	21,132.0
USGS	5480500	8/28/1981	1330	FALSE	FALSE	FALSE	FALSE	TRUE	660	87.0	89.83	639.1	15,450.2	670	87.0	89.83	643.8	15,450.2	1,282.9	30,788.4
USGS	5480500	8/29/1981	2170	FALSE	FALSE	FALSE	FALSE	TRUE	670	87.0	89.83	643.8	15,450.2	670	87.0	89.83	643.8	15,450.2	1,287.5	30,900.3
USGS	5480500	8/30/1981	2270	FALSE	FALSE	FALSE	FALSE	TRUE	670	87.0	89.83	643.8	15,450.2	670	87.0	89.83	643.8	15,450.2	1,287.5	30,900.3
USGS	5480500	8/31/1981	2290	FALSE	FALSE	FALSE	FALSE	TRUE	670	87.0	89.83	643.8	15,450.2	670	87.0	89.83	643.8	15,450.2	1,287.5	30,900.3
		Total- August 1981											**413,818.2**					**358,371.9**		**772,190.1**

(*continued*)

TABLE 12.13 (*Continued*)

Source	Station	Date	Q	F1	F2	F3	F4	C9	Temp	C11	C12	C13	C14	Temp2	C16	C17	C18	C19	C20
USGS	5480500	9/1/1981	2330	FALSE	FALSE	FALSE	TRUE	670	87.0	89.83	643.8	15,450.2	670	87.0	89.83	643.8	15,450.2	1,287.5	30,900.3
USGS	5480500	9/2/1981	2390	FALSE	FALSE	FALSE	TRUE	670	87.0	89.83	643.8	15,450.2	670	87.0	89.83	643.8	15,450.2	1,287.5	30,900.3
USGS	5480500	9/3/1981	2420	FALSE	FALSE	FALSE	TRUE	670	87.0	89.83	643.8	15,450.2	670	87.0	89.83	643.8	15,450.2	1,287.5	30,900.3
USGS	5480500	9/4/1981	2310	FALSE	FALSE	FALSE	TRUE	670	87.0	89.83	643.8	15,450.2	670	87.0	89.83	643.8	15,450.2	1,287.5	30,900.3
USGS	5480500	9/5/1981	2050	FALSE	FALSE	FALSE	TRUE	670	87.0	89.83	643.8	15,450.2	670	87.0	89.83	643.8	15,450.2	1,287.5	30,900.3
USGS	5480500	9/6/1981	1730	FALSE	FALSE	FALSE	TRUE	670	87.0	89.83	643.8	15,450.2	670	87.0	89.83	643.8	15,450.2	1,287.5	30,900.3
USGS	5480500	9/7/1981	1510	FALSE	FALSE	FALSE	TRUE	670	87.0	89.83	643.8	15,450.2	670	87.0	89.83	643.8	15,450.2	1,287.5	30,900.3
USGS	5480500	9/8/1981	1230	FALSE	FALSE	TRUE	FALSE	560	92.1	89.83	569.2	13,661.9	670	87.0	89.83	643.8	15,450.2	1,213.0	29,112.1
USGS	5480500	9/9/1981	1030	FALSE	FALSE	FALSE	FALSE	360	87.6	89.83	348.1	8,354.9	670	87.0	89.83	643.8	15,450.2	991.9	23,805.0
USGS	5480500	9/10/1981	907	FALSE	TRUE	TRUE	FALSE	639	89.0	89.83	627.8	15,067.3	268	79.6	89.83	235.5	5,653.1	863.3	20,720.4
USGS	5480500	9/11/1981	804	FALSE	TRUE	TRUE	FALSE	536	92.5	89.83	547.2	13,132.1	268	79.6	89.83	235.5	5,653.1	782.7	18,785.2
USGS	5480500	9/12/1981	736	FALSE	TRUE	TRUE	FALSE	468	90.2	89.83	476.4	11,433.0	268	79.6	89.83	235.5	5,653.1	711.9	17,086.1
USGS	5480500	9/13/1981	673	FALSE	FALSE	TRUE	FALSE	405	90.1	89.83	403.0	9,672.3	268	79.6	89.83	235.5	5,653.1	638.6	15,325.4
USGS	5480500	9/14/1981	609	FALSE	FALSE	FALSE	FALSE	609	90.5	89.83	608.4	14,601.0	0	0.0	0.0	0.0	0.0	608.4	14,601.0
USGS	5480500	9/15/1981	552	FALSE	FALSE	FALSE	FALSE	552	92.2	89.83	562.1	13,490.0	0	0.0	0.0	0.0	0.0	562.1	13,490.0
USGS	5480500	9/16/1981	522	FALSE	FALSE	FALSE	FALSE	522	92.6	89.83	533.5	12,804.5	0	0.0	0.0	0.0	0.0	533.5	12,804.5
USGS	5480500	9/17/1981	491	FALSE	FALSE	FALSE	FALSE	491	92.5	89.83	501.5	12,036.3	0	0.0	0.0	0.0	0.0	501.5	12,036.3
USGS	5480500	9/18/1981	448	FALSE	FALSE	FALSE	FALSE	448	91.7	89.83	453.7	10,889.1	0	0.0	0.0	0.0	0.0	453.7	10,889.1
USGS	5480500	9/19/1981	410	FALSE	FALSE	FALSE	FALSE	410	90.4	89.83	409.0	9,816.5	0	0.0	0.0	0.0	0.0	409.0	9,816.5
USGS	5480500	9/20/1981	391	FALSE	FALSE	FALSE	FALSE	391	89.5	89.83	386.1	9,265.9	0	0.0	0.0	0.0	0.0	386.1	9,265.9
USGS	5480500	9/21/1981	379	FALSE	FALSE	FALSE	FALSE	379	88.8	89.83	371.4	8,914.6	0	0.0	0.0	0.0	0.0	371.4	8,914.6
USGS	5480500	9/22/1981	364	FALSE	FALSE	FALSE	FALSE	364	87.9	89.83	353.0	8,473.0	0	0.0	0.0	0.0	0.0	353.0	8,473.0
USGS	5480500	9/23/1981	344	FALSE	FALSE	FALSE	FALSE	344	86.5	89.83	328.4	7,881.4	0	0.0	0.0	0.0	0.0	328.4	7,881.4
USGS	5480500	9/24/1981	344	FALSE	FALSE	FALSE	FALSE	344	86.5	89.83	328.4	7,881.4	0	0.0	0.0	0.0	0.0	328.4	7,881.4
USGS	5480500	9/25/1981	404	FALSE	FALSE	FALSE	FALSE	404	90.1	89.83	401.8	9,643.4	0	0.0	0.0	0.0	0.0	401.8	9,643.4
USGS	5480500	9/26/1981	385	FALSE	FALSE	FALSE	FALSE	385	89.1	89.83	378.8	9,090.5	0	0.0	0.0	0.0	0.0	378.8	9,090.5
USGS	5480500	9/27/1981	369	FALSE	FALSE	FALSE	FALSE	369	88.2	89.83	359.2	8,620.5	0	0.0	0.0	0.0	0.0	359.2	8,620.5
USGS	5480500	9/28/1981	344	FALSE	FALSE	FALSE	FALSE	344	86.5	89.83	328.4	7,881.4	0	0.0	0.0	0.0	0.0	328.4	7,881.4
USGS	5480500	9/29/1981	342	FALSE	FALSE	FALSE	FALSE	342	86.3	89.83	325.9	7,822.2	0	0.0	0.0	0.0	0.0	325.9	7,822.2
USGS	5480500	9/30/1981	336	FALSE	FALSE	FALSE	FALSE	336	85.9	89.83	318.5	7,644.5	0	0.0	0.0	0.0	0.0	318.5	7,644.5
												Total- September 1981 346,228.9					161,663.8		507,892.7

(continued)

USGS	5480500	10/1/1981	326	FALSE	FALSE	FALSE	TRUE	FALSE	326	85.1	89.83	306.2	7,348.6	0	0.0	89.83	0.0	0.0	306.2	7,348.6
USGS	5480500	10/2/1981	314	FALSE	FALSE	FALSE	TRUE	FALSE	314	84.1	89.83	291.4	6,994.1	0	0.0	89.83	0.0	0.0	291.4	6,994.1
USGS	5480500	10/3/1981	337	FALSE	FALSE	FALSE	TRUE	FALSE	337	86.0	89.83	319.8	7,674.2	0	0.0	89.83	0.0	0.0	319.8	7,674.2
USGS	5480500	10/4/1981	454	FALSE	FALSE	FALSE	TRUE	FALSE	454	91.9	89.83	460.6	11,053.9	0	0.0	89.83	0.0	0.0	460.6	11,053.9
USGS	5480500	10/5/1981	630	FALSE	FALSE	FALSE	TRUE	FALSE	630	89.5	89.83	622.4	14,936.8	0	0.0	89.83	0.0	0.0	622.4	14,936.8
USGS	5480500	10/6/1981	661	FALSE	FALSE	TRUE	FALSE	FALSE	661	87.7	89.83	639.6	15,350.0	268	79.6	89.83	235.5	5,653.1	639.6	15,350.0
USGS	5480500	10/7/1981	687	FALSE	FALSE	TRUE	FALSE	FALSE	419	90.8	89.83	419.8	10,074.4	268	79.6	89.83	235.5	5,653.1	655.3	15,727.5
USGS	5480500	10/8/1981	718	FALSE	FALSE	TRUE	FALSE	FALSE	450	91.8	89.83	456.0	10,944.2	268	79.6	89.83	235.5	5,653.1	691.6	16,597.3
USGS	5480500	10/9/1981	697	FALSE	FALSE	TRUE	FALSE	FALSE	429	91.1	89.83	431.6	10,358.4	268	79.6	89.83	235.5	5,653.1	667.1	16,011.5
USGS	5480500	10/10/1981	687	FALSE	FALSE	FALSE	TRUE	FALSE	419	90.8	89.83	419.8	10,074.4	0	0.0	89.83	0.0	0.0	655.3	15,727.5
USGS	5480500	10/11/1981	657	FALSE	FALSE	FALSE	TRUE	FALSE	657	87.9	89.83	637.6	15,302.5	0	0.0	89.83	0.0	0.0	637.6	15,302.5
USGS	5480500	10/12/1981	619	FALSE	FALSE	FALSE	TRUE	FALSE	619	90.0	89.83	615.3	14,766.3	0	0.0	89.83	0.0	0.0	615.3	14,766.3
USGS	5480500	10/13/1981	609	FALSE	FALSE	FALSE	TRUE	FALSE	609	90.5	89.83	608.4	14,601.0	0	0.0	89.83	0.0	0.0	608.4	14,601.0
USGS	5480500	10/14/1981	593	FALSE	FALSE	FALSE	TRUE	FALSE	593	91.1	89.83	596.5	14,317.2	0	0.0	89.83	0.0	0.0	596.5	14,317.2
USGS	5480500	10/15/1981	629	FALSE	FALSE	FALSE	TRUE	FALSE	629	89.5	89.83	621.7	14,921.8	0	0.0	89.83	0.0	0.0	621.7	14,921.8
USGS	5480500	10/16/1981	665	FALSE	FALSE	FALSE	TRUE	FALSE	665	87.4	89.83	641.5	15,395.6	0	0.0	89.83	0.0	0.0	641.5	15,395.6
USGS	5480500	10/17/1981	739	FALSE	FALSE	TRUE	FALSE	FALSE	471	92.3	89.83	479.7	11,513.1	268	79.6	89.83	235.5	5,653.1	715.3	17,166.2
USGS	5480500	10/18/1981	779	FALSE	FALSE	TRUE	FALSE	FALSE	511	92.6	89.83	522.4	12,538.3	268	79.6	89.83	235.5	5,653.1	758.0	18,191.4
USGS	5480500	10/19/1981	1000	FALSE	TRUE	FALSE	FALSE	FALSE	330	85.4	89.83	311.1	7,466.9	670	87.0	89.83	643.8	15,450.2	954.9	22,917.1
USGS	5480500	10/20/1981	1190	FALSE	TRUE	FALSE	FALSE	FALSE	520	92.6	89.83	531.5	12,756.7	670	87.0	89.83	643.8	15,450.2	1,175.3	28,206.8
USGS	5480500	10/21/1981	1230	FALSE	TRUE	FALSE	FALSE	FALSE	560	92.1	89.83	569.2	13,661.9	670	87.0	89.83	643.8	15,450.2	1,213.0	29,112.1
USGS	5480500	10/22/1981	1220	FALSE	TRUE	FALSE	FALSE	FALSE	550	92.3	89.83	560.3	13,446.2	670	87.0	89.83	643.8	15,450.2	1,204.0	28,896.4
USGS	5480500	10/23/1981	1180	FALSE	TRUE	FALSE	FALSE	FALSE	510	92.6	89.83	521.4	12,513.8	670	87.0	89.83	643.8	15,450.2	1,165.2	27,963.9
USGS	5480500	10/24/1981	1110	FALSE	TRUE	FALSE	FALSE	FALSE	440	91.5	89.83	444.5	10,667.1	670	87.0	89.83	643.8	15,450.2	1,088.2	26,117.3
USGS	5480500	10/25/1981	1070	FALSE	TRUE	FALSE	FALSE	FALSE	340	89.9	89.83	397.0	9,527.6	670	87.0	89.83	643.8	15,450.2	1,040.7	24,977.8
USGS	5480500	10/26/1981	1010	FALSE	TRUE	FALSE	FALSE	FALSE	316	86.2	89.83	323.5	7,763.0	670	87.0	89.83	643.8	15,450.2	967.2	23,213.2
USGS	5480500	10/27/1981	986	FALSE	TRUE	FALSE	FALSE	FALSE	302	84.2	89.83	293.9	7,053.1	670	87.0	89.83	643.8	15,450.2	937.6	22,503.3
USGS	5480500	10/28/1981	972	FALSE	TRUE	FALSE	FALSE	FALSE	275	83.0	89.83	276.7	6,641.0	670	87.0	89.83	643.8	15,450.2	920.5	22,091.2
USGS	5480500	10/29/1981	945	FALSE	TRUE	FALSE	FALSE	FALSE		80.4	89.83	243.9	5,854.5	670	87.0	89.83	643.8	15,450.2	887.7	21,304.7
USGS	5480500	10/30/1981	930	FALSE	FALSE	TRUE	FALSE	FALSE	662	87.6	89.83	640.1	15,361.5	268	79.6	89.83	235.5	5,653.1	875.6	21,014.6
USGS	5480500	10/31/1981	930	FALSE	FALSE	TRUE	FALSE	FALSE	662	87.6	89.83	640.1	15,361.5	268	79.6	89.83	235.5	5,653.1	875.6	21,014.6
												Total-October 1981	356,239.8					215,176.5		571,416.3

TABLE 12.13 (Continued)

USGS	5480500	11/1/1981	880	FALSE	FALSE	TRUE	FALSE	612	90.4	89.83	610.5	14,651.6	268	79.6	89.83	235.5	5,653.1	846.0	20,304.7
USGS	5480500	11/2/1981	853	FALSE	FALSE	TRUE	FALSE	585	91.4	89.83	590.3	14,166.7	268	79.6	89.83	235.5	5,653.1	825.8	19,819.8
USGS	5480500	11/3/1981	824	FALSE	FALSE	TRUE	FALSE	556	92.2	89.83	565.7	13,576.5	268	79.6	89.83	235.5	5,653.1	801.2	19,229.6
USGS	5480500	11/4/1981	799	FALSE	FALSE	TRUE	FALSE	531	92.5	89.83	542.4	13,016.6	268	79.6	89.83	235.5	5,653.1	777.9	18,669.7
USGS	5480500	11/5/1981	798	FALSE	FALSE	TRUE	FALSE	530	92.5	89.83	541.4	12,993.3	268	79.6	89.83	235.5	5,653.1	776.9	18,646.4
USGS	5480500	11/6/1981	774	FALSE	FALSE	TRUE	FALSE	506	92.6	89.83	517.3	12,415.0	268	79.6	89.83	235.5	5,653.1	752.8	18,068.0
USGS	5480500	11/7/1981	752	FALSE	FALSE	TRUE	FALSE	484	92.5	89.83	494.0	11,855.4	268	79.6	89.83	235.5	5,653.1	729.5	17,508.5
USGS	5480500	11/8/1981	738	FALSE	FALSE	TRUE	FALSE	470	92.2	89.83	478.6	11,486.4	268	79.6	89.83	235.5	5,653.1	714.1	17,139.5
USGS	5480500	11/9/1981	701	FALSE	FALSE	TRUE	FALSE	433	91.3	89.83	436.3	10,471.1	268	79.6	89.83	235.5	5,653.1	671.8	16,124.2
USGS	5480500	11/10/1981	675	FALSE	FALSE	TRUE	FALSE	407	90.2	89.83	405.4	9,730.1	268	79.6	89.83	235.5	5,653.1	641.0	15,383.2
USGS	5480500	11/11/1981	729	FALSE	FALSE	TRUE	FALSE	461	92.1	89.83	468.5	11,244.5	268	79.6	89.83	235.5	5,653.1	704.1	16,897.6
USGS	5480500	11/12/1981	645	FALSE	TRUE	FALSE	FALSE	645	88.7	89.83	631.2	15,149.5	0	0.0	89.83	0.0	0.0	631.2	15,149.5
USGS	5480500	11/13/1981	631	FALSE	TRUE	FALSE	FALSE	631	89.4	89.83	623.0	14,951.7	0	0.0	89.83	0.0	0.0	623.0	14,951.7
USGS	5480500	11/14/1981	618	FALSE	TRUE	FALSE	FALSE	618	90.1	89.83	614.6	14,750.2	0	0.0	89.83	0.0	0.0	614.6	14,750.2
USGS	5480500	11/15/1981	613	FALSE	TRUE	FALSE	FALSE	613	90.3	89.83	611.2	14,668.3	0	0.0	89.83	0.0	0.0	611.2	14,668.3
USGS	5480500	11/16/1981	608	FALSE	TRUE	FALSE	FALSE	608	90.5	89.83	607.7	14,584.0	0	0.0	89.83	0.0	0.0	607.7	14,584.0
USGS	5480500	11/17/1981	601	FALSE	TRUE	FALSE	FALSE	601	90.8	89.83	602.6	14,462.0	0	0.0	89.83	0.0	0.0	602.6	14,462.0
USGS	5480500	11/18/1981	584	FALSE	TRUE	FALSE	FALSE	584	91.4	89.83	589.5	14,147.5	0	0.0	89.83	0.0	0.0	589.5	14,147.5
USGS	5480500	11/19/1981	595	FALSE	TRUE	FALSE	FALSE	595	91.1	89.83	598.1	14,353.9	0	0.0	89.83	0.0	0.0	598.1	14,353.9
USGS	5480500	11/20/1981	577	FALSE	TRUE	FALSE	FALSE	577	91.7	89.83	583.8	14,010.7	0	0.0	89.83	0.0	0.0	583.8	14,010.7
USGS	5480500	11/21/1981	516	FALSE	TRUE	FALSE	FALSE	516	92.6	89.83	527.5	12,660.3	0	0.0	89.83	0.0	0.0	527.5	12,660.3
USGS	5480500	11/22/1981	515	FALSE	TRUE	FALSE	FALSE	515	92.6	89.83	526.5	12,636.0	0	0.0	89.83	0.0	0.0	526.5	12,636.0
USGS	5480500	11/23/1981	587	FALSE	TRUE	FALSE	FALSE	587	91.3	89.83	591.9	14,204.8	0	0.0	89.83	0.0	0.0	591.9	14,204.8
USGS	5480500	11/24/1981	566	FALSE	TRUE	FALSE	FALSE	566	91.9	89.83	574.5	13,787.6	0	0.0	89.83	0.0	0.0	574.5	13,787.6
USGS	5480500	11/25/1981	578	FALSE	TRUE	FALSE	FALSE	578	91.6	89.83	584.6	14,030.5	0	0.0	89.83	0.0	0.0	584.6	14,030.5
USGS	5480500	11/26/1981	594	FALSE	TRUE	FALSE	FALSE	594	91.1	89.83	597.3	14,335.6	0	0.0	89.83	0.0	0.0	597.3	14,335.6
USGS	5480500	11/27/1981	593	FALSE	TRUE	FALSE	FALSE	593	91.1	89.83	596.5	14,317.2	0	0.0	89.83	0.0	0.0	596.5	14,317.2
USGS	5480500	11/28/1981	591	FALSE	TRUE	FALSE	FALSE	591	91.2	89.83	595.0	14,280.1	0	0.0	89.83	0.0	0.0	595.0	14,280.1
USGS	5480500	11/29/1981	583	FALSE	TRUE	FALSE	FALSE	583	91.5	89.83	588.7	14,128.2	0	0.0	89.83	0.0	0.0	588.7	14,128.2
USGS	5480500	11/30/1981	609	FALSE	TRUE	FALSE	FALSE	609	90.5	89.83	608.4	14,601.0	0	0.0	89.83	0.0	0.0	608.4	14,601.0
											Total- November 1981	405,666.3					62,184.1		467,850.4

Table of daily streamflow data (USGS site 5480500), December 1981.

Agency	Site No.	Date	(cfs)	flag1	flag2	flag3	flag4	flag5	(cfs)	(%)	elev	(ac-ft)	(ac-ft)	(cfs)	(cfs)	elev	(ac-ft)	(ac-ft)	(ac-ft)	(ac-ft)
USGS	5480500	12/1/1981	638	FALSE	FALSE	TRUE	FALSE	FALSE	638	89.1	89.83	627.2	15,053.2	0	0.0	89.83	0.0	0.0	627.2	15,053.2
USGS	5480500	12/2/1981	520	FALSE	FALSE	TRUE	FALSE	FALSE	520	92.6	89.83	531.5	12,756.7	0	0.0	89.83	0.0	0.0	531.5	12,756.7
USGS	5480500	12/3/1981	470	FALSE	FALSE	TRUE	FALSE	FALSE	470	92.2	89.83	478.6	11,486.4	0	0.0	89.83	0.0	0.0	478.6	11,486.4
USGS	5480500	12/4/1981	630	FALSE	FALSE	TRUE	FALSE	FALSE	630	89.5	89.83	622.4	14,936.8	0	0.0	89.83	0.0	0.0	622.4	14,936.8
USGS	5480500	12/5/1981	610	FALSE	FALSE	TRUE	FALSE	FALSE	610	90.5	89.83	609.1	14,618.0	0	0.0	89.83	0.0	0.0	609.1	14,618.0
USGS	5480500	12/6/1981	600	FALSE	FALSE	FALSE	TRUE	FALSE	600	90.9	89.83	601.8	14,444.2	0	0.0	89.83	0.0	0.0	601.8	14,444.2
USGS	5480500	12/7/1981	710	FALSE	FALSE	TRUE	FALSE	FALSE	442	91.6	89.83	446.8	10,722.8	268	79.6	89.83	235.5	5,653.1	682.3	16,375.9
USGS	5480500	12/8/1981	700	FALSE	FALSE	TRUE	FALSE	FALSE	432	91.2	89.83	435.1	10,443.0	268	79.6	89.83	235.5	5,653.1	670.7	16,096.1
USGS	5480500	12/9/1981	540	FALSE	FALSE	TRUE	FALSE	FALSE	540	92.4	89.83	551.0	13,223.3	0	0.0	89.83	0.0	0.0	551.0	13,223.3
USGS	5480500	12/10/1981	410	FALSE	FALSE	TRUE	FALSE	FALSE	410	90.4	89.83	409.0	9,816.5	0	0.0	89.83	0.0	0.0	409.0	9,816.5
USGS	5480500	12/11/1981	460	FALSE	FALSE	TRUE	FALSE	FALSE	460	92.0	89.83	467.4	11,217.4	0	0.0	89.83	0.0	0.0	467.4	11,217.4
USGS	5480500	12/12/1981	610	FALSE	FALSE	TRUE	FALSE	FALSE	610	90.5	89.83	609.1	14,618.0	0	0.0	89.83	0.0	0.0	609.1	14,618.0
USGS	5480500	12/13/1981	640	FALSE	FALSE	TRUE	FALSE	FALSE	640	88.9	89.83	628.4	15,081.2	0	0.0	89.83	0.0	0.0	628.4	15,081.2
USGS	5480500	12/14/1981	500	FALSE	FALSE	TRUE	FALSE	FALSE	500	92.6	89.83	511.0	12,265.0	0	0.0	89.83	0.0	0.0	511.0	12,265.0
USGS	5480500	12/15/1981	330	FALSE	FALSE	TRUE	FALSE	FALSE	330	85.4	89.83	311.1	7,466.9	0	0.0	89.83	0.0	0.0	311.1	7,466.9
USGS	5480500	12/16/1981	390	FALSE	FALSE	TRUE	FALSE	FALSE	390	89.4	89.83	384.9	9,236.7	0	0.0	89.83	0.0	0.0	384.9	9,236.7
USGS	5480500	12/17/1981	400	FALSE	FALSE	TRUE	FALSE	FALSE	400	89.9	89.83	397.0	9,527.6	0	0.0	89.83	0.0	0.0	397.0	9,527.6
USGS	5480500	12/18/1981	450	FALSE	FALSE	TRUE	FALSE	FALSE	450	91.8	89.83	456.0	10,944.2	0	0.0	89.83	0.0	0.0	456.0	10,944.2
USGS	5480500	12/19/1981	430	FALSE	FALSE	TRUE	FALSE	FALSE	430	91.2	89.83	432.8	10,386.6	0	0.0	89.83	0.0	0.0	432.8	10,386.6
USGS	5480500	12/20/1981	500	FALSE	FALSE	TRUE	FALSE	FALSE	500	92.6	89.83	511.0	12,265.0	0	0.0	89.83	0.0	0.0	511.0	12,265.0
USGS	5480500	12/21/1981	450	FALSE	FALSE	TRUE	FALSE	FALSE	450	91.8	89.83	456.0	10,944.2	0	0.0	89.83	0.0	0.0	456.0	10,944.2
USGS	5480500	12/22/1981	400	FALSE	FALSE	TRUE	FALSE	FALSE	400	89.9	89.83	397.0	9,527.6	0	0.0	89.83	0.0	0.0	397.0	9,527.6
USGS	5480500	12/23/1981	416	FALSE	FALSE	TRUE	FALSE	FALSE	416	90.6	89.83	416.2	9,988.7	0	0.0	89.83	0.0	0.0	416.2	9,988.7
USGS	5480500	12/24/1981	380	FALSE	FALSE	TRUE	FALSE	FALSE	380	88.8	89.83	372.7	8,944.0	0	0.0	89.83	0.0	0.0	372.7	8,944.0
USGS	5480500	12/25/1981	370	FALSE	FALSE	TRUE	FALSE	FALSE	370	88.2	89.83	360.4	8,649.9	0	0.0	89.83	0.0	0.0	360.4	8,649.9
USGS	5480500	12/26/1981	350	FALSE	FALSE	TRUE	FALSE	FALSE	350	86.9	89.83	335.8	8,059.1	0	0.0	89.83	0.0	0.0	335.8	8,059.1
USGS	5480500	12/27/1981	330	FALSE	FALSE	TRUE	FALSE	FALSE	330	85.4	89.83	311.1	7,466.9	0	0.0	89.83	0.0	0.0	311.1	7,466.9
USGS	5480500	12/28/1981	320	FALSE	FALSE	TRUE	FALSE	FALSE	320	84.6	89.83	298.8	7,171.2	0	0.0	89.83	0.0	0.0	298.8	7,171.2
USGS	5480500	12/29/1981	310	FALSE	FALSE	TRUE	FALSE	FALSE	310	83.7	89.83	286.5	6,876.2	0	0.0	89.83	0.0	0.0	286.5	6,876.2
USGS	5480500	12/30/1981	280	FALSE	FALSE	TRUE	FALSE	FALSE	280	80.9	89.83	250.0	5,999.1	0	0.0	89.83	0.0	0.0	250.0	5,999.1
USGS	5480500	12/31/1981	260	TRUE	FALSE	FALSE	FALSE	TRUE	0	0.0	89.83	0.0	0.0	0	0.0	89.83	0.0	0.0	0.0	0.0
		Total- December 1981											324,136.4					11,306.2		335,442.6

(continued)

TABLE 12.13 (*Continued*)

USGS	5480500	1/1/1982	250	TRUE	FALSE	FALSE	FALSE	FALSE	0	0.0	0.0	89.83	0.0	0.0	0.0	89.83	0.0	0.0
USGS	5480500	1/2/1982	245	TRUE	FALSE	FALSE	FALSE	FALSE	0	0.0	0.0	89.83	0.0	0.0	0.0	89.83	0.0	0.0
USGS	5480500	1/3/1982	240	TRUE	FALSE	FALSE	FALSE	FALSE	0	0.0	0.0	89.83	0.0	0.0	0.0	89.83	0.0	0.0
USGS	5480500	1/4/1982	250	TRUE	FALSE	FALSE	FALSE	FALSE	0	0.0	0.0	89.83	0.0	0.0	0.0	89.83	0.0	0.0
USGS	5480500	1/5/1982	240	TRUE	FALSE	FALSE	FALSE	FALSE	0	0.0	0.0	89.83	0.0	0.0	0.0	89.83	0.0	0.0
USGS	5480500	1/6/1982	230	TRUE	FALSE	FALSE	FALSE	FALSE	0	0.0	0.0	89.83	0.0	0.0	0.0	89.83	0.0	0.0
USGS	5480500	1/7/1982	220	TRUE	FALSE	FALSE	FALSE	FALSE	0	0.0	0.0	89.83	0.0	0.0	0.0	89.83	0.0	0.0
USGS	5480500	1/8/1982	210	TRUE	FALSE	FALSE	FALSE	FALSE	0	0.0	0.0	89.83	0.0	0.0	0.0	89.83	0.0	0.0
USGS	5480500	1/9/1982	200	TRUE	FALSE	FALSE	FALSE	FALSE	0	0.0	0.0	89.83	0.0	0.0	0.0	89.83	0.0	0.0
USGS	5480500	1/10/1982	180	TRUE	FALSE	FALSE	FALSE	FALSE	0	0.0	0.0	89.83	0.0	0.0	0.0	89.83	0.0	0.0
USGS	5480500	1/11/1982	180	TRUE	FALSE	FALSE	FALSE	FALSE	0	0.0	0.0	89.83	0.0	0.0	0.0	89.83	0.0	0.0
USGS	5480500	1/12/1982	190	TRUE	FALSE	FALSE	FALSE	FALSE	0	0.0	0.0	89.83	0.0	0.0	0.0	89.83	0.0	0.0
USGS	5480500	1/13/1982	180	TRUE	FALSE	FALSE	FALSE	FALSE	0	0.0	0.0	89.83	0.0	0.0	0.0	89.83	0.0	0.0
USGS	5480500	1/14/1982	168	TRUE	FALSE	FALSE	FALSE	FALSE	0	0.0	0.0	89.83	0.0	0.0	0.0	89.83	0.0	0.0
USGS	5480500	1/15/1982	165	TRUE	FALSE	FALSE	FALSE	FALSE	0	0.0	0.0	89.83	0.0	0.0	0.0	89.83	0.0	0.0
USGS	5480500	1/16/1982	160	TRUE	FALSE	FALSE	FALSE	FALSE	0	0.0	0.0	89.83	0.0	0.0	0.0	89.83	0.0	0.0
USGS	5480500	1/17/1982	162	TRUE	FALSE	FALSE	FALSE	FALSE	0	0.0	0.0	89.83	0.0	0.0	0.0	89.83	0.0	0.0
USGS	5480500	1/18/1982	165	TRUE	FALSE	FALSE	FALSE	FALSE	0	0.0	0.0	89.83	0.0	0.0	0.0	89.83	0.0	0.0
USGS	5480500	1/19/1982	165	TRUE	FALSE	FALSE	FALSE	FALSE	0	0.0	0.0	89.83	0.0	0.0	0.0	89.83	0.0	0.0
USGS	5480500	1/20/1982	165	TRUE	FALSE	FALSE	FALSE	FALSE	0	0.0	0.0	89.83	0.0	0.0	0.0	89.83	0.0	0.0
USGS	5480500	1/21/1982	168	TRUE	FALSE	FALSE	FALSE	FALSE	0	0.0	0.0	89.83	0.0	0.0	0.0	89.83	0.0	0.0
USGS	5480500	1/22/1982	172	TRUE	FALSE	FALSE	FALSE	FALSE	0	0.0	0.0	89.83	0.0	0.0	0.0	89.83	0.0	0.0
USGS	5480500	1/23/1982	160	TRUE	FALSE	FALSE	FALSE	FALSE	0	0.0	0.0	89.83	0.0	0.0	0.0	89.83	0.0	0.0
USGS	5480500	1/24/1982	140	TRUE	FALSE	FALSE	FALSE	FALSE	0	0.0	0.0	89.83	0.0	0.0	0.0	89.83	0.0	0.0
USGS	5480500	1/25/1982	125	TRUE	FALSE	FALSE	FALSE	FALSE	0	0.0	0.0	89.83	0.0	0.0	0.0	89.83	0.0	0.0
USGS	5480500	1/26/1982	125	TRUE	FALSE	FALSE	FALSE	FALSE	0	0.0	0.0	89.83	0.0	0.0	0.0	89.83	0.0	0.0
USGS	5480500	1/27/1982	128	TRUE	FALSE	FALSE	FALSE	FALSE	0	0.0	0.0	89.83	0.0	0.0	0.0	89.83	0.0	0.0
USGS	5480500	1/28/1982	135	TRUE	FALSE	FALSE	FALSE	FALSE	0	0.0	0.0	89.83	0.0	0.0	0.0	89.83	0.0	0.0
USGS	5480500	1/29/1982	148	TRUE	FALSE	FALSE	FALSE	FALSE	0	0.0	0.0	89.83	0.0	0.0	0.0	89.83	0.0	0.0
USGS	5480500	1/30/1982	146	TRUE	FALSE	FALSE	FALSE	FALSE	0	0.0	0.0	89.83	0.0	0.0	0.0	89.83	0.0	0.0
USGS	5480500	1/31/1982	140	TRUE	FALSE	FALSE	FALSE	FALSE	0	0.0	0.0	89.83	0.0	0.0	0.0	89.83	0.0	0.0
										Total- January 1982	0.0				0.0		0.0	0.0

Agency	Site	Date	Q	F1	F2	F3	F4	F5	c1	c2	c3	c4	c5	c6	c7	c8	c9	c10	c11	c12
USGS	5480500	2/1/1982	140	TRUE	FALSE	FALSE	FALSE	FALSE	0	0.0	89.83	0.0	0.0	0	0.0	89.83	0.0	0.0	0.0	0.0
USGS	5480500	2/2/1982	146	TRUE	FALSE	FALSE	FALSE	FALSE	0	0.0	89.83	0.0	0.0	0	0.0	89.83	0.0	0.0	0.0	0.0
USGS	5480500	2/3/1982	140	TRUE	FALSE	FALSE	FALSE	FALSE	0	0.0	89.83	0.0	0.0	0	0.0	89.83	0.0	0.0	0.0	0.0
USGS	5480500	2/4/1982	138	TRUE	FALSE	FALSE	FALSE	FALSE	0	0.0	89.83	0.0	0.0	0	0.0	89.83	0.0	0.0	0.0	0.0
USGS	5480500	2/5/1982	142	TRUE	FALSE	FALSE	FALSE	FALSE	0	0.0	89.83	0.0	0.0	0	0.0	89.83	0.0	0.0	0.0	0.0
USGS	5480500	2/6/1982	135	TRUE	FALSE	FALSE	FALSE	FALSE	0	0.0	89.83	0.0	0.0	0	0.0	89.83	0.0	0.0	0.0	0.0
USGS	5480500	2/7/1982	140	TRUE	FALSE	FALSE	FALSE	FALSE	0	0.0	89.83	0.0	0.0	0	0.0	89.83	0.0	0.0	0.0	0.0
USGS	5480500	2/8/1982	145	TRUE	FALSE	FALSE	FALSE	FALSE	0	0.0	89.83	0.0	0.0	0	0.0	89.83	0.0	0.0	0.0	0.0
USGS	5480500	2/9/1982	138	TRUE	FALSE	FALSE	FALSE	FALSE	0	0.0	89.83	0.0	0.0	0	0.0	89.83	0.0	0.0	0.0	0.0
USGS	5480500	2/10/1982	134	TRUE	FALSE	FALSE	FALSE	FALSE	0	0.0	89.83	0.0	0.0	0	0.0	89.83	0.0	0.0	0.0	0.0
USGS	5480500	2/11/1982	136	TRUE	FALSE	FALSE	FALSE	FALSE	0	0.0	89.83	0.0	0.0	0	0.0	89.83	0.0	0.0	0.0	0.0
USGS	5480500	2/12/1982	140	TRUE	FALSE	FALSE	FALSE	FALSE	0	0.0	89.83	0.0	0.0	0	0.0	89.83	0.0	0.0	0.0	0.0
USGS	5480500	2/13/1982	142	TRUE	FALSE	FALSE	FALSE	FALSE	0	0.0	89.83	0.0	0.0	0	0.0	89.83	0.0	0.0	0.0	0.0
USGS	5480500	2/14/1982	148	TRUE	FALSE	FALSE	FALSE	FALSE	0	0.0	89.83	0.0	0.0	0	0.0	89.83	0.0	0.0	0.0	0.0
USGS	5480500	2/15/1982	160	TRUE	FALSE	FALSE	FALSE	FALSE	0	0.0	89.83	0.0	0.0	0	0.0	89.83	0.0	0.0	0.0	0.0
USGS	5480500	2/16/1982	170	TRUE	FALSE	FALSE	FALSE	FALSE	0	0.0	89.83	0.0	0.0	0	0.0	89.83	0.0	0.0	0.0	0.0
USGS	5480500	2/17/1982	175	TRUE	FALSE	FALSE	FALSE	FALSE	0	0.0	89.83	0.0	0.0	0	0.0	89.83	0.0	0.0	0.0	0.0
USGS	5480500	2/18/1982	180	TRUE	FALSE	FALSE	FALSE	FALSE	0	0.0	89.83	0.0	0.0	0	0.0	89.83	0.0	0.0	0.0	0.0
USGS	5480500	2/19/1982	230	TRUE	FALSE	FALSE	FALSE	FALSE	0	0.0	89.83	0.0	0.0	0	0.0	89.83	0.0	0.0	0.0	0.0
USGS	5480500	2/20/1982	235	TRUE	TRUE	FALSE	FALSE	FALSE	0	0.0	89.83	0.0	0.0	0	0.0	89.83	0.0	0.0	0.0	0.0
USGS	5480500	2/21/1982	430	FALSE	FALSE	TRUE	FALSE	FALSE	430	91.2	89.83	432.8	10,386.6	268	79.6	89.83	235.5	5,653.1	432.8	10,386.6
USGS	5480500	2/22/1982	880	FALSE	FALSE	FALSE	FALSE	TRUE	512	90.4	89.83	610.5	14,651.6	670	87.0	89.83	643.8	15,450.2	846.0	20,304.7
USGS	5480500	2/23/1982	2250	FALSE	FALSE	FALSE	FALSE	TRUE	670	87.0	89.83	643.8	15,450.2	670	87.0	89.83	643.8	15,450.2	1,287.5	30,900.3
USGS	5480500	2/24/1982	2650	FALSE	FALSE	FALSE	FALSE	TRUE	670	87.0	89.83	643.8	15,450.2	670	87.0	89.83	643.8	15,450.2	1,287.5	30,900.3
USGS	5480500	2/25/1982	3100	FALSE	FALSE	FALSE	FALSE	TRUE	670	87.0	89.83	643.8	15,450.2	670	87.0	89.83	643.8	15,450.2	1,287.5	30,900.3
USGS	5480500	2/26/1982	3420	FALSE	FALSE	FALSE	FALSE	TRUE	670	87.0	89.83	643.8	15,450.2	670	87.0	89.83	643.8	15,450.2	1,287.5	30,900.3
USGS	5480500	2/27/1982	3600	FALSE	FALSE	FALSE	FALSE	TRUE	670	87.0	89.83	643.8	15,450.2	670	87.0	89.83	643.8	15,450.2	1,287.5	30,900.3
USGS	5480500	2/28/1982	3500	FALSE	FALSE	FALSE	FALSE	TRUE	670	87.0	89.83	643.8	15,450.2	670	87.0	89.83	643.8	15,450.2	1,287.5	30,900.3
		Total- February 1982											117,739.2					98,354.0		216,093.2

(continued)

TABLE 12.13 (*Continued*)

USGS	5480500	3/1/1982	3700	FALSE	FALSE	FALSE	TRUE	670	87.0	89.83	643.8	15,450.2	670	87.0	89.83	643.8	15,450.2	1,287.5	30,900.3
USGS	5480500	3/2/1982	3350	FALSE	FALSE	FALSE	TRUE	670	87.0	89.83	643.8	15,450.2	670	87.0	89.83	643.8	15,450.2	1,287.5	30,900.3
USGS	5480500	3/3/1982	3100	FALSE	FALSE	FALSE	TRUE	670	87.0	89.83	643.8	15,450.2	670	87.0	89.83	643.8	15,450.2	1,287.5	30,900.3
USGS	5480500	3/4/1982	2800	FALSE	FALSE	FALSE	TRUE	670	87.0	89.83	643.8	15,450.2	670	87.0	89.83	643.8	15,450.2	1,287.5	30,900.3
USGS	5480500	3/5/1982	2700	FALSE	FALSE	FALSE	TRUE	670	87.0	89.83	643.8	15,450.2	670	87.0	89.83	643.8	15,450.2	1,287.5	30,900.3
USGS	5480500	3/6/1982	2520	FALSE	FALSE	FALSE	TRUE	670	87.0	89.83	643.8	15,450.2	670	87.0	89.83	643.8	15,450.2	1,287.5	30,900.3
USGS	5480500	3/7/1982	2320	FALSE	FALSE	FALSE	TRUE	670	87.0	89.83	643.8	15,450.2	670	87.0	89.83	643.8	15,450.2	1,287.5	30,900.3
USGS	5480500	3/8/1982	2200	FALSE	FALSE	FALSE	TRUE	670	87.0	89.83	643.8	15,450.2	670	87.0	89.83	643.8	15,450.2	1,287.5	30,900.3
USGS	5480500	3/9/1982	2000	FALSE	FALSE	FALSE	TRUE	670	87.0	89.83	643.8	15,450.2	670	87.0	89.83	643.8	15,450.2	1,287.5	30,900.3
USGS	5480500	3/10/1982	2000	FALSE	FALSE	FALSE	TRUE	670	87.0	89.83	643.8	15,450.2	670	87.0	89.83	643.8	15,450.2	1,287.5	30,900.3
USGS	5480500	3/11/1982	2140	FALSE	FALSE	FALSE	TRUE	670	87.0	89.83	643.8	15,450.2	670	87.0	89.83	643.8	15,450.2	1,287.5	30,900.3
USGS	5480500	3/12/1982	2390	FALSE	FALSE	FALSE	TRUE	670	87.0	89.83	643.8	15,450.2	670	87.0	89.83	643.8	15,450.2	1,287.5	30,900.3
USGS	5480500	3/13/1982	2540	FALSE	FALSE	FALSE	TRUE	670	87.0	89.83	643.8	15,450.2	670	87.0	89.83	643.8	15,450.2	1,287.5	30,900.3
USGS	5480500	3/14/1982	2860	FALSE	FALSE	FALSE	TRUE	670	87.0	89.83	643.8	15,450.2	670	87.0	89.83	643.8	15,450.2	1,287.5	30,900.3
USGS	5480500	3/15/1982	3060	FALSE	FALSE	FALSE	TRUE	670	87.0	89.83	643.8	15,450.2	670	87.0	89.83	643.8	15,450.2	1,287.5	30,900.3
USGS	5480500	3/16/1982	3180	FALSE	FALSE	FALSE	TRUE	670	87.0	89.83	643.8	15,450.2	670	87.0	89.83	643.8	15,450.2	1,287.5	30,900.3
USGS	5480500	3/17/1982	3060	FALSE	FALSE	FALSE	TRUE	670	87.0	89.83	643.8	15,450.2	670	87.0	89.83	643.8	15,450.2	1,287.5	30,900.3
USGS	5480500	3/18/1982	3280	FALSE	FALSE	FALSE	TRUE	670	87.0	89.83	643.8	15,450.2	670	87.0	89.83	643.8	15,450.2	1,287.5	30,900.3
USGS	5480500	3/19/1982	4740	FALSE	FALSE	FALSE	TRUE	670	87.0	89.83	643.8	15,450.2	670	87.0	89.83	643.8	15,450.2	1,287.5	30,900.3
USGS	5480500	3/20/1982	6860	FALSE	FALSE	FALSE	TRUE	670	87.0	89.83	643.8	15,450.2	670	87.0	89.83	643.8	15,450.2	1,287.5	30,900.3
USGS	5480500	3/21/1982	7190	FALSE	FALSE	FALSE	TRUE	670	87.0	89.83	643.8	15,450.2	670	87.0	89.83	643.8	15,450.2	1,287.5	30,900.3
USGS	5480500	3/22/1982	6790	FALSE	FALSE	FALSE	TRUE	670	87.0	89.83	643.8	15,450.2	670	87.0	89.83	643.8	15,450.2	1,287.5	30,900.3
USGS	5480500	3/23/1982	6040	FALSE	FALSE	FALSE	TRUE	670	87.0	89.83	643.8	15,450.2	670	87.0	89.83	643.8	15,450.2	1,287.5	30,900.3
USGS	5480500	3/24/1982	5420	FALSE	FALSE	FALSE	TRUE	670	87.0	89.83	643.8	15,450.2	670	87.0	89.83	643.8	15,450.2	1,287.5	30,900.3
USGS	5480500	3/25/1982	4960	FALSE	FALSE	FALSE	TRUE	670	87.0	89.83	643.8	15,450.2	670	87.0	89.83	643.8	15,450.2	1,287.5	30,900.3
USGS	5480500	3/26/1982	4470	FALSE	FALSE	FALSE	TRUE	670	87.0	89.83	643.8	15,450.2	670	87.0	89.83	643.8	15,450.2	1,287.5	30,900.3
USGS	5480500	3/27/1982	4090	FALSE	FALSE	FALSE	TRUE	670	87.0	89.83	643.8	15,450.2	670	87.0	89.83	643.8	15,450.2	1,287.5	30,900.3
USGS	5480500	3/28/1982	3660	FALSE	FALSE	FALSE	TRUE	670	87.0	89.83	643.8	15,450.2	670	87.0	89.83	643.8	15,450.2	1,287.5	30,900.3
USGS	5480500	3/29/1982	3390	FALSE	FALSE	FALSE	TRUE	670	87.0	89.83	643.8	15,450.2	670	87.0	89.83	643.8	15,450.2	1,287.5	30,900.3
USGS	5480500	3/30/1982	3150	FALSE	FALSE	FALSE	TRUE	670	87.0	89.83	643.8	15,450.2	670	87.0	89.83	643.8	15,450.2	1,287.5	30,900.3
USGS	5480500	3/31/1982	2960	FALSE	FALSE	FALSE	TRUE	670	87.0	89.83	643.8	15,450.2	670	87.0	89.83	643.8	15,450.2	1,287.5	30,900.3
												Total- March 1982 478,954.9					478,954.9		957,909.8

USGS	5480500	4/1/1982	2800	FALSE	FALSE	FALSE	FALSE	TRUE	670	87.0	89.83	643.8	15,450.2	670	87.0	89.83	643.8	15,450.2	1,287.5	30,900.3
USGS	5480500	4/2/1982	2700	FALSE	FALSE	FALSE	FALSE	TRUE	670	87.0	89.83	643.8	15,450.2	670	87.0	89.83	643.8	15,450.2	1,287.5	30,900.3
USGS	5480500	4/3/1982	2830	FALSE	FALSE	FALSE	FALSE	TRUE	670	87.0	89.83	643.8	15,450.2	670	87.0	89.83	643.8	15,450.2	1,287.5	30,900.3
USGS	5480500	4/4/1982	2640	FALSE	FALSE	FALSE	FALSE	TRUE	670	87.0	89.83	643.8	15,450.2	670	87.0	89.83	643.8	15,450.2	1,287.5	30,900.3
USGS	5480500	4/5/1982	3080	FALSE	FALSE	FALSE	FALSE	TRUE	670	87.0	89.83	643.8	15,450.2	670	87.0	89.83	643.8	15,450.2	1,287.5	30,900.3
USGS	5480500	4/6/1982	3090	FALSE	FALSE	FALSE	FALSE	TRUE	670	87.0	89.83	643.8	15,450.2	670	87.0	89.83	643.8	15,450.2	1,287.5	30,900.3
USGS	5480500	4/7/1982	2940	FALSE	FALSE	FALSE	FALSE	TRUE	670	87.0	89.83	643.8	15,450.2	670	87.0	89.83	643.8	15,450.2	1,287.5	30,900.3
USGS	5480500	4/8/1982	2820	FALSE	FALSE	FALSE	FALSE	TRUE	670	87.0	89.83	643.8	15,450.2	670	87.0	89.83	643.8	15,450.2	1,287.5	30,900.3
USGS	5480500	4/9/1982	2820	FALSE	FALSE	FALSE	FALSE	TRUE	670	87.0	89.83	643.8	15,450.2	670	87.0	89.83	643.8	15,450.2	1,287.5	30,900.3
USGS	5480500	4/10/1982	2810	FALSE	FALSE	FALSE	FALSE	TRUE	670	87.0	89.83	643.8	15,450.2	670	87.0	89.83	643.8	15,450.2	1,287.5	30,900.3
USGS	5480500	4/11/1982	2950	FALSE	FALSE	FALSE	FALSE	TRUE	670	87.0	89.83	643.8	15,450.2	670	87.0	89.83	643.8	15,450.2	1,287.5	30,900.3
USGS	5480500	4/12/1982	3250	FALSE	FALSE	FALSE	FALSE	TRUE	670	87.0	89.83	643.8	15,450.2	670	87.0	89.83	643.8	15,450.2	1,287.5	30,900.3
USGS	5480500	4/13/1982	3360	FALSE	FALSE	FALSE	FALSE	TRUE	670	87.0	89.83	643.8	15,450.2	670	87.0	89.83	643.8	15,450.2	1,287.5	30,900.3
USGS	5480500	4/14/1982	3220	FALSE	FALSE	FALSE	FALSE	TRUE	670	87.0	89.83	643.8	15,450.2	670	87.0	89.83	643.8	15,450.2	1,287.5	30,900.3
USGS	5480500	4/15/1982	3280	FALSE	FALSE	FALSE	FALSE	TRUE	670	87.0	89.83	643.8	15,450.2	670	87.0	89.83	643.8	15,450.2	1,287.5	30,900.3
USGS	5480500	4/16/1982	4820	FALSE	FALSE	FALSE	FALSE	TRUE	670	87.0	89.83	643.8	15,450.2	670	87.0	89.83	643.8	15,450.2	1,287.5	30,900.3
USGS	5480500	4/17/1982	4710	FALSE	FALSE	FALSE	FALSE	TRUE	670	87.0	89.83	643.8	15,450.2	670	87.0	89.83	643.8	15,450.2	1,287.5	30,900.3
USGS	5480500	4/18/1982	4550	FALSE	FALSE	FALSE	FALSE	TRUE	670	87.0	89.83	643.8	15,450.2	670	87.0	89.83	643.8	15,450.2	1,287.5	30,900.3
USGS	5480500	4/19/1982	4410	FALSE	FALSE	FALSE	FALSE	TRUE	670	87.0	89.83	643.8	15,450.2	670	87.0	89.83	643.8	15,450.2	1,287.5	30,900.3
USGS	5480500	4/20/1982	4270	FALSE	FALSE	FALSE	FALSE	TRUE	670	87.0	89.83	643.8	15,450.2	670	87.0	89.83	643.8	15,450.2	1,287.5	30,900.3
USGS	5480500	4/21/1982	4030	FALSE	FALSE	FALSE	FALSE	TRUE	670	87.0	89.83	643.8	15,450.2	670	87.0	89.83	643.8	15,450.2	1,287.5	30,900.3
USGS	5480500	4/22/1982	3800	FALSE	FALSE	FALSE	FALSE	TRUE	670	87.0	89.83	643.8	15,450.2	670	87.0	89.83	643.8	15,450.2	1,287.5	30,900.3
USGS	5480500	4/23/1982	3590	FALSE	FALSE	FALSE	FALSE	TRUE	670	87.0	89.83	643.8	15,450.2	670	87.0	89.83	643.8	15,450.2	1,287.5	30,900.3
USGS	5480500	4/24/1982	3420	FALSE	FALSE	FALSE	FALSE	TRUE	670	87.0	89.83	643.8	15,450.2	670	87.0	89.83	643.8	15,450.2	1,287.5	30,900.3
USGS	5480500	4/25/1982	3170	FALSE	FALSE	FALSE	FALSE	TRUE	670	87.0	89.83	643.8	15,450.2	670	87.0	89.83	643.8	15,450.2	1,287.5	30,900.3
USGS	5480500	4/26/1982	2900	FALSE	FALSE	FALSE	FALSE	TRUE	670	87.0	89.83	643.8	15,450.2	670	87.0	89.83	643.8	15,450.2	1,287.5	30,900.3
USGS	5480500	4/27/1982	2850	FALSE	FALSE	FALSE	FALSE	TRUE	670	87.0	89.83	643.8	15,450.2	670	87.0	89.83	643.8	15,450.2	1,287.5	30,900.3
USGS	5480500	4/28/1982	2450	FALSE	FALSE	FALSE	FALSE	TRUE	670	87.0	89.83	643.8	15,450.2	670	87.0	89.83	643.8	15,450.2	1,287.5	30,900.3
USGS	5480500	4/29/1982	2290	FALSE	FALSE	FALSE	FALSE	TRUE	670	87.0	89.83	643.8	15,450.2	670	87.0	89.83	643.8	15,450.2	1,287.5	30,900.3
USGS	5480500	4/30/1982	2160	FALSE	FALSE	FALSE	FALSE	TRUE	670	87.0	89.83	643.8	15,450.2	670	87.0	89.83	643.8	15,450.2	1,287.5	30,900.3
													Total- April 1982 463,504.8					463,504.8		927,009.5

(continued)

TABLE 12.13 (Continued)

USGS	5480500	5/1/1982	1970	FALSE	FALSE	FALSE	TRUE	670	87.0	89.83	643.8	15,450.2	670	87.0	89.83	643.8	15,450.2	1,287.5	30,900.3
USGS	5480500	5/2/1982	1870	FALSE	FALSE	FALSE	TRUE	670	87.0	89.83	643.8	15,450.2	670	87.0	89.83	643.8	15,450.2	1,287.5	30,900.3
USGS	5480500	5/3/1982	1810	FALSE	FALSE	FALSE	TRUE	670	87.0	89.83	643.8	15,450.2	670	87.0	89.83	643.8	15,450.2	1,287.5	30,900.3
USGS	5480500	5/4/1982	1760	FALSE	FALSE	FALSE	TRUE	670	87.0	89.83	643.8	15,450.2	670	87.0	89.83	643.8	15,450.2	1,287.5	30,900.3
USGS	5480500	5/5/1982	2030	FALSE	FALSE	FALSE	TRUE	670	87.0	89.83	643.8	15,450.2	670	87.0	89.83	643.8	15,450.2	1,287.5	30,900.3
USGS	5480500	5/6/1982	2710	FALSE	FALSE	FALSE	TRUE	670	87.0	89.83	643.8	15,450.2	670	87.0	89.83	643.8	15,450.2	1,287.5	30,900.3
USGS	5480500	5/7/1982	3180	FALSE	FALSE	FALSE	TRUE	670	87.0	89.83	643.8	15,450.2	670	87.0	89.83	643.8	15,450.2	1,287.5	30,900.3
USGS	5480500	5/8/1982	3190	FALSE	FALSE	FALSE	TRUE	670	87.0	89.83	643.8	15,450.2	670	87.0	89.83	643.8	15,450.2	1,287.5	30,900.3
USGS	5480500	5/9/1982	3070	FALSE	FALSE	FALSE	TRUE	670	87.0	89.83	643.8	15,450.2	670	87.0	89.83	643.8	15,450.2	1,287.5	30,900.3
USGS	5480500	5/10/1982	2880	FALSE	FALSE	FALSE	TRUE	670	87.0	89.83	643.8	15,450.2	670	87.0	89.83	643.8	15,450.2	1,287.5	30,900.3
USGS	5480500	5/11/1982	2720	FALSE	FALSE	FALSE	TRUE	670	87.0	89.83	643.8	15,450.2	670	87.0	89.83	643.8	15,450.2	1,287.5	30,900.3
USGS	5480500	5/12/1982	2970	FALSE	FALSE	FALSE	TRUE	670	87.0	89.83	643.8	15,450.2	670	87.0	89.83	643.8	15,450.2	1,287.5	30,900.3
USGS	5480500	5/13/1982	3680	FALSE	FALSE	FALSE	TRUE	670	87.0	89.83	643.8	15,450.2	670	87.0	89.83	643.8	15,450.2	1,287.5	30,900.3
USGS	5480500	5/14/1982	4370	FALSE	FALSE	FALSE	TRUE	670	87.0	89.83	643.8	15,450.2	670	87.0	89.83	643.8	15,450.2	1,287.5	30,900.3
USGS	5480500	5/15/1982	5020	FALSE	FALSE	FALSE	TRUE	670	87.0	89.83	643.8	15,450.2	670	87.0	89.83	643.8	15,450.2	1,287.5	30,900.3
USGS	5480500	5/16/1982	5930	FALSE	FALSE	FALSE	TRUE	670	87.0	89.83	643.8	15,450.2	670	87.0	89.83	643.8	15,450.2	1,287.5	30,900.3
USGS	5480500	5/17/1982	6000	FALSE	FALSE	FALSE	TRUE	670	87.0	89.83	643.8	15,450.2	670	87.0	89.83	643.8	15,450.2	1,287.5	30,900.3
USGS	5480500	5/18/1982	6490	FALSE	FALSE	FALSE	TRUE	670	87.0	89.83	643.8	15,450.2	670	87.0	89.83	643.8	15,450.2	1,287.5	30,900.3
USGS	5480500	5/19/1982	6190	FALSE	FALSE	FALSE	TRUE	670	87.0	89.83	643.8	15,450.2	670	87.0	89.83	643.8	15,450.2	1,287.5	30,900.3
USGS	5480500	5/20/1982	6360	FALSE	FALSE	FALSE	TRUE	670	87.0	89.83	643.8	15,450.2	670	87.0	89.83	643.8	15,450.2	1,287.5	30,900.3
USGS	5480500	5/21/1982	6510	FALSE	FALSE	FALSE	TRUE	670	87.0	89.83	643.8	15,450.2	670	87.0	89.83	643.8	15,450.2	1,287.5	30,900.3
USGS	5480500	5/22/1982	6650	FALSE	FALSE	FALSE	TRUE	670	87.0	89.83	643.8	15,450.2	670	87.0	89.83	643.8	15,450.2	1,287.5	30,900.3
USGS	5480500	5/23/1982	6290	FALSE	FALSE	FALSE	TRUE	670	87.0	89.83	643.8	15,450.2	673	87.0	89.83	643.8	15,450.2	1,287.5	30,900.3
USGS	5480500	5/24/1982	5870	FALSE	FALSE	FALSE	TRUE	670	87.0	89.83	643.8	15,450.2	670	87.0	89.83	643.8	15,450.2	1,287.5	30,900.3
USGS	5480500	5/25/1982	5430	FALSE	FALSE	FALSE	TRUE	670	87.0	89.83	643.8	15,450.2	670	87.0	89.83	643.8	15,450.2	1,287.5	30,900.3
USGS	5480500	5/26/1982	7490	FALSE	FALSE	FALSE	TRUE	670	87.0	89.83	643.8	15,450.2	670	87.0	89.83	643.8	15,450.2	1,287.5	30,900.3
USGS	5480500	5/27/1982	8950	FALSE	FALSE	FALSE	TRUE	670	87.0	89.83	643.8	15,450.2	670	87.0	89.83	643.8	15,450.2	1,287.5	30,900.3
USGS	5480500	5/28/1982	8790	FALSE	FALSE	FALSE	TRUE	670	87.0	89.83	643.8	15,450.2	670	87.0	89.83	643.8	15,450.2	1,287.5	30,900.3
USGS	5480500	5/29/1982	8200	FALSE	FALSE	FALSE	TRUE	670	87.0	89.83	643.8	15,450.2	670	87.0	89.83	643.8	15,450.2	1,287.5	30,900.3
USGS	5480500	5/30/1982	7290	FALSE	FALSE	FALSE	TRUE	670	87.0	89.83	643.8	15,450.2	670	87.0	89.83	643.8	15,450.2	1,287.5	30,900.3
USGS	5480500	5/31/1982	6560	FALSE	FALSE	FALSE	TRUE	670	87.0	89.83	643.8	15,450.2	670	87.0	89.83	643.8	15,450.2	1,287.5	30,900.3
							Total- May 1982					478,954.9					478,954.9		957,906.8

USGS	5480500	6/1/1982	6040	FALSE	FALSE	FALSE	FALSE	TRUE	670	87.0	89.83	643.8	15,450.2	670	87.0	89.83	643.8	15,450.2	1,287.5	30,900.3
USGS	5480500	6/2/1982	5580	FALSE	FALSE	FALSE	FALSE	TRUE	670	87.0	89.83	643.8	15,450.2	670	87.0	89.83	643.8	15,450.2	1,287.5	30,900.3
USGS	5480500	6/3/1982	5100	FALSE	FALSE	FALSE	FALSE	TRUE	670	87.0	89.83	643.8	15,450.2	670	87.0	89.83	643.8	15,450.2	1,287.5	30,900.3
USGS	5480500	6/4/1982	4680	FALSE	FALSE	FALSE	FALSE	TRUE	670	87.0	89.83	643.8	15,450.2	670	87.0	89.83	643.8	15,450.2	1,287.5	30,900.3
USGS	5480500	6/5/1982	4300	FALSE	FALSE	FALSE	FALSE	TRUE	670	87.0	89.83	643.8	15,450.2	670	87.0	89.83	643.8	15,450.2	1,287.5	30,900.3
USGS	5480500	6/6/1982	4180	FALSE	FALSE	FALSE	FALSE	TRUE	670	87.0	89.83	643.8	15,450.2	670	87.0	89.83	643.8	15,450.2	1,287.5	30,900.3
USGS	5480500	6/7/1982	4210	FALSE	FALSE	FALSE	FALSE	TRUE	670	87.0	89.83	643.8	15,450.2	670	87.0	89.83	643.8	15,450.2	1,287.5	30,900.3
USGS	5480500	6/8/1982	5250	FALSE	FALSE	FALSE	FALSE	TRUE	670	87.0	89.83	643.8	15,450.2	670	87.0	89.83	643.8	15,450.2	1,287.5	30,900.3
USGS	5480500	6/9/1982	6230	FALSE	FALSE	FALSE	FALSE	TRUE	670	87.0	89.83	643.8	15,450.2	670	87.0	89.83	643.8	15,450.2	1,287.5	30,900.3
USGS	5480500	6/10/1982	6160	FALSE	FALSE	FALSE	FALSE	TRUE	670	87.0	89.83	643.8	15,450.2	670	87.0	89.83	643.8	15,450.2	1,287.5	30,900.3
USGS	5480500	6/11/1982	5800	FALSE	FALSE	FALSE	FALSE	TRUE	670	87.0	89.83	643.8	15,450.2	670	87.0	89.83	643.8	15,450.2	1,287.5	30,900.3
USGS	5480500	6/12/1982	5450	FALSE	FALSE	FALSE	FALSE	TRUE	670	87.0	89.83	643.8	15,450.2	670	87.0	89.83	643.8	15,450.2	1,287.5	30,900.3
USGS	5480500	6/13/1982	5040	FALSE	FALSE	FALSE	FALSE	TRUE	670	87.0	89.83	643.8	15,450.2	670	87.0	89.83	643.8	15,450.2	1,287.5	30,900.3
USGS	5480500	6/14/1982	4610	FALSE	FALSE	FALSE	FALSE	TRUE	670	87.0	89.83	643.8	15,450.2	670	87.0	89.83	643.8	15,450.2	1,287.5	30,900.3
USGS	5480500	6/15/1982	4330	FALSE	FALSE	FALSE	FALSE	TRUE	670	87.0	89.83	643.8	15,450.2	670	87.0	89.83	643.8	15,450.2	1,287.5	30,900.3
USGS	5480500	6/16/1982	3890	FALSE	FALSE	FALSE	FALSE	TRUE	670	87.0	89.83	643.8	15,450.2	670	87.0	89.83	643.8	15,450.2	1,287.5	30,900.3
USGS	5480500	6/17/1982	3680	FALSE	FALSE	FALSE	FALSE	TRUE	670	87.0	89.83	643.8	15,450.2	670	87.0	89.83	643.8	15,450.2	1,287.5	30,900.3
USGS	5480500	6/18/1982	3480	FALSE	FALSE	FALSE	FALSE	TRUE	670	87.0	89.83	643.8	15,450.2	670	87.0	89.83	643.8	15,450.2	1,287.5	30,900.3
USGS	5480500	6/19/1982	3410	FALSE	FALSE	FALSE	FALSE	TRUE	670	87.0	89.83	643.8	15,450.2	670	87.0	89.83	643.8	15,450.2	1,287.5	30,900.3
USGS	5480500	6/20/1982	3400	FALSE	FALSE	FALSE	FALSE	TRUE	670	87.0	89.83	643.8	15,450.2	670	87.0	89.83	643.8	15,450.2	1,287.5	30,900.3
USGS	5480500	6/21/1982	3330	FALSE	FALSE	FALSE	FALSE	TRUE	670	87.0	89.83	643.8	15,450.2	670	87.0	89.83	643.8	15,450.2	1,287.5	30,900.3
USGS	5480500	6/22/1982	3180	FALSE	FALSE	FALSE	FALSE	TRUE	670	87.0	89.83	643.8	15,450.2	670	87.0	89.83	643.8	15,450.2	1,287.5	30,900.3
USGS	5480500	6/23/1982	3000	FALSE	FALSE	FALSE	FALSE	TRUE	670	87.0	89.83	643.8	15,450.2	670	87.0	89.83	643.8	15,450.2	1,287.5	30,900.3
USGS	5480500	6/24/1982	2800	FALSE	FALSE	FALSE	FALSE	TRUE	670	87.0	89.83	643.8	15,450.2	670	87.0	89.83	643.8	15,450.2	1,287.5	30,900.3
USGS	5480500	6/25/1982	2690	FALSE	FALSE	FALSE	FALSE	TRUE	670	87.0	89.83	643.8	15,450.2	670	87.0	89.83	643.8	15,450.2	1,287.5	30,900.3
USGS	5480500	6/26/1982	2760	FALSE	FALSE	FALSE	FALSE	TRUE	670	87.0	89.83	643.8	15,450.2	670	87.0	89.83	643.8	15,450.2	1,287.5	30,900.3
USGS	5480500	6/27/1982	2640	FALSE	FALSE	FALSE	FALSE	TRUE	670	87.0	89.83	643.8	15,450.2	670	87.0	89.83	643.8	15,450.2	1,287.5	30,900.3
USGS	5480500	6/28/1982	2480	FALSE	FALSE	FALSE	FALSE	TRUE	670	87.0	89.83	643.8	15,450.2	670	87.0	89.83	643.8	15,450.2	1,287.5	30,900.3
USGS	5480500	6/29/1982	2290	FALSE	FALSE	FALSE	FALSE	TRUE	670	87.0	89.83	643.8	15,450.2	670	87.0	89.83	643.8	15,450.2	1,287.5	30,900.3
USGS	5480500	6/30/1982	2160	FALSE	FALSE	FALSE	FALSE	TRUE	670	87.0	89.83	643.8	15,450.2	670	87.0	89.83	643.8	15,450.2	1,287.5	30,900.3
								Total-June 1992					463,504.8					463,504.8		927,009.5

(continued)

TABLE 12.13 (*Continued*)

USGS	5480500	7/1/1982	2000	FALSE	FALSE	FALSE	TRUE	670	87.0	89.83	643.8	15,450.2	670	87.0	89.83	643.8	15,450.2	1,287.5	30,900.3
USGS	5480500	7/2/1982	1860	FALSE	FALSE	FALSE	TRUE	670	87.0	89.83	643.8	15,450.2	670	87.0	89.83	643.8	15,450.2	1,287.5	30,900.3
USGS	5480500	7/3/1982	1760	FALSE	FALSE	FALSE	TRUE	670	87.0	89.83	643.8	15,450.2	670	87.0	89.83	643.8	15,450.2	1,287.5	30,900.3
USGS	5480500	7/4/1982	1640	FALSE	FALSE	FALSE	TRUE	670	87.0	89.83	643.8	15,450.2	670	87.0	89.83	643.8	15,450.2	1,287.5	30,900.3
USGS	5480500	7/5/1982	1510	FALSE	FALSE	FALSE	TRUE	670	87.0	89.83	643.8	15,450.2	670	87.0	89.83	643.8	15,450.2	1,287.5	30,900.3
USGS	5480500	7/6/1982	1680	FALSE	FALSE	FALSE	TRUE	670	87.0	89.83	643.8	15,450.2	670	87.0	89.83	643.8	15,450.2	1,287.5	30,900.3
USGS	5480500	7/7/1982	3780	FALSE	FALSE	FALSE	TRUE	670	87.0	89.83	643.8	15,450.2	670	87.0	89.83	643.8	15,450.2	1,287.5	30,900.3
USGS	5480500	7/8/1982	4430	FALSE	FALSE	FALSE	TRUE	670	87.0	89.83	643.8	15,450.2	670	87.0	89.83	643.8	15,450.2	1,287.5	30,900.3
USGS	5480500	7/9/1982	5420	FALSE	FALSE	FALSE	TRUE	670	87.0	89.83	643.8	15,450.2	670	87.0	89.83	643.8	15,450.2	1,287.5	30,900.3
USGS	5480500	7/10/1982	8370	FALSE	FALSE	FALSE	TRUE	670	87.0	89.83	643.8	15,450.2	670	87.0	89.83	643.8	15,450.2	1,287.5	30,900.3
USGS	5480500	7/11/1982	9310	FALSE	FALSE	FALSE	TRUE	670	87.0	89.83	643.8	15,450.2	670	87.0	89.83	643.8	15,450.2	1,287.5	30,900.3
USGS	5480500	7/12/1982	9110	FALSE	FALSE	FALSE	TRUE	670	87.0	89.83	643.8	15,450.2	670	87.0	89.83	643.8	15,450.2	1,287.5	30,900.3
USGS	5480500	7/13/1982	7760	FALSE	FALSE	FALSE	TRUE	670	87.0	89.83	643.8	15,450.2	670	87.0	89.83	643.8	15,450.2	1,287.5	30,900.3
USGS	5480500	7/14/1982	6410	FALSE	FALSE	FALSE	TRUE	670	87.0	89.83	643.8	15,450.2	670	87.0	89.83	643.8	15,450.2	1,287.5	30,900.3
USGS	5480500	7/15/1982	5470	FALSE	FALSE	FALSE	TRUE	670	87.0	89.83	643.8	15,450.2	670	87.0	89.83	643.8	15,450.2	1,287.5	30,900.3
USGS	5480500	7/16/1982	4760	FALSE	FALSE	FALSE	TRUE	670	87.0	89.83	643.8	15,450.2	670	87.0	89.83	643.8	15,450.2	1,287.5	30,900.3
USGS	5480500	7/17/1982	4230	FALSE	FALSE	FALSE	TRUE	670	87.0	89.83	643.8	15,450.2	670	87.0	89.83	643.8	15,450.2	1,287.5	30,900.3
USGS	5480500	7/18/1982	4850	FALSE	FALSE	FALSE	TRUE	670	87.0	89.83	643.8	15,450.2	670	87.0	89.83	643.8	15,450.2	1,287.5	30,900.3
USGS	5480500	7/19/1982	4220	FALSE	FALSE	FALSE	TRUE	670	87.0	89.83	643.8	15,450.2	670	87.0	89.83	643.8	15,450.2	1,287.5	30,900.3
USGS	5480500	7/20/1982	3680	FALSE	FALSE	FALSE	TRUE	670	87.0	89.83	643.8	15,450.2	670	87.0	89.83	643.8	15,450.2	1,287.5	30,900.3
USGS	5480500	7/21/1982	3460	FALSE	FALSE	FALSE	TRUE	670	87.0	89.83	643.8	15,450.2	670	87.0	89.83	643.8	15,450.2	1,287.5	30,900.3
USGS	5480500	7/22/1982	3200	FALSE	FALSE	FALSE	TRUE	670	87.0	89.83	643.8	15,450.2	670	87.0	89.83	643.8	15,450.2	1,287.5	30,900.3
USGS	5480500	7/23/1982	2840	FALSE	FALSE	FALSE	TRUE	670	87.0	89.83	643.8	15,450.2	670	87.0	89.83	643.8	15,450.2	1,287.5	30,900.3
USGS	5480500	7/24/1982	2560	FALSE	FALSE	FALSE	TRUE	670	87.0	89.83	643.8	15,450.2	670	87.0	89.83	643.8	15,450.2	1,287.5	30,900.3
USGS	5480500	7/25/1982	2280	FALSE	FALSE	FALSE	TRUE	670	87.0	89.83	643.8	15,450.2	670	87.0	89.83	643.8	15,450.2	1,287.5	30,900.3
USGS	5480500	7/26/1982	2000	FALSE	FALSE	FALSE	TRUE	670	87.0	89.83	643.8	15,450.2	670	87.0	89.83	643.8	15,450.2	1,287.5	30,900.3
USGS	5480500	7/27/1982	1820	FALSE	FALSE	FALSE	TRUE	670	87.0	89.83	643.8	15,450.2	670	87.0	89.83	643.8	15,450.2	1,287.5	30,900.3
USGS	5480500	7/28/1982	1800	FALSE	FALSE	FALSE	TRUE	670	87.0	89.83	643.8	15,450.2	670	87.0	89.83	643.8	15,450.2	1,287.5	30,900.3
USGS	5480500	7/29/1982	1850	FALSE	FALSE	FALSE	TRUE	670	87.0	89.83	643.8	15,450.2	670	87.0	89.83	643.8	15,450.2	1,287.5	30,900.3
USGS	5480500	7/30/1982	1780	FALSE	FALSE	FALSE	TRUE	670	87.0	89.83	643.8	15,450.2	670	87.0	89.83	643.8	15,450.2	1,287.5	30,900.3
USGS	5480500	7/31/1982	1580	FALSE	FALSE	FALSE	TRUE	670	87.0	89.83	643.8	15,450.2	670	87.0	89.83	643.8	15,450.2	1,287.5	30,900.3
							Total- July 1982					478,954.9					478,954.9		957,909.8

USGS	5480500	8/1/1982	1390	FALSE	FALSE	FALSE	FALSE	TRUE	670	87.0	89.83	643.8	15,450.2	670	87.0	89.83	643.8	15,450.2	1,287.5	30,900.3
USGS	5480500	8/2/1982	1260	FALSE	FALSE	FALSE	TRUE	FALSE	590	91.2	89.83	594.2	14,261.4	670	87.0	89.83	643.8	15,450.2	1,238.0	29,711.6
USGS	5480500	8/3/1982	1100	FALSE	FALSE	FALSE	TRUE	FALSE	430	91.2	89.83	432.8	10,386.6	670	87.0	89.83	643.8	15,450.2	1,076.5	25,836.8
USGS	5480500	8/4/1982	1090	FALSE	FALSE	FALSE	TRUE	FALSE	420	90.8	89.83	421.0	10,102.9	670	87.0	89.83	643.8	15,450.2	1,064.7	25,553.1
USGS	5480500	8/5/1982	1060	FALSE	FALSE	FALSE	TRUE	FALSE	390	89.4	89.83	384.9	9,236.7	670	87.0	89.83	643.8	15,450.2	1,028.6	24,686.8
USGS	5480500	8/6/1982	1080	FALSE	FALSE	FALSE	TRUE	FALSE	410	90.4	89.83	409.0	9,816.5	670	87.0	89.83	643.8	15,450.2	1,052.8	25,266.7
USGS	5480500	8/7/1982	1080	FALSE	FALSE	FALSE	TRUE	FALSE	410	90.4	89.83	409.0	9,816.5	670	87.0	89.83	643.8	15,450.2	1,052.8	25,266.7
USGS	5480500	8/8/1982	1100	FALSE	FALSE	TRUE	FALSE	FALSE	430	91.2	89.83	432.8	10,386.6	670	87.0	89.83	643.8	15,450.2	1,076.5	25,836.8
USGS	5480500	8/9/1982	985	FALSE	FALSE	TRUE	FALSE	FALSE	315	84.2	89.83	292.7	7,023.6	670	87.0	89.83	643.8	15,450.2	936.4	22,473.8
USGS	5480500	8/10/1982	897	FALSE	FALSE	TRUE	FALSE	FALSE	629	89.5	89.83	621.7	14,921.8	268	79.6	89.83	235.5	5,653.1	857.3	20,574.9
USGS	5480500	8/11/1982	817	FALSE	TRUE	FALSE	FALSE	FALSE	549	92.3	89.83	559.3	13,424.3	268	79.6	89.83	235.5	5,653.1	794.9	19,077.4
USGS	5480500	8/12/1982	751	FALSE	TRUE	FALSE	FALSE	FALSE	483	92.4	89.83	492.9	11,829.4	268	79.6	89.83	235.5	5,653.1	728.4	17,482.5
USGS	5480500	8/13/1982	754	FALSE	TRUE	FALSE	FALSE	FALSE	486	92.5	89.83	496.1	11,907.3	268	79.6	89.83	235.5	5,653.1	731.7	17,560.4
USGS	5480500	8/14/1982	694	FALSE	TRUE	FALSE	FALSE	FALSE	426	91.0	89.83	428.1	10,273.5	268	79.6	89.83	235.5	5,653.1	663.6	15,926.6
USGS	5480500	8/15/1982	670	FALSE	TRUE	FALSE	FALSE	FALSE	670	87.0	89.83	643.8	15,450.2	0	0.0	89.83	0.0	0.0	643.8	15,450.2
USGS	5480500	8/16/1982	620	FALSE	TRUE	FALSE	FALSE	FALSE	620	90.0	89.83	615.9	14,782.3	0	0.0	89.83	0.0	0.0	615.9	14,782.3
USGS	5480500	8/17/1982	586	FALSE	TRUE	FALSE	FALSE	FALSE	586	91.4	89.83	591.1	14,185.8	0	0.0	89.83	0.0	0.0	591.1	14,185.8
USGS	5480500	8/18/1982	554	FALSE	TRUE	FALSE	FALSE	FALSE	554	92.2	89.83	563.9	13,533.4	0	0.0	89.83	0.0	0.0	563.9	13,533.4
USGS	5480500	8/19/1982	520	FALSE	TRUE	FALSE	FALSE	FALSE	520	92.6	89.83	531.5	12,756.7	0	0.0	89.83	0.0	0.0	531.5	12,756.7
USGS	5480500	8/20/1982	495	FALSE	TRUE	FALSE	FALSE	FALSE	495	92.6	89.83	505.8	12,138.4	0	0.0	89.83	0.0	0.0	505.8	12,138.4
USGS	5480500	8/21/1982	462	FALSE	TRUE	FALSE	FALSE	FALSE	462	92.1	89.83	469.6	11,271.5	0	0.0	89.83	0.0	0.0	469.6	11,271.5
USGS	5480500	8/22/1982	448	FALSE	TRUE	FALSE	FALSE	FALSE	448	91.7	89.83	453.7	10,889.1	0	0.0	89.83	0.0	0.0	453.7	10,889.1
USGS	5480500	8/23/1982	427	FALSE	TRUE	FALSE	FALSE	FALSE	427	91.1	89.83	429.2	10,301.8	0	0.0	89.83	0.0	0.0	429.2	10,301.8
USGS	5480500	8/24/1982	421	FALSE	TRUE	FALSE	FALSE	FALSE	421	90.8	89.83	422.1	10,131.4	0	0.0	89.83	0.0	0.0	422.1	10,131.4
USGS	5480500	8/25/1982	403	FALSE	TRUE	FALSE	FALSE	FALSE	403	90.1	89.83	400.6	9,614.5	0	0.0	89.83	0.0	0.0	400.6	9,614.5
USGS	5480500	8/26/1982	405	FALSE	TRUE	FALSE	FALSE	FALSE	405	90.1	89.83	403.0	9,672.3	0	0.0	89.83	0.0	0.0	403.0	9,672.3
USGS	5480500	8/27/1982	409	FALSE	TRUE	FALSE	FALSE	FALSE	409	90.3	89.83	407.8	9,787.7	0	0.0	89.83	0.0	0.0	407.8	9,787.7
USGS	5480500	8/28/1982	389	FALSE	TRUE	FALSE	FALSE	FALSE	389	89.3	89.83	383.6	9,207.5	0	0.0	89.83	0.0	0.0	383.6	9,207.5
USGS	5480500	8/29/1982	423	FALSE	TRUE	FALSE	FALSE	FALSE	423	90.9	89.83	424.5	10,188.3	0	0.0	89.83	0.0	0.0	424.5	10,188.3
USGS	5480500	8/30/1982	478	FALSE	TRUE	FALSE	FALSE	FALSE	478	92.4	89.83	487.4	11,698.4	0	0.0	89.83	0.0	0.0	487.4	11,698.4
USGS	5480500	8/31/1982	601	FALSE	TRUE	FALSE	FALSE	FALSE	601	90.8	89.83	602.6	14,462.0	0	0.0	89.83	0.0	0.0	602.6	14,462.0
													Total- August 1982 358,908.7					**167,316.9**		**526,225.6**

(continued)

340

TABLE 12.13 (Continued)

		Date																			
USGS	5480500	10/1/1982	2560	FALSE	FALSE	FALSE	FALSE	TRUE	670	87.0	89.83	643.8	15,450.2	670	87.0	89.83	643.8	15,450.2	1,287.5	30,900.3
USGS	5480500	10/2/1982	4250	FALSE	FALSE	FALSE	FALSE	TRUE	670	87.0	89.83	643.8	15,450.2	670	87.0	89.83	643.8	15,450.2	1,287.5	30,900.3
USGS	5480500	10/3/1982	5310	FALSE	FALSE	FALSE	FALSE	TRUE	670	87.0	89.83	643.8	15,450.2	670	87.0	89.83	643.8	15,450.2	1,287.5	30,900.0
USGS	5480500	10/4/1982	5980	FALSE	FALSE	FALSE	FALSE	TRUE	670	87.0	89.83	643.8	15,450.2	670	87.0	89.83	643.8	15,450.2	1,287.5	30,900.3
USGS	5480500	10/5/1982	5910	FALSE	FALSE	FALSE	FALSE	TRUE	670	87.0	89.83	643.8	15,450.2	670	87.0	89.83	643.8	15,450.2	1,287.5	30,900.3
USGS	5480500	10/6/1982	5460	FALSE	FALSE	FALSE	FALSE	TRUE	670	87.0	89.83	643.8	15,450.2	670	87.0	89.83	643.8	15,450.2	1,287.5	30,900.3
USGS	5480500	10/7/1982	5010	FALSE	FALSE	FALSE	FALSE	TRUE	670	87.0	89.83	643.8	15,450.2	670	87.0	89.83	643.8	15,450.2	1,287.5	30,900.3
USGS	5480500	10/8/1982	4710	FALSE	FALSE	FALSE	FALSE	TRUE	670	87.0	89.83	643.8	15,450.2	670	87.0	89.83	643.8	15,450.2	1,287.5	30,900.3
USGS	5480500	10/9/1982	4840	FALSE	FALSE	FALSE	FALSE	TRUE	670	87.0	89.83	643.8	15,450.2	670	87.0	89.83	643.8	15,450.2	1,287.5	30,900.3
USGS	5480500	10/10/1982	5440	FALSE	FALSE	FALSE	FALSE	TRUE	670	87.0	89.83	643.8	15,450.2	670	87.0	89.83	643.8	15,450.2	1,287.5	30,900.3
USGS	5480500	10/11/1982	5700	FALSE	FALSE	FALSE	FALSE	TRUE	670	87.0	89.83	643.8	15,450.2	670	87.0	89.83	643.8	15,450.2	1,287.5	30,900.3
USGS	5480500	10/12/1982	5570	FALSE	FALSE	FALSE	FALSE	TRUE	670	87.0	89.83	643.8	15,450.2	670	87.0	89.83	643.8	15,450.2	1,287.5	30,900.3
USGS	5480500	10/13/1982	5270	FALSE	FALSE	FALSE	FALSE	TRUE	670	87.0	89.83	643.8	15,450.2	670	87.0	89.83	643.8	15,450.2	1,287.5	30,900.3
USGS	5480500	10/14/1982	4890	FALSE	FALSE	FALSE	FALSE	TRUE	670	87.0	89.83	643.8	15,450.2	670	87.0	89.83	643.8	15,450.2	1,287.5	30,900.3
USGS	5480500	10/15/1982	4560	FALSE	FALSE	FALSE	FALSE	TRUE	670	87.0	89.83	643.8	15,450.2	670	87.0	89.83	643.8	15,450.2	1,287.5	30,900.3
USGS	5480500	10/16/1982	4230	FALSE	FALSE	FALSE	FALSE	TRUE	670	87.0	89.83	643.8	15,450.2	670	87.0	89.83	643.8	15,450.2	1,287.5	30,900.3
USGS	5480500	10/17/1982	3950	FALSE	FALSE	FALSE	FALSE	TRUE	670	87.0	89.83	643.8	15,450.2	670	87.0	89.83	643.8	15,450.2	1,287.5	30,900.3
USGS	5480500	10/18/1982	3740	FALSE	FALSE	FALSE	FALSE	TRUE	670	87.0	89.83	643.8	15,450.2	670	87.0	89.83	643.8	15,450.2	1,287.5	30,900.3
USGS	5480500	10/19/1982	3660	FALSE	FALSE	FALSE	FALSE	TRUE	670	87.0	89.83	643.8	15,450.2	670	87.0	89.83	643.8	15,450.2	1,287.5	30,900.3
USGS	5480500	10/20/1982	3820	FALSE	FALSE	FALSE	FALSE	TRUE	670	87.0	89.83	643.8	15,450.2	670	87.0	89.83	643.8	15,450.2	1,287.5	30,900.3
USGS	5480500	10/21/1982	4460	FALSE	FALSE	FALSE	FALSE	TRUE	670	87.0	89.83	643.8	15,450.2	670	87.0	89.83	643.8	15,450.2	1,287.5	30,900.3
USGS	5480500	10/22/1982	4800	FALSE	FALSE	FALSE	FALSE	TRUE	670	87.0	89.83	643.8	15,450.2	670	87.0	89.83	643.8	15,450.2	1,287.5	30,900.3
USGS	5480500	10/23/1982	4770	FALSE	FALSE	FALSE	FALSE	TRUE	670	87.0	89.83	643.8	15,450.2	670	87.0	89.83	643.8	15,450.2	1,287.5	30,900.3
USGS	5480500	10/24/1982	4610	FALSE	FALSE	FALSE	FALSE	TRUE	670	87.0	89.83	643.8	15,450.2	670	87.0	89.83	643.8	15,450.2	1,287.5	30,900.3
USGS	5480500	10/25/1982	4440	FALSE	FALSE	FALSE	FALSE	TRUE	670	87.0	89.83	643.8	15,450.2	670	87.0	89.83	643.8	15,450.2	1,287.5	30,900.3
USGS	5480500	10/26/1982	4290	FALSE	FALSE	FALSE	FALSE	TRUE	670	87.0	89.83	643.8	15,450.2	670	87.0	89.83	643.8	15,450.2	1,287.5	30,900.3
USGS	5480500	10/27/1982	4130	FALSE	FALSE	FALSE	FALSE	TRUE	670	87.0	89.83	643.8	15,450.2	670	87.0	89.83	643.8	15,450.2	1,287.5	30,900.3
USGS	5480500	10/28/1982	4000	FALSE	FALSE	FALSE	FALSE	TRUE	670	87.0	89.83	643.8	15,450.2	670	87.0	89.83	643.8	15,450.2	1,287.5	30,900.3
USGS	5480500	10/29/1982	3860	FALSE	FALSE	FALSE	FALSE	TRUE	670	87.0	89.83	643.8	15,450.2	670	87.0	89.83	643.8	15,450.2	1,287.5	30,900.3
USGS	5480500	10/30/1982	3680	FALSE	FALSE	FALSE	FALSE	TRUE	670	87.0	89.83	643.8	15,450.2	670	87.0	89.83	643.8	15,450.2	1,287.5	30,900.3
USGS	5480500	10/31/1982	3510	FALSE	FALSE	FALSE	FALSE	TRUE	670	87.0	89.83	643.8	15,450.2	670	87.0	89.83	643.8	15,450.2	1,287.5	30,900.3
												Total- October 1982	478,954.9					478,954.9		957,909.8

USGS	5480500	11/1/1982	3380	FALSE	FALSE	FALSE	TRUE	670	87.0	89.83	643.8	15,450.2	670	87.0	89.83	643.8	15,450.2	89.83	643.8	15,450.2	1,287.5	30,900.3
USGS	5480500	11/2/1982	3250	FALSE	FALSE	FALSE	TRUE	670	87.0	89.83	643.8	15,450.2	670	87.0	89.83	643.8	15,450.2	89.83	643.8	15,450.2	1,287.5	30,900.3
USGS	5480500	11/3/1982	3020	FALSE	FALSE	FALSE	TRUE	670	87.0	89.83	643.8	15,450.2	670	87.0	89.83	643.8	15,450.2	89.83	643.8	15,450.2	1,287.5	30,900.3
USGS	5480500	11/4/1982	2940	FALSE	FALSE	FALSE	TRUE	670	87.0	89.83	643.8	15,450.2	670	87.0	89.83	643.8	15,450.2	89.83	643.8	15,450.2	1,287.5	30,900.3
USGS	5480500	11/5/1982	2810	FALSE	FALSE	FALSE	TRUE	670	87.0	89.83	643.8	15,450.2	670	87.0	89.83	643.8	15,450.2	89.83	643.8	15,450.2	1,287.5	30,900.3
USGS	5480500	11/6/1982	2690	FALSE	FALSE	FALSE	TRUE	670	87.0	89.83	643.8	15,450.2	670	87.0	89.83	643.8	15,450.2	89.83	643.8	15,450.2	1,287.5	30,900.3
USGS	5480500	11/7/1982	2610	FALSE	FALSE	FALSE	TRUE	670	87.0	89.83	643.8	15,450.2	670	87.0	89.83	643.8	15,450.2	89.83	643.8	15,450.2	1,287.5	30,900.3
USGS	5480500	11/8/1982	2510	FALSE	FALSE	FALSE	TRUE	670	87.0	89.83	643.8	15,450.2	670	87.0	89.83	643.8	15,450.2	89.83	643.8	15,450.2	1,287.5	30,900.3
USGS	5480500	11/9/1982	2460	FALSE	FALSE	FALSE	TRUE	670	87.0	89.83	643.8	15,450.2	670	87.0	89.83	643.8	15,450.2	89.83	643.8	15,450.2	1,287.5	30,900.3
USGS	5480500	11/10/1982	2780	FALSE	FALSE	FALSE	TRUE	670	87.0	89.83	643.8	15,450.2	670	87.0	89.83	643.8	15,450.2	89.83	643.8	15,450.2	1,287.5	30,900.3
USGS	5480500	11/11/1982	4190	FALSE	FALSE	FALSE	TRUE	670	87.0	89.83	643.8	15,450.2	670	87.0	89.83	643.8	15,450.2	89.83	643.8	15,450.2	1,287.5	30,900.3
USGS	5480500	11/12/1982	6210	FALSE	FALSE	FALSE	TRUE	670	87.0	89.83	643.8	15,450.2	670	87.0	89.83	643.8	15,450.2	89.83	643.8	15,450.2	1,287.5	30,900.3
USGS	5480500	11/13/1982	7160	FALSE	FALSE	FALSE	TRUE	670	87.0	89.83	643.8	15,450.2	670	87.0	89.83	643.8	15,450.2	89.83	643.8	15,450.2	1,287.5	30,900.3
USGS	5480500	11/14/1982	7060	FALSE	FALSE	FALSE	TRUE	670	87.0	89.83	643.8	15,450.2	670	87.0	89.83	643.8	15,450.2	89.83	643.8	15,450.2	1,287.5	30,900.3
USGS	5480500	11/15/1982	6740	FALSE	FALSE	FALSE	TRUE	670	87.0	89.83	643.8	15,450.2	670	87.0	89.83	643.8	15,450.2	89.83	643.8	15,450.2	1,287.5	30,900.3
USGS	5480500	11/16/1982	6320	FALSE	FALSE	FALSE	TRUE	670	87.0	89.83	643.8	15,450.2	670	87.0	89.83	643.8	15,450.2	89.83	643.8	15,450.2	1,287.5	30,900.3
USGS	5480500	11/17/1982	5900	FALSE	FALSE	FALSE	TRUE	670	87.0	89.83	643.8	15,450.2	670	87.0	89.83	643.8	15,450.2	89.83	643.8	15,450.2	1,287.5	30,900.3
USGS	5480500	11/18/1982	5620	FALSE	FALSE	FALSE	TRUE	670	87.0	89.83	643.8	15,450.2	670	87.0	89.83	643.8	15,450.2	89.83	643.8	15,450.2	1,287.5	30,900.3
USGS	5480500	11/19/1982	5500	FALSE	FALSE	FALSE	TRUE	670	87.0	89.83	643.8	15,450.2	670	87.0	89.83	643.8	15,450.2	89.83	643.8	15,450.2	1,287.5	30,900.3
USGS	5480500	11/20/1982	5410	FALSE	FALSE	FALSE	TRUE	670	87.0	89.83	643.8	15,450.2	670	87.0	89.83	643.8	15,450.2	89.83	643.8	15,450.2	1,287.5	30,900.3
USGS	5480500	11/21/1982	5340	FALSE	FALSE	FALSE	TRUE	670	87.0	89.83	643.8	15,450.2	670	87.0	89.83	643.8	15,450.2	89.83	643.8	15,450.2	1,287.5	30,900.3
USGS	5480500	11/22/1982	5200	FALSE	FALSE	FALSE	TRUE	670	87.0	89.83	643.8	15,450.2	670	87.0	89.83	643.8	15,450.2	89.83	643.8	15,450.2	1,287.5	30,900.3
USGS	5480500	11/23/1982	5020	FALSE	FALSE	FALSE	TRUE	670	87.0	89.83	643.8	15,450.2	670	87.0	89.83	643.8	15,450.2	89.83	643.8	15,450.2	1,287.5	30,900.3
USGS	5480500	11/24/1982	4780	FALSE	FALSE	FALSE	TRUE	670	87.0	89.83	643.8	15,450.2	670	87.0	89.83	643.8	15,450.2	89.83	643.8	15,450.2	1,287.5	30,900.3
USGS	5480500	11/25/1982	4550	FALSE	FALSE	FALSE	TRUE	670	87.0	89.83	643.8	15,450.2	670	87.0	89.83	643.8	15,450.2	89.83	643.8	15,450.2	1,287.5	30,900.3
USGS	5480500	11/26/1982	4260	FALSE	FALSE	FALSE	TRUE	670	87.0	89.83	643.8	15,450.2	670	87.0	89.83	643.8	15,450.2	89.83	643.8	15,450.2	1,287.5	30,900.3
USGS	5480500	11/27/1982	3810	FALSE	FALSE	FALSE	TRUE	670	87.0	89.83	643.8	15,450.2	670	87.0	89.83	643.8	15,450.2	89.83	643.8	15,450.2	1,287.5	30,900.3
USGS	5480500	11/28/1982	3920	FALSE	FALSE	FALSE	TRUE	670	87.0	89.83	643.8	15,450.2	670	87.0	89.83	643.8	15,450.2	89.83	643.8	15,450.2	1,287.5	30,900.3
USGS	5480500	11/29/1982	4010	FALSE	FALSE	FALSE	TRUE	670	87.0	89.83	643.8	15,450.2	670	87.0	89.83	643.8	15,450.2	89.83	643.8	15,450.2	1,287.5	30,900.3
USGS	5480500	11/30/1982	3960	FALSE	FALSE	FALSE	TRUE	670	87.0	89.83	643.8	15,450.2	670	87.0	89.83	643.8	15,450.2	89.83	643.8	15,450.2	1,287.5	30,900.3
												Total- November 1982 463,504.8					463,504.8					927,009.5

(continued)

341

TABLE 12.13 (*Continued*)

USGS	5480500	12/1/1982	3940	FALSE	FALSE	FALSE	TRUE	670	87.0	89.83	643.8	15,450.2	670	87.0	89.83	643.8	15,450.2	1,287.5	30,900.3
USGS	5480500	12/2/1982	4010	FALSE	FALSE	FALSE	TRUE	670	87.0	89.83	643.8	15,450.2	670	87.0	89.83	643.8	15,450.2	1,287.5	30,900.3
USGS	5480500	12/3/1982	4140	FALSE	FALSE	FALSE	TRUE	670	87.0	89.83	643.8	15,450.2	670	87.0	89.83	643.8	15,450.2	1,287.5	30,900.3
USGS	5480500	12/4/1982	4260	FALSE	FALSE	FALSE	TRUE	670	87.0	89.83	643.8	15,450.2	670	87.0	89.83	643.8	15,450.2	1,287.5	30,900.3
USGS	5480500	12/5/1982	4510	FALSE	FALSE	FALSE	TRUE	670	87.0	89.83	643.8	15,450.2	670	87.0	89.83	643.8	15,450.2	1,287.5	30,900.3
USGS	5480500	12/6/1982	4610	FALSE	FALSE	FALSE	TRUE	670	87.0	89.83	643.8	15,450.2	670	87.0	89.83	643.8	15,450.2	1,287.5	30,900.3
USGS	5480500	12/7/1982	4510	FALSE	FALSE	FALSE	TRUE	670	87.0	89.83	643.8	15,450.2	670	87.0	89.83	643.8	15,450.2	1,287.5	30,900.3
USGS	5480500	12/8/1982	4220	FALSE	FALSE	FALSE	TRUE	670	87.0	89.83	643.8	15,450.2	670	87.0	89.83	643.8	15,450.2	1,287.5	30,900.3
USGS	5480500	12/9/1982	3580	FALSE	FALSE	FALSE	TRUE	670	87.0	89.83	643.8	15,450.2	670	87.0	89.83	643.8	15,450.2	1,287.5	30,900.3
USGS	5480500	12/10/1982	3280	FALSE	FALSE	FALSE	TRUE	670	87.0	89.83	643.8	15,450.2	670	87.0	89.83	643.8	15,450.2	1,287.5	30,900.3
USGS	5480500	12/11/1982	2920	FALSE	FALSE	FALSE	TRUE	670	87.0	89.83	643.8	15,450.2	670	87.0	89.83	643.8	15,450.2	1,287.5	30,900.3
USGS	5480500	12/12/1982	2430	FALSE	FALSE	FALSE	TRUE	670	87.0	89.83	643.8	15,450.2	670	87.0	89.83	643.8	15,450.2	1,287.5	30,900.3
USGS	5480500	12/13/1982	2270	FALSE	FALSE	FALSE	TRUE	670	87.0	89.83	643.8	15,450.2	670	87.0	89.83	643.8	15,450.2	1,287.5	30,900.3
USGS	5480500	12/14/1982	2740	FALSE	FALSE	FALSE	TRUE	670	87.0	89.83	643.8	15,450.2	670	87.0	89.83	643.8	15,450.2	1,287.5	30,900.3
USGS	5480500	12/15/1982	3010	FALSE	FALSE	FALSE	TRUE	670	87.0	89.83	643.8	15,450.2	670	87.0	89.83	643.8	15,450.2	1,287.5	30,900.3
USGS	5480500	12/16/1982	3160	FALSE	FALSE	FALSE	TRUE	670	87.0	89.83	643.8	15,450.2	670	87.0	89.83	643.8	15,450.2	1,287.5	30,900.3
USGS	5480500	12/17/1982	3230	FALSE	FALSE	FALSE	TRUE	670	87.0	89.83	643.8	15,450.2	670	87.0	89.83	643.8	15,450.2	1,287.5	30,900.3
USGS	5480500	12/18/1982	3200	FALSE	FALSE	FALSE	TRUE	670	87.0	89.83	643.8	15,450.2	670	87.0	89.83	643.8	15,450.2	1,287.5	30,900.3
USGS	5480500	12/19/1982	3180	FALSE	FALSE	FALSE	TRUE	670	87.0	89.83	643.8	15,450.2	670	87.0	89.83	643.8	15,450.2	1,287.5	30,900.3
USGS	5480500	12/20/1982	3010	FALSE	FALSE	FALSE	TRUE	670	87.0	89.83	643.8	15,450.2	670	87.0	89.83	643.8	15,450.2	1,287.5	30,900.3
USGS	5480500	12/21/1982	2760	FALSE	FALSE	FALSE	TRUE	670	87.0	89.83	643.8	15,450.2	670	87.0	89.83	643.8	15,450.2	1,287.5	30,900.3
USGS	5480500	12/22/1982	2640	FALSE	FALSE	FALSE	TRUE	670	87.0	89.83	643.8	15,450.2	670	87.0	89.83	643.8	15,450.2	1,287.5	30,900.3
USGS	5480500	12/23/1982	2660	FALSE	FALSE	FALSE	TRUE	670	87.0	89.83	643.8	15,450.2	670	87.0	89.83	643.8	15,450.2	1,287.5	30,900.3
USGS	5480500	12/24/1982	2910	FALSE	FALSE	FALSE	TRUE	670	87.0	89.83	643.8	15,450.2	670	87.0	89.83	643.8	15,450.2	1,287.5	30,900.3
USGS	5480500	12/25/1982	4160	FALSE	FALSE	FALSE	TRUE	670	87.0	89.83	643.8	15,450.2	670	87.0	89.83	643.8	15,450.2	1,287.5	30,900.3
USGS	5480500	12/26/1982	5710	FALSE	FALSE	FALSE	TRUE	670	87.0	89.83	643.8	15,450.2	670	87.0	89.83	643.8	15,450.2	1,287.5	30,900.3
USGS	5480500	12/27/1982	5990	FALSE	FALSE	FALSE	TRUE	670	87.0	89.83	643.8	15,450.2	670	87.0	89.83	643.8	15,450.2	1,287.5	30,900.3
USGS	5480500	12/28/1982	5680	FALSE	FALSE	FALSE	TRUE	670	87.0	89.83	643.8	15,450.2	670	87.0	89.83	643.8	15,450.2	1,287.5	30,900.3
USGS	5480500	12/29/1982	3840	FALSE	FALSE	FALSE	TRUE	670	87.0	89.83	643.8	15,450.2	670	87.0	89.83	643.8	15,450.2	1,287.5	30,900.3
USGS	5480500	12/30/1982	3900	FALSE	FALSE	FALSE	TRUE	670	87.0	89.83	643.8	15,450.2	670	87.0	89.83	643.8	15,450.2	1,287.5	30,900.3
USGS	5480500	12/31/1982	4190	FALSE	FALSE	FALSE	TRUE	670	87.0	89.83	643.8	15,450.2	670	87.0	89.83	643.8	15,450.2	1,287.5	30,900.3
							Total- December 1982					478,954.9					478,954.9		957,909.6

Agency	Site	Date																		
USGS	5480500	1/1/1983	3970	FALSE	FALSE	FALSE	TRUE	FALSE	670	87.0	89.83	643.8	15,450.2	670	87.0	89.83	643.8	15,450.2	1,287.5	30,900.3
USGS	5480500	1/2/1983	3940	FALSE	FALSE	FALSE	TRUE	FALSE	670	87.0	89.83	643.8	15,450.2	670	87.0	89.83	643.8	15,450.2	1,287.5	30,900.3
USGS	5480500	1/3/1983	3640	FALSE	FALSE	FALSE	TRUE	FALSE	670	87.0	89.83	643.8	15,450.2	670	87.0	89.83	643.8	15,450.2	1,287.5	30,900.3
USGS	5480500	1/4/1983	3480	FALSE	FALSE	FALSE	TRUE	FALSE	670	87.0	89.83	643.8	15,450.2	670	87.0	89.83	643.8	15,450.2	1,287.5	30,900.3
USGS	5480500	1/5/1983	3420	FALSE	FALSE	FALSE	TRUE	FALSE	670	87.0	89.83	643.8	15,450.2	670	87.0	89.83	643.8	15,450.2	1,287.5	30,900.3
USGS	5480500	1/6/1983	3240	FALSE	FALSE	FALSE	TRUE	FALSE	670	87.0	89.83	643.8	15,450.2	670	87.0	89.83	643.8	15,450.2	1,287.5	30,900.3
USGS	5480500	1/7/1983	2980	FALSE	FALSE	FALSE	TRUE	FALSE	670	87.0	89.83	643.8	15,450.2	670	87.0	89.83	643.8	15,450.2	1,287.5	30,900.3
USGS	5480500	1/8/1983	2980	FALSE	FALSE	FALSE	TRUE	FALSE	670	87.0	89.83	643.8	15,450.2	670	87.0	89.83	643.8	15,450.2	1,287.5	30,900.3
USGS	5480500	1/9/1983	2960	FALSE	FALSE	FALSE	TRUE	FALSE	670	87.0	89.83	643.8	15,450.2	670	87.0	89.83	643.8	15,450.2	1,287.5	30,900.3
USGS	5480500	1/10/1983	2600	FALSE	FALSE	FALSE	TRUE	FALSE	670	87.0	89.83	643.8	15,450.2	670	87.0	89.83	643.8	15,450.2	1,287.5	30,900.3
USGS	5480500	1/11/1983	1800	FALSE	FALSE	FALSE	TRUE	FALSE	670	87.0	89.83	643.8	15,450.2	670	87.0	89.83	643.8	15,450.2	1,287.5	30,900.3
USGS	5480500	1/12/1983	2260	FALSE	FALSE	FALSE	TRUE	FALSE	670	87.0	89.83	643.8	15,450.2	670	87.0	89.83	643.8	15,450.2	1,287.5	30,900.3
USGS	5480500	1/13/1983	2300	FALSE	FALSE	FALSE	TRUE	FALSE	670	87.0	89.83	643.8	15,450.2	670	87.0	89.83	643.8	15,450.2	1,287.5	30,900.3
USGS	5480500	1/14/1983	2000	FALSE	FALSE	FALSE	TRUE	FALSE	670	87.0	89.83	643.8	15,450.2	670	87.0	89.83	643.8	15,450.2	1,287.5	30,900.3
USGS	5480500	1/15/1983	1720	FALSE	FALSE	FALSE	TRUE	FALSE	670	87.0	89.83	643.8	15,450.2	670	87.0	89.83	643.8	15,450.2	1,287.5	30,900.3
USGS	5480500	1/16/1983	1660	FALSE	FALSE	FALSE	TRUE	FALSE	670	87.0	89.83	643.8	15,450.2	670	87.0	89.83	643.8	15,450.2	1,287.5	30,900.3
USGS	5480500	1/17/1983	1600	FALSE	FALSE	FALSE	TRUE	FALSE	670	87.0	89.83	643.8	15,450.2	670	87.0	89.83	643.8	15,450.2	1,287.5	30,900.3
USGS	5480500	1/18/1983	1700	FALSE	FALSE	FALSE	TRUE	FALSE	670	87.0	89.83	643.8	15,450.2	670	87.0	89.83	643.8	15,450.2	1,287.5	30,900.3
USGS	5480500	1/19/1983	1760	FALSE	FALSE	FALSE	TRUE	FALSE	670	87.0	89.83	643.8	15,450.2	670	87.0	89.83	643.8	15,450.2	1,287.5	30,900.3
USGS	5480500	1/20/1983	1800	FALSE	FALSE	FALSE	TRUE	FALSE	670	87.0	89.83	643.8	15,450.2	670	87.0	89.83	643.8	15,450.2	1,287.5	30,900.3
USGS	5480500	1/21/1983	1840	FALSE	FALSE	FALSE	TRUE	FALSE	670	87.0	89.83	643.8	15,450.2	670	87.0	89.83	643.8	15,450.2	1,287.5	30,900.3
USGS	5480500	1/22/1983	1760	FALSE	FALSE	FALSE	TRUE	FALSE	670	87.0	89.83	643.8	15,450.2	670	87.0	89.83	643.8	15,450.2	1,287.5	30,900.3
USGS	5480500	1/23/1983	1630	FALSE	FALSE	FALSE	TRUE	FALSE	670	87.0	89.83	643.8	15,450.2	670	87.0	89.83	643.8	15,450.2	1,287.5	30,900.3
USGS	5480500	1/24/1983	1500	FALSE	FALSE	FALSE	TRUE	FALSE	670	87.0	89.83	643.8	15,450.2	670	87.0	89.83	643.8	15,450.2	1,287.5	30,900.3
USGS	5480500	1/25/1983	1500	FALSE	FALSE	FALSE	TRUE	FALSE	670	87.0	89.83	643.8	15,450.2	670	87.0	89.83	643.8	15,450.2	1,287.5	30,900.3
USGS	5480500	1/26/1983	1330	FALSE	FALSE	FALSE	TRUE	FALSE	670	87.0	89.83	643.8	15,450.2	670	87.0	89.83	643.8	15,450.2	1,287.5	30,900.3
USGS	5480500	1/27/1983	1150	FALSE	FALSE	FALSE	FALSE	TRUE	660	87.7	89.83	639.1	15,338.3	670	87.0	89.83	643.8	15,450.2	1,282.9	30,788.4
USGS	5480500	1/28/1983	1260	FALSE	FALSE	FALSE	FALSE	TRUE	480	92.4	89.83	489.6	11,750.9	670	87.0	89.83	643.8	15,450.2	1,133.4	27,201.1
USGS	5480500	1/29/1983	1500	FALSE	FALSE	FALSE	FALSE	TRUE	590	91.2	89.83	594.2	14,261.4	670	87.0	89.83	643.8	15,450.2	1,238.0	29,711.6
USGS	5480500	1/30/1983	1360	FALSE	FALSE	FALSE	TRUE	FALSE	670	87.0	89.83	643.8	15,450.2	670	87.0	89.83	643.8	15,450.2	1,287.5	30,900.3
USGS	5480500	1/31/1983	1190	FALSE	FALSE	FALSE	FALSE	FALSE	520	92.6	89.83	531.5	12,756.7	670	87.0	89.83	643.8	15,450.2	1,175.3	28,206.8
									Total- January 1983				471,261.6					478,954.9		960,216.5

(*continued*)

TABLE 12.13 (*Continued*)

Agency	Site No.	Date	Q																		
USGS	5480500	2/1/1983	1150	FALSE	FALSE	FALSE	TRUE	FALSE	480	92.4	89.83	489.6	11,750.9	670	87.0	89.83	643.8	15,450.2	1,133.4	27,201.1
USGS	5480500	2/2/1983	1210	FALSE	FALSE	FALSE	TRUE	FALSE	540	92.4	89.83	551.0	13,223.3	670	87.0	89.83	643.8	15,450.2	1,194.7	28,673.4
USGS	5480500	2/3/1983	981	FALSE	FALSE	FALSE	TRUE	FALSE	311	83.8	89.83	287.7	6,905.7	670	87.0	89.83	643.8	15,450.2	931.5	22,355.8
USGS	5480500	2/4/1983	854	FALSE	FALSE	TRUE	FALSE	FALSE	586	81.4	89.83	591.1	14,185.8	268	79.6	89.83	235.5	5,653.1	826.6	19,838.9
USGS	5480500	2/5/1983	955	FALSE	FALSE	TRUE	FALSE	FALSE	285	81.4	89.83	256.0	6,144.2	670	87.0	89.83	643.8	15,450.2	899.8	21,594.4
USGS	5480500	2/6/1983	929	FALSE	FALSE	TRUE	FALSE	FALSE	661	87.7	89.83	639.6	15,350.0	268	79.6	89.83	235.5	5,653.1	875.1	21,003.0
USGS	5480500	2/7/1983	866	FALSE	FALSE	TRUE	FALSE	FALSE	598	90.9	89.83	600.3	14,408.4	268	79.6	89.83	235.5	5,653.1	835.9	20,061.5
USGS	5480500	2/8/1983	877	FALSE	FALSE	TRUE	FALSE	FALSE	609	90.5	89.83	608.4	14,601.0	268	79.6	89.83	235.5	5,653.1	843.9	20,254.1
USGS	5480500	2/9/1983	880	FALSE	FALSE	TRUE	FALSE	FALSE	612	90.4	89.83	610.5	14,651.6	268	79.6	89.83	235.5	5,653.1	846.0	20,304.7
USGS	5480500	2/10/1983	883	FALSE	FALSE	TRUE	FALSE	FALSE	615	90.2	89.83	612.6	14,701.3	268	79.6	89.83	235.5	5,653.1	848.1	20,354.4
USGS	5480500	2/11/1983	886	FALSE	FALSE	TRUE	FALSE	FALSE	618	90.1	89.83	614.6	14,750.2	268	79.6	89.83	235.5	5,653.1	850.1	20,403.3
USGS	5480500	2/12/1983	890	FALSE	FALSE	TRUE	FALSE	FALSE	622	89.9	89.83	617.3	14,814.0	268	79.6	89.83	235.5	5,653.1	852.8	20,467.1
USGS	5480500	2/13/1983	929	FALSE	FALSE	TRUE	FALSE	FALSE	661	87.7	89.83	639.6	15,350.0	268	79.6	89.83	235.5	5,653.1	875.1	21,003.0
USGS	5480500	2/14/1983	1320	FALSE	FALSE	TRUE	FALSE	TRUE	650	88.4	89.83	634.0	15,215.2	670	87.0	89.83	643.8	15,450.2	1,277.7	30,665.3
USGS	5480500	2/15/1983	2120	FALSE	FALSE	FALSE	FALSE	TRUE	670	87.0	89.83	643.8	15,450.2	670	87.0	89.83	643.8	15,450.2	1,287.5	30,900.3
USGS	5480500	2/16/1983	2980	FALSE	FALSE	FALSE	FALSE	TRUE	670	87.0	89.83	643.8	15,450.2	670	87.0	89.83	643.8	15,450.2	1,287.5	30,900.3
USGS	5480500	2/17/1983	3500	FALSE	FALSE	FALSE	FALSE	TRUE	670	87.0	89.83	643.8	15,450.2	670	87.0	89.83	643.8	15,450.2	1,287.5	30,900.3
USGS	5480500	2/18/1983	5200	FALSE	FALSE	FALSE	FALSE	TRUE	670	87.0	89.83	643.8	15,450.2	670	87.0	89.83	643.8	15,450.2	1,287.5	30,900.3
USGS	5480500	2/19/1983	7790	FALSE	FALSE	FALSE	FALSE	TRUE	670	87.0	89.83	643.8	15,450.2	670	87.0	89.83	643.8	15,450.2	1,287.5	30,900.3
USGS	5480500	2/20/1983	8350	FALSE	FALSE	FALSE	FALSE	TRUE	670	87.0	89.83	643.8	15,450.2	670	87.0	89.83	643.8	15,450.2	1,287.5	30,900.3
USGS	5480500	2/21/1983	9670	FALSE	FALSE	FALSE	FALSE	TRUE	670	87.0	89.83	643.8	15,450.2	670	87.0	89.83	643.8	15,450.2	1,287.5	30,900.3
USGS	5480500	2/22/1983	10300	FALSE	FALSE	FALSE	FALSE	TRUE	670	87.0	89.83	643.8	15,450.2	670	87.0	89.83	643.8	15,450.2	1,287.5	30,900.3
USGS	5480500	2/23/1983	11700	FALSE	FALSE	FALSE	FALSE	TRUE	670	87.0	89.83	643.8	15,450.2	670	87.0	89.83	643.8	15,450.2	1,287.5	30,900.3
USGS	5480500	2/24/1983	11100	FALSE	FALSE	FALSE	FALSE	TRUE	670	87.0	89.83	643.8	15,450.2	670	87.0	89.83	643.8	15,450.2	1,287.5	30,900.3
USGS	5480500	2/25/1983	10200	FALSE	FALSE	FALSE	FALSE	TRUE	670	87.0	89.83	643.8	15,450.2	670	87.0	89.83	643.8	15,450.2	1,287.5	30,900.3
USGS	5480500	2/26/1983	10100	FALSE	FALSE	FALSE	FALSE	TRUE	670	87.0	89.83	643.8	15,450.2	670	87.0	89.83	643.8	15,450.2	1,287.5	30,900.3
USGS	5480500	2/27/1983	10100	FALSE	FALSE	FALSE	FALSE	TRUE	670	87.0	89.83	643.8	15,450.2	670	87.0	89.83	643.8	15,450.2	1,287.5	30,900.3
USGS	5480500	2/28/1983	10800	FALSE	FALSE	FALSE	FALSE	TRUE	670	87.0	89.83	643.8	15,450.2	670	87.0	89.83	643.8	15,450.2	1,287.5	30,900.3
									Total–February 1983				401,312.0					334,633.8		735,945.8

344

USGS	5480500	3/1/1983	10900	FALSE	FALSE	FALSE	FALSE	TRUE	670	87.0	89.83	643.8	15,450.2	670	87.0	89.83	643.8	15,450.2	1,287.5	30,900.3
USGS	5480500	3/2/1983	11100	FALSE	FALSE	FALSE	FALSE	TRUE	670	87.0	89.83	643.8	15,450.2	670	87.0	89.83	643.8	15,450.2	1,287.5	30,900.3
USGS	5480500	3/3/1983	11400	FALSE	FALSE	FALSE	FALSE	TRUE	670	87.0	89.83	643.8	15,450.2	670	87.0	89.83	643.8	15,450.2	1,287.5	30,900.3
USGS	5480500	3/4/1983	11500	FALSE	FALSE	FALSE	FALSE	TRUE	670	87.0	89.83	643.8	15,450.2	670	87.0	89.83	643.8	15,450.2	1,287.5	30,900.3
USGS	5480500	3/5/1983	11900	FALSE	FALSE	FALSE	FALSE	TRUE	670	87.0	89.83	643.8	15,450.2	670	87.0	89.83	643.8	15,450.2	1,287.5	30,900.3
USGS	5480500	3/6/1983	14600	FALSE	FALSE	FALSE	FALSE	TRUE	670	87.0	89.83	643.8	15,450.2	670	87.0	89.83	643.8	15,450.2	1,287.5	30,900.3
USGS	5480500	3/7/1983	18100	FALSE	FALSE	FALSE	FALSE	TRUE	670	87.0	89.83	643.8	15,450.2	670	87.0	89.83	643.8	15,450.2	1,287.5	30,900.3
USGS	5480500	3/8/1983	17800	FALSE	FALSE	FALSE	FALSE	TRUE	670	87.0	89.83	643.8	15,450.2	670	87.0	89.83	643.8	15,450.2	1,287.5	30,900.3
USGS	5480500	3/9/1983	15300	FALSE	FALSE	FALSE	FALSE	TRUE	670	87.0	89.83	643.8	15,450.2	670	87.0	89.83	643.8	15,450.2	1,287.5	30,900.3
USGS	5480500	3/10/1983	13800	FALSE	FALSE	FALSE	FALSE	TRUE	670	87.0	89.83	643.8	15,450.2	670	87.0	89.83	643.8	15,450.2	1,287.5	30,900.3
USGS	5480500	3/11/1983	12500	FALSE	FALSE	FALSE	FALSE	TRUE	670	87.0	89.83	643.8	15,450.2	670	87.0	89.83	643.8	15,450.2	1,287.5	30,900.3
USGS	5480500	3/12/1983	12000	FALSE	FALSE	FALSE	FALSE	TRUE	670	87.0	89.83	643.8	15,450.2	670	87.0	89.83	643.8	15,450.2	1,287.5	30,900.3
USGS	5480500	3/13/1983	11500	FALSE	FALSE	FALSE	FALSE	TRUE	670	87.0	89.83	643.8	15,450.2	670	87.0	89.83	643.8	15,450.2	1,287.5	30,900.3
USGS	5480500	3/14/1983	10700	FALSE	FALSE	FALSE	FALSE	TRUE	670	87.0	89.83	643.8	15,450.2	670	87.0	89.83	643.8	15,450.2	1,287.5	30,900.3
USGS	5480500	3/15/1983	10400	FALSE	FALSE	FALSE	FALSE	TRUE	670	87.0	89.83	643.8	15,450.2	670	87.0	89.83	643.8	15,450.2	1,287.5	30,900.3
USGS	5480500	3/16/1983	12000	FALSE	FALSE	FALSE	FALSE	TRUE	670	87.0	89.83	643.8	15,450.2	670	87.0	89.83	643.8	15,450.2	1,287.5	30,900.3
USGS	5480500	3/17/1983	13200	FALSE	FALSE	FALSE	FALSE	TRUE	670	87.0	89.83	643.8	15,450.2	670	87.0	89.83	643.8	15,450.2	1,287.5	30,900.3
USGS	5480500	3/18/1983	12800	FALSE	FALSE	FALSE	FALSE	TRUE	670	87.0	89.83	643.8	15,450.2	670	87.0	89.83	643.8	15,450.2	1,287.5	30,900.3
USGS	5480500	3/19/1983	12200	FALSE	FALSE	FALSE	FALSE	TRUE	670	87.0	89.83	643.8	15,450.2	670	87.0	89.83	643.8	15,450.2	1,287.5	30,900.3
USGS	5480500	3/20/1983	11500	FALSE	FALSE	FALSE	FALSE	TRUE	670	87.0	89.83	643.8	15,450.2	670	87.0	89.83	643.8	15,450.2	1,287.5	30,900.3
USGS	5480500	3/21/1983	10800	FALSE	FALSE	FALSE	FALSE	TRUE	670	87.0	89.83	643.8	15,450.2	670	87.0	89.83	643.8	15,450.2	1,287.5	30,900.3
USGS	5480500	3/22/1983	9810	FALSE	FALSE	FALSE	FALSE	TRUE	670	87.0	89.83	643.8	15,450.2	670	87.0	89.83	643.8	15,450.2	1,287.5	30,900.3
USGS	5480500	3/23/1983	9090	FALSE	FALSE	FALSE	FALSE	TRUE	670	87.0	89.83	643.8	15,450.2	670	87.0	89.83	643.8	15,450.2	1,287.5	30,900.3
USGS	5480500	3/24/1983	8520	FALSE	FALSE	FALSE	FALSE	TRUE	670	87.0	89.83	643.8	15,450.2	670	87.0	89.83	643.8	15,450.2	1,287.5	30,900.3
USGS	5480500	3/25/1983	8000	FALSE	FALSE	FALSE	FALSE	TRUE	670	87.0	89.83	643.8	15,450.2	670	87.0	89.83	643.8	15,450.2	1,287.5	30,900.3
USGS	5480500	3/26/1983	7640	FALSE	FALSE	FALSE	FALSE	TRUE	670	87.0	89.83	643.8	15,450.2	670	87.0	89.83	643.8	15,450.2	1,287.5	30,900.3
USGS	5480500	3/27/1983	6920	FALSE	FALSE	FALSE	FALSE	TRUE	670	87.0	89.83	643.8	15,450.2	670	87.0	89.83	643.8	15,450.2	1,287.5	30,900.3
USGS	5480500	3/28/1983	6460	FALSE	FALSE	FALSE	FALSE	TRUE	670	87.0	89.83	643.8	15,450.2	670	87.0	89.83	643.8	15,450.2	1,287.5	30,900.3
USGS	5480500	3/29/1983	6690	FALSE	FALSE	FALSE	FALSE	TRUE	670	87.0	89.83	643.8	15,450.2	670	87.0	89.83	643.8	15,450.2	1,287.5	30,900.3
USGS	5480500	3/30/1983	6630	FALSE	FALSE	FALSE	FALSE	TRUE	670	87.0	89.83	643.8	15,450.2	670	87.0	89.83	643.8	15,450.2	1,287.5	30,900.3
USGS	5480500	3/31/1983	7890	FALSE	FALSE	FALSE	FALSE	TRUE	670	87.0	89.83	643.8	15,450.2	670	87.0	89.83	643.8	15,450.2	1,287.5	30,900.3
		Total- March 1983											478,954.9					478,954.9		957,909.8

(continued)

TABLE 12.13 (Continued)

USGS	5480500	4/1/1983	10100	FALSE	FALSE	TRUE	670	87.0	89.83	643.8	15,450.2	670	87.0	89.83	643.8	15,450.2	1,287.5	30,900.3
USGS	5480500	4/2/1983	11300	FALSE	FALSE	TRUE	670	87.0	89.83	643.8	15,450.2	670	87.0	89.83	643.8	15,450.2	1,287.5	30,900.3
USGS	5480500	4/3/1983	11700	FALSE	FALSE	TRUE	670	87.0	89.83	643.8	15,450.2	670	87.0	89.83	643.8	15,450.2	1,287.5	30,900.3
USGS	5480500	4/4/1983	11200	FALSE	FALSE	TRUE	670	87.0	89.83	643.8	15,450.2	670	87.0	89.83	643.8	15,450.2	1,287.5	30,900.3
USGS	5480500	4/5/1983	11200	FALSE	FALSE	TRUE	670	87.0	89.83	643.8	15,450.2	670	87.0	89.83	643.8	15,450.2	1,287.5	30,900.3
USGS	5480500	4/6/1983	11400	FALSE	FALSE	TRUE	670	87.0	89.83	643.8	15,450.2	670	87.0	89.83	643.8	15,450.2	1,287.5	30,900.3
USGS	5480500	4/7/1983	11700	FALSE	FALSE	TRUE	670	87.0	89.83	643.8	15,450.2	670	87.0	89.83	643.8	15,450.2	1,287.5	30,900.3
USGS	5480500	4/8/1983	11600	FALSE	FALSE	TRUE	670	87.0	89.83	643.8	15,450.2	670	87.0	89.83	643.8	15,450.2	1,287.5	30,900.3
USGS	5480500	4/9/1983	11600	FALSE	FALSE	TRUE	670	87.0	89.83	643.8	15,450.2	670	87.0	89.83	643.8	15,450.2	1,287.5	30,900.3
USGS	5480500	4/10/1983	11900	FALSE	FALSE	TRUE	670	87.0	89.83	643.8	15,450.2	670	87.0	89.83	643.8	15,450.2	1,287.5	30,900.3
USGS	5480500	4/11/1983	12500	FALSE	FALSE	TRUE	670	87.0	89.83	643.8	15,450.2	670	87.0	89.83	643.8	15,450.2	1,287.5	30,900.3
USGS	5480500	4/12/1983	13100	FALSE	FALSE	TRUE	670	87.0	89.83	643.8	15,450.2	670	87.0	89.83	643.8	15,450.2	1,287.5	30,900.3
USGS	5480500	4/13/1983	15400	FALSE	FALSE	TRUE	670	87.0	89.83	643.8	15,450.2	670	87.0	89.83	643.8	15,450.2	1,287.5	30,900.3
USGS	5480500	4/14/1983	17800	FALSE	FALSE	TRUE	670	87.0	89.83	643.8	15,450.2	670	87.0	89.83	643.8	15,450.2	1,287.5	30,900.3
USGS	5480500	4/15/1983	18000	FALSE	FALSE	TRUE	670	87.0	89.83	643.8	15,450.2	670	87.0	89.83	643.8	15,450.2	1,287.5	30,900.3
USGS	5480500	4/16/1983	19100	FALSE	FALSE	TRUE	670	87.0	89.83	643.8	15,450.2	670	87.0	89.83	643.8	15,450.2	1,287.5	30,900.3
USGS	5480500	4/17/1983	19300	FALSE	FALSE	TRUE	670	87.0	89.83	643.8	15,450.2	670	87.0	89.83	643.8	15,450.2	1,287.5	30,900.3
USGS	5480500	4/18/1983	18300	FALSE	FALSE	TRUE	670	87.0	89.83	643.8	15,450.2	670	87.0	89.83	643.8	15,450.2	1,287.5	30,900.3
USGS	5480500	4/19/1983	17300	FALSE	FALSE	TRUE	670	87.0	89.83	643.8	15,450.2	670	87.0	89.83	643.8	15,450.2	1,287.5	30,900.3
USGS	5480500	4/20/1983	16200	FALSE	FALSE	TRUE	670	87.0	89.83	643.8	15,450.2	670	87.0	89.83	643.8	15,450.2	1,287.5	30,900.3
USGS	5480500	4/21/1983	15100	FALSE	FALSE	TRUE	670	87.0	89.83	643.8	15,450.2	670	87.0	89.83	643.8	15,450.2	1,287.5	30,900.3
USGS	5480500	4/22/1983	14100	FALSE	FALSE	TRUE	670	87.0	89.83	643.8	15,450.2	670	87.0	89.83	643.8	15,450.2	1,287.5	30,900.3
USGS	5480500	4/23/1983	13400	FALSE	FALSE	TRUE	670	87.0	89.83	643.8	15,450.2	670	87.0	89.83	643.8	15,450.2	1,287.5	30,900.3
USGS	5480500	4/24/1983	12600	FALSE	FALSE	TRUE	670	87.0	89.83	643.8	15,450.2	670	87.0	89.83	643.8	15,450.2	1,287.5	30,900.3
USGS	5480500	4/25/1983	11900	FALSE	FALSE	TRUE	670	87.0	89.83	643.8	15,450.2	670	87.0	89.83	643.8	15,450.2	1,287.5	30,900.3
USGS	5480500	4/26/1983	11200	FALSE	FALSE	TRUE	670	87.0	89.83	643.8	15,450.2	673	87.0	89.83	643.8	15,450.2	1,287.5	30,900.3
USGS	5480500	4/27/1983	10500	FALSE	FALSE	TRUE	670	87.0	88.83	643.8	15,450.2	670	87.0	88.83	643.8	15,450.2	1,287.5	30,900.3
USGS	5480500	4/28/1983	9820	FALSE	FALSE	TRUE	670	87.0	89.83	643.8	15,450.2	670	87.0	89.83	643.8	15,450.2	1,287.5	30,900.3
USGS	5480500	4/29/1983	9260	FALSE	FALSE	TRUE	670	87.0	89.83	643.8	15,450.2	670	87.0	89.83	643.8	15,450.2	1,287.5	30,900.3
USGS	5480500	4/30/1983	8730	FALSE	FALSE	TRUE	670	87.0	89.83	643.8	15,450.2	670	87.0	89.83	643.8	15,450.2	1,287.5	30,900.3
										Total- April 1983	463,504.8					463,504.8		927,009.5

USGS	5480500	Date	Flow																	
USGS	5480500	5/1/1983	8440	FALSE	FALSE	FALSE	FALSE	TRUE	670	87.0	89.83	643.8	15,450.2	670	87.0	89.83	643.8	15,450.2	1,287.5	30,900.3
USGS	5480500	5/2/1983	8970	FALSE	FALSE	FALSE	FALSE	TRUE	670	87.0	89.83	643.8	15,450.2	670	87.0	89.83	643.8	15,450.2	1,287.5	30,900.3
USGS	5480500	5/3/1983	10600	FALSE	FALSE	FALSE	FALSE	TRUE	670	87.0	89.83	643.8	15,450.2	670	87.0	89.83	643.8	15,450.2	1,287.5	30,900.3
USGS	5480500	5/4/1983	10800	FALSE	FALSE	FALSE	FALSE	TRUE	670	87.0	89.83	643.8	15,450.2	670	87.0	89.83	643.8	15,450.2	1,287.5	30,900.3
USGS	5480500	5/5/1983	10300	FALSE	FALSE	FALSE	FALSE	TRUE	670	87.0	89.83	643.8	15,450.2	670	87.0	89.83	643.8	15,450.2	1,287.5	30,900.3
USGS	5480500	5/6/1983	9890	FALSE	FALSE	FALSE	FALSE	TRUE	670	87.0	89.83	643.8	15,450.2	670	87.0	89.83	643.8	15,450.2	1,287.5	30,900.3
USGS	5480500	5/7/1983	9790	FALSE	FALSE	FALSE	FALSE	TRUE	670	87.0	89.83	643.8	15,450.2	670	87.0	89.83	643.8	15,450.2	1,287.5	30,900.3
USGS	5480500	5/8/1983	9140	FALSE	FALSE	FALSE	FALSE	TRUE	670	87.0	89.83	643.8	15,450.2	670	87.0	89.83	643.8	15,450.2	1,287.5	30,900.3
USGS	5480500	5/9/1983	8470	FALSE	FALSE	FALSE	FALSE	TRUE	670	87.0	89.83	643.8	15,450.2	670	87.0	89.83	643.8	15,450.2	1,287.5	30,900.3
USGS	5480500	5/10/1983	7870	FALSE	FALSE	FALSE	FALSE	TRUE	670	87.0	89.83	643.8	15,450.2	670	87.0	89.83	643.8	15,450.2	1,287.5	30,900.3
USGS	5480500	5/11/1983	7550	FALSE	FALSE	FALSE	FALSE	TRUE	670	87.0	89.83	643.8	15,450.2	670	87.0	89.83	643.8	15,450.2	1,287.5	30,900.3
USGS	5480500	5/12/1983	7340	FALSE	FALSE	FALSE	FALSE	TRUE	670	87.0	89.83	643.8	15,450.2	670	87.0	89.83	643.8	15,450.2	1,287.5	30,900.3
USGS	5480500	5/13/1983	7290	FALSE	FALSE	FALSE	FALSE	TRUE	670	87.0	89.83	643.8	15,450.2	670	87.0	89.83	643.8	15,450.2	1,287.5	30,900.3
USGS	5480500	5/14/1983	7220	FALSE	FALSE	FALSE	FALSE	TRUE	670	87.0	89.83	643.8	15,450.2	670	87.0	89.83	643.8	15,450.2	1,287.5	30,900.3
USGS	5480500	5/15/1983	7150	FALSE	FALSE	FALSE	FALSE	TRUE	670	87.0	89.83	643.8	15,450.2	670	87.0	89.83	643.8	15,450.2	1,287.5	30,900.3
USGS	5480500	5/16/1983	7050	FALSE	FALSE	FALSE	FALSE	TRUE	670	87.0	89.83	643.8	15,450.2	670	87.0	89.83	643.8	15,450.2	1,287.5	30,900.3
USGS	5480500	5/17/1983	6930	FALSE	FALSE	FALSE	FALSE	TRUE	670	87.0	89.83	643.8	15,450.2	670	87.0	89.83	643.8	15,450.2	1,287.5	30,900.3
USGS	5480500	5/18/1983	7120	FALSE	FALSE	FALSE	FALSE	TRUE	670	87.0	89.83	643.8	15,450.2	670	87.0	89.83	643.8	15,450.2	1,287.5	30,900.3
USGS	5480500	5/19/1983	7990	FALSE	FALSE	FALSE	FALSE	TRUE	670	87.0	89.83	643.8	15,450.2	670	87.0	89.83	643.8	15,450.2	1,287.5	30,900.3
USGS	5480500	5/20/1983	8300	FALSE	FALSE	FALSE	FALSE	TRUE	670	87.0	89.83	643.8	15,450.2	670	87.0	89.83	643.8	15,450.2	1,287.5	30,900.3
USGS	5480500	5/21/1983	7840	FALSE	FALSE	FALSE	FALSE	TRUE	670	87.0	89.83	643.8	15,450.2	670	87.0	89.83	643.8	15,450.2	1,287.5	30,900.3
USGS	5480500	5/22/1983	7340	FALSE	FALSE	FALSE	FALSE	TRUE	670	87.0	89.83	643.8	15,450.2	670	87.0	89.83	643.8	15,450.2	1,287.5	30,900.3
USGS	5480500	5/23/1983	6790	FALSE	FALSE	FALSE	FALSE	TRUE	670	87.0	89.83	643.8	15,450.2	670	87.0	89.83	643.8	15,450.2	1,287.5	30,900.3
USGS	5480500	5/24/1983	6350	FALSE	FALSE	FALSE	FALSE	TRUE	670	87.0	89.83	643.8	15,450.2	670	87.0	89.83	643.8	15,450.2	1,287.5	30,900.3
USGS	5480500	5/25/1983	5920	FALSE	FALSE	FALSE	FALSE	TRUE	670	87.0	89.83	643.8	15,450.2	670	87.0	89.83	643.8	15,450.2	1,287.5	30,900.3
USGS	5480500	5/26/1983	5610	FALSE	FALSE	FALSE	FALSE	TRUE	670	87.0	89.83	643.8	15,450.2	670	87.0	89.83	643.8	15,450.2	1,287.5	30,900.3
USGS	5480500	5/27/1983	5480	FALSE	FALSE	FALSE	FALSE	TRUE	670	87.0	89.83	643.8	15,450.2	670	87.0	89.83	643.8	15,450.2	1,287.5	30,900.3
USGS	5480500	5/28/1983	5140	FALSE	FALSE	FALSE	FALSE	TRUE	670	87.0	89.83	643.8	15,450.2	670	87.0	89.83	643.8	15,450.2	1,287.5	30,900.3
USGS	5480500	5/29/1983	4780	FALSE	FALSE	FALSE	FALSE	TRUE	670	87.0	89.83	643.8	15,450.2	670	87.0	89.83	643.8	15,450.2	1,287.5	30,900.3
USGS	5480500	5/30/1983	4510	FALSE	FALSE	FALSE	FALSE	TRUE	670	87.0	89.83	643.8	15,450.2	670	87.0	89.83	643.8	15,450.2	1,287.5	30,900.3
USGS	5480500	5/31/1983	4340	FALSE	FALSE	FALSE	FALSE	TRUE	670	87.0	89.83	643.8	15,450.2	670	87.0	89.83	643.8	15,450.2	1,287.5	30,900.3
											Total- May 1983		478,954.9					478,954.9		
																				957,909.8

(*continued*)

TABLE 12.13 (*Continued*)

USGS	5480500	6/1/1983	4150	FALSE	FALSE	FALSE	TRUE	670	87.0	89.83	643.8	15,450.2	670	87.0	89.83	643.8	15,450.2	1,287.5	30,900.3
USGS	5480500	6/2/1983	3980	FALSE	FALSE	FALSE	TRUE	670	87.0	89.83	643.8	15,450.2	670	87.0	89.83	643.8	15,450.2	1,287.5	30,900.3
USGS	5480500	6/3/1983	3830	FALSE	FALSE	FALSE	TRUE	670	87.0	89.83	643.8	15,450.2	670	87.0	89.83	643.8	15,450.2	1,287.5	30,900.3
USGS	5480500	6/4/1983	3670	FALSE	FALSE	FALSE	TRUE	670	87.0	89.83	643.8	15,450.2	670	87.0	89.83	643.8	15,450.2	1,287.5	30,900.3
USGS	5480500	6/5/1983	3500	FALSE	FALSE	FALSE	TRUE	670	87.0	89.83	643.8	15,450.2	670	87.0	89.83	643.8	15,450.2	1,287.5	30,900.3
USGS	5480500	6/6/1983	3330	FALSE	FALSE	FALSE	TRUE	670	87.0	89.83	643.8	15,450.2	670	87.0	89.83	643.8	15,450.2	1,287.5	30,900.3
USGS	5480500	6/7/1983	3170	FALSE	FALSE	FALSE	TRUE	670	87.0	89.83	643.8	15,450.2	670	87.0	89.83	643.8	15,450.2	1,287.5	30,900.3
USGS	5480500	6/8/1983	3020	FALSE	FALSE	FALSE	TRUE	670	87.0	89.83	643.8	15,450.2	670	87.0	89.83	643.8	15,450.2	1,287.5	30,900.3
USGS	5480500	6/9/1983	2880	FALSE	FALSE	FALSE	TRUE	670	87.0	89.83	643.8	15,450.2	670	87.0	89.83	643.8	15,450.2	1,287.5	30,900.3
USGS	5480500	6/10/1983	2710	FALSE	FALSE	FALSE	TRUE	670	87.0	89.83	643.8	15,450.2	670	87.0	89.83	643.8	15,450.2	1,287.5	30,900.3
USGS	5480500	6/11/1983	2500	FALSE	FALSE	FALSE	TRUE	670	87.0	89.83	643.8	15,450.2	670	87.0	89.83	643.8	15,450.2	1,287.5	30,900.3
USGS	5480500	6/12/1983	2450	FALSE	FALSE	FALSE	TRUE	670	87.0	89.83	643.8	15,450.2	670	87.0	89.83	643.8	15,450.2	1,287.5	30,900.3
USGS	5480500	6/13/1983	2380	FALSE	FALSE	FALSE	TRUE	670	87.0	89.83	643.8	15,450.2	670	87.0	89.83	643.8	15,450.2	1,287.5	30,900.3
USGS	5480500	6/14/1983	3200	FALSE	FALSE	FALSE	TRUE	670	87.0	89.83	643.8	15,450.2	670	87.0	89.83	643.8	15,450.2	1,287.5	30,900.3
USGS	5480500	6/15/1983	4900	FALSE	FALSE	FALSE	TRUE	670	87.0	89.83	643.8	15,450.2	670	87.0	89.83	643.8	15,450.2	1,287.5	30,900.3
USGS	5480500	6/16/1983	5420	FALSE	FALSE	FALSE	TRUE	670	87.0	89.83	643.8	15,450.2	670	87.0	89.83	643.8	15,450.2	1,287.5	30,900.3
USGS	5480500	6/17/1983	5170	FALSE	FALSE	FALSE	TRUE	670	87.0	89.83	643.8	15,450.2	670	87.0	89.83	643.8	15,450.2	1,287.5	30,900.3
USGS	5480500	6/18/1983	5230	FALSE	FALSE	FALSE	TRUE	670	87.0	89.83	643.8	15,450.2	670	87.0	89.83	643.8	15,450.2	1,287.5	30,900.3
USGS	5480500	6/19/1983	5540	FALSE	FALSE	FALSE	TRUE	670	87.0	89.83	643.8	15,450.2	670	87.0	89.83	643.8	15,450.2	1,287.5	30,900.3
USGS	5480500	6/20/1983	6410	FALSE	FALSE	FALSE	TRUE	670	87.0	89.83	643.8	15,450.2	670	87.0	89.83	643.8	15,450.2	1,287.5	30,900.3
USGS	5480500	6/21/1983	9260	FALSE	FALSE	FALSE	TRUE	670	87.0	89.83	643.8	15,450.2	670	87.0	89.83	643.8	15,450.2	1,287.5	30,900.3
USGS	5480500	6/22/1983	11600	FALSE	FALSE	FALSE	TRUE	670	87.0	89.83	643.8	15,450.2	670	87.0	89.83	643.8	15,450.2	1,287.5	30,900.3
USGS	5480500	6/23/1983	10700	FALSE	FALSE	FALSE	TRUE	670	87.0	89.83	643.8	15,450.2	670	87.0	89.83	643.8	15,450.2	1,287.5	30,900.3
USGS	5480500	6/24/1983	9830	FALSE	FALSE	FALSE	TRUE	670	87.0	89.83	643.8	15,450.2	670	87.0	89.83	643.8	15,450.2	1,287.5	30,900.3
USGS	5480500	6/25/1983	9160	FALSE	FALSE	FALSE	TRUE	670	87.0	89.83	643.8	15,450.2	670	87.0	89.83	643.8	15,450.2	1,287.5	30,900.3
USGS	5480500	6/26/1983	8660	FALSE	FALSE	FALSE	TRUE	670	87.0	89.83	643.8	15,450.2	670	87.0	89.83	643.8	15,450.2	1,287.5	30,900.3
USGS	5480500	6/27/1983	8730	FALSE	FALSE	FALSE	TRUE	670	87.0	89.83	643.8	15,450.2	670	87.0	89.83	643.8	15,450.2	1,287.5	30,900.3
USGS	5480500	6/28/1983	9280	FALSE	FALSE	FALSE	TRUE	670	87.0	89.83	643.8	15,450.2	670	87.0	89.83	643.8	15,450.2	1,287.5	30,900.3
USGS	5480500	6/29/1983	9840	FALSE	FALSE	FALSE	TRUE	670	87.0	89.83	643.8	15,450.2	670	87.0	89.83	643.8	15,450.2	1,287.5	30,900.3
USGS	5480500	6/30/1983	11100	FALSE	FALSE	FALSE	TRUE	670	87.0	89.83	643.8	15,450.2	670	87.0	89.83	643.8	15,450.2	1,287.5	30,900.3
		Total- June 1983										463,504.8					463,504.8		927,009.5

USGS	5480500	7/1/1983	13200	FALSE	FALSE	FALSE	FALSE	TRUE	670	87.0	89.83	643.8	15,450.2	670	87.0	89.83	643.8	15,450.2	1,287.5	30,900.3
USGS	5480500	7/2/1983	16200	FALSE	FALSE	FALSE	FALSE	TRUE	670	87.0	89.83	643.8	15,450.2	670	87.0	89.83	643.8	15,450.2	1,287.5	30,900.3
USGS	5480500	7/3/1983	19300	FALSE	FALSE	FALSE	FALSE	TRUE	670	87.0	89.83	643.8	15,450.2	670	87.0	89.83	643.8	15,450.2	1,287.5	30,900.3
USGS	5480500	7/4/1983	17900	FALSE	FALSE	FALSE	FALSE	TRUE	670	87.0	89.83	643.8	15,450.2	670	87.0	89.83	643.8	15,450.2	1,287.5	30,900.3
USGS	5480500	7/5/1983	15800	FALSE	FALSE	FALSE	FALSE	TRUE	670	87.0	89.83	643.8	15,450.2	670	87.0	89.83	643.8	15,450.2	1,287.5	30,900.3
USGS	5480500	7/6/1983	14100	FALSE	FALSE	FALSE	FALSE	TRUE	670	87.0	89.83	643.8	15,450.2	670	87.0	89.83	643.8	15,450.2	1,287.5	30,900.3
USGS	5480500	7/7/1983	12700	FALSE	FALSE	FALSE	FALSE	TRUE	670	87.0	89.83	643.8	15,450.2	670	87.0	89.83	643.8	15,450.2	1,287.5	30,900.3
USGS	5480500	7/8/1983	11700	FALSE	FALSE	FALSE	FALSE	TRUE	670	87.0	89.83	643.8	15,450.2	670	87.0	89.83	643.8	15,450.2	1,287.5	30,900.3
USGS	5480500	7/9/1983	10900	FALSE	FALSE	FALSE	FALSE	TRUE	670	87.0	89.83	643.8	15,450.2	670	87.0	89.83	643.8	15,450.2	1,287.5	30,900.3
USGS	5480500	7/10/1983	10400	FALSE	FALSE	FALSE	FALSE	TRUE	670	87.0	89.83	643.8	15,450.2	670	87.0	89.83	643.8	15,450.2	1,287.5	30,900.3
USGS	5480500	7/11/1983	9940	FALSE	FALSE	FALSE	FALSE	TRUE	670	87.0	89.83	643.8	15,450.2	670	87.0	89.83	643.8	15,450.2	1,287.5	30,900.3
USGS	5480500	7/12/1983	9400	FALSE	FALSE	FALSE	FALSE	TRUE	670	87.0	89.83	643.8	15,450.2	670	87.0	89.83	643.8	15,450.2	1,287.5	30,900.3
USGS	5480500	7/13/1983	8630	FALSE	FALSE	FALSE	FALSE	TRUE	670	87.0	89.83	643.8	15,450.2	670	87.0	89.83	643.8	15,450.2	1,287.5	30,900.3
USGS	5480500	7/14/1983	7910	FALSE	FALSE	FALSE	FALSE	TRUE	670	87.0	89.83	643.8	15,450.2	670	87.0	89.83	643.8	15,450.2	1,287.5	30,900.3
USGS	5480500	7/15/1983	7220	FALSE	FALSE	FALSE	FALSE	TRUE	670	87.0	89.83	643.8	15,450.2	670	87.0	89.83	643.8	15,450.2	1,287.5	30,900.3
USGS	5480500	7/16/1983	6820	FALSE	FALSE	FALSE	FALSE	TRUE	670	87.0	89.83	643.8	15,450.2	670	87.0	89.83	643.8	15,450.2	1,287.5	30,900.3
USGS	5480500	7/17/1983	6020	FALSE	FALSE	FALSE	FALSE	TRUE	670	87.0	89.83	643.8	15,450.2	670	87.0	89.83	643.8	15,450.2	1,287.5	30,900.3
USGS	5480500	7/18/1983	5590	FALSE	FALSE	FALSE	FALSE	TRUE	670	87.0	89.83	643.8	15,450.2	670	87.0	89.83	643.8	15,450.2	1,287.5	30,900.3
USGS	5480500	7/19/1983	5270	FALSE	FALSE	FALSE	FALSE	TRUE	670	87.0	89.83	643.8	15,450.2	670	87.0	89.83	643.8	15,450.2	1,287.5	30,900.3
USGS	5480500	7/20/1983	4870	FALSE	FALSE	FALSE	FALSE	TRUE	670	87.0	89.83	643.8	15,450.2	670	87.0	89.83	643.8	15,450.2	1,287.5	30,900.3
USGS	5480500	7/21/1983	4470	FALSE	FALSE	FALSE	FALSE	TRUE	670	87.0	89.83	643.8	15,450.2	670	87.0	89.83	643.8	15,450.2	1,287.5	30,900.3
USGS	5480500	7/22/1983	4010	FALSE	FALSE	FALSE	FALSE	TRUE	670	87.0	89.83	643.8	15,450.2	670	87.0	89.83	643.8	15,450.2	1,287.5	30,900.3
USGS	5480500	7/23/1983	3820	FALSE	FALSE	FALSE	FALSE	TRUE	670	87.0	89.83	643.8	15,450.2	670	87.0	89.83	643.8	15,450.2	1,287.5	30,900.3
USGS	5480500	7/24/1983	3640	FALSE	FALSE	FALSE	FALSE	TRUE	670	87.0	89.83	643.8	15,450.2	670	87.0	89.83	643.8	15,450.2	1,287.5	30,900.3
USGS	5480500	7/25/1983	3500	FALSE	FALSE	FALSE	FALSE	TRUE	670	87.0	89.83	643.8	15,450.2	670	87.0	89.83	643.8	15,450.2	1,287.5	30,900.3
USGS	5480500	7/26/1983	3270	FALSE	FALSE	FALSE	FALSE	TRUE	670	87.0	89.83	643.8	15,450.2	670	87.0	89.83	643.8	15,450.2	1,287.5	30,900.3
USGS	5480500	7/27/1983	3060	FALSE	FALSE	FALSE	FALSE	TRUE	670	87.0	89.83	643.8	15,450.2	670	87.0	89.83	643.8	15,450.2	1,287.5	30,900.3
USGS	5480500	7/28/1983	2900	FALSE	FALSE	FALSE	FALSE	TRUE	670	87.0	89.83	643.8	15,450.2	670	87.0	89.83	643.8	15,450.2	1,287.5	30,900.3
USGS	5480500	7/29/1983	2720	FALSE	FALSE	FALSE	FALSE	TRUE	670	87.0	89.83	643.8	15,450.2	670	87.0	89.83	643.8	15,450.2	1,287.5	30,900.3
USGS	5480500	7/30/1983	2550	FALSE	FALSE	FALSE	FALSE	TRUE	670	87.0	89.83	643.8	15,450.2	670	87.0	89.83	643.8	15,450.2	1,287.5	30,900.3
USGS	5480500	7/31/1983	2370	FALSE	FALSE	FALSE	FALSE	TRUE	670	87.0	89.83	643.8	15,450.2	670	87.0	89.83	643.8	15,450.2	1,287.5	30,900.3
													Total- July 1983 478,954.9					478,954.9		957,909.8

(continued)

TABLE 12.13 (Continued)

USGS	5480500	8/1/1983	2160	FALSE	FALSE	FALSE	FALSE	TRUE	670	87.0	643.8	89.83	15,450.2	670	87.0	643.8	89.83	15,450.2	1,287.5	30,900.3
USGS	5480500	8/2/1983	1970	FALSE	FALSE	FALSE	FALSE	TRUE	670	87.0	643.8	89.83	15,450.2	670	87.0	643.8	89.83	15,450.2	1,287.5	30,900.3
USGS	5480500	8/3/1983	1830	FALSE	FALSE	FALSE	FALSE	TRUE	670	87.0	643.8	89.83	15,450.2	670	87.0	643.8	89.83	15,450.2	1,287.5	30,900.3
USGS	5480500	8/4/1983	1690	FALSE	FALSE	FALSE	FALSE	TRUE	670	87.0	643.8	89.83	15,450.2	670	87.0	643.8	89.83	15,450.2	1,287.5	30,900.3
USGS	5480500	8/5/1983	1530	FALSE	FALSE	FALSE	FALSE	TRUE	670	87.0	643.8	89.83	15,450.2	670	87.0	643.8	89.83	15,450.2	1,287.5	30,900.3
USGS	5480500	8/6/1983	1430	FALSE	FALSE	FALSE	FALSE	TRUE	670	87.0	643.8	89.83	15,450.2	670	87.0	643.8	89.83	15,450.2	1,287.5	30,900.3
USGS	5480500	8/7/1983	1330	FALSE	FALSE	FALSE	TRUE	FALSE	660	87.7	639.1	89.83	15,338.3	670	87.0	643.8	89.83	15,450.2	1,282.9	30,788.4
USGS	5480500	8/8/1983	1220	FALSE	FALSE	FALSE	TRUE	FALSE	550	92.3	560.3	89.83	13,446.2	670	87.0	643.8	89.83	15,450.2	1,204.0	28,896.4
USGS	5480500	8/9/1983	1110	FALSE	FALSE	FALSE	TRUE	FALSE	440	91.5	444.5	89.83	10,667.1	670	87.0	643.8	89.83	15,450.2	1,088.2	26,117.3
USGS	5480500	8/10/1983	1040	FALSE	FALSE	FALSE	TRUE	FALSE	370	88.2	360.4	89.83	8,649.9	670	87.0	643.8	89.83	15,450.2	1,004.2	24,100.1
USGS	5480500	8/11/1983	964	FALSE	FALSE	TRUE	FALSE	FALSE	294	82.3	266.9	89.83	6,406.6	670	87.0	643.8	89.83	15,450.2	910.7	21,856.8
USGS	5480500	8/12/1983	894	FALSE	FALSE	TRUE	FALSE	FALSE	626	89.7	619.8	89.83	14,876.2	268	79.6	235.5	89.83	5,653.1	855.4	20,529.3
USGS	5480500	8/13/1983	849	FALSE	FALSE	TRUE	FALSE	FALSE	581	91.5	587.1	89.83	14,089.4	268	79.6	235.5	89.83	5,653.1	822.6	19,742.5
USGS	5480500	8/14/1983	814	FALSE	FALSE	TRUE	FALSE	FALSE	546	92.3	556.6	89.83	13,357.9	268	79.6	235.5	89.83	5,653.1	792.1	19,011.0
USGS	5480500	8/15/1983	771	FALSE	FALSE	TRUE	FALSE	FALSE	503	92.6	514.2	89.83	12,340.2	268	79.6	235.5	89.83	5,653.1	749.7	17,993.3
USGS	5480500	8/16/1983	735	FALSE	FALSE	TRUE	FALSE	FALSE	467	92.2	475.3	89.83	11,406.2	268	79.6	235.5	89.83	5,653.1	710.8	17,059.3
USGS	5480500	8/17/1983	701	FALSE	TRUE	FALSE	FALSE	FALSE	433	91.3	436.3	89.83	10,471.1	268	79.6	235.5	89.83	5,653.1	671.8	16,124.2
USGS	5480500	8/18/1983	664	FALSE	TRUE	FALSE	FALSE	FALSE	664	87.5	641.0	89.83	15,384.4	0	0.0	0.0	89.83	0.0	641.0	15,384.4
USGS	5480500	8/19/1983	636	FALSE	TRUE	FALSE	FALSE	FALSE	636	89.2	626.0	89.83	15,024.7	0	0.0	0.0	89.83	0.0	626.0	15,024.7
USGS	5480500	8/20/1983	611	FALSE	TRUE	FALSE	FALSE	FALSE	611	90.4	609.8	89.83	14,634.8	0	0.0	0.0	89.83	0.0	609.8	14,634.8
USGS	5480500	8/21/1983	634	FALSE	TRUE	FALSE	FALSE	FALSE	634	89.3	624.8	89.83	14,995.8	0	0.0	0.0	89.83	0.0	624.8	14,995.8
USGS	5480500	8/22/1983	621	FALSE	TRUE	FALSE	FALSE	FALSE	621	89.9	616.6	89.83	14,798.2	0	0.0	0.0	89.83	0.0	616.6	14,798.2
USGS	5480500	8/23/1983	591	FALSE	TRUE	FALSE	FALSE	FALSE	591	91.2	595.0	89.83	14,280.1	0	0.0	0.0	89.83	0.0	595.0	14,280.1
USGS	5480500	8/24/1983	567	FALSE	TRUE	FALSE	FALSE	FALSE	567	91.9	575.3	89.83	13,808.3	0	0.0	0.0	89.83	0.0	575.3	13,808.3
USGS	5480500	8/25/1983	554	FALSE	TRUE	FALSE	FALSE	FALSE	554	92.2	563.9	89.83	13,533.4	0	0.0	0.0	89.83	0.0	563.9	13,533.4
USGS	5480500	8/26/1983	547	FALSE	TRUE	FALSE	FALSE	FALSE	547	92.3	557.5	89.83	13,380.1	0	0.0	0.0	89.83	0.0	557.5	13,380.1
USGS	5480500	8/27/1983	774	TRUE	FALSE	FALSE	FALSE	FALSE	506	92.6	517.3	89.83	12,415.0	268	79.6	235.5	89.83	5,653.1	752.8	18,068.0
USGS	5480500	8/28/1983	764	TRUE	FALSE	FALSE	FALSE	FALSE	496	92.6	506.8	89.83	12,163.9	268	79.6	235.5	89.83	5,653.1	742.4	17,817.0
USGS	5480500	8/29/1983	650	FALSE	TRUE	FALSE	FALSE	FALSE	650	88.4	634.0	89.83	15,215.2	0	0.0	0.0	89.83	0.0	634.0	15,215.2
USGS	5480500	8/30/1983	636	FALSE	TRUE	FALSE	FALSE	FALSE	636	89.2	626.0	89.83	15,024.7	0	0.0	0.0	89.83	0.0	626.0	15,024.7
USGS	5480500	8/31/1983	588	FALSE	TRUE	FALSE	FALSE	FALSE	588	91.3	592.7	89.83	14,223.8	0	0.0	0.0	89.83	0.0	592.7	14,223.8
													Total- August 1983 422,632.5					215,176.5		637,809.0

Agency	Site	Date	Value																		
USGS	5480500	9/1/1983	569	FALSE	TRUE	FALSE	FALSE	FALSE	569	91.9	89.83	577.1	13,849.4	0	0.0	89.83	0.0	0.0	577.1	13,849.4
USGS	5480500	9/2/1983	542	FALSE	TRUE	FALSE	FALSE	FALSE	542	92.4	89.83	552.9	13,268.4	0	0.0	89.83	0.0	0.0	552.9	13,268.4
USGS	5480500	9/3/1983	516	FALSE	TRUE	FALSE	FALSE	FALSE	516	92.6	89.83	527.5	12,660.3	0	0.0	89.83	0.0	0.0	527.5	12,660.3
USGS	5480500	9/4/1983	491	FALSE	TRUE	FALSE	FALSE	FALSE	491	92.5	89.83	501.5	12,036.3	0	0.0	89.83	0.0	0.0	501.5	12,036.3
USGS	5480500	9/5/1983	488	FALSE	TRUE	FALSE	FALSE	FALSE	488	92.2	89.83	498.3	11,959.1	0	0.0	89.83	0.0	0.0	498.3	11,959.1
USGS	5480500	9/6/1983	552	FALSE	TRUE	FALSE	FALSE	FALSE	552	92.5	89.83	562.1	13,490.0	0	0.0	89.83	0.0	0.0	562.1	13,490.0
USGS	5480500	9/7/1983	532	FALSE	TRUE	FALSE	FALSE	FALSE	532	92.5	89.83	543.3	13,039.8	0	0.0	89.83	0.0	0.0	543.3	13,039.8
USGS	5480500	9/8/1983	489	FALSE	TRUE	FALSE	FALSE	FALSE	489	92.2	89.83	499.4	11,984.8	0	0.0	89.83	0.0	0.0	499.4	11,984.8
USGS	5480500	9/9/1983	465	FALSE	TRUE	FALSE	FALSE	FALSE	465	92.2	89.83	473.0	11,352.5	0	0.0	89.83	0.0	0.0	473.0	11,352.5
USGS	5480500	9/10/1983	465	FALSE	TRUE	FALSE	FALSE	FALSE	465	92.2	89.83	473.0	11,352.5	0	0.0	89.83	0.0	0.0	473.0	11,352.5
USGS	5480500	9/11/1983	444	FALSE	TRUE	FALSE	FALSE	FALSE	444	91.6	89.83	449.1	10,778.4	0	0.0	89.83	0.0	0.0	449.1	10,778.4
USGS	5480500	9/12/1983	426	FALSE	TRUE	FALSE	FALSE	FALSE	426	91.0	89.83	428.1	10,273.5	0	0.0	89.83	0.0	0.0	428.1	10,273.5
USGS	5480500	9/13/1983	408	FALSE	TRUE	FALSE	FALSE	FALSE	408	90.3	89.83	406.6	9,758.9	0	0.0	89.83	0.0	0.0	406.6	9,758.9
USGS	5480500	9/14/1983	398	FALSE	TRUE	FALSE	FALSE	FALSE	398	89.8	89.83	394.6	9,469.6	0	0.0	89.83	0.0	0.0	394.6	9,469.6
USGS	5480500	9/15/1983	431	FALSE	TRUE	FALSE	FALSE	FALSE	431	91.2	89.83	433.9	10,414.8	0	0.0	89.83	0.0	0.0	433.9	10,414.8
USGS	5480500	9/16/1983	427	FALSE	TRUE	FALSE	FALSE	FALSE	427	91.1	89.83	429.2	10,301.8	0	0.0	89.83	0.0	0.0	429.2	10,301.8
USGS	5480500	9/17/1983	426	FALSE	TRUE	FALSE	FALSE	FALSE	426	91.0	89.83	428.1	10,273.5	0	0.0	89.83	0.0	0.0	428.1	10,273.5
USGS	5480500	9/18/1983	453	FALSE	TRUE	FALSE	FALSE	FALSE	453	91.9	89.83	459.4	11,026.6	0	0.0	89.83	0.0	0.0	459.4	11,026.6
USGS	5480500	9/19/1983	742	FALSE	FALSE	TRUE	FALSE	FALSE	474	92.3	89.83	483.0	11,592.8	268	79.6	89.83	235.5	5,653.1	718.6	17,245.9
USGS	5480500	9/20/1983	2210	FALSE	FALSE	FALSE	FALSE	TRUE	670	87.0	89.83	643.8	15,450.2	670	87.0	89.83	643.8	15,450.2	1,287.5	30,900.3
USGS	5480500	9/21/1983	2790	FALSE	FALSE	FALSE	FALSE	TRUE	670	87.0	89.83	643.8	15,450.2	670	87.0	89.83	643.8	15,450.2	1,287.5	30,900.3
USGS	5480500	9/22/1983	2170	FALSE	FALSE	FALSE	FALSE	TRUE	670	87.0	89.83	643.8	15,450.2	670	87.0	89.83	643.8	15,450.2	1,287.5	30,900.3
USGS	5480500	9/23/1983	1650	FALSE	FALSE	FALSE	FALSE	TRUE	670	87.0	89.83	643.8	15,450.2	670	87.0	89.83	643.8	15,450.2	1,287.5	30,900.3
USGS	5480500	9/24/1983	1360	FALSE	FALSE	FALSE	FALSE	TRUE	670	87.0	89.83	643.8	15,450.2	670	87.0	89.83	643.8	15,450.2	1,287.5	30,900.3
USGS	5480500	9/25/1983	1180	FALSE	FALSE	FALSE	TRUE	FALSE	510	88.2	89.83	521.4	12,513.8	670	87.0	89.83	643.8	15,450.2	1,165.2	27,963.9
USGS	5480500	9/26/1983	1040	FALSE	FALSE	FALSE	TRUE	FALSE	370	87.5	89.83	360.4	8,649.9	670	87.0	89.83	643.8	15,450.2	1,004.2	24,100.1
USGS	5480500	9/27/1983	932	FALSE	FALSE	TRUE	FALSE	FALSE	664	91.5	89.83	641.0	15,384.4	268	79.6	89.83	235.5	5,653.1	876.6	21,037.5
USGS	5480500	9/28/1983	849	FALSE	FALSE	TRUE	FALSE	FALSE	581	91.9	89.83	587.1	14,089.4	268	79.6	89.83	235.5	5,653.1	822.6	19,742.5
USGS	5480500	9/29/1983	799	FALSE	FALSE	TRUE	FALSE	FALSE	531	92.5	89.83	542.4	13,016.6	268	79.6	89.83	235.5	5,653.1	777.9	18,669.7
USGS	5480500	9/30/1983	795	FALSE	FALSE	TRUE	FALSE	FALSE	527	92.6	89.83	538.5	12,923.0	268	79.6	89.83	235.5	5,653.1	774.0	18,576.1
								Total - September 1983					372,710.6					136,416.6		509,127.2

(continued)

TABLE 12.13 (*Continued*)

Agency	Station	Date																			
USGS	5480500	10/1/1983	824	FALSE	TRUE	FALSE	FALSE	FALSE	556	92.2	89.83	565.7	13,576.5	268	79.6	89.83	235.5	5,653.1	801.2	19,229.6
USGS	5480500	10/2/1983	806	FALSE	TRUE	FALSE	FALSE	FALSE	538	92.5	89.83	549.1	13,177.8	268	79.6	89.83	235.5	5,653.1	784.6	18,830.9
USGS	5480500	10/3/1983	848	FALSE	TRUE	FALSE	FALSE	FALSE	580	91.6	89.83	586.2	14,069.8	268	79.6	89.83	235.5	5,653.1	821.8	19,722.9
USGS	5480500	10/4/1983	783	FALSE	TRUE	FALSE	FALSE	FALSE	515	92.6	89.83	526.5	12,636.0	268	79.6	89.83	235.5	5,653.1	762.0	18,289.1
USGS	5480500	10/5/1983	749	FALSE	TRUE	FALSE	FALSE	FALSE	481	92.4	89.83	490.7	11,777.1	268	79.6	89.83	235.5	5,653.1	726.3	17,430.2
USGS	5480500	10/6/1983	706	FALSE	FALSE	TRUE	FALSE	FALSE	438	91.4	89.83	442.1	10,611.3	268	79.6	89.83	235.5	5,653.1	677.7	16,264.4
USGS	5480500	10/7/1983	668	FALSE	FALSE	TRUE	FALSE	FALSE	668	87.2	89.83	642.9	15,428.7	0	0.0	89.83	0.0	0.0	642.9	15,428.7
USGS	5480500	10/8/1983	641	FALSE	FALSE	TRUE	FALSE	FALSE	641	88.9	89.83	629.0	15,095.1	0	0.0	89.83	0.0	0.0	629.0	15,095.1
USGS	5480500	10/9/1983	612	FALSE	FALSE	FALSE	FALSE	FALSE	612	90.4	89.83	610.5	14,651.6	0	0.0	89.83	0.0	0.0	610.5	14,651.6
USGS	5480500	10/10/1983	603	FALSE	FALSE	FALSE	FALSE	FALSE	603	90.7	89.83	604.1	14,497.3	0	0.0	89.83	0.0	0.0	604.1	14,497.3
USGS	5480500	10/11/1983	893	FALSE	FALSE	TRUE	TRUE	FALSE	625	89.8	89.83	619.2	14,860.8	268	79.6	89.83	235.5	5,653.1	854.7	20,513.9
USGS	5480500	10/12/1983	1210	FALSE	FALSE	FALSE	FALSE	TRUE	540	92.4	89.83	551.0	13,223.3	670	87.0	89.83	643.8	15,450.2	1,194.7	28,673.4
USGS	5480500	10/13/1983	1420	FALSE	FALSE	FALSE	FALSE	TRUE	670	87.0	89.83	643.8	15,450.2	670	87.0	89.83	643.8	15,450.2	1,287.5	30,900.3
USGS	5480500	10/14/1983	1360	FALSE	FALSE	FALSE	TRUE	FALSE	670	87.0	89.83	643.8	15,450.2	670	87.0	89.83	643.8	15,450.2	1,287.5	30,900.3
USGS	5480500	10/15/1983	1260	FALSE	FALSE	FALSE	TRUE	FALSE	590	91.2	89.83	594.2	14,261.4	670	87.0	89.83	643.8	15,450.2	1,238.0	29,711.6
USGS	5480500	10/16/1983	1150	FALSE	FALSE	FALSE	TRUE	TRUE	480	92.4	89.83	489.6	11,750.9	670	87.0	89.83	643.8	15,450.2	1,133.4	27,201.1
USGS	5480500	10/17/1983	1060	FALSE	FALSE	FALSE	TRUE	TRUE	390	89.4	89.83	384.9	9,236.7	670	87.0	89.83	643.8	15,450.2	1,028.6	24,686.8
USGS	5480500	10/18/1983	990	FALSE	FALSE	FALSE	FALSE	TRUE	320	84.6	89.83	298.8	7,171.2	670	87.0	89.83	643.8	15,450.2	942.6	22,621.4
USGS	5480500	10/19/1983	1220	FALSE	FALSE	FALSE	FALSE	FALSE	550	92.3	89.83	560.3	13,446.2	670	87.0	89.83	643.8	15,450.2	1,204.0	28,896.4
USGS	5480500	10/20/1983	1600	FALSE	FALSE	FALSE	FALSE	FALSE	670	87.0	89.83	643.8	15,450.2	670	87.0	89.83	643.8	15,450.2	1,287.5	30,900.3
USGS	5480500	10/21/1983	1750	FALSE	FALSE	FALSE	TRUE	TRUE	670	87.0	89.83	643.8	15,450.2	670	87.0	89.83	643.8	15,450.2	1,287.5	30,900.3
USGS	5480500	10/22/1983	1790	FALSE	FALSE	FALSE	TRUE	TRUE	670	87.0	89.83	643.8	15,450.2	670	87.0	89.83	643.8	15,450.2	1,287.5	30,900.3
USGS	5480500	10/23/1983	1720	FALSE	FALSE	FALSE	TRUE	TRUE	670	87.0	89.83	643.8	15,450.2	670	87.0	89.83	643.8	15,450.2	1,287.5	30,900.3
USGS	5480500	10/24/1983	1630	FALSE	FALSE	FALSE	TRUE	TRUE	670	87.0	89.83	643.8	15,450.2	670	87.0	89.83	643.8	15,450.2	1,287.5	30,900.3
USGS	5480500	10/25/1983	1510	FALSE	FALSE	FALSE	FALSE	TRUE	670	87.0	89.83	643.8	15,450.2	670	87.0	89.83	643.8	15,450.2	1,287.5	30,900.3
USGS	5480500	10/26/1983	1410	FALSE	FALSE	FALSE	FALSE	FALSE	670	87.0	89.83	643.8	15,450.2	670	87.0	89.83	643.8	15,450.2	1,287.5	30,900.3
USGS	5480500	10/27/1983	1340	FALSE	FALSE	FALSE	FALSE	FALSE	670	87.0	89.83	643.8	15,450.2	670	87.0	89.83	643.8	15,450.2	1,287.5	30,900.3
USGS	5480500	10/28/1983	1260	FALSE	FALSE	FALSE	TRUE	FALSE	590	91.2	89.83	594.2	14,261.4	670	87.0	89.83	643.8	15,450.2	1,238.0	29,711.6
USGS	5480500	10/29/1983	1150	FALSE	FALSE	FALSE	FALSE	TRUE	480	92.4	89.83	489.6	11,750.9	670	87.0	89.83	643.8	15,450.2	1,133.4	27,201.1
USGS	5480500	10/30/1983	1080	FALSE	FALSE	FALSE	TRUE	TRUE	410	90.4	89.83	409.0	9,816.5	670	87.0	89.83	643.8	15,450.2	1,052.8	25,266.7
USGS	5480500	10/31/1983	1050	FALSE	FALSE	FALSE	TRUE	FALSE	380	88.8	89.83	372.7	8,944.0	670	87.0	89.83	643.8	15,450.2	1,016.4	24,394.2
											Total- October 1983		**418,746.3**					**348,574.9**		**767,321.1**

Agency	Site	Date																						
USGS	5480500	11/1/1983	1030	FALSE	FALSE	FALSE	TRUE	FALSE	360	87.6	89.83	348.1	670	87.0	89.83	643.8	8,354.9	15,450.2	643.8	89.83	87.0	670	991.9	23,805.0
USGS	5480500	11/2/1983	998	FALSE	FALSE	FALSE	TRUE	FALSE	328	85.2	89.83	308.7	670	87.0	89.83	643.8	7,407.7	15,450.2	643.8	89.83	87.0	670	952.4	22,857.9
USGS	5480500	11/3/1983	987	FALSE	FALSE	FALSE	TRUE	FALSE	297	82.5	89.83	270.6	670	87.0	89.83	643.8	6,494.4	15,450.2	643.8	89.83	87.0	670	914.4	21,944.6
USGS	5480500	11/4/1983	988	FALSE	FALSE	FALSE	FALSE	TRUE	318	84.4	89.83	296.5	670	87.0	89.83	643.8	7,112.2	15,450.2	643.8	89.83	87.0	670	940.1	22,562.3
USGS	5480500	11/5/1983	1500	FALSE	FALSE	FALSE	FALSE	TRUE	670	87.0	89.83	643.8	670	87.0	89.83	643.8	15,450.2	15,450.2	643.8	89.83	87.0	670	1,287.5	30,900.3
USGS	5480500	11/6/1983	1600	FALSE	FALSE	FALSE	FALSE	TRUE	670	87.0	89.83	643.8	670	87.0	89.83	643.8	15,450.2	15,450.2	643.8	89.83	87.0	670	1,287.5	30,900.3
USGS	5480500	11/7/1983	1440	FALSE	FALSE	FALSE	TRUE	TRUE	670	87.0	89.83	643.8	670	87.0	89.83	643.8	15,450.2	15,450.2	643.8	89.83	87.0	670	1,287.5	30,900.3
USGS	5480500	11/8/1983	1340	FALSE	FALSE	FALSE	FALSE	TRUE	670	87.0	89.83	643.8	670	87.0	89.83	643.8	15,450.2	15,450.2	643.8	89.83	87.0	670	1,287.5	30,900.3
USGS	5480500	11/9/1983	1380	FALSE	FALSE	FALSE	FALSE	TRUE	670	87.0	89.83	643.8	670	87.0	89.83	643.8	15,450.2	15,450.2	643.8	89.83	87.0	670	1,287.5	30,900.3
USGS	5480500	11/10/1983	1620	FALSE	FALSE	FALSE	FALSE	TRUE	670	87.0	89.83	643.8	670	87.0	89.83	643.8	15,450.2	15,450.2	643.8	89.83	87.0	670	1,287.5	30,900.3
USGS	5480500	11/11/1983	1920	FALSE	FALSE	FALSE	FALSE	TRUE	670	87.0	89.83	643.8	670	87.0	89.83	643.8	15,450.2	15,450.2	643.8	89.83	87.0	670	1,287.5	30,900.3
USGS	5480500	11/12/1983	1890	FALSE	FALSE	FALSE	FALSE	TRUE	670	87.0	89.83	643.8	670	87.0	89.83	643.8	15,450.2	15,450.2	643.8	89.83	87.0	670	1,287.5	30,900.3
USGS	5480500	11/13/1983	1790	FALSE	FALSE	FALSE	FALSE	TRUE	670	87.0	89.83	643.8	670	87.0	89.83	643.8	15,450.2	15,450.2	643.8	89.83	87.0	670	1,287.5	30,900.3
USGS	5480500	11/14/1983	1840	FALSE	FALSE	FALSE	FALSE	TRUE	670	87.0	89.83	643.8	670	87.0	89.83	643.8	15,450.2	15,450.2	643.8	89.83	87.0	670	1,287.5	30,900.3
USGS	5480500	11/15/1983	2330	FALSE	FALSE	FALSE	FALSE	TRUE	670	87.0	89.83	643.8	670	87.0	89.83	643.8	15,450.2	15,450.2	643.8	89.83	87.0	670	1,287.5	30,900.3
USGS	5480500	11/16/1983	2600	FALSE	FALSE	FALSE	FALSE	TRUE	670	87.0	89.83	643.8	670	87.0	89.83	643.8	15,450.2	15,450.2	643.8	89.83	87.0	670	1,287.5	30,900.3
USGS	5480500	11/17/1983	2480	FALSE	FALSE	FALSE	FALSE	TRUE	670	87.0	89.83	643.8	670	87.0	89.83	643.8	15,450.2	15,450.2	643.8	89.83	87.0	670	1,287.5	30,900.3
USGS	5480500	11/18/1983	2390	FALSE	FALSE	FALSE	FALSE	TRUE	670	87.0	89.83	643.8	670	87.0	89.83	643.8	15,450.2	15,450.2	643.8	89.83	87.0	670	1,287.5	30,900.3
USGS	5480500	11/19/1983	2420	FALSE	FALSE	FALSE	FALSE	TRUE	670	87.0	89.83	643.8	670	87.0	89.83	643.8	15,450.2	15,450.2	643.8	89.83	87.0	670	1,287.5	30,900.3
USGS	5480500	11/20/1983	2870	FALSE	FALSE	FALSE	FALSE	TRUE	670	87.0	89.83	643.8	670	87.0	89.83	643.8	15,450.2	15,450.2	643.8	89.83	87.0	670	1,287.5	30,900.3
USGS	5480500	11/21/1983	3270	FALSE	FALSE	FALSE	FALSE	TRUE	670	87.0	89.83	643.8	670	87.0	89.83	643.8	15,450.2	15,450.2	643.8	89.83	87.0	670	1,287.5	30,900.3
USGS	5480500	11/22/1983	3180	FALSE	FALSE	FALSE	FALSE	TRUE	670	87.0	89.83	643.8	670	87.0	89.83	643.8	15,450.2	15,450.2	643.8	89.83	87.0	670	1,287.5	30,900.3
USGS	5480500	11/23/1983	3190	FALSE	FALSE	FALSE	FALSE	TRUE	670	87.0	89.83	643.8	670	87.0	89.83	643.8	15,450.2	15,450.2	643.8	89.83	87.0	670	1,287.5	30,900.3
USGS	5480500	11/24/1983	3110	FALSE	FALSE	FALSE	FALSE	TRUE	670	87.0	89.83	643.8	670	87.0	89.83	643.8	15,450.2	15,450.2	643.8	89.83	87.0	670	1,287.5	30,900.3
USGS	5480500	11/25/1983	2830	FALSE	FALSE	FALSE	FALSE	TRUE	670	87.0	89.83	643.8	670	87.0	89.83	643.8	15,450.2	15,450.2	643.8	89.83	87.0	670	1,287.5	30,900.3
USGS	5480500	11/26/1983	2550	FALSE	FALSE	FALSE	FALSE	TRUE	670	87.0	89.83	643.8	670	87.0	89.83	643.8	15,450.2	15,450.2	643.8	89.83	87.0	670	1,287.5	30,900.3
USGS	5480500	11/27/1983	2490	FALSE	FALSE	FALSE	FALSE	TRUE	670	87.0	89.83	643.8	670	87.0	89.83	643.8	15,450.2	15,450.2	643.8	89.83	87.0	670	1,287.5	30,900.3
USGS	5480500	11/28/1983	2220	FALSE	FALSE	FALSE	FALSE	TRUE	670	87.0	89.83	643.8	670	87.0	89.83	643.8	15,450.2	15,450.2	643.8	89.83	87.0	670	1,287.5	30,900.3
USGS	5480500	11/29/1983	1640	FALSE	FALSE	FALSE	FALSE	TRUE	670	87.0	89.83	643.8	670	87.0	89.83	643.8	15,450.2	15,450.2	643.8	89.83	87.0	670	1,287.5	30,900.3
USGS	5480500	11/30/1983	1800	FALSE	FALSE	FALSE	FALSE	TRUE	670	87.0	89.83	643.8	670	87.0	89.83	643.8	15,450.2	15,450.2	643.8	89.83	87.0	670	1,287.5	30,900.3
		1										Total: November 1983					431,073.3	463,504.8						894,578.0

(continued)

353

TABLE 12.13 (*Continued*)

Agency	Station	Date																			
USGS	5480500	12/1/1983	1900	1	FALSE	FALSE	FALSE	FALSE	TRUE	670	87.0	89.83	643.8	15,450.2	670	87.0	89.83	643.8	15,450.2	1,287.5	30,900.3
USGS	5480500	12/2/1983	1950	1	FALSE	FALSE	FALSE	FALSE	TRUE	670	87.0	89.83	643.8	15,450.2	670	87.0	89.83	643.8	15,450.2	1,287.5	30,900.3
USGS	5480500	12/3/1983	1950	1	FALSE	FALSE	FALSE	FALSE	TRUE	670	87.0	89.83	643.8	15,450.2	670	87.0	89.83	643.8	15,450.2	1,287.5	30,900.3
USGS	5480500	12/4/1983	1900	1	FALSE	FALSE	FALSE	FALSE	TRUE	670	87.0	89.83	643.8	15,450.2	670	87.0	89.83	643.8	15,450.2	1,287.5	30,900.3
USGS	5480500	12/5/1983	1800	1	FALSE	FALSE	FALSE	FALSE	TRUE	670	87.0	89.83	643.8	15,450.2	670	87.0	89.83	643.8	15,450.2	1,287.5	30,900.3
USGS	5480500	12/6/1983	1650	1	FALSE	FALSE	FALSE	FALSE	TRUE	670	87.0	89.83	643.8	15,450.2	670	87.0	89.83	643.8	15,450.2	1,287.5	30,900.3
USGS	5480500	12/7/1983	1450	1	FALSE	FALSE	FALSE	FALSE	TRUE	670	87.0	89.83	643.8	15,450.2	670	87.0	89.83	643.8	15,450.2	1,287.5	30,900.3
USGS	5480500	12/8/1983	1300	1	FALSE	FALSE	FALSE	TRUE	FALSE	630	89.5	89.83	622.4	14,936.8	268	79.6	89.83	235.5	5,653.1	1,266.1	30,387.0
USGS	5480500	12/9/1983	1100	1	FALSE	FALSE	FALSE	TRUE	FALSE	430	91.2	89.83	432.8	10,386.6	268	79.6	89.83	235.5	5,653.1	1,076.5	25,836.8
USGS	5480500	12/10/1983	900	1	FALSE	FALSE	FALSE	TRUE	FALSE	632	89.4	89.83	623.6	14,966.5	268	79.6	89.83	235.5	5,653.1	859.2	20,619.6
USGS	5480500	12/11/1983	850	1	FALSE	FALSE	FALSE	TRUE	FALSE	582	91.5	89.83	587.9	14,108.8	268	79.6	89.83	235.5	5,653.1	823.4	19,761.9
USGS	5480500	12/12/1983	800	1	FALSE	FALSE	FALSE	TRUE	FALSE	532	92.5	89.83	543.3	13,039.8	268	79.6	89.83	235.5	5,653.1	778.9	18,692.9
USGS	5480500	12/13/1983	770	1	FALSE	FALSE	FALSE	TRUE	FALSE	502	92.6	89.83	513.1	12,315.2	268	79.6	89.83	235.5	5,653.1	748.7	17,968.3
USGS	5480500	12/14/1983	740	1	FALSE	FALSE	FALSE	TRUE	FALSE	472	92.3	89.83	480.8	11,539.7	268	79.6	89.83	235.5	5,653.1	716.4	17,192.8
USGS	5480500	12/15/1983	730	1	FALSE	FALSE	FALSE	TRUE	FALSE	462	92.1	89.83	469.6	11,271.5	268	79.6	89.83	235.5	5,653.1	705.2	16,924.6
USGS	5480500	12/16/1983	720	1	FALSE	FALSE	FALSE	TRUE	FALSE	452	91.9	89.83	458.3	10,999.1	268	79.6	89.83	235.5	5,653.1	693.8	16,652.2
USGS	5480500	12/17/1983	730	1	FALSE	FALSE	FALSE	TRUE	FALSE	462	92.1	89.83	469.6	11,271.5	268	79.6	89.83	235.5	5,653.1	705.2	16,924.6
USGS	5480500	12/18/1983	750	1	FALSE	FALSE	FALSE	TRUE	FALSE	482	92.4	89.83	491.8	11,803.3	268	79.6	89.83	235.5	5,653.1	727.3	17,456.4
USGS	5480500	12/19/1983	770	1	FALSE	FALSE	FALSE	TRUE	FALSE	502	92.6	89.83	513.1	12,315.2	268	79.6	89.83	235.5	5,653.1	748.7	17,968.3
USGS	5480500	12/20/1983	800	1	FALSE	FALSE	FALSE	TRUE	FALSE	532	92.5	89.83	543.3	13,039.8	268	79.6	89.83	235.5	5,653.1	778.9	18,692.9
USGS	5480500	12/21/1983	830	1	FALSE	FALSE	FALSE	TRUE	FALSE	562	92.0	89.83	571.0	13,704.1	268	79.6	89.83	235.5	5,653.1	806.6	19,357.2
USGS	5480500	12/22/1983	860	1	FALSE	FALSE	FALSE	TRUE	FALSE	592	91.2	89.83	595.8	14,298.7	268	79.6	89.83	235.5	5,653.1	831.3	19,951.8
USGS	5480500	12/23/1983	880	1	FALSE	FALSE	FALSE	TRUE	FALSE	612	90.4	89.83	610.5	14,651.6	268	79.6	89.83	235.5	5,653.1	846.0	20,304.7
USGS	5480500	12/24/1983	890	1	FALSE	FALSE	FALSE	TRUE	FALSE	622	89.9	89.83	617.3	14,814.0	268	79.6	89.83	235.5	5,653.1	852.8	20,467.1
USGS	5480500	12/25/1983	890	1	FALSE	FALSE	FALSE	TRUE	FALSE	622	89.9	89.83	617.3	14,814.0	268	79.6	89.83	235.5	5,653.1	852.8	20,467.1
USGS	5480500	12/26/1983	890	1	FALSE	FALSE	FALSE	TRUE	FALSE	622	89.9	89.83	617.3	14,814.0	268	79.6	89.83	235.5	5,653.1	852.8	20,467.1
USGS	5480500	12/27/1983	890	1	FALSE	FALSE	FALSE	TRUE	FALSE	622	89.9	89.83	617.3	14,814.0	268	79.6	89.83	235.5	5,653.1	852.8	20,467.1
USGS	5480500	12/28/1983	890	1	FALSE	FALSE	FALSE	TRUE	FALSE	622	89.9	89.83	617.3	14,814.0	268	79.6	89.83	235.5	5,653.1	852.8	20,467.1
USGS	5480500	12/29/1983	890	1	FALSE	FALSE	FALSE	TRUE	FALSE	622	89.9	89.83	617.3	14,814.0	268	79.6	89.83	235.5	5,653.1	852.8	20,467.1
USGS	5480500	12/30/1983	890	1	FALSE	FALSE	FALSE	TRUE	FALSE	622	89.9	89.83	617.3	14,814.0	268	79.6	89.83	235.5	5,653.1	852.8	20,467.1
USGS	5480500	12/31/1983	900	1	FALSE	FALSE	FALSE	TRUE	FALSE	632	89.4	89.83	623.6	14,966.5	268	79.6	89.83	235.5	5,653.1	859.2	20,619.6
		Total- December 1983												**431,464.1**					**263,419.6**		**694,883.6**

(continued)

Agency	Site	Date			L1	L2	L3	L4												
USGS	5480500	1/1/1984	910	1	FALSE	TRUE	FALSE	FALSE	642	88.8	89.83	629.5	15,108.9	268	79.6	89.83	235.5	5,653.1	865.1	20,762.0
USGS	5480500	1/2/1984	920	1	FALSE	TRUE	FALSE	FALSE	652	88.2	89.83	635.0	15,240.7	268	79.6	89.83	235.5	5,653.1	870.6	20,893.8
USGS	5480500	1/3/1984	930	1	FALSE	FALSE	FALSE	FALSE	662	87.6	89.83	640.1	15,361.5	268	79.6	89.83	235.5	5,653.1	875.6	21,014.6
USGS	5480500	1/4/1984	940	1	FALSE	FALSE	FALSE	FALSE	270	79.8	89.83	237.9	5,710.9	670	87.0	89.83	643.8	15,450.2	881.7	21,160.7
USGS	5480500	1/5/1984	950	1	FALSE	TRUE	FALSE	FALSE	280	80.9	89.83	250.0	5,999.1	670	87.0	89.83	643.8	15,450.2	893.7	21,449.3
USGS	5480500	1/6/1984	940	1	FALSE	TRUE	FALSE	FALSE	270	79.8	89.83	237.9	5,710.5	670	87.0	89.83	643.8	15,450.2	881.7	21,160.7
USGS	5480500	1/7/1984	940	1	FALSE	TRUE	FALSE	FALSE	270	79.8	89.83	237.9	5,710.5	670	87.0	89.83	643.8	15,450.2	881.7	21,160.7
USGS	5480500	1/8/1984	900	1	FALSE	FALSE	FALSE	FALSE	632	89.4	89.83	623.6	14,966.5	268	79.6	89.83	235.5	5,653.1	859.2	20,619.6
USGS	5480500	1/9/1984	840	1	FALSE	TRUE	FALSE	FALSE	572	91.8	89.83	579.6	13,910.5	268	79.6	89.83	235.5	5,653.1	815.2	19,563.6
USGS	5480500	1/10/1984	780	1	FALSE	TRUE	FALSE	FALSE	512	92.6	89.83	523.5	12,562.8	268	79.6	89.83	235.5	5,653.1	759.0	18,215.9
USGS	5480500	1/11/1984	750	1	FALSE	TRUE	FALSE	FALSE	482	92.4	89.83	491.8	11,803.3	268	79.6	89.83	235.5	5,653.1	727.3	17,456.4
USGS	5480500	1/12/1984	730	1	FALSE	TRUE	FALSE	FALSE	462	92.1	89.83	469.6	11,271.5	268	79.6	89.83	235.5	5,653.1	705.2	16,924.6
USGS	5480500	1/13/1984	710	1	FALSE	TRUE	FALSE	FALSE	442	91.6	89.83	446.8	10,722.8	268	79.6	89.83	235.5	5,653.1	682.3	16,375.9
USGS	5480500	1/14/1984	700	1	FALSE	TRUE	FALSE	FALSE	432	91.2	89.83	435.1	10,443.0	268	79.6	89.83	235.5	5,653.1	670.7	16,096.1
USGS	5480500	1/15/1984	690	1	FALSE	TRUE	FALSE	FALSE	422	90.9	89.83	423.3	10,159.9	268	79.6	89.83	235.5	5,653.1	658.9	15,813.0
USGS	5480500	1/16/1984	680	1	FALSE	FALSE	FALSE	FALSE	412	90.5	89.83	411.4	9,674.0	268	79.6	89.83	235.5	5,653.1	647.0	15,527.1
USGS	5480500	1/17/1984	670	1	FALSE	FALSE	TRUE	TRUE	670	87.0	89.83	643.8	15,450.2	0	0.0	89.83	0.0	0.0	643.8	15,450.2
USGS	5480500	1/18/1984	660	1	FALSE	FALSE	TRUE	TRUE	660	87.7	89.83	639.1	15,338.3	0	0.0	89.83	0.0	0.0	639.1	15,338.3
USGS	5480500	1/19/1984	660	1	FALSE	FALSE	TRUE	TRUE	660	87.7	89.83	639.1	15,338.3	0	0.0	89.83	0.0	0.0	639.1	15,338.3
USGS	5480500	1/20/1984	650	1	FALSE	FALSE	TRUE	TRUE	650	88.4	89.83	634.0	15,215.2	0	0.0	89.83	0.0	0.0	634.0	15,215.2
USGS	5480500	1/21/1984	630	1	FALSE	FALSE	TRUE	TRUE	630	89.5	89.83	622.4	14,936.8	0	0.0	89.83	0.0	0.0	622.4	14,936.8
USGS	5480500	1/22/1984	600	1	FALSE	FALSE	TRUE	TRUE	600	90.9	89.83	601.8	14,444.2	0	0.0	89.83	0.0	0.0	601.8	14,444.2
USGS	5480500	1/23/1984	560	1	FALSE	FALSE	TRUE	TRUE	560	92.1	89.83	569.2	13,661.9	0	0.0	89.83	0.0	0.0	569.2	13,661.9
USGS	5480500	1/24/1984	530	1	FALSE	FALSE	TRUE	TRUE	530	92.5	89.83	541.4	12,993.3	0	0.0	89.83	0.0	0.0	541.4	12,993.3
USGS	5480500	1/25/1984	510	1	FALSE	FALSE	TRUE	TRUE	510	92.6	89.83	521.4	12,513.8	0	0.0	89.83	0.0	0.0	521.4	12,513.8
USGS	5480500	1/26/1984	490	1	FALSE	FALSE	TRUE	TRUE	490	92.5	89.83	500.4	12,010.6	0	0.0	89.83	0.0	0.0	500.4	12,010.6
USGS	5480500	1/27/1984	480	1	FALSE	FALSE	TRUE	TRUE	480	92.4	89.83	489.6	11,750.9	0	0.0	89.83	0.0	0.0	489.6	11,750.9
USGS	5480500	1/28/1984	470	1	FALSE	FALSE	TRUE	TRUE	470	92.2	89.83	478.6	11,486.4	0	0.0	89.83	0.0	0.0	478.6	11,486.4
USGS	5480500	1/29/1984	460	1	FALSE	FALSE	TRUE	TRUE	460	92.0	89.83	467.4	11,217.4	0	0.0	89.83	0.0	0.0	467.4	11,217.4
USGS	5480500	1/30/1984	450	1	FALSE	FALSE	TRUE	TRUE	450	91.8	89.83	456.0	10,944.2	0	0.0	89.83	0.0	0.0	456.0	10,944.2
USGS	5480500	1/31/1984	450	1	FALSE	FALSE	TRUE	TRUE	450	91.8	89.83	456.0	10,944.2	0	0.0	89.83	0.0	0.0	456.0	10,944.2
												Total- January 1984	372,801.7					129,637.8		502,439.5

TABLE 12.13 (Continued)

Agency	Station	Date	Flow																		
USGS	5480500	2/1/1984	440	1	FALSE	TRUE	FALSE	FALSE	FALSE	440	91.5	89.83	444.5	10,667.1	0	0.0	89.83	0.0	0.0	444.5	10,667.1
USGS	5480500	2/2/1984	430	1	FALSE	TRUE	FALSE	FALSE	FALSE	430	91.2	89.83	432.8	10,386.6	0	0.0	89.83	0.0	0.0	432.8	10,386.6
USGS	5480500	2/3/1984	420	1	FALSE	TRUE	FALSE	FALSE	FALSE	420	90.8	89.83	421.0	10,102.9	0	0.0	89.83	0.0	0.0	421.0	10,102.9
USGS	5480500	2/4/1984	410	1	FALSE	TRUE	FALSE	FALSE	FALSE	410	90.4	89.83	409.0	9,816.5	0	0.0	89.83	0.0	0.0	409.0	9,816.5
USGS	5480500	2/5/1984	400	1	FALSE	TRUE	FALSE	FALSE	FALSE	400	89.9	89.83	397.0	9,527.6	0	0.0	89.83	0.0	0.0	397.0	9,527.6
USGS	5480500	2/6/1984	410	1	FALSE	TRUE	FALSE	FALSE	FALSE	410	90.4	89.83	409.0	9,816.5	0	0.0	89.83	0.0	0.0	409.0	9,816.5
USGS	5480500	2/7/1984	380	1	FALSE	TRUE	FALSE	FALSE	FALSE	380	88.8	89.83	372.7	8,944.0	0	0.0	89.83	0.0	0.0	372.7	8,944.0
USGS	5480500	2/8/1984	370	1	FALSE	TRUE	FALSE	FALSE	FALSE	370	88.2	89.83	360.4	8,649.9	0	0.0	89.83	0.0	0.0	360.4	8,649.9
USGS	5480500	2/9/1984	380	1	FALSE	TRUE	FALSE	FALSE	FALSE	380	88.8	89.83	372.7	8,944.0	0	0.0	89.83	0.0	0.0	372.7	8,944.0
USGS	5480500	2/10/1984	410	1	FALSE	TRUE	FALSE	FALSE	FALSE	410	90.4	89.83	409.0	9,816.5	0	0.0	89.83	0.0	0.0	409.0	9,816.5
USGS	5480500	2/11/1984	430	1	FALSE	TRUE	FALSE	FALSE	FALSE	430	91.2	89.83	432.8	10,386.6	0	0.0	89.83	0.0	0.0	432.8	10,386.6
USGS	5480500	2/12/1984	520	1	FALSE	TRUE	FALSE	FALSE	FALSE	520	92.6	89.83	531.5	12,756.7	0	0.0	89.83	0.0	0.0	531.5	12,756.7
USGS	5480500	2/13/1984	720	1	FALSE	FALSE	FALSE	TRUE	FALSE	452	91.9	89.83	458.3	10,999.1	268	79.6	89.83	235.5	5,653.1	693.8	16,652.2
USGS	5480500	2/14/1984	920	1	FALSE	FALSE	TRUE	FALSE	FALSE	652	88.2	89.83	635.0	15,240.7	268	79.6	89.83	235.5	5,653.1	870.6	20,893.8
USGS	5480500	2/15/1984	2200	1	FALSE	FALSE	FALSE	FALSE	TRUE	670	87.0	89.83	643.8	15,450.2	670	87.0	89.83	643.8	15,450.2	1,287.5	30,900.3
USGS	5480500	2/16/1984	5400		FALSE	FALSE	FALSE	FALSE	TRUE	670	87.0	89.83	643.8	15,450.2	670	87.0	89.83	643.8	15,450.2	1,287.5	30,900.3
USGS	5480500	2/17/1984	13000		FALSE	FALSE	FALSE	FALSE	TRUE	670	87.0	89.83	643.8	15,450.2	670	87.0	89.83	643.8	15,450.2	1,287.5	30,900.3
USGS	5480500	2/18/1984	10000		FALSE	FALSE	FALSE	FALSE	TRUE	670	87.0	89.83	643.8	15,450.2	670	87.0	89.83	643.8	15,450.2	1,287.5	30,900.3
USGS	5480500	2/19/1984	7400		FALSE	FALSE	FALSE	FALSE	TRUE	670	87.0	89.83	643.8	15,450.2	670	87.0	89.83	643.8	15,450.2	1,287.5	30,900.3
USGS	5480500	2/20/1984	6840		FALSE	FALSE	FALSE	FALSE	TRUE	670	87.0	89.83	643.8	15,450.2	670	87.0	89.83	643.8	15,450.2	1,287.5	30,900.3
USGS	5480500	2/21/1984	6880		FALSE	FALSE	FALSE	FALSE	TRUE	670	87.0	89.83	643.8	15,450.2	670	87.0	89.83	643.8	15,450.2	1,287.5	30,900.3
USGS	5480500	2/22/1984	8240		FALSE	FALSE	FALSE	FALSE	TRUE	670	87.0	89.83	643.8	15,450.2	670	87.0	89.83	643.8	15,450.2	1,287.5	30,900.3
USGS	5480500	2/23/1984	10100		FALSE	FALSE	FALSE	FALSE	TRUE	670	87.0	89.83	643.8	15,450.2	670	87.0	89.83	643.8	15,450.2	1,287.5	30,900.3
USGS	5480500	2/24/1984	9670		FALSE	FALSE	FALSE	FALSE	TRUE	670	87.0	89.83	643.8	15,450.2	670	87.0	89.83	643.8	15,450.2	1,287.5	30,900.3
USGS	5480500	2/25/1984	8990		FALSE	FALSE	FALSE	FALSE	TRUE	670	87.0	89.83	643.8	15,450.2	670	87.0	89.83	643.8	15,450.2	1,287.5	30,900.3
USGS	5480500	2/26/1984	8750		FALSE	FALSE	FALSE	FALSE	TRUE	670	87.0	89.83	643.8	15,450.2	670	87.0	89.83	643.8	15,450.2	1,287.5	30,900.3
USGS	5480500	2/27/1984	8540		FALSE	FALSE	FALSE	FALSE	TRUE	670	87.0	89.83	643.8	15,450.2	670	87.0	89.83	643.8	15,450.2	1,287.5	30,900.3
USGS	5480500	2/28/1984	7360		FALSE	FALSE	FALSE	FALSE	TRUE	670	87.0	89.83	643.8	15,450.2	670	87.0	89.83	643.8	15,450.2	1,287.5	30,900.3
USGS	5480500	2/29/1984	6190		FALSE	FALSE	FALSE	FALSE	TRUE	670	87.0	89.83	643.8	15,450.2	670	87.0	89.83	643.8	15,450.2	1,287.5	30,900.3
		Total- February 1984												**377,807.1**					**243,058.6**		**620,865.7**

USGS	5480500	3/1/1984	5690		FALSE	FALSE	FALSE	FALSE	TRUE	670	87.0	89.83	643.8	15,450.2	670	87.0	89.83	643.8	15,450.2	1,287.5	30,900.3
USGS	5480500	3/2/1984	5120		FALSE	FALSE	FALSE	FALSE	TRUE	670	87.0	89.83	643.8	15,450.2	670	87.0	89.83	643.8	15,450.2	1,287.5	30,900.3
USGS	5480500	3/3/1984	4610		FALSE	FALSE	FALSE	FALSE	TRUE	670	87.0	89.83	643.8	15,450.2	670	87.0	89.83	643.8	15,450.2	1,287.5	30,900.3
USGS	5480500	3/4/1984	4330		FALSE	FALSE	FALSE	FALSE	TRUE	670	87.0	89.83	643.8	15,450.2	670	87.0	89.83	643.8	15,450.2	1,287.5	30,900.3
USGS	5480500	3/5/1984	4020		FALSE	FALSE	FALSE	FALSE	TRUE	670	87.0	89.83	643.8	15,450.2	670	87.0	89.83	643.8	15,450.2	1,287.5	30,900.3
USGS	5480500	3/6/1984	3380		FALSE	FALSE	FALSE	FALSE	TRUE	670	87.0	89.83	643.8	15,450.2	670	87.0	89.83	643.8	15,450.2	1,287.5	30,900.3
USGS	5480500	3/7/1984	2820		FALSE	FALSE	FALSE	FALSE	TRUE	670	87.0	89.83	643.8	15,450.2	670	87.0	89.83	643.8	15,450.2	1,287.5	30,900.3
USGS	5480500	3/8/1984	2600		FALSE	FALSE	FALSE	FALSE	TRUE	670	87.0	89.83	643.8	15,450.2	670	87.0	89.83	643.8	15,450.2	1,287.5	30,900.3
USGS	5480500	3/9/1984	2650	1	FALSE	FALSE	FALSE	FALSE	TRUE	670	87.0	89.83	643.8	15,450.2	670	87.0	89.83	643.8	15,450.2	1,287.5	30,900.3
USGS	5480500	3/10/1984	2300		FALSE	FALSE	FALSE	FALSE	TRUE	670	87.0	89.83	643.8	15,450.2	670	87.0	89.83	643.8	15,450.2	1,287.5	30,900.3
USGS	5480500	3/11/1984	2000	1	FALSE	FALSE	FALSE	FALSE	TRUE	670	87.0	89.83	643.8	15,450.2	670	87.0	89.83	643.8	15,450.2	1,287.5	30,900.3
USGS	5480500	3/12/1984	1900	1	FALSE	FALSE	FALSE	FALSE	TRUE	670	87.0	89.83	643.8	15,450.2	670	87.0	89.83	643.8	15,450.2	1,287.5	30,900.3
USGS	5480500	3/13/1984	2250	1	FALSE	FALSE	FALSE	FALSE	TRUE	670	87.0	89.83	643.8	15,450.2	670	87.0	89.83	643.8	15,450.2	1,287.5	30,900.3
USGS	5480500	3/14/1984	2100	1	FALSE	FALSE	FALSE	FALSE	TRUE	670	87.0	89.83	643.8	15,450.2	670	87.0	89.83	643.8	15,450.2	1,287.5	30,900.3
USGS	5480500	3/15/1984	2400		FALSE	FALSE	FALSE	FALSE	TRUE	670	87.0	89.83	643.8	15,450.2	670	87.0	89.83	643.8	15,450.2	1,287.5	30,900.3
USGS	5480500	3/16/1984	2600	1	FALSE	FALSE	FALSE	FALSE	TRUE	670	87.0	89.83	643.8	15,450.2	670	87.0	89.83	643.8	15,450.2	1,287.5	30,900.3
USGS	5480500	3/17/1984	2400		FALSE	FALSE	FALSE	FALSE	TRUE	670	87.0	89.83	643.8	15,450.2	670	87.0	89.83	643.8	15,450.2	1,287.5	30,900.3
USGS	5480500	3/18/1984	2200	1	FALSE	FALSE	FALSE	FALSE	TRUE	670	87.0	89.83	643.8	15,450.2	670	87.0	89.83	643.8	15,450.2	1,287.5	30,900.3
USGS	5480500	3/19/1984	2100	1	FALSE	FALSE	FALSE	FALSE	TRUE	670	87.0	89.83	643.8	15,450.2	670	87.0	89.83	643.8	15,450.2	1,287.5	30,900.3
USGS	5480500	3/20/1984	2080		FALSE	FALSE	FALSE	FALSE	TRUE	670	87.0	89.83	643.8	15,450.2	670	87.0	89.83	643.8	15,450.2	1,287.5	30,900.3
USGS	5480500	3/21/1984	2000		FALSE	FALSE	FALSE	FALSE	TRUE	670	87.0	89.83	643.8	15,450.2	670	87.0	89.83	643.8	15,450.2	1,287.5	30,900.3
USGS	5480500	3/22/1984	2260		FALSE	FALSE	FALSE	FALSE	TRUE	670	87.0	89.83	643.8	15,450.2	670	87.0	89.83	643.8	15,450.2	1,287.5	30,900.3
USGS	5480500	3/23/1984	3780		FALSE	FALSE	FALSE	FALSE	TRUE	670	87.0	89.83	643.8	15,450.2	670	87.0	89.83	643.8	15,450.2	1,287.5	30,900.3
USGS	5480500	3/24/1984	4930		FALSE	FALSE	FALSE	FALSE	TRUE	670	87.0	89.83	643.8	15,450.2	670	87.0	89.83	643.8	15,450.2	1,287.5	30,900.3
USGS	5480500	3/25/1984	5710		FALSE	FALSE	FALSE	FALSE	TRUE	670	87.0	89.83	643.8	15,450.2	670	87.0	89.83	643.8	15,450.2	1,287.5	30,900.3
USGS	5480500	3/26/1984	6520		FALSE	FALSE	FALSE	FALSE	TRUE	670	87.0	89.83	643.8	15,450.2	670	87.0	89.83	643.8	15,450.2	1,287.5	30,900.3
USGS	5480500	3/27/1984	6920		FALSE	FALSE	FALSE	FALSE	TRUE	670	87.0	89.83	643.8	15,450.2	670	87.0	89.83	643.8	15,450.2	1,287.5	30,900.3
USGS	5480500	3/28/1984	7350		FALSE	FALSE	FALSE	FALSE	TRUE	670	87.0	89.83	643.8	15,450.2	670	87.0	89.83	643.8	15,450.2	1,287.5	30,900.3
USGS	5480500	3/29/1984	7690		FALSE	FALSE	FALSE	FALSE	TRUE	670	87.0	89.83	643.8	15,450.2	670	87.0	89.83	643.8	15,450.2	1,287.5	30,900.3
USGS	5480500	3/30/1984	7710		FALSE	FALSE	FALSE	FALSE	TRUE	670	87.0	89.83	643.8	15,450.2	670	87.0	89.83	643.8	15,450.2	1,287.5	30,900.3
USGS	5480500	3/31/1984	7490		FALSE	FALSE	FALSE	FALSE	TRUE	670	87.0	89.83	643.8	15,450.2	670	87.0	89.83	643.8	15,450.2	1,287.5	30,900.3
													Total- March 1984	**478,954.9**					**478,954.9**		**957,909.8**

(continued)

TABLE 12.13 (*Continued*)

USGS	5480500	4/1/1984	7190	FALSE	FALSE	FALSE	TRUE	670	87.0	89.83	643.8	15,450.2	670	87.0	89.83	643.8	15,450.2	1,287.5	30,900.3
USGS	5480500	4/2/1984	6900	FALSE	FALSE	FALSE	TRUE	670	87.0	89.83	643.8	15,450.2	670	87.0	89.83	643.8	15,450.2	1,287.5	30,900.3
USGS	5480500	4/3/1984	7090	FALSE	FALSE	FALSE	TRUE	670	87.0	89.83	643.8	15,450.2	670	87.0	89.83	643.8	15,450.2	1,287.5	30,900.3
USGS	5480500	4/4/1984	8220	FALSE	FALSE	FALSE	TRUE	670	87.0	89.83	643.8	15,450.2	670	87.0	89.83	643.8	15,450.2	1,287.5	30,900.3
USGS	5480500	4/5/1984	8840	FALSE	FALSE	FALSE	TRUE	670	87.0	89.83	643.8	15,450.2	670	87.0	89.83	643.8	15,450.2	1,287.5	30,900.3
USGS	5480500	4/6/1984	8790	FALSE	FALSE	FALSE	TRUE	670	87.0	89.83	643.8	15,450.2	670	87.0	89.83	643.8	15,450.2	1,287.5	30,900.3
USGS	5480500	4/7/1984	8870	FALSE	FALSE	FALSE	TRUE	670	87.0	89.83	643.8	15,450.2	670	87.0	89.83	643.8	15,450.2	1,287.5	30,900.3
USGS	5480500	4/8/1984	9620	FALSE	FALSE	FALSE	TRUE	670	87.0	89.83	643.8	15,450.2	670	87.0	89.83	643.8	15,450.2	1,287.5	30,900.3
USGS	5480500	4/9/1984	11400	FALSE	FALSE	FALSE	TRUE	670	87.0	89.83	643.8	15,450.2	670	87.0	89.83	643.8	15,450.2	1,287.5	30,900.3
USGS	5480500	4/10/1984	12000	FALSE	FALSE	FALSE	TRUE	670	87.0	89.83	643.8	15,450.2	670	87.0	89.83	643.8	15,450.2	1,287.5	30,900.3
USGS	5480500	4/11/1984	13100	FALSE	FALSE	FALSE	TRUE	670	87.0	89.83	643.8	15,450.2	670	87.0	89.83	643.8	15,450.2	1,287.5	30,900.3
USGS	5480500	4/12/1984	13300	FALSE	FALSE	FALSE	TRUE	670	87.0	89.83	643.8	15,450.2	670	87.0	89.83	643.8	15,450.2	1,287.5	30,900.3
USGS	5480500	4/13/1984	13900	FALSE	FALSE	FALSE	TRUE	670	87.0	89.83	643.8	15,450.2	670	87.0	89.83	643.8	15,450.2	1,287.5	30,900.3
USGS	5480500	4/14/1984	14500	FALSE	FALSE	FALSE	TRUE	670	87.0	89.83	643.8	15,450.2	670	87.0	89.83	643.8	15,450.2	1,287.5	30,900.3
USGS	5480500	4/15/1984	14800	FALSE	FALSE	FALSE	TRUE	670	87.0	89.83	643.8	15,450.2	670	87.0	89.83	643.8	15,450.2	1,287.5	30,900.3
USGS	5480500	4/16/1984	14900	FALSE	FALSE	FALSE	TRUE	670	87.0	89.83	643.8	15,450.2	670	87.0	89.83	643.8	15,450.2	1,287.5	30,900.3
USGS	5480500	4/17/1984	15200	FALSE	FALSE	FALSE	TRUE	670	87.0	89.83	643.8	15,450.2	670	87.0	89.83	643.8	15,450.2	1,287.5	30,900.3
USGS	5480500	4/18/1984	15000	FALSE	FALSE	FALSE	TRUE	670	87.0	89.83	643.8	15,450.2	670	87.0	89.83	643.8	15,450.2	1,287.5	30,900.3
USGS	5480500	4/19/1984	14100	FALSE	FALSE	FALSE	TRUE	670	87.0	89.83	643.8	15,450.2	670	87.0	89.83	643.8	15,450.2	1,287.5	30,900.3
USGS	5480500	4/20/1984	13200	FALSE	FALSE	FALSE	TRUE	670	87.0	89.83	643.8	15,450.2	670	87.0	89.83	643.8	15,450.2	1,287.5	30,900.3
USGS	5480500	4/21/1984	12400	FALSE	FALSE	FALSE	TRUE	670	87.0	89.83	643.8	15,450.2	670	87.0	89.83	643.8	15,450.2	1,287.5	30,900.3
USGS	5480500	4/22/1984	12500	FALSE	FALSE	FALSE	TRUE	670	87.0	89.83	643.8	15,450.2	670	87.0	89.83	643.8	15,450.2	1,287.5	30,900.3
USGS	5480500	4/23/1984	13400	FALSE	FALSE	FALSE	TRUE	670	87.0	89.83	643.8	15,450.2	670	87.0	89.83	643.8	15,450.2	1,287.5	30,900.3
USGS	5480500	4/24/1984	13700	FALSE	FALSE	FALSE	TRUE	670	87.0	89.83	643.8	15,450.2	670	87.0	89.83	643.8	15,450.2	1,287.5	30,900.3
USGS	5480500	4/25/1984	12500	FALSE	FALSE	FALSE	TRUE	670	87.0	89.83	643.8	15,450.2	670	87.0	89.83	643.8	15,450.2	1,287.5	30,900.3

From the computations tabulated from daily flow data (Table 12.13), some examples will illustrate the computational process.

Example 1: From Table 12.13, page 1, date 1/1/1981, run of river flow is 198 cfs, i.e., $Q = 198$. Testing this flow with the five conditions:

Condition 1   IF(198 < 268, $Q_1 = 0$, $Q_2 = 0$) TRUE; thus proceed with computations to next columns and with $Q_1 = 0$ cfs; then power from Turbine 1 is 0 kW and energy produced from Turbine 1 is 0 kWh; and with $Q_2 = 0$ cfs, the power for Turbine 2 is 0 kW and 0 kWh.

Example 2: Consider date, 2/17/1981, run of river flow is 300 cfs. Testing this flow with the five conditions:

Condition 1   IF($Q < q_{min}$, $Q_1 = 0$, $Q_2 = 0$)
              IF(300 < 268, $Q_1 = 0$, $Q_2 = 0$) FALSE
Condition 2   IF(AND($Q \geq q_{min}$, $Q \leq q_{max}$), $Q_1 = Q$, $Q_2 = 0$)
              IF(AND(300 ≥ 268, 300 ≤ 670, $Q_1 = 300$, $Q_2 = 0$) TRUE
Condition 3   IF(AND($Q > q_{max}$, $Q \leq q_{max} + q_{min}$), $Q_1 = Q - q_{min}$, $Q_2 = q_{min}$
              IF(AND(300 > 670, 300 ≤ 670 + 268), $Q_1 = 300 - 268$, $Q_2 = 268$ FALSE
Condition 4   IF(AND(Q > $q_{max} + q_{min}$, $Q \leq q_{max} + q_{max}$), $Q_1 = Q - q_{max}$, $Q_2 = q_{max}$
              IF(AND(300 > 670 + 268, 300 ≤ 670 + 268), $Q_1 = 300 - 670$, $Q_2 = 670$
              FALSE
Condition 5   IF($Q > q_{max} + q_{max}$, $Q_1 = q_{max}$, $Q_2 = q_{max}$)
              IF(300 > 670 + 670, $Q_1 = 670$, $Q_2 = 670$) FALSE

Thus, computations are done according to Condition 2 being TRUE. $Q_1 = 300$ cfs; E_t of Turbine 1 is 82.8% from the turbine efficiency table above at $Q_1 = 300$ cfs.

E_{gtot} is computed from the related electromechanical efficiencies present above as $E_{gen} \times E_{gear} \times E_{trans}$, which is (94.5/100 × 97.0/100 × 98.0/100 = 0.8983).

$H = 14.5$ feet.

From Equation (12.1), we have KW = 300 cfs × 14.5 feet × 82.8/00 × 0.8983/ 11.8 = 274.3 kW. For 24 hours in a day the energy production is 6,582 kWh from Turbine 1. Turbine-2 output is 0 kWh, since the flow to this turbine is 0 cfs.

Total energy from Turbines 1 and 2 = 6,582 + 0 = 6,582 kWh.

Example 3: Consider date, 6/3/1981, run of river flow is 935 cfs. Testing this flow with the five conditions:

Condition 1   IF($Q < q_{min}$, $Q_1 = 0$, $Q_2 = 0$)
              IF(935 < 268, $Q_1 = 0$, $Q_2 = 0$) FALSE
Condition 2   IF(AND($Q \geq q_{min}$, $Q \leq q_{max}$), $Q_1 = Q$, $Q_2 = 0$
              IF(AND(935>=268, 935 ≤ 670, Q1 = 935, $Q_2 = 0$) FALSE
Condition 3   IF(AND($Q > q_{max}$, $Q \leq q_{max} + q_{min}$), $Q_1 = Q - q_{min}$, $Q_2 = q_{min}$)
              IF(AND(935 > 670, 935 ≤ 670 + 268), $Q_1 = 935 - 268$, $Q_2 = 268$ TRUE
Condition 4   IF(AND($Q > q_{max} + q_{min}$, $Q \leq q_{max} + q_{max}$), $Q_1 = Q - q_{max}$, $Q_2 = q_{max}$
              IF(AND(935 > 670 + 268, 935 ≤ 670 + 268), $Q_1 = 935 - 670$, $Q_2 = 670$
              FALSE

Condition 5   IF($Q > q_{max} + q_{max}$, $Q_1 = q_{max}$, $Q_2 = q_{max}$)
　　　　　　　IF(935 > 670 + 670, $Q_1 = 670$, $Q_2 = 670$) FALSE

Thus, computations are done according to Condition 3 being TRUE. $Q_1 = 935 - 268 = 667$ cfs; E_t of Turbine 1 is 87.3% from the turbine efficiency table above at $Q_1 = 667$ cfs. $Q_2 = 268$ cfs and E_t of Turbine 2 is 79.6%. E_{gtot} is computed from the related electromechanical efficiencies present above as $E_{gen} \times E_{gear} \times E_{trans}$, which is (94.5/100 × 97.0/100 × 98.0/100 = 0.8983).

H = 14.5 feet.

From Equation (12.1), we have for Turbine 1, KW = 667 cfs × 14.5 feet ×87.3/100 × 0.8983/11.8 = 642.4 kW. For 24 hours in a day, the energy production is 15,418 kWh from Turbine 1.

From Equation (12.1), we have for Turbine 2, KW = 268 cfs × 14.5 feet × 79.6/100 × 0.8983/11.8 = 235.5 kW. For 24 hours in a day, the energy production is 5,652 kWh from Turbine 2. Total energy from Turbines 1 nd 2 = 15,418 + 5,652 = 21,070 kWh.

Computations like the above examples were done on all the daily flow data from 01-01-1981 to 12-31-2000. Details are shown in spreadsheet in Table 12.13, Fort Dodge Mill Dam Flow and Production Computations, Fort Dodge, Iowa.

A summary of energy production from the plant with monthly and annual production is shown in Table 12.12, Summary of Potential Energy Generation of Fort Dodge Mill Dam, Fort Dodge, Iowa, for the years 1981 to 2000. This summary table was generated by selecting the energy production (kWh) for a particular month and year from Table 12.13 and recording it in the proper place in Table 12.12.

REFERENCES

1. Bunch, J., "Most Promising Source of Power, Wind, As Old As Civilization," The Denver Post, April 20, 2003.
2. DeWitte, D., "Wind Farm Weathers Failure," The Gazette, May 8, 2003.
3. GE Energy, 1.5 MW Series Wind Turbine, www.gepower.com, June 22, 2009.
4. U.S. Department of Energy, "Energy Efficiency and Renewable Energy Wind and Hydropower Technologies Program, How Wind Turbines Work," wwwl.eere.energy.gov, June 6, 2009.
5. Young, G. C., "Why Big Wind Wins Out," Public Utilities Fortnightly, April 2004.
6. Young, G.C., "The Economics of Wind: Looking at MidAmerican Energy," Public Utilities Fortnightly, August 2003.
7. Boshart, R., "Wind Plant to Generate Revenue," The Gazette, March 26, 2003.
8. Private Communication, Dave Dewitte, The Gazette, April 22, 2003.
9. Phillips, C., "Private Communications, Financial and Economic Review," Universal Electric Power, Inc., May 2003.
10. Sinclair, K., "A Population Study of Golden Eagles in the Altamont Pass Wind Resource Area: Population Trend Analysis 1994-1997," NREL/SR-500-26092, June 1999.
11. Hunt, W.G., Jackman, R.E., Hunt, T.L., Driscoll, D.E., and Culp, L., 1998, "A Population Study of Golden Eagles in the Altamont Pass Wind Resource Area: Population Trend

Analysis," 1997. Report To National Renewable Energy Laboratory, Subcontract XAT-6-16459-01. Predatory Bird Research Group, University of California, Santa Cruz, NREL/SR-500-26092, June 1999.

12. Hunt, G., "Golden Eagles in a Perilous Landscape: Predicting the Effects of Mitigation for Wind Turbine Blade-Strike Mortality," P500-02-043F, July 2002, www.energy.ca.gov.

13. Sinclair, K.C., "Status of Avian Research at the National Renewable Energy Laboratory," National Renewable Energy Laboratory (NREL), NREL/CP-500-30546, September 2001.

14. Hunt, G. and Hunt, T., "The Trend of Golden Eagle Territory Occupancy in the Vicinity of the Altamont Pass Wind Resource Area: 2005 Survey," California Energy Commission, PIER Energy-Related Environmental Research. CEC-500-2006-056, June 2006.

15. National Renewable Energy Laboratory, NREL, Photographic Information Exchange, http://www.nrel.gov/data/pix/, June 2009.

16. USGS, "Sources of electricity in the USA 2006," Data from http://www.eia.doe.gov/cneaf/electricity/epa/eapt1p1.html, 07/09/2009.

17. USGS, "Hydroelectric Power," http://ga.water.usgs/edu/wuhy.html, 07/09/2009.

18. U.S. Army Corps of Engineers & USGS, http://ga.water.usgs.gov/edu/hyhowworks.html, 07/09/2009.

19. Noyes, R., "Small And Micro Hydroelectric Power Plants, Technology and Feasibility," Noyes Data Corporation, Park Ridge, New Jersey, U.S.A., 1980.

20. Energy Policy Act of 2005, 109th Congress, United States, July 27, 2005.

21. American Recovery and Reinvestment Act of 2009, Payments for Specified Energy Property in Lieu of Tax Credits, U.S. Treasury Department, Office of the Fiscal Assistant Secretary, July 2009.

22. Young, G.C., Rieck, K.V., and Phillips, C., "The Economics of Low-Head Dams," Public Utilities Fortnightly, December 2005.

23. Sinclair, D., "Private Communications," Advanced Hydro Solutions, LLP, 2004.

CHAPTER 13

Waste Energy to Recycled Energy

INTRODUCTION

Of interest today is the cost of energy and particularly the efforts to reduce energy costs by using less energy in a system. Many have studied the transfer of energy from one system to another. This field of study is called thermodynamics. From the past studies, it was concluded that energy is conserved. Thus, energy can be transferred from one form to another or from one body to another body, but it is neither created nor destroyed. This conclusion was expressed as the first law of thermodynamics or sometimes referring to as the law of conservation of energy. However in recent history, Einstein studied the interrelation of energy and mass and its application to the atomic energy field and pointed out that energy is not conserved. However, in the normal chemical and physical processes, the loss or gain in mass is undetectable, and the independent conservation of mass and energy can be assumed with confidence.[1]

The energy balance makes use of the "first law of thermodynamics," which states that energy is conserved, i.e., mass and energy are neither created nor destroyed. Consequently, an energy balance is typically determined about a boundary within a process, which is called the "system." The area outside the system boundary is called the "surroundings." The system and surroundings are therefore called the universe. Thus, the sum of the energy of the system and surroundings is constant, i.e., it remains conserved according to the first law of thermodynamics. Therefore after the "system" is defined by a boundary, the balance of energy entering the system across the boundary, the energy leaving the system across the boundary, and the accumulation of energy within the system is constant. The application of the first law of thermodynamics regarding the energy about a boundary within a system can be determined by Equation (13.1).

$$\text{Energy In} = \text{Energy Out} + \text{Energy Accumulation} \qquad (13.1)$$

The application, (Eq. (13.1)), of the first law of thermodynamics about a boundary within a process describing a system is called an "energy balance" of the system. All

Municipal Solid Waste to Energy Conversion Processes: Economic, Technical, and Renewable Comparisons By Gary C. Young
Copyright © 2010 John Wiley & Sons, Inc.

forms of energy must be considered in the energy balance, which for a typical chemical process involving a flow of fluid can be kinetic energy (KE), potential energy (PE), internal energy (E), heat energy (Q), and work energy (W). Energy of a moving object is "kinetic" energy. "Potential" energy is determined by the position in the earth's gravitational field. The motion of molecules of a substance imparts internal energy within the substance and is called "internal" energy. The temperature of a material is a measure of its internal energy. The "second law of thermodynamics" states that for a process involving heat transfer alone, energy is only transferred from a higher temperature to a lower temperature.

For a conventional flow system at steady state, where accumulation is zero, Equation (13.1) becomes the energy balance for a system at steady state and is presented as Equation (13.2):

$$\text{Energy In} = \text{Energy Out}$$

$$E_1 + KE_1 + PE_1 + Q' = E_2 + KE_2 + PE_2 + W' \tag{13.2}$$

where subscripts 1 represents into the system and 2 represents out of the system. The primes designate heat and work removed from or appearing in the surroundings.

Substitution of more specific terms per pound of fluid into Equation (13.2) and rearranging as a difference, the result is:

$$\Delta E + \Delta(PV) + \Delta(v^2)/2g_c + \Delta zg/g_c = Q' - W_s \tag{13.3}$$

where:

$P =$ the force exerted upon the fluid,
$V =$ the specific volume of the fluid,
$v =$ velocity,
$z =$ height above a reference level,
$Q' =$ heat energy neglecting loss due to frictional effects,
$W_s =$ work neglecting loss due to frictional effects,
$g_c =$ dimensional conversion factor, and
$g =$ acceleration of gravity.

Equation (13.3) represents a typical expression of a total energy balance in a flow system.

By arranging a group as $(E + PV)$ and labeling enthalpy as (H), H becomes a state function much like E. Consequently, terms in Equation (13.3) become as follows:

$$\Delta H + \Delta(v^2)/2g_c + \Delta zg/g_c = Q' - W_s \tag{13.4}$$

In chemical process applications, the changes in velocity and height are small; thus, the kinetic energy term, $\Delta(v^2)/2g_c$, and the potential energy term, $\Delta zg/g_c$, are negligible, resulting in an energy balance. Equation (13.4) reduces to Equation (13.5):

$$\Delta H = Q' - W_s \tag{13.5}$$

In a similar manner for a constant-pressure process, which is true for many systems in a chemical process, Equation (13.5) can be shown as:

$$\Delta H = Q' \tag{13.6}$$

for which Equation (13.6) states that the increase in the enthalpy of a system in a constant-pressure process is equal to the heat added to the system. With many processes in the chemical industry as constant-pressure processes, Equation (13.6) will be shown as a useful tool in solving process energy problems.

A precise and exceptional development of energy balance equations as applied to systems can be found in this reference.[1] This educational material is priceless for the academic professional desiring a true understanding of energy balances and proper application to systems.

Energy transferred between a system and the surroundings is either "heat" or "work." Heat and work are considered energy in transit. Work depends upon the path followed by the system.

Heat represents energy in transition under the influence of temperature differences. Energy flows as "heat" from a hot body to a cold body. Heat is usually expressed in units of "Btu" or "gram-calorie."

When the various forms of energy are added algebraically for a total energy balance at steady state, i.e., for "no" accumulation, Equation (13.1) becomes "energy in" equals "energy out" and various energy terms can be rearranged and grouped together. In so doing, a group is placed together consisting of internal energy plus pressure times specific volume per unit mass of the system, which is called "enthalpy." The enthalpy is a widely used function and is a state function. The usefulness of enthalpy is that for a constant-pressure process, the change in enthalpy is equal to the heat added to the system. Because so many constant-pressure processes are so common in chemical processes, the enthalpy function becomes important. The useful relationship therefore becomes Equation (13.6) and is used widely for solving many energy process problems. A more detailed and definitive understanding of the "energy balance" can be found in reference materials.[1,2]

The mass balance is based upon the "law of conservation of mass." The "law of conservation of mass" states that mass can be neither created nor destroyed. Thus, the mass of a closed system must be constant or conserved. Mass can change form but the total mass must not be changed. From the "law of conservation of mass," Equation (13.7) applies.[1]

$$\text{Mass In} = \text{Mass Out} + \text{Mass Accumulation} \tag{13.7}$$

Many solutions in the industrial and chemical processes today are derived from the use of these laws: "law of conservation of mass," "first law of thermodynamics," and "second law of thermodynamics." A real and practical application using the "law of conservation of mass," Equation (13.7), the "first law of thermodynamics,"

Equation (13.1), and "second law of thermodynamics," will be illustrated involving the recovery and recycling of heat in a system.

The heat investigated is atmospheric waste heat from a concentrator emitted into the atmosphere. The concentrator exists in a chemical plant. This concentrator is an agitated vessel with steam coils using plant boiler steam in the coils. The concentrator concentrates a stream for further use in the chemical process. The concentrator discharges about 9,000 lbs/hour of steam to the atmosphere at a temperature of about 212°F. This process information of the concentrator vessel operation was determined and known for years but without a solution. The problem was: What can be done with this low atmospheric pressure steam (0.0 psig, 14.7 psia) that is wasted to the atmosphere? A representation of the agitated concentrator is shown in Fig. 13.1. It had been concluded in the past by others for this particular chemical process plant that nothing could be done technically and/or economically with this atmospheric steam, i.e., low-pressure steam with essentially no pressure driving force to be used elsewhere in the plant.

Assessing a process plant that is nearly 50 + years old presents a challenge but is not daunting. The most difficult aspect of the problem initially is the attitude of the vast majority of personnel at the plant; it has been investigated many times before in the past by others, both professionals and nonprofessionals, and concluded to be an "energy waste" problem with no solution. However one elderly chemist, a son of the founder of the original process and company, was unique in that he could discuss and accept technical process ideas based upon sound and scientific reasoning. In other words, he could, "think outside the box on scientific matters." This was the key ingredient that the new process engineer needed, since he had much of the same technical characteristics. The new process engineer was experienced and a

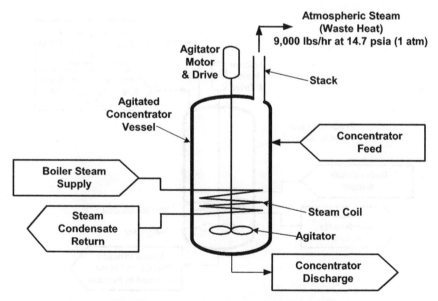

FIGURE 13.1 Agitated Concentrator Vessel.

professional, i.e., this individual could be labeled respectfully as "the engineer." Thus, the new process engineer had support from the top and could discuss freely on several occasions with this wise elderly son of the founder, i.e., label him with respect as a "chemist." Thus, a true technical team had developed with the proper professional skills and management support for potential future success with a "can do" attitude. The team was simply composed of a "chemist" and an "engineer."

The chemist and engineer, who comprised the team, discussed the chemistry and process operations throughout the plant for several months. The engineer asked about operations with process streams at temperatures below the concentrator stack temperature of 212°F. After several months of discussions, one operation of the plant appeared to have some process streams at temperatures below 212°F. This operation is shown in Fig. 13.2 associated with the process liquor preheat vessel. Process streams come from the crystallizer operation, stream (1) and stream (2). Process liquor feed stream (1) is an aqueous stream at 129.2°F and at a flow rate of 90 gallons per minute (GPM). Process liquor feed stream (2) is an aqueous stream at 96.8°F and at a flow rate of 636 GPM.

These two process liquor feed streams are heated in the agitated process liquor preheat vessel to a temperature of 111.2°F and are called the process discharge liquor. Note, the two process liquor feed streams are at temperatures below 212°F, which is the concentrator stack atmospheric steam temperature. With this process information, the next step was to do a further process evaluation to formulate a potential waste heat recovery system that was both technically and economically feasible.

Referring to Fig. 13.2, a process evaluation will be done first by determining a mass balance about the process liquor preheat vessel according to Equation (13.7). Secondly, an energy balance according to Equations (13.1) and (13.6) is done.

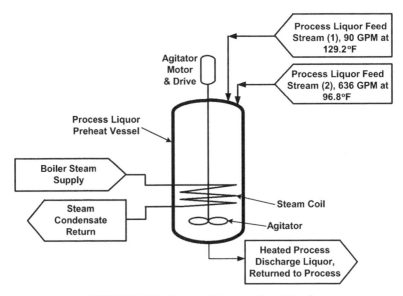

FIGURE 13.2 Process Liquor Preheat Vessel.

A mass balance using Equation (13.7) is conducted on numbered streams encircled as streams 5, 6, and 7 as shown in Fig. 13.3. The computations below illustrate the mass balance of the liquor streams entering and leaving this vessel to determine individual stream mass flow rates:

$$Q_m(\text{lbs/hour}) = Q_v \times 60 \times \text{Sp.gr.} \times 8.34 \tag{13.8}$$

where

Q_m = mass flow rate, lbs/hour

Q_v = volumetric flow rate, gallons/minute,

60 = constant, minutes/hour,

Sp.gr. = specific gravity of liquid at about 68°F in reference to liquid water,

8.34 = constant, lbs/gallon.

Stream 5 is an aqueous stream having the properties: 90 GPM and 1.20 specific gravity. Substitution into Equation (13.8) yields the following:

$$\begin{aligned}Q_m \text{ of stream } 5 &= 90\, \text{gallons/minute} \times 60\, \text{minutes/hour} \times 1.20 \times 8.34\, \text{lbs/gallon} \\ &= 54{,}043\, \text{lbs/hour}\end{aligned}$$

For stream 6 with properties: 636 GPM and 1.20 specific gravity, substitution into Equation (13.8) yields:

$$\begin{aligned}Q_m \text{ of stream } 6 &= 636\, \text{gallons/minute} \times 60\, \text{minutes/hour} \times 1.20 \times 8.34\, \text{lbs/gallon} \\ &= 381{,}905\, \text{lbs/hour}\end{aligned}$$

From the mass balance about the process liquor streams 5, 6, and 7 as shown encircled in Fig. 13.3 and using the mass balance equation, Equation (13.7), at steady state whereby the accumulation is zero and stream 7 has a Sp.gr. of 1.20:

$$\text{Mass (stream 5)} + \text{Mass (stream 6)} = \text{Mass (stream 7)}$$

Therefore, 54,043 lbs/hour + 381,905 lbs/hour = Q_m (stream 7) × 60 × 1.20 × 8.34

or, Q_m (stream 7) = 726 GPM or 435,948 lbs/hour.

Secondly, the energy balance is computed according to Equations (13.1) and (13.6). According to Equation (13.6), the energy balance about the process liquor vessel shown by the encirclement of streams 5, 6, and 7 is:

$$\Delta H \text{ (stream 5)} + \Delta H \text{ (stream 6)} = \Delta H \text{ (stream 7)} \tag{13.9}$$

For the aqueous streams, enthalpy change is computed by Equation (13.10).

$$\Delta H \text{ (aqueous stream)} = Q_m \times C_p \times (T_o - T_i) \tag{13.10}$$

where,

ΔH (aqueous stream) = enthalpy change of the aqueous stream, Btu/hour

C_p = heat capacity of the aqueous stream, Btu/lb-°F

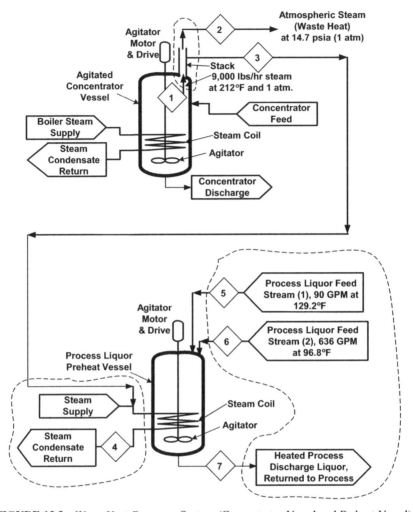

FIGURE 13.3   Waste Heat Recovery System (Concentrator Vessel and Preheat Vessel).

T_i = inlet temperature of the aqueous stream

T_o = outlet temperature of the aqueous stream

Consequently for stream 5 where the heat capacity is 1.0 Btu/lb-°F, inlet temperature T_i is 129.2°F, and outlet temperature of 111.2°F, computations using Equation (13.10) become:

$$\Delta H \text{ (stream 5)} = 54,043 \text{ lbs/hour} \times 1.0 \text{ Btu/lb-°F} \times (111.2°F\text{-}129.2°F)$$
$$= -972,774 \text{ Btu/hour}$$

For stream 6 where the heat capacity is 1.0 Btu/lb-°F, inlet temperature T_i is 96.8°F, and outlet temperature is 111.2°F, enthalpy change computed by

Equation (13.10) becomes:

$$\Delta H \text{ (stream 6)} = 381,905 \text{ lbs/hour} \times 1.0 \text{ Btu/lb-}°\text{F} \times (111.2°\text{F-}96.8°\text{F})$$
$$= 5,499,432 \text{ Btu/hour}$$

Substituting into Equation (13.9), the enthalpy change about streams 5, 6, and 7 becomes:

$$5,499,432 \text{ Btu/hour} + (-972,774) \text{ Btu/hour} = 4,526,658 \text{ Btu/hour}$$

From an overall energy balance about the process liquor preheat vessel, the change in enthalpy of streams 5, 6, and 7 equals the heat input from the boiler steam supply/ steam condensate as noted by stream 4 in Fig. 13.3. Thus, if the latent heat of condensation of the boiler steam to the steam coil in the process liquor preheat vessel is about 1,000 Btu/lb, then the estimate for the boiler steam supply to the coil is approximately:[3]

$$4,526,658 \text{ Btu/hour}/1,000 \text{ Btu/lb condensate} = 4,527 \text{ lbs/hour (condensate)}$$

Thus, the steam supply to the process liquor preheat vessel is estimated to be 4,527 lbs/hour.

With a supply of atmospheric steam from the concentrator stack at 9,000 lbs/hr, sufficient steam is available from the concentrator stack, i.e., (surplus of 9,000 lbs/ hr–4,527 lbs hr = 4,473 lbs/hour steam surplus). The surplus is important as will be discussed later. Also, the concentrator and the process liquor preheat units are in operation at the same time continuously. In conclusion from the preliminary energy recovery discussions and analysis, a recovery system is possible provided a technical and practical energy recovery system is developed for using the atmospheric waste heat from the concentrator stack to preheat the process liquor feed streams 5 and 6 to 111.2°F for return to the process as heated process discharge liquor, stream 7.

Now, the question arises, can a heat recovery system be developed that can technically and economically recover the potential waste heat of about 4,526,658 lbs/ hr at 212°F and 1 atmosphere from the concentrator system and be used for the process liquor preheat system as envisioned in Fig. 13.3.? It helps to summarize what is known and what is not as follows:

- Waste energy source (concentrator stack) has steam at 9,000 lbs/hour at 212°F and 1 atmosphere.
- Use for the waste energy source is potentially the process liquor preheat system requiring 4,526,658 Btu/hour steam to preheat mixed aqueous process streams, stream 5 at 90 GPM and 129.2°F and stream 6 at 636 GPM and 96.8°F, to a final mixture, heated process discharge liquor, for return to the plant process at 111.2°F.
- The source of waste heat, concentrator vessel and end use of this waste heat, process liquor preheat vessel, are located at far ends and opposite to each other in the large plant.

- Both the concentrator system and the preheat liquor system operate simultaneously and continuously so that waste heat supply and waste heat end use are fully compatible.
- Finally, what heat recovery system, both technically and economically, can be developed as illustrated in Fig. 13.3 when the waste heat steam supply at atmospheric pressure is to be used by another operation at considerable distance from the low-pressure steam source?

After considerable thought, the engineer went through the following analyses to determine the technical feasibility for the waste heat process described in Fig. 13.3.

1. What size of steam coil is required for the process liquor preheat vessel and, in addition, will it fit into that vessel?

   a. Determine the overall heat transfer coefficient for the steam coil according to heat transfer Equation (13.11).[4]

$$1/U_o = 1/(1/h_1 + 1/h_o + 1/h_d) \qquad (13.11)$$

where

U_o = overall heat transfer coefficient, Btu/hr-ft^2-°F

h_i = inside film heat transfer coefficient, Btu/hr-ft^2-°F

h_o = outside film heat transfer coefficient, Btu/hr-ft^2-°F

and h_d = fouling heat transfer coefficient, Btu/hr-ft^2-°F

h_o is determined by Equation (13.12).[4]

$$h_o = 0.00285(L^2 N\rho/\mu)^{0.7} \qquad (13.12)$$

where

L = length of agitator paddle, diameter, feet

N = agitator speed, rev/hr

ρ = fluid density, lbm/ft^3

μ = viscosity of fluid, lbm/ft-hr

The properties of the fluid and vessel are:

L = 6.125 feet

N = 6000 rev/hr

ρ = 74.9 lb/ft^3

μ = 1.694 lbm/ft-hr

and substitution into Equation (13.12),

$$h_o = 0.00285 \,(6.1252^2 \times 6,000 \times 74.9/1.694)^{0.7} = 226 \text{ Btu/hr-ft}^2\text{-°F}$$

h_i equals 1,500 Btu/hr-ft^2-°F for steam condensation inside a steam coil.[2]

h_i equals 1,500 Btu/hr-ft^2-°F for an aqueous system.[4]

Substitution of h_i, h_o, and h_d into the overall heat transfer Equation (13.11), becomes:

$$1/U_o = 1/(1/1,500 + 1/226 + 1/1,500)$$

or $U_o = 174$ Btu/hr-ft^2-°F, which is the overall heat transfer coefficient for the steam coil in the process liquor preheat vessel.

b. Determine the steam condensation temperature inside the steam coil of the process liquor preheat vessel and various coil sizes using the heat transfer equation (Eq. (13.13)).

$$Q = U_o A_o \Delta T_{lm} \tag{13.13}$$

where

Q = the energy transferred from the waste steam through the coil to the process liquor, Btu/hour,

U_o = overall heat transfer coefficient, Btu/hr-ft^2-°F

A_o = outside heat transfer area of coil, ft^2

ΔT_{lm} = logarithmic mean temperature difference, LMTD, °F

For the steam coil heat transfer, $U_o = 174$ Btu/hr-ft^2-°F, $Q = 4,526,658$ Btu/hour. The logarithmic mean temperature difference is determined from Equation (13.13) but with a modification to Equation (13.14) because the process liquor in the preheat vessel will be considered well mixed and isothermal as well as the steam condensate inside the coil.

$$\Delta T_{lm} = \Delta T = (T_s - T_o) \tag{13.14}$$

where

ΔT = temperature difference between mixed process liquor in vessel, °F

T_s = saturated temperature of steam condensate in coil, °F

T_o = temperature of process liquor in vessel, °F.

$T_o = 111.2$°F and T_s and A_o are unknowns after substitution of the known data into Equations (13.13) and (13.14) Equation (13.15) results.

$$4,526,658 \text{ Btu/hr} = 174 \text{ Btu/hr-ft}^2\text{-°F} \times A_o(\text{ft}^2) \times (T_s - 111.2) \text{ (°F)} \tag{13.15}$$

In Equation (13.15), A_o and T_s are unknowns. However, if A_o is selected, a corresponding T_s can be computed. An example of a few computations with Equation (13.15) is shown in Table 13.1.

2. From Table 13.1, a selected coil size (heat transfer surface area) will determine the necessary waste steam coil condensation temperature (°F) and the corresponding saturated pressure (psia). For example, a steam coil surface area of 300 ft^2 equates to a corresponding steam coil waste steam condensation temperature of 197.7°F and saturation pressure of 11.035 psia. With the concentrator waste heat stack at 1 atmosphere (14.7 psia), the pressure differential between the concentrator waste heat stack and the steam coil in the liquor preheat vessel is 3.665 psi, i.e. (14.7−11.035 = 3.665 psi). Then the 3.665 psi

TABLE 13.1   Corresponding Coil Surface Area and Steam Coil Condensation Temperature

Selected Area (A_o)	Computed Steam Coil Condensation Temperature	Corresponding Steam Saturation Pressure
300 ft^2	197.9°F	11.035 psia
380 ft^2	179.7°F	7.462 psia
400 ft^2	176.2°F	6.901 psia

differential becomes the driving force to move waste heat steam in the concentrator stack to the steam coil in the liquor preheat vessel. For the conditions shown in Table 13.1, the driving force (pressure differential) for moving waste heat to the steam coil is presented in Table 13.2.

3. In summation, a heat recovery system is technically feasible if the following conditions are satisfied:
   - Waste heat concentrator stack has at least 4,526,658 Btu/hour of steam at 212°F and 1 atmosphere (14.7 psia).
   - Steam coil in process liquor preheat vessel is about 380 ft^2 heat transfer area for a waste heat steam condensation temperature of about 179.7°F corresponding to a saturated steam condensation temperature of 7.462 psia.
   - Differential pressure available to transport steam from the concentrator waste heat stack to the steam coil of the process liquor preheat vessel is about 7.462 psia.

4. The next step is to determine whether a steam coil with about 380 ft^2 of heat transfer area can fit into the process liquor preheat vessel.

   The process liquor preheat vessel has physical limitations due to the space available in the agitated tank for baffles and available liquid submergence due to the height of the vessel. Consequently, it was logical to consider a steam coil of 10 ft 2 in. O.D. and about 7 ft 0 in. overall height. For practicality, a 4 in. pipe coil, 4.50 in. O.D., schedule-40 carbon steel pipe with 1.5 coil turns/foot, or 8 in. centerline-to-centerline distance between pipe coils was used. This permitted a 3.5 in. open space between coils to avoid pluggage from a crystalline buildup, since the process liquor can have carryover solids from upstream process units. The effective heat transfer area based upon the outside surface area of the coil

TABLE 13.2   Corresponding Coil Surface Area and Steam Differential Pressure

Selected Area (A_o)	Computed Steam Condensation Pressure	Corresponding Steam Pressure Differential (Between Waste Heat Stack And Process Liquor Preheat Coil)
300 ft^2	11.035 psia	3.665 psi
380 ft^2	7.462 psia	7.238 psi
400 ft^2	6.901 psia	7.799 psi

Note: Waste Heat Stack at 1 atmosphere (14.7 psia).

can be determined by Equation (13.16).[5]

$$A_{co} = 3.14 \times d_{co} \times H_c \times n[(3.14 \times D_c)^2 + n^{-2}]^{0.5} \qquad (13.16)$$

where

A_{co} = effective heat transfer area of the coil, outside area, ft^2

d_{co} = outside diameter of pipe coil, ft

n = number of coil turns per foot of coil height, turns/ft, (which equals the inverse of the coil pitch

H_c = total height of coil, ft

D_c = mean or centerline diameter of internal coil helix, ft

On substitution of the following coil parameters into Equation (13.16),

$n = 1.5$ coil turns per foot

$d_{co} = 0.3750$ ft, O.D. pipe coil (4.5/12 ft)

$H_c = 7.0$ ft, total height of coil

$D_c = 9.7917$ ft, centerline diameter of coil helix (10 ft 2 inches – 4.5 inches),

$A_{co} = 3.14 \times 0.3750 \times 7.0 \times 1.5[(3.14 \times 9.7917)^2 + 1.5^{-2}]^{0.5} = 380$ ft^2

The approximate length of the pipe coil can be determined by Equation (13.17).

$$L = A_{co}/A_o \qquad (13.17)$$

where

$$A_o = 3.14 \, d_{co} \qquad (13.18)$$

A_o = heat transfer area of pipe coil per unit length, ft^2/ft

For the 4-in. schedule 40 pipe, $A_o = 3.14 \times 0.375 = 1.1775$ ft^2/ft from Equation (13.18). Thus, from Equation (13.17), L = 380 ft^2/1.1775 ft^2/ft = 323 feet, approximate length of pipe coil. A simplified illustration of a helical pipe coil is shown in Fig. 13.4. Since the steam coil needs 380 ft^2 outside heat transfer area and the coil investigated can fit into the process liquor preheat vessel, which is 323 feet of 4—in. schedule 40 carbon steel pipe having outside surface area of 380 ft^2, the heat recovery system envisioned in Fig. 13.3 is technically feasible, i. e., a technical success. As a final comment, many computations were done before coming to the valuable process feasibility so clearly presented. The project involved a multitude of computations and engineering design before the final concept was conceived and progress finally made. Some idea of the multitude of work done for the final system will be apparent in the following economic feasibility analysis for the heat recovery system as presented in Figs. 13.3 and 13.5. As a final comment, the amount of process liquor to the preheat vessel could vary considerably due to plant process upsets so that the steam rate to the process liquor preheat coil could vary between about 600 and 6,700 lbs/hour. This variation in the process heat load on the process liquor preheat coil was taken into account in the final design of the heat recovery and recycle system.

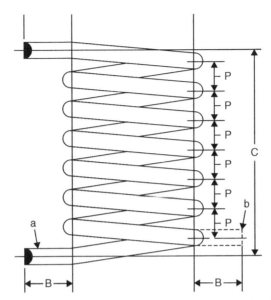

FIGURE 13.4 Simple Illustration of Helical Pipe Coil.[6]

5. Concentrator waste steam from the stack can be used to heat process liquor in
   the process liquor preheat vessel as illustrated in Fig. 13.3, i.e., technically
   feasible. Now, the next question is: is the technically sound proposed heat
   recovery system economical?

   For this waste heat recovery system, a more detailed engineering study and
   project design work was completed as illustrated in the simplified process
   and instrumentation drawing (P&ID) shown in Fig. 13.5. As an explanation
   of the system basic concepts, a large pipe carried the waste heat steam from
   the concentrator stack to the waste heat coil in the process liquor preheat
   vessel. Flow control valve, FCV-1, on the large steam line, modulated the
   waste heat steam based upon the temperature of the vessel liquor contents as
   transmitted by temperature element, TI-1, to temperature indicating controller, TIC-1. Condensate from the waste heat coil was collected in the
   condensate tank and recycled to the plant as either plant process water
   and/or boiler makeup water.

   The boiler steam coil was maintained as a backup option if needed for any
   unforeseen process downtime and supplied by boiler steam if and only if
   needed. The boiler steam coil operated using flow control valve, TCV-2, and
   temperature element, TI-1, with temperature indicating controller, TIC-1.

   To understand the capital costs for the heat recovery system, a list of major
   system components is presented below with basic information:

   • Waste heat steam coil, 4—in. pipe, carbon steel, schedule 40, 4.5 in. O.D.,
     ASTM A106B seamless approximately 318 feet coil length, 7 ft 0 in. height,
     9 ft 7.5 inches centerline diameter and 10 ft 0 in. O.D. coil, 8 in. pitch

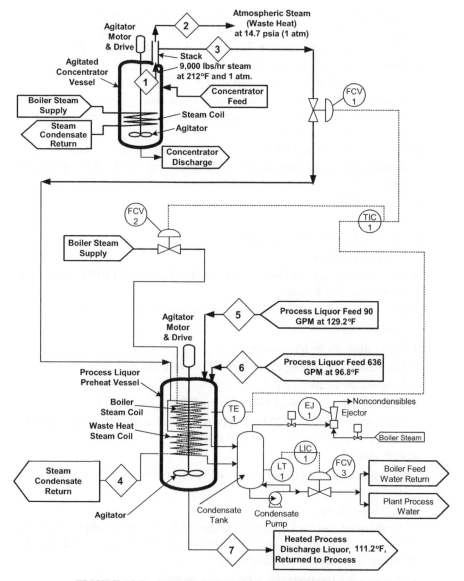

FIGURE 13.5 Simplified P&ID, Waste Heat Recovery System.

- Boiler steam coil, 4—in. pipe, carbon steel, schedule 40, ASTM A106B seamless approximately 108 feet coil length, 3 ft 0 in. approximate height, 7 ft 7.5 inches centerline diameter, 8 ft 0 in. O.D. coil, 8 in. pitch
- Pipe supports for coils
- Waste heat steam coil split in two coil sections to accommodate condensate drainage
- Check valves on condensate lines

- Ejector, EJ-1, for noncondensible removal for start-up and as needed when necessary, 2 in. diameter suction, 6 lb/hr air at 1.56 psia and 117°F, requires 145 lbs/hour motive steam at 120 psig and discharges at 14.7 psia
- Flow control valve, FCV-2, 2-in. control valve for steam service, steam valve coefficient 3.02 at 10% open and 99.0 at 100% open
- Flow control valve, FCV-1, concentrator waste heat steam from stack, 6 inch control valve for saturated steam service, steam valve coefficient 1.51 at 10% open and 855 at 60% open
- Level control valve, FCV-3, for level control in condensate tank, 1—inch control valve, $^3/_4$-in. port diameter, 5% open C_v or 0.276, 100% open C_v of 8.84
- Steam shutoff valves
- Gages
- Level indicating controller, LIC-1
- Level transmitters,
- Condensate pump, 30 GPM at 120 T.D.H., 5 hp × 1750 RPM, for hot condensate at 252°F max.
- Pressure reducing valve for steam line on ejector steam supply
- Line strainers
- Steam traps on boiler steam lines
- Temperature indicating controller, TIC-1, with alarms
- Temperature element, TE-1
- Metal flexible corrugated hose expansion joints for steam lines
- Steel condensate tank, ASME flanged and dished heads, 30 in. O.D. × 3/8 in. thickness, 2ft 10 inches tangent to tangent
- Ball valves
- Steel pipe and fittings, which included 126 feet of 6—in. schedule 40 carbon steel pipe, and
- Installation costs.

Total capital cost for year 1994 was about $90,000.00 including installation and start-up costs. The basic economic analysis can now be conducted using the following economic specifications:

- Capital cost: $90,000.00 at 6.00% interest for 11 years
- Waste heat steam saved at 4,527 lbs/hour valued at $5.30/1,000 lbs steam
- Operation of plant at 7,446 hours/year
- O&M for this project well suited for being absorbed into existing plant budget, and
- Savings for using the waste heat condensate of 4,527 lbs/hr (9.05 GPM) as process water and/or boiler makeup water were not considered even though for this plant the savings would be appreciable.

Cost of capital,

$90,000.00 at 6.00% for 11 years, two payments annually : -$11,294,53/year

Energy savings,

4,527 lbs/hr × $5.30/1,000 lbs × 7,446 hours/year : $179,132.48/year

Net annual savings (before taxes): $167,837.95/year

Obviously, this energy conservation project is economically viable. In addition, with the project being operational in the year 1994, within a short time after start-up, the boiler operator called over to the plant and asked what had happened. The boiler operator said that the export steam meter on the boilers had dropped by about 5000 lbs/hour export steam rate.

Conclusion: the waste heat energy recovery and recycle system was technically feasible and economical.

Note, since the project was operational in year -1994, the 15 years time span to year 2009 has accumulated a total energy savings of nearly $2,517,569. Energy recovery and recycle can be very lucrative for a business.

6. Operation of the Concentrator Waste Heat/Steam Recovery System

This project covers many fundamental and key aspects of engineering in order to be successful as discussed previously. These basic engineering principles will be discussed with regard to how the waste heat recovery system is started up and operates. An engineer needs to be aware of these fundamentals involving plant processes, since such understanding insures a proper design, construction, and start-up for success, both professionally and economically for the stockholders of a corporation. With this in mind, the following discussion of the waste heat recovery system operation is presented for clarity so that fundamentals of science can be learned by the process and project design engineers to be successful.

With reference to Fig. 13.5, the following sequence for operation of the waste heat recovery system is presented.

a. Boiler steam is supplied to ejector, EJ-1, to remove noncondensables from the condensate tank.

b. Flow control valve, FCV-1, is cracked open to allow atmospheric heat atmospheric steam from the concentrator stack to displace any noncondensables in the 6-in. diameter steam line from the concentrator stack to the process liquor preheat waste heat steam coil. These noncondensables will pass through the steam coil into the condensate tank and finally exit the system to the atmosphere via the ejector, EJ-1.

c. Condensate will build up a level in the condensate tank at which the condensate can be turned on with recirculation to the condensate tank. As more condensate builds up in the condensate tank, the condensate will be sent back to the plant as process water and/or as boiler feed water.

d. Waste heat steam from the concentrator stack will soon be sufficient to heat the recirculating process liquor stream 5 and stream 6 to 111.2°F. From process evaluations conducted for the design of the heat recovery system, the heat load of the waste heat steam coil for heating process liquor stream 5 and stream 6 can vary from 600 to 6700 lbs/hour of steam. Consequently, the flow control valve, FCV-1, was specified to have a steam valve coefficient for stable operation over this range of steam flow rates. In addition, note, the waste heat steam available in the concentrator stack is 9,000 lbs/hour, which is above the maximum steam required to operate the process liquor preheat waste heat steam coil. Thus, noncondensables are not sucked into the 6—in.-diameter steam line via the atmospheric exit of the concentrator stack. This is important because noncondensables can rapidly reduce heat transfer in the coils and make the system nearly inoperable.

e. The ejector, EJ-1, is isolated with shutoff valve to the condensate tank and boiler steam shutoff to the ejector. This ejector can be used if necessary to remove noncondensables building up in the waste heat steam coils as indicated by a loss in heat transfer by a problem of maintaining the heated process discharge liquor to 111.2°F.

f. When necessary if unknown process fluctuations or some unknown problem does not permit the temperature control of the heated process discharge liquor to the desired process temperature of 111.2°F, boiler steam can be supplied via flow control valve, FCV-2, to the boiler steam coil in the process liquor preheat vessel. This backup heating feature adds to the plant flexibility and efficiency when unforeseen process conditions arise.

This case study of recovering atmospheric waste heat steam from the concentrator stack is an example of what a chemist and an engineer can accomplish with a team effort based upon scientific principles. Prior to this heat recovery project, much energy was lost to the atmosphere via the concentrator stack. By understanding the basic fundamentals of engineering and applying them to the real world of plant operations, much efficiency and profits to the bottom line can be achieved for a company. Needless to say, much hard work went into this project and much tenacity was used to stay on course by the key team players. As a final comment, this heat recovery case study was a real-life project and adventure. After much sweat, hard work, and tenacity in applying the laws of science, success is the likely outcome.

REFERENCES

1. Andersen, B.L. and Wenzel, L.A., *"Introduction to Chemical Engineering,"* McGraw-Hill Book Company, Inc. , New York, 1961.

2. Hougen, O.A., Watson, K.M., and Ragatz, R.A., *"Chemical Process Principles, Part I, Material and Energy Balances,"* 2nd edition, Wiley. , New York, 1954.

3. Steam Tables, Properties of Saturated and Superheated Steam—From 0.08865 to 15,500 lb per sq in. absolute pressure, Seventh Printing, Values Reprinted from 1967 ASME STEAM TABLES, The American Society of Mechanical Engineers, 1967.

4. Kern, D.Q., *"Process Heat Transfer,"* McGraw-Hill Book Company, New York, 1959.

5. Chemical Engineering, April 4, 1983, p. 69.

Printed and bound by CPI Group (UK) Ltd, Croydon, CR0 4YY

16/04/2025

14658606-0002